P9-APP-137

MAGNETO-SOLID MECHANICS

MAGNETO-SOLID MECHANICS

FRANCIS C. MOON

Theoretical and Applied Mechanics
Cornell University
Ithaca, New York

A Wiley-Interscience Publication

JOHN WILEY & SONS

New York · Chichester · Brisbane · Toronto · Singapore

***Library of Congress Cataloging in Publication Data*:**

Moon, F. C., 1939-
Magneto-solid mechanics.

"A Wiley-Interscience publication."
Includes index.
1. Magnetic devices. 2. Electromagnets. 3. Superconductors. I. Title.

TK454.4.E5M66 1984 621.34 83-23372
ISBN 0-471-88536-3

Printed in the United States of America

10 9 8 7 6 5 4 3 2 1

To Lee
and in memory of
Frank Moon, Sr.

"*Mechanics is the paradise of the mathematical sciences because by means of it one comes to the fruit of mathematics.*"

– LEONARDO DA VINCI, NOTEBOOKS

"*Thus it is settled by nature, not without reason, that the parts nigher the pole shall have the greatest attractive force; and that the pole itself shall be the seat, the throne as it were, of a high and splendid power.*"

– W. GILBERT, DE MAGNETE (1600)

Preface

There is a growing class of problems in solid mechanics related to new electromagnetic devices that involve large magnetic forces and new materials such as superconductors. These devices include magnets for particle physics research, magnets for medical applications (NMR), large magnets for magnetohydrodynamic (MHD) devices and fusion reactors, magnetically levitated trains, magnetic forming tools, and magnetic ore and solid-waste separation systems. While the engineer can turn to dozens of texts and monographs on magneto-fluid mechanics when faced with a related problem, there is no summary of knowledge in monograph or text form for reference to problems in magneto-solid mechanics. I have written this book for the engineer or applied physicist who wishes to understand the stresses, dynamics, or elastic stability of magnetic devices, as well as for the novitiate researcher in magneto-solid mechanics.

This project began as two separate literature reviews on magneto-solid mechanics and vibration problems in magnetic levitation and propulsion of vehicles. It contains more tutorial discussion of principles and over 60 illustrative examples, as well as introductory material on superconductivity. An extensive literature review of the field includes over 280 references. Also included is an appendix of electromagnetic properties of materials and a collection of magnetic force and couple formulas.

Most of the problems treated in this book are approached from methods similar to those in strength of materials. That is, I have posited as simple a magneto-mechanical model as possible to describe the phenomena. While the foundations of the subject are certainly to be found in continuum physics, I have often bypassed the strict deductive paths along continuum solid mechanics and electromagnetics to use simpler integrated equations of beams, plates, and circuits. I have done this to provide a more intuitive or inductive approach to magnetomechanics and to avoid the pomposity of including details that do not contribute to an understanding of the phenomena. These simple models of magnetomechanical problems are designed to illustrate classic problems of magnetic devices. It is recognized that more sophisticated numerical analyses are required to actually design these devices.

In the opening chapter I introduce some physical concepts, such as magnetic stresses and skin depth, and indicate potential application of magneto-solid mechanics to magnetic forming devices, fusion reactors, and levitation of trains. A brief historical review concludes Chapter 1. In Chapters 2 and 3 the basic equations of the subject of magneto-solid mechanics are summarized and a discussion is given on magnetic forces and energy. Stress analysis of current-carrying structures is introduced in Chapter 4. Chapter 5 presents a discussion of magnetoelastic instabilities in magnetomechanical devices and the implications of Earnshaw's theorem for the design of full-scale magnets.

The next three chapters deal with specific problem areas—superconducting structures, ferromagnetic devices, and eddy-current and dynamically induced magnetic forces in solids. While there are other introductory texts on superconductivity, a brief review of the terminology and elementary physics of the subject was thought necessary so that the structural engineer could appreciate the problems attendant to superconducting devices.

Finally, in Chapter 9, I present a review of some experimental techniques in magneto-solid mechanics with the hope of encouraging experimental mechanicians to get over their phobia of things magnetic.

I have not attempted to summarize solid-mechanics problems related to rotating electromagnetic devices, since much of that research was done in the late 19th and early 20th century. Nor have I included material on electromagnetic solid-state phenomena, such as piezoelectricity, magnetostrictive devices, or solid-state plasmas, since modern summaries of these areas have been written.

My early interest in interdisciplinary research was sparked by several professors at Pratt Institute, Brooklyn, New York, around 1960, particularly, Irving Shames (now at State University of New York at Buffalo) and Charles Mischke (now at Iowa State University). My studies into magnetoelastic stability were nurtured by my mentor, Yih-Hsing Pao at Cornell University, who became an early collaborator. The many and varied magnetomechanical problems and analyses which are the substance of this book are in part the result of the hard work of a dozen students and postdoctoral colleagues. In particular, I wish to recognize the contributions of Somnath Chattopadhyay, Donald Chu, Eddie Chian, Kosei Hara, B. S. Kim, Claude Swanson, and Kuan-Ya Yuan. This book would not have been written were it not for the continued support given by Clifford Astill of the National Science Foundation and, more recently, Nicholas Basdekas of the Office of Naval Research. The four-year pregnancy and delivery of the book has had the patient midwifery of my spouse, Lee Rumstich Moon. I also want to thank Lynne Vrooman for her skillful typing of the final manuscript.

FRANCIS C. MOON

Ithaca, New York
January 1984

Contents

1

Introduction

1.1 WHAT IS MAGNETO-SOLID MECHANICS?

Magneto-solid mechanics is the study of the effect of magnetic forces on the deformation, stresses, motion, and stability of solid bodies. Early studies of magnetism focused on the forces produced by electric currents and magnetic bodies. But in modern times so much attention has been given to the field nature of electromagnetism and the informational applications of electromagnetic fields, that it is difficult to find an up-to-date treatment of magnetic forces and the internal stresses and deformations in magnetic solids.

In mechanics, however, a great number of general theories of electromagnetism and deformable bodies have been posited but applications have been limited to special geometries and special field configurations such as uniform magnetic fields. Rarely have mechanicians dealt with problems related to actual magnetic devices.

It is the belief of the author that a combined treatment of principles and problems related to magnetic fields and solid bodies is needed. There are many texts and monographs related to magnetic fields and fluids (e.g., magnetohydrodynamics or MHD) as well as the magnetomechanics of rotating machinery. Deformable solids, however, demand a special treatment because of the complications that arise when both shear and normal stresses act on the body.

Aside from the obvious dearth of texts on magnetic forces and solids, why another hyphenated subject? The need for magneto-solid mechanics arises from the increasing applications and devices that employ high magnetic fields and high electric currents. Conventional devices with high magnetic fields and currents include switches, motors, generators, and electromagnets. However, recent advances in superconductivity have extended this list to include superconducting motors, transmission lines, inductive energy storage devices, and large high-field magnets for magnetic levitation of trains, MHD systems, fusion reactors, high-energy particle accelerators, magnetic separation devices, and medical nuclear-magnetic-resonance scanners (NMR).

It is the author's belief that electromagnetic forces will play an increasing role in moving, levitating, controlling, and forming solids. While electric machines have played a major role in the last century in metal forming and machining, the actual cutting and forming forces have been transmitted by mechanical devices driven by rotary electric motors. In the 1960s direct forming of sheetlike metal parts was developed under the name of "magnetic forming," which used pulsed magnetic fields. Direct forming means that there is no mechanical connection between the source of the magnetic field and the solid workpiece. In the 1970s the development of high-power laser beams resulted in manufacturing applications for drilling, sawing, and surface heat treating of solids.

There have been many research papers and reports published on the state of stress and motion of solids under magnetic fields. Except for the general theories proposed by the mechanics community, these analyses arose out of specific applications in magnetic devices such as electric contacts, Holm (1967), or electromagnet design, Montgomery (1980). There are, however, other problems which until recently have received almost no attention, such as the interaction of high current arcs with solids and the elastic stability of superconducting magnets.

The purpose of this monograph is to organize the existing literature on magneto-solid mechanics and to give a tutorial presentation of the basic principles and some useful methods of analysis. Of course some reviews of the field already exist, such as Paria (1967) on continuum magnetoelasticity, Hutter and van der Ven (1978), Pao (1978) on theories of magnetic forces, Moon (1978), on problems in magneto-solid mechanics and a review by Ambartsumyan (1982).

In recent years contemporary mechanicians have emphasized the deductive method based on the balance laws of mechanics and electromagnetics. The problems in this monograph do not always follow this deductive path. Since many of these problems have never been posed before, we proceed inductively and sometimes intuitively to "guess" at an appropriate magnetomechanical model to "explain" the experimental observation. There is no doubt that by careful analysis and fortune of hindsight, the solution of these problems can be derived from the general theories of continuum mechanics. But the author assumes that the average reader's experience in mechanics will allow the reader to draw analogies more easily than to follow the minutia of perturbation analyses of the general field equations of mechanics and electromagnetism. In most of these problems the emphasis will be to identify the principal forces and the electromagnetic and mechanical resistances that the forces must work against.

Most engineers generalize from example and analogy and from the study of simple problems to complex. It is hoped that this monograph will provide paradigms for future work in the field. In typical engineering design procedures for magnetomechanical devices, the problem is often split into a magnetic field problem and an elasticity or structures problem, with each group applying powerful but complicated and expensive numerical computer codes such as finite elements to achieve a design. In this mode of design, interaction problems such as magnetoelastic instabilities are often overlooked. While we recognize the importance and even the necessity of using large computer codes to solve these problems, we also believe it important for the designer to understand the magnetomechanical phenomena through simple models. By understanding simple models for the macroscopic interaction of magnetic fields with structural solids such as beams, plates, and shells, which have stood the test of experimental verification, the designer or analyst can proceed with some confidence to more complex structures and field configurations.

1.2 PHYSICAL QUANTITIES AND NONDIMENSIONAL GROUPS

The basic set of observables in magneto-solid mechanics includes displacement (**u**) and strains, electric current (**J**), magnetization (**M**), and temperatures (θ). These fields are associated with the material or mass of the body and have no meaning outside the body. Electromagnetic bodies interact with other bodies by either creating forces or inducing additional charges, currents, polarization, and magnetization in them. To account for these "long-range" effects, the concept of magnetic and electric fields, **B** and **E**, respectively, were introduced. Unlike **u**, **J**, and **M**, the fields **B** and **E** permeate both material space and space between material bodies.†

Electric currents in a solid are associated with the motion of electrons in the solid. The rate of charge flow is called current I, and the current crossing a unit area is called the current density **J**. The negative charge of the free electrons is balanced by the net positive charge on the stationary ions. In good conductors the time for any accumulated net change to diffuse is so short ($\cong 10^{-19}$ sec) that for practical purposes we can assume that the net charge density is zero within the body. However, charge can accumulate on the surface of the body.

Another source of magnetic forces in solids is the *magnetization density* **M**. These forces exist even when no currents are flowing. Far from a magnetized body the magnetic effects are the same as if a loop of persistent current were circulating in the body. Some theories of magnetization imagine the body to be composed of a distribution of small current loops

†There are many introductory and advanced texts on electromagnetism. Three of my favorites are Stratton (1941), Reitz and Milford (1960), and Jackson (1962).

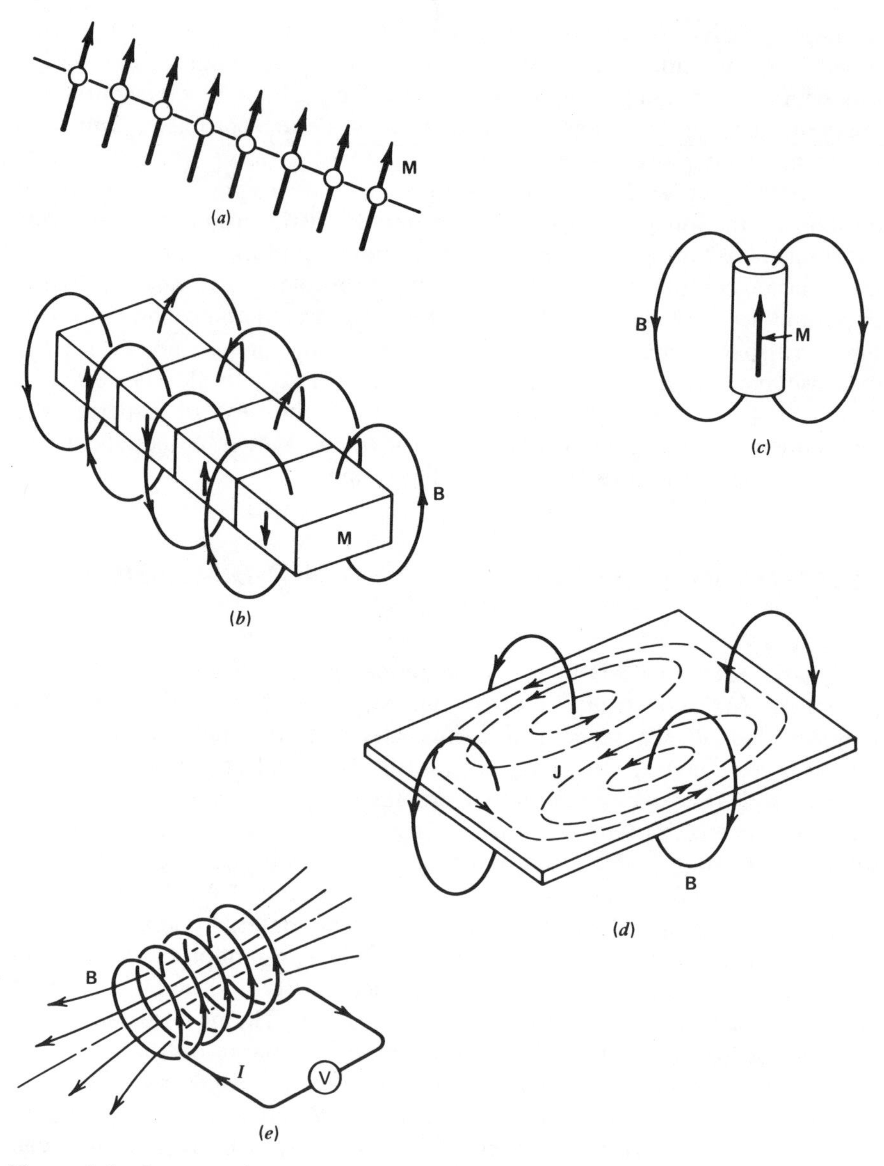

Figure 1-1 Sources of magnetic fields: (*a*) alignment of spins in a linear array of atoms, (*b*) magnetic domains, (*c*) magnetic dipole, (*d*) electric current distributions, and (*e*) electric circuit.

called *Ampere currents*. In fact, some small magnetization can be ascribed to the motion of electrons around the nucleus of each atom. However, in ferromagnetic materials this effect is small and the magnetization is attributed to a property of atoms called the "spin." Locally, the spins can align themselves in regions of net magnetization called domains (see Fig. 1-1). These domains may align themselves to produce a bulk magnetization and the body is called "permanently magnetized" or a "hard magnetic material." On the other hand, the magnetic domains may be randomly oriented to produce zero bulk magnetic effect when there is no external magnetic field. However, the spins may align themselves to produce net effects in an external field. These materials are referred to as "soft magnetic materials."

When the body is assumed to be rigid, only the displacements, angular motions, and linear and angular velocities are of interest as in rotating electric motors, generators, and levitated vehicles. But when the deformations of the body are of concern then the displacement gradient or strains become important as well as the resultant stresses they induce. In the classical macroscopic or continuum view of deformable bodies, the strains are related to the internal stresses. One pictures a body as a collection of small rectangular blocks which have normal and tangential forces acting on each of their six sides. The ratio of these forces to the area over which they act is a quantity called *stress*. Solid bodies are distinguished from fluids by their resistance to shear and tendency to return to original shape after shear stresses are removed (of course, some materials, characterized as viscoelastic, can exhibit fluidlike and solidlike behavior). For isotropic elastic materials, there are two elastic constants Y† and G relating normal and shear stresses to their respective strains.

Magnetic Stresses

An idea of great importance in magnetomechanics is that of electromagnetic stresses. The mathematical definition of these stresses is presented in Chapters 2 and 3. The electromagnetic stress concept is attributed to Faraday, who imagined that lines of tension act along field lines and pressure forces act normal to field lines. For example, consider the attraction of north and south poles of two magnets as shown in Figure 1-2. Faraday imagined that the effect of each magnet on the other could be represented by lines of magnetic force (i.e., magnetic field lines) and that the attractive force was equivalent to tension forces acting along these lines. Thus one replaces the forces acting between the bodies by "stresses" acting in the field between them. The total force on the one pole face is given by the integration of these "stresses" over the pole face; that is,

†In standard mechanics texts the symbol E is used for Young's modulus. However, to avoid confusion with the electric field, we will use Y in this book.

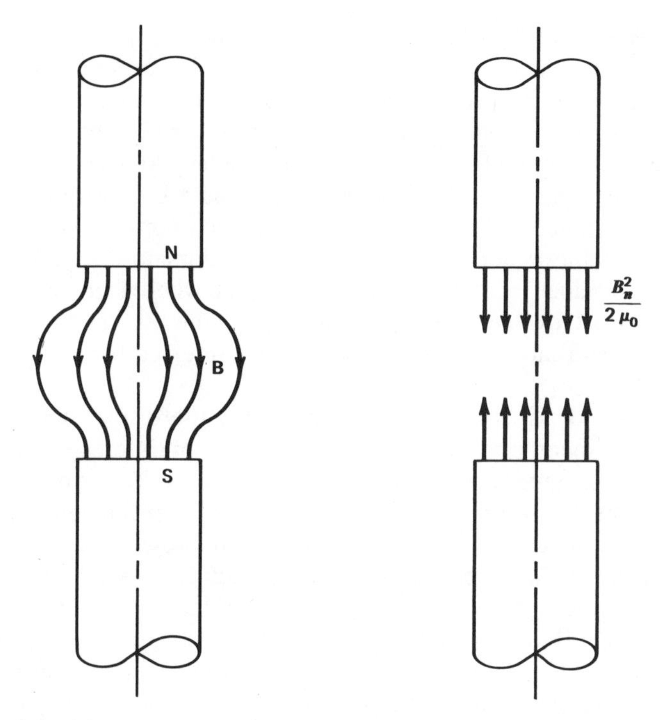

Figure 1-2 Magnetic field lines and magnetic stresses near magnetic poles.

$$F = \int_A \frac{B_n^2}{2\mu_0} da \tag{1-2.1}$$

where B_n is the normal component of the field to the surface.

Another illustration is shown in Figure 1-3, where current in a long cylinder with circumferential currents, called a solenoid, produces very high axial magnetic fields inside the cylinder and low-density fields outside the cylinder. The body forces on the currents near the center of the solenoid produce radial magnetic forces. However, these forces are replaced by a magnetic "pressure" acting on the inside surface of the cylinder of strength

$$P_m = \frac{B_t^2}{2\mu_0} \tag{1-2.2}$$

where B_t is the tangential component of the magnetic field to the cylinder surface. Thus a solenoid which produces a 1-T magnetic field will produce $10^3/8\pi$ N/cm^2 (57.7 psi) pressure on the inside wall of the cylinder. Of course, the actual forces are distributed over the current filaments in the cylinder wall. However, where the details of the stress distribution in the wall are not important, as in a thin-walled solenoid, the magnetic pressure concept is useful.

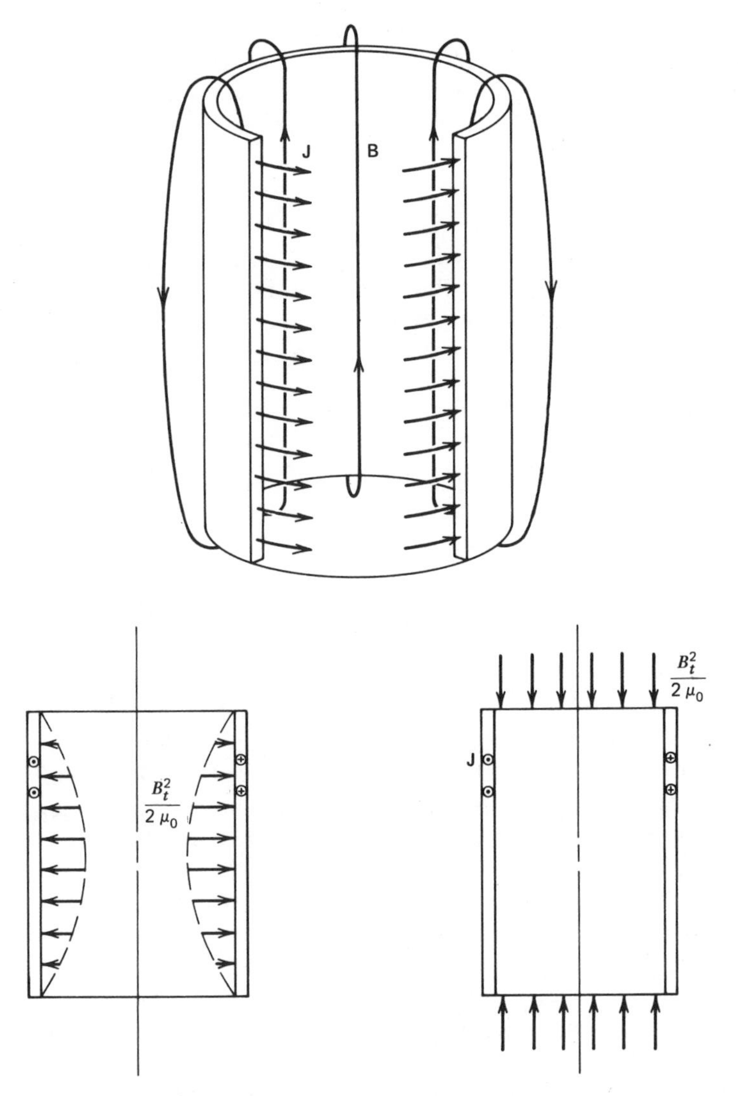

Figure 1-3 Magnetic field lines and magnetic stresses in a solenoid.

Nondimensional Groups

For isotropic elastic, magnetic, and thermal materials there are at least seven or more material constants, not to mention over seven primary and secondary observables such as displacement and stress or current and magnetic field. In any parameter study of a magnetomechanical problem it is useful to reduce the number of independent variables by introducing nondimensional groups such as is done in thermal-fluid mechanics. (See Chapter 8 for a more extended discussion of nondimensional groups.)

One important nondimensional group is the ratio of magnetic to mechanical stresses. As noted in the previous section, B^2/μ_0 has units of stress or energy density. This can be nondimensionalized by the elastic modulus Y or by one of the six stress components t_{ij} forming the groups

$$\frac{B^2}{\mu_0 Y} \quad \text{or} \quad \frac{B^2}{\mu_0 t_{ij}}$$

These groups will appear where applied magnetic fields induce either currents or magnetization in the solid. For these problems the stresses and strains are proportional to the *square* of the applied magnetic field.

A nondimensional group that appears in electric current problems is

$$\frac{\mu_0 I^2}{Y\Delta^2}$$

where $\mu_0 I^2$ has units of force and Δ is a thickness or length variable. When more than one current appears, such as the bending of an elastic coil with current I_1 in the field of another coil with current I_2, the following group will occur:

$$\frac{\mu_0 I_1 I_2}{Y\Delta^2}$$

When time enters the problem, such as frequency of an oscillating field ω or the length of a current pulse τ, the concept of *skin depth* becomes important. The skin depth is the depth to which an oscillating magnetic field can penetrate a conductor (see Chapter 8 or Reitz and Milford, 1960). This length is given by

$$\delta_m = \left(\frac{2}{\mu_0 \sigma \omega}\right)^{1/2} \tag{1-2.3}$$

A natural nondimensional group is the ratio of δ_m to some geometric length such as the thickness of a plate, that is, δ_m/Δ. When $\delta_m/\Delta \ll 1$, we can sometimes assume that the magnetic field is zero inside the conductor.

When electric currents generate heat in a solid, one can define a thermal skin depth

$$\delta_t = \left(\frac{2k}{c\omega}\right)^{1/2} \tag{1-2.4}$$

where k is the thermal conductivity and c is the heat capacity. Another natural nondimensional group is the ratio

$$\frac{\delta_m}{\delta_t} = \left(\frac{c}{\mu_0 \sigma k}\right)^{1/2} \tag{1-2.5}$$

When both magnetic field and temperature oscillate outside a good conductor, the magnetic field penetrates much deeper into the solid than the temperature.

When a conductor moves in a magnetic field with velocity v, there exists a nondimensional group similar to the Reynolds number in viscous-fluid mechanics. Hence we call this group a *magnetic Reynolds number*,

$$R_m = \frac{v\mu_0\sigma\Delta}{2} = \frac{v}{V_m}$$

where Δ is a length parameter. The quantity $V_m = 2/\mu_0\sigma\Delta$ represents a characteristic velocity parameter. For oscillating field problems, we can replace v by $\omega\Delta$ or

$$R_m = \frac{\omega\mu_0\sigma\Delta^2}{2} = \frac{\Delta^2}{\delta_m^2} \tag{1-2.6}$$

Finally, we consider the nondimensional group

$$\frac{\varepsilon_0 E^2}{B^2/\mu_0}$$

where ε_0 is the electric permittivity. Since B^2/μ_0 represents a magnetic stress, then $\varepsilon_0 E^2$ must also have units of stress. The electric "stress" $\varepsilon_0 E^2$ is associated with forces produced by charges on the surface of a conductor in an electric field. Since $1/\mu_0\varepsilon_0$ represents the square of the speed of light, the field E must reach very high values for the ratio of magnetic to electric forces to reach unity. Recognition of this fact allows us to neglect electric-charge-type forces in magnetic field problems. Of course, when $B = 0$, or when B has negligible influence on the material, such as in a dielectric, the forces associated with E or $\varepsilon_0 E^2$ must be retained (see, e.g., Melcher, 1981).

1.3 APPLICATIONS

As noted at the beginning of this chapter, applications of high magnetic fields have advanced far beyond conventional motors and electromagnets, especially in the areas of pulsed magnetic fields, linear motors, and superconductivity. Magnetic devices have been built which can form and bend metals in a fraction of a second, guide and accelerate electrons and charged particles in high-energy physics experiments, contain plasmas at temperatures of $10^8\,^\circ$C and levitate and accelerate 50-ton vehicles. Three of these applications are discussed below.

Magnetic Forming

The intuitive concepts of skin depth and magnetic pressure introduced above can be combined in a creative way to describe a very useful

Figure 1-4 Photograph of magnetically deformed aluminum plates and forming coil.

application in which transient magnetic fields can deform a solid. We imagine a coil placed near a conducting solid of thickness Δ and a transient current pulse placed in the coil with pulse time τ. The resulting transient magnetic field produced by the coil will induce circulating or eddy currents in the nearby conductor. If the ratio of skin depth to thickness is less than unity, that is,

$$\left(\frac{\tau}{\sigma\mu_0\pi}\right)^{1/2}\frac{1}{\Delta}\ll 1$$

the magnetic field will be almost zero on one side of the solid slab. Then we can think of a magnetic pressure on the coil side of the solid as acting to push the solid away from the coil. If a die is placed behind the accelerating solid, the metal can be formed around the die. An example of this is shown in Figure 1-4 where a flat coil was used to punch a hole through a thin aluminum plate.

Magnetic forming was developed as part of the space research in the 1960s (see, e.g., Furth, 1961). A review of this field will be given in Chapter 8. Magnetic forming can be used as a magnetic hammer to take dents out of thin conducting structures, as a swagging tool, and even for riveting applications.

Magnetic Fusion and MHD Magnets

Two applications in which magneto-solid mechanics plays an essential technical role are the design of magnets for the containment of fusion reactions and for magnetohydrodynamic electric generators (MHD).

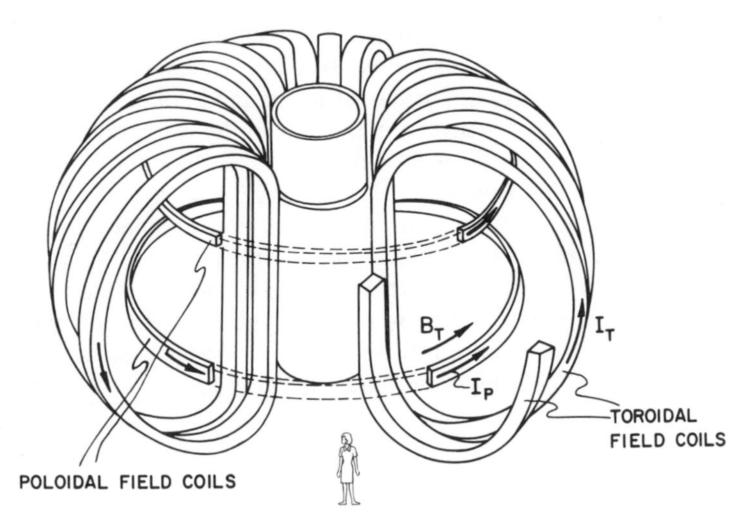

Figure 1-5 Magnets in a toroidal magnetic fusion reactor (tokamak).

Nuclear fusion reactions require that ionized hydrogen atoms be held at extremely high temperatures for sufficient time so that hydrogen can combine to form helium and release energy. At these temperatures magnetic fields are used to keep the plasma from cooling off on the reactor vessel walls. Many shapes for this so called "magnetic bottle" have been proposed. One reactor concept called the *tokamak* is shown in Figure 1-5. The magnetic fields for these devices are created by magnets in which wire conductors are wound onto structural forms. These structures serve to hold the windings in place and to withstand the magnetic forces when current flows in the windings.

Full-scale reactors will have current of the order of $(1–10)\times 10^6$ At (ampere turns) per magnet. The forces on these magnets can be determined with the aid of the magnetic pressure concept. In the toroidal reactor (see Fig. 1-5) the plasma and the magnetic field are confined principally to a torus or "donut"-shaped region inside the toroidal field magnets. Thus inside these magnets there exists a large magnetic pressure while outside the magnetic field is small. The magnets act as a kind of magnetic pressure vessel. The magnetic forces which act on these magnets act normal to the windings, tending to increase the length of the coil, thereby producing tension in the windings. The "pressure" on the inner or vertical lengths of the coils will place very high compressive stresses on the central support cylinder.

In the late 1970s, the conductors were normal copper windings which heat up when current is present and require cooling to prevent thermal damage and thermoelastic stresses. Magnetic fusion magnets for the 1980s are expected to be wound from superconducting wire. Superconducting magnets require temperatures below 10°K (see Chapter 6). To minimize heat leaks from the warm environment to the cold conductors, the number of structural connections between magnets and the warm environment must be kept small. Such magnets will be more flexible than normal coils and are subject to bending and torsional deformations.

One might recall from elementary physics that parallel currents attract. In the case of toroidal field magnets one observes that all the outer parts of the magnets carry parallel currents and so that each coil attracts its neighbor. The net lateral force on each magnet is zero if all the coils are perfectly aligned, but small misalignments will always exist which can lead to a critical current at which the magnets can buckle (see Chapters 5 and 6).

Figure 1-6 Construction of yin–yang magnets for a tandem mirror fusion reactor. (Photo courtesy of Lawrence Livermore Laboratory, Livermore, California.)

Another magnet configuration in magnetic fusion is the so-called yin-yang or "baseball" magnets (see Fig. 1-6). These complex out-of-plane coils require finite-element computer codes to perform both the magnetic and structural design.

MHD magnets shown in Figure 1-7 are designed to provide a steady magnetic field transverse to the flow of ionized gas. These devices could be employed as topping cycles to fossil-fuel power plants and might increase the efficiency of these plants by as much as 30–40%. The shape of these coils are often called "saddle" coils or dipole magnets. They are usually out of plane and require numerical codes for the magnetic and structural design. A recent monograph on the design of both MHD and fusion magnets has been published by Thome and Tarrh (1982).

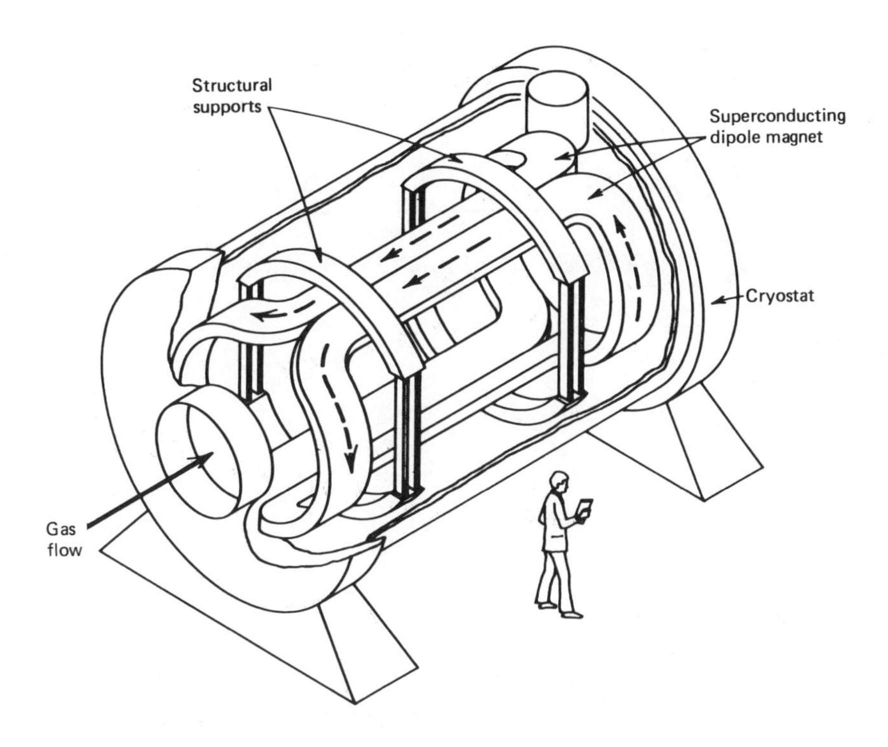

Figure 1-7 Saddle or dipole magnets for a magnetohydrodynamics (MHD) electric generator. (Illustration courtesy of F. Bitter, National Magnet Laboratory, Cambridge, Massachusetts.)

Magnetically Levitated Vehicles

To skim over the ground with no visible means of suspension has long been a dream of science fiction writers. Now that dream is reality, at least at a few research centers, for people-carrying vehicles can be suspended, guided, and propelled by electromagnetic fields. Designs for revenue systems range from airport to city-center service to 480 km/h (300 mi/h) intercity vehicles (see, e.g., Laithwaite, 1977; Rhodes and Mulhall, 1981). While the steel-wheeled vehicle on steel rails has long been the most efficient way to travel on land at low speeds, at high speeds (>300 km/h) dynamic problems arise causing increased noise, vibration, and high maintenance costs for the guideway alignment.

There are two principal methods for magnetically levitating vehicles. The first, shown in Figure 1-8*a*, called electromagnetic (EML) or attracting levitation, uses ordinary electromagnets which suspend the vehicle

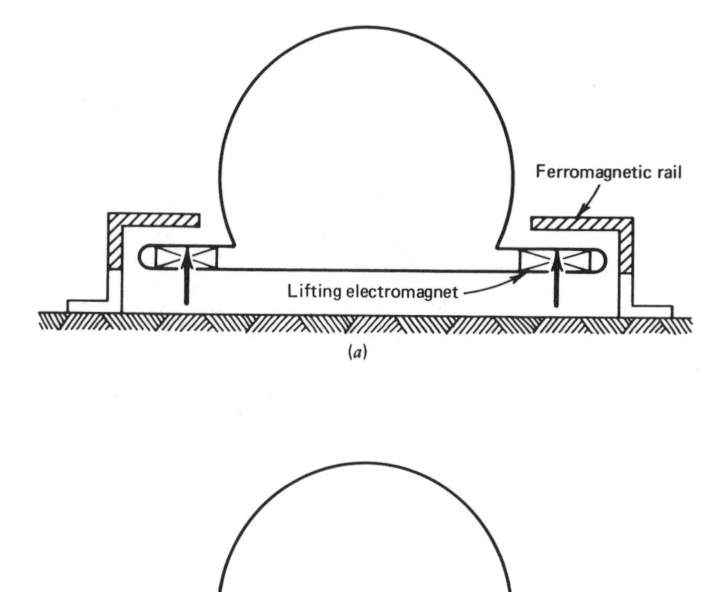

Figure 1-8 Two magnetic levitation methods: (*a*) electromagnetic levitation (EML) or attractive method and (*b*) electrodynamic levitation (EDL) or repulsive method.

below ferromagnetic rails. The natural tendency of magnets to slam up against the steel rail is overcome by feedback control of the currents. The suspension gap is of the order of 1 cm (see Chapter 7).

The second method, shown in Figure 1-8*b*, called electrodynamic (EDL) or repulsive levitation, employs large magnets on the vehicle which generate eddy currents in an aluminum track below the vehicle. Lift is developed when the vehicle moves, reaching an asymptotic limit at high speeds (see Fig. 1-9). The associated magnetic drag increases at first with speed and then decreases with increasing velocities. The magnetic lift-to-drag forces ratio for passenger-vehicle designs have ranged from 20 to 100. The dependence of lift and drag forces on speed can be put in nondimensional form by plotting

$$\frac{F}{\mu_0 I^2} \quad \text{versus} \quad R_m = \frac{v\mu_0\sigma\Delta}{2}$$

where I is the total current in the magnet, R_m is the magnetic Reynolds number, and Δ is the track thickness.

The magnets proposed for the EDL method are made from superconducting wire. The large fields generated by superconducting magnets (~2–3 T) can create a levitation gap of up to 30 cm. Permanent magnets

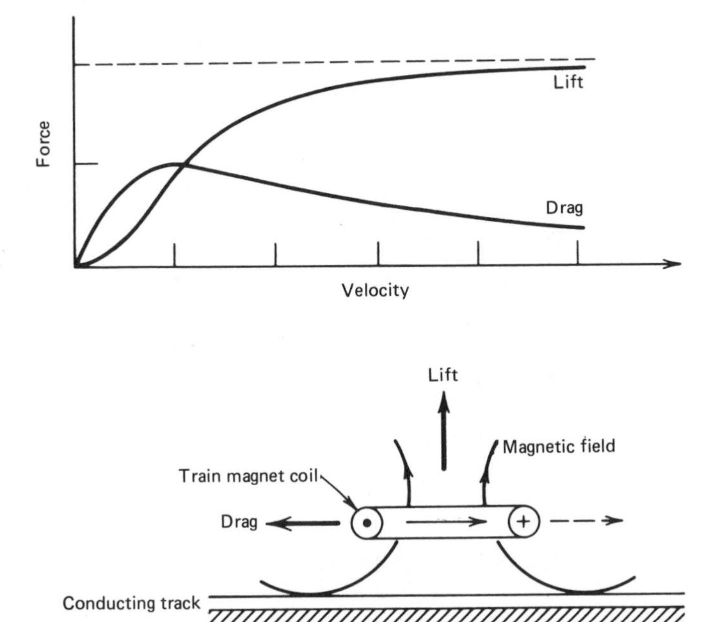

Figure 1-9 Magnetic lift and drag forces vs. velocity for the electrodynamic levitation method.

can also be used on the vehicle for EDL, but they have a large weight penalty for full-sized vehicles.

Mechanics-related problems which arise in magnetic levitation concern the transmission of forces from the cryogenic environment of the superconductor to the warm-vehicle structure, rigid-body dynamics and instabilities of the vehicle itself, and the dynamic interaction between a high-speed vehicle and flexible guideway, especially when the guideway is above ground. Dynamic instabilities of EDL magnetic-levitation vehicles have been observed in laboratory models and speeds near the high magnetic drag portion of the force–velocity regime. This phenomenon is analogous to dynamic flutter in aircraft.

Another interesting phenomenon in magnetic-levitation mechanics is negative damping. When a magnet vibrates above a conducting sheet, positive damping forces occur due to induced eddy-current flow in the conductor. However, if the magnet moves over the conductor while it vibrates, then damping can become negative and can produce growing oscillations. These problems and others are discussed in Chapter 8.

Magnetic Cannons and Medical Magnets (NMR)

Two recent applications of magnetic fields span the human experience of life and death. In the former, new techniques in nuclear-magnetic-resonance measurements (NMR) have resulted in the invention of a system that can scan the inside organs of a human body without the need for dangerous X-rays, or injection of radioactive tracers. However, these NMR devices require that the patient be placed in a large solenoidal magnet. At present these magnets are superconducting and of the order of 1-m diameter with a 1-T central field.

A natural extension of the magnetic levitation and propulsion of vehicles is the development of magnetic cannons, sometimes called euphemistically, "mass drivers" or "rail guns." The object here is to take stored energy in a capacitance bank or rotating generator and dump it into a linear magnetic propulsion device to accelerate a mass or projectile to very high velocities. In one proposal by O'Neill and Kolm (see O'Neill, 1977) a mass driver is proposed to launch material off the moon or other extraterrestrial objects in space, while in another design (Marshall, 1982) a more military use is envisioned. Such devices have also been proposed for inertial fusion and high-pressure physics research (see Chapter 8).

1.4 BRIEF HISTORICAL REVIEW

Magnetism has always fascinated mankind from the moment its magical powers were discovered in the lodestone to the present plunge into the mysteries of the electron and nuclear spin. The history of magneto-solid mechanics is intertwined with the history of electromagnetism and electrical engineering. This was especially true in the 19th century when magnetic and electromagnetic devices were primarily used to produce motions, forces, and torques. In the present century electromagnetism has been associated more closely with communication and information. Thus contemporary research in electromagnetic solids has, until recently, been focused on transducers such as piezoelectric, ferroelectric, electrostrictive, and magnetostrictive devices for translating mechanical information into electrical signals and vice versa.

However, with the development of high-energy pulsed magnetic field devices, high-field superconducting magnets, lasers, and large electromagnetic linear motors and levitators, problems involving the motion and deformation of solids and structures in high magnetic fields have become important again.

Early Applications

The use of magnetism for compasses dates back to the Chinese (between 400 BC and 1000 AD). These natural magnets were found in iron ore, Fe_3O_4, and were later called lodestones after the Saxon word "leaden"—to lead, or leading stones. The name magnet is attributed to a region in Greece, Magnesia, where many natural magnets were found.

But the use of the lodestone in the West for navigation did not develop until many centuries later. A French crusader, de Maricourt, in 1269 described the use of the compass for navigation and also is credited with the use of the term poles as a result of experiments with the lodestone and small pieces of iron.

There are many claims to the invention of the first electric motor, depending on one's source.† The decade of the 1830s seems to have witnessed a number of magnetomotive devices, more than two centuries after Gilbert's treatise on magnetism. Henry created a rocking motor in 1831, and dal Negro of Italy invented an oscillating motor in the same year. A rotating motor is credited to H. Pixii in Paris in 1832. In England, Clarke and Saxton each invented rotating machines for converting mechanical energy into electrical energy. A Belgian, Gramme, in 1870, built an alternating-current generator. In 1888 an Italian, G. Ferrari, and N. Tesla in the United States built induction motors. These devices used a rotating magnetic field to induce eddy currents and resulting torque on an armature. This created a device for converting magnetic power into mechanical power without the use of sliding contacts or brushes.

Another important magnetomechanical device to appear in the 19th century was the magnetic relay or switch. This invention is credited to Henry around 1831. It was used by Samuel B. Morse in his famous

†See, for example, Dunsheath (1962).

telegraph demonstration in 1843 between Washington, D.C. and Baltimore, Maryland. Morse's original system had one circuit, battery, and switch and was limited in distance by the resistance of the wire. Using Henry's relay an infinite line of circuits could be connected; one circuit provides the magnetic field to close the switch in the next circuit and so on.

W. Sturgeon is credited with the first electromagnet. This magnet, built in 1825 from 16 turns of wire around a 1.27-cm-diam ($\frac{1}{2}$-in.) soft iron rod, was able to lift 40 N (9 lbf). Joseph Henry was able to improve the lift capability of electromagnets in 1832 with a force of 125 N (28 lbf) and then later with a force of 15.5 kN (3500 lbf) for a magnet weighing only 445 N (100 lbf). These were spectacular advances at the time and these lifting devices were a source of wonder to those who witnessed them. Electromagnets were used for magnetic separation of iron ore and for cleaning grain, such as removing nails from oats fed to horses.

Basic Laws of Electromagnetism

Socrates is reputed by Plato to have been aware of the attractive properties of the lodestone for pieces of iron. The Romans were also believed to have observed magnetic repelling forces between lodestones. However, the first extensive treatise on magnetism did not appear until 1600 when Gilbert, a court physician to Queen Elizabeth I, wrote *De Magnete*.

For over 200 years knowledge of magnetism did not advance much beyond Gilbert's work. In 1820 Oersted discovered that electric currents can produce magnetic fields and Ampere found that coils of current-carrying wire would act as a magnet. Ampere also discovered the laws of force between currents, that is, parallel currents attract and antiparallel currents repel.

Finally, in the 1830s, Faraday in England and Henry in the United States discovered electromagnetic induction, namely, that a changing magnetic field would induce a voltage or current in a nearby conductor not in physical contact with the source of magnetic field.

Maxwell credits Faraday with the concept of magnetic tension and pressures, thus endowing the magnetic field with a more physical reality. Finally, Maxwell added the missing displacement current to the laws discovered by Faraday, Ampere, and Henry which resulted in his celebrated "Maxwell's equations," from which the wave nature of the electric and magnetic fields results. Except for laser–solid interactions, however, the wave nature of the magnetic field does not play a crucial part in magneto-mechanical problems.

Later studies by Lorentz and others laid the foundations for the connection between Maxwell's theory of electromagnetism and relativity. In particular, it was discovered that if one changed from one moving reference frame to another the electric and magnetic fields must change, or that

Maxwell's equations were not Galilean-invariant. This important discovery is relevant to the study of electromagnetism in moving media, such as MHD or deformable solids.

Magneto-Solid Mechanics

With the development of Maxwell's equations and the theory of elasticity, the basic tenets of magneto-solid mechanics would appear to have been established by the late 19th century. The interrelationship between strain and magnetization in solids was a subject of great interest in the 19th century as may be seen from reading Todhunter and Pearson's history of elasticity (1886, 1893). Werthiem (1848), Matteucci (1850), and Wiedemann (1860), to mention only a few, investigated the effects of plastic strain, vibrations, and shock on the ability to magnetize or retain magnetization in magnetic materials such as iron and steel. The change of shape of a body when placed in a magnetic field, called magnetostriction, was another subject of great study (see, e.g., Bozorth, 1951).

However, until recently, the fact that for magnetizable materials the representation of internal magnetic forces was not unique led to much confusion. Brown (1966), for example, has shown that the total force on a magnetized body in a magnetic field could be expressed in terms of different forms of body and surface force densities, each producing the same total force on a body, but leading to different internal stresses. Since Brown's paper other authors have shown that to uniquely determine the internal forces in a magnetized body, the constitutive relations between strain, stress, and magnetic fields must be rigorously formulated using a nonlinear energy approach (see, e.g., Penfield and Haus, 1967; Pao 1978).

In the last decade intense interest in nonlinear continuum mechanics and the axiomatic method led researchers to a reexamination of the interaction of electromagnetic fields with solids as illustrated by Toupin (1963) and Eringen (1963) in their phenomenological theory of elastic dielectrics. Many of these works were motivated primarily by an interest in the reciprocal effects of electric and magnetic fields on the constitutive or stress–strain relations and did not arise out of a specific engineering application.

Aside from general theories of magneto-solid mechanics, many of the specific problems may be grouped around three areas: static stresses in magnet structures (Chapters 4, 6, and 7), magnetoelastic stability (Chapters 5 and 6), and dynamic magnetic forces (Chapter 8). A review of the many research papers in each of these specific areas of magneto-solid mechanics is presented in the above-cited chapters of this book. In the following we review some of the earlier work on particular subjects of magnetomechanics.

Stresses in Magnetic Structures

The study of forces and stresses in electromagnetic devices was mainly pursued by electrical engineers in the first half of the century. An almost forgotten book by Dwight (1945) on induction coils includes a summary of forces between current-carrying circular coils, solenoids, and filaments. Some of these forces are also summarized in Grover (1946). Forces in rotating machinery, for example, between current conductors and ferromagnetic materials, may be found in a text by Hague (1929). The forces and dynamics of solenoids, relays, and other electromechanical devices may be found in Roters (1941).

Another problem of importance to electrical engineering is the electric contact. The flow of current between two convex solids depends on the contact area and the force between them. The phenomenon involves plastic deformation, heating, thermoelastic stresses, and magnetic forces which act to separate the two solids during current flow (see Snowden, 1961). Of importance to high-power circuit-breaker designs, some of these problems are discussed in a text by Holm (1967).

Besides electrical generation, motors, and switches, magnets for physics research have been a source for new problems in magnetoelasticity. A classic problem has been the determination of stresses in a solenoid or circular cylinder. The early work of Kapitza in the 1930s on the use of pulsed current solenoids to produce fields of 30 T developed magnetic pressures in the magnets over $30\,10^3$ N/cm^2 (see Bitter, 1959). In 1936 Francis Bitter designed a water-cooled magnet, now referred to as a Bitter solenoid, which produced 10 T with an equivalent magnetic pressure of over 4×10^3 N/cm^2.

Magnetoelastic Stability

Two classes of magnetoelastic stability problems have received attention in the last decade. One involves ferromagnetic structures in steady magnetic fields, while the other set of problems concerns nonferromagnetic conductors carrying high electric currents, such as superconducting magnets. Both sets of instabilities are of the buckling, or bifurcation type; that is, multiple equilibria appear for some value of the magnetic field or electric current. In this sense they are analogous to the classic Euler column buckling. These problems fall into the more general class of problems in applied mathematics called *catastrophe theory*. Dynamic instabilities can also occur in magnetoelastic problems, but these involve either time-varying magnetic fields, feedback-controlled magnetic forces, or moving conductors in magnetic fields. Two reviews on the subject of magnetoelastic stability may be found in Moon (1978) and Ambartsumyan (1982).

Buckling of ferromagnetic whiskers in the form of elastica has been observed by DeBlois (1967). In this phenomenon the whiskers are initially magnetized and are placed in an antiparallel magnetic field. At a critical value of the applied field, the magnetization in the whisker tries to align itself with the applied field, which induces bending in the elastic whisker.

Buckling of a beam between the poles of an electromagnet was observed by Mozniker (1959) and buckling of a beamlike plate in a transverse uniform magnetic field was studied by Moon and Pao (1968).

Buckling of structures with high electric currents was first proposed by Leontovich and Shafronov (1961), who predicted a kink-type instability similar to one occurring in plasma physics for a flexible wire carrying current. Experimental studies have demonstrated that mutual magnetic forces between magnets in a toroidal set such as a fusion tokamak reactor might lead to buckling (Moon and Swanson, 1976).

A classic dynamic instability involving magnetic forces is the parametric instability of a pendulum in an oscillating magnetic field observed by Bethonod (Rayleigh 1894; McLachlan, 1947). This same phenomenon was observed for a ferromagnetic beam in an oscillating uniform magnetic field (Moon and Pao, 1969). Magnetic feedback-controlled forces have been used to control a beam undergoing flutter-type oscillations (Kalmbach et al., 1974; Horakawa et al., 1978). Magnetic feedback forces have been used to stabilize magnetically levitated trains (Meisenholder and Wang, 1972), and superconducting magnets in a magnetic field (Tenney, 1969).

A recent review of papers dealing with flutter of elastic plates in aerodynamic flows and in magnetic fields has been published by Librescu (1977). This problem may have application to MHD devices, though no experimental work on the subject has appeared.

Dynamic Magnetic Forces

Dynamical problems on magnetoelasticity have been grouped into two classes—dynamics of solids in steady magnetic fields and dynamical behavior produced by time-varying magnetic fields.

In bulk solids the effect of steady magnetic fields on elastic wave propagation has received much attention. This was motivated in part by the question of whether the Earth's magnetic field affects the propagation of seismic waves. Knopoff (1955) and Chadwick (1956) concluded, however, that the effect was insignificant. This result did not temper interest in this problem because of the development of ultrasonic devices for delay lines and communication using magnetoelastic waves. Tiersten (1964), for example, investigated waves in saturated magnetoelastic materials. Theoretical analysis of wave propagation in nonlinear elastic solids in magnetic fields has been made by McCarthy (1966*a*, *b*).

Similar papers appeared on the study of elastic and plastic wave motion in the presence of a large static magnetic field. Many of these papers originated in the Polish journals by Kaliski (1959/1960) and co-workers. A review of many of these papers was given by Paria (1967).

Measurements of the effect of steady magnetic fields on wave propagation in conducting solids, such as wave attenuation in magnetic crystals, were made by Levy and Truel (1953) and on wave speeds by Galkin and Koroliuk (1958), and Alers and Fleury (1963). The effect of a steady magnetic field on the wave speed was found to be very small.

Wave and vibratory motion due to pulsed magnetic fields have been studied in connection with magnetic forming devices. In these problems a transient magnetic field induces circulating or eddy currents in a nearby solid. The interaction of the eddy currents with the magnetic field can produce a wave motion in the solid. The eddy currents can also generate dynamic thermoelastic strains. Pulsed eddy-current-induced elastic waves in bars and plates have been investigated by the author and co-workers (see Chapter 8).

2

Review of Electromagnetics

2.1 INTRODUCTION

To many trained in the mechanical sciences, electromagnetics often appears as a subject full of mystery and complexity. However, electromagnetics and mechanics have important similarities that are useful to recognize and understand. Many ideas from elasticity and continuum mechanics are also utilized in electromagnetics, such as balance laws, constitutive relations, and boundary conditions.

A rigorous mathematical development of electromagnetic theory and continuum solid mechanics will not be attempted in this book. There are a number of excellent monographs which discuss the general theory (e.g., Brown, 1966; Penfield and Haus, 1967; Hutter and van der Ven, 1978 to mention just a few). We want, however, to study specific problems, phenomena, and procedures for predicting forces, stresses, and deformations in solids in electromagnetic fields.

There are many excellent pedagogical treatments of electromagnetics and we shall not attempt to do more than summarize the basic equations for low-frequency, low-velocity electromagnetics. In this chapter we shall review the basic relations between the electric charges, currents, and magnetization, and the electromagnetic fields, forces, and moments.

In Chapter 3 a more detailed discussion on electromagnetic forces, stress tensors, and energy is given, especially as it pertains to magnetizable bodies. In Chapter 3 we will also discuss the basic equations of solid mechanics, and the incorporation of electromagnetic forces into Lagrange's equations.

In a continuum description of physical material, we use charge density q (C/m^3), current density $\mathbf{J}$ (A/m^2), polarization density $\mathbf{P}$ (C $\cdot$ m/m^3), and magnetization density $\mathbf{M}$ (A $\cdot$ m/m^2), where the MKSC system of units is used. The long-range interactions between material objects, separated in space, are described with the aid of auxiliary variables **E**, **B**, **D**, and **H**, which one assumes permeates all space inside and outside of the material objects of interest. This, in effect, replaces an essentially nonlocal physics—that is, what happens "here" depends on what is going on "everywhere else"—with a local theory where the force on the charge Q at position $\mathbf{r}^0$ depends on $\mathbf{E}(\mathbf{r}^0)$. [Note that $\mathbf{E}(\mathbf{r}^0)$ depends on all the charges at all other $\mathbf{r} \neq \mathbf{r}^0$.] This device leads to local differential forms of the laws of electromagnetics, but adds the complication that one must deal with up to eight different physical variables instead of the original four. In solving practical problems it means that the fields **E** and **B** must be determined outside as well as inside the material body of interest.

The difference between the **B** and **H** is related to the physical magnetization density in material bodies; that is,

$$\mathbf{B} = \mu_0(\mathbf{H} + \mathbf{M})$$

where $\mu_0 = 4\pi \times 10^{-7}$ in MKS units and is called the permeability of a vacuum. Thus, outside a magnetizable body, $\mathbf{M} = 0$ and $\mathbf{B} = \mu_0\mathbf{H}$. When $\mathbf{M} = 0$, there is no real difference in the two fields **B** and **H** except the constant μ_0.

In the same way the fields **D** and **E** are related to the electric polarization density **P**; that is,

$$\mathbf{D} = \varepsilon_0\mathbf{P} + \mathbf{E}$$

where $\varepsilon_0 = 8.854 \times 10^{-12}$ F/m. In this book we assume that $\mathbf{P} = 0$. This term becomes important in problems involving piezoelectric or ferroelectric materials (see, e.g., Tiersten, 1969).

An elementary model of an electric dipole **p** consists of two charges of equal magnitude $|Q|$ and opposite sign separated by a distance **d**. The vector **d**, which points toward the positive charge, defines the direction of **p** such that (see Fig. 2-1*a*).

$$\mathbf{p} = Q\mathbf{d}$$

This concept is not important in good conductors, but is used in the theory of dielectrics.

The magnetic dipole model is often represented as the limit of a small circuit of radius r and current I as $r \to 0$ and $I \to \infty$. The magnetic dipole **m** is defined by the equation (see Fig. 2-1*b*)

$$\mathbf{m} = \frac{1}{2}\oint \mathbf{r} \times \mathbf{I}\, ds$$

or for a circular circuit

$$\mathbf{m} = AI\mathbf{n}$$

where $A = \pi r^2$ and **n** is the normal in the direction of $\mathbf{r} \times \mathbf{I}$. Another magnetic

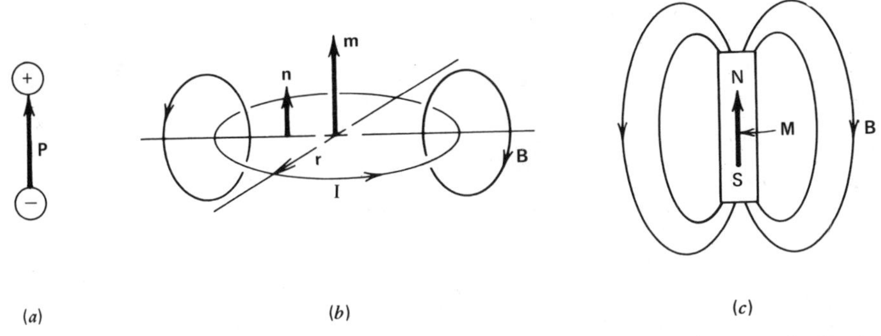

Figure 2-1 (*a*) Electric dipole, (*b*) electric current dipole, and (*c*) magnetic dipole.

dipole model is that of a small bar magnet. In analogy to the electric dipole, we have two magnetic poles of equal magnitude and opposite sense separated by a distance **d**. However, in materials with high magnetization neither circulating microcurrents nor small bar magnets are responsible for the magnetization but a quantity called *spin* which resides in the electrons. Although the current and pole models for elementary magnetism are physically artificial, they are useful concepts for mathematical models of magnetic materials.

2.2 MAXWELL'S EQUATIONS

The differential forms of the laws of electromagnetics are known as Maxwell's equations (Maxwell, 1891). There are numerous formulations of electromagnetics. A comparison of different formulations of electromagnetics and representations for magnetic force laws has been given by Pao (1978). One used in this text is due to Minkowski (see, e.g., Sommerfeld, 1952). These equations are summarized below:

Conservation of charge

$$\nabla \cdot \mathbf{J} + \frac{\partial q}{\partial t} = 0 \tag{2-2.1}$$

Conservation of flux

$$\nabla \cdot \mathbf{B} = 0 \tag{2-2.2}$$

Gauss's law

$$\nabla \cdot \mathbf{D} = q \tag{2-2.3}$$

Maxwell's generalization of Ampere's law

$$\nabla \times \mathbf{H} = \mathbf{J} + \frac{\partial \mathbf{D}}{\partial t} \tag{2-2.4}$$

Faraday's law of induction

$$\nabla \times \mathbf{E} + \frac{\partial \mathbf{B}}{\partial t} = 0 \tag{2-2.5}$$

These equations are not all independent since Eqs. (2-2.1) and (2-2.3) imply that $\mathbf{J} + \partial\mathbf{D}/\partial t$ is solenoidal as expressed in Eq. (2-2.4).

The above relations, known generally as Maxwell's equations, can be written in the form of balance laws:

Conservation of charge

$$\int_S \mathbf{J} \cdot d\mathbf{a} = -\frac{\partial}{\partial t}\int_V q\,dv \tag{2-2.6}$$

Conservation of flux

$$\int_S \mathbf{B} \cdot d\mathbf{a} = 0 \tag{2-2.7}$$

Gauss's law

$$\int_S \mathbf{D} \cdot d\mathbf{a} = \int_V q\,dv \tag{2-2.8}$$

Ampere's law

$$\oint_C \mathbf{H} \cdot d\mathbf{l} = \int_S \mathbf{J} \cdot d\mathbf{a} + \frac{\partial}{\partial t}\int_S \mathbf{D} \cdot d\mathbf{a} \tag{2-2.9}$$

Faraday's law

$$\oint_C \mathbf{E} \cdot d\mathbf{l} = -\frac{\partial}{\partial t}\int_S \mathbf{B} \cdot d\mathbf{a} \tag{2-2.10}$$

In the last two expressions, the area S is defined as that enclosed by the closed curve C (see, e.g., Sommerfeld, 1952).

Low-Frequency Electromagnetics

The full set of Maxwell's equations leads to hyperbolic differential equations for the field variables with propagating wave solutions. Such wave-type solutions are important in the study of waveguides, antennaes, and electromagnetic wave propagation and scattering problems. However, for most magneto-solid mechanics problems, wave-type solutions in electromagnetic field variables are not required. This is because for frequencies of importance to structural problems (less than 10^7 Hz) the wavelengths associated with such wave solutions are much longer than conventional structures of interest. For example, the wavelength of a 10^6-Hz wave in air is 300 m ($c = 3 \times 10^8$ m/sec). The essential term in Maxwell's equations which leads to wave propagation is the displacement current $\partial \mathbf{D}/\partial t$. When this term is dropped, the equations take on the characteristics of either a diffusion equation or elliptic equation. Neglect of $\partial \mathbf{D}/\partial t$ in Ampere's law is sometimes called the *quasistatic approximation*, even though it is valid for very high frequencies in structural mechanics.

2.3 ELECTROMAGNETIC BOUNDARY CONDITIONS

Understanding of the behavior of electromagnetic fields at the interface between different materials is very important in the study of structures in magnetic fields. Because drastic changes of field can occur at a boundary, deformation or movement of a surface provides a primary coupling between the magnetic field and the deformation of the structure. In many problems this coupling is more important than constitutive coupling between field and strain.

Electromagnetic boundary conditions are best understood by applying the integral forms of Maxwell's equations [Eqs. (2-2.6)–(2-2.10)] to an infinitesimal volume containing two media, with an interface dA, and normal $\mathbf{n}$ directed from material 1 to material 2 (see Fig. 2-2). The electric and

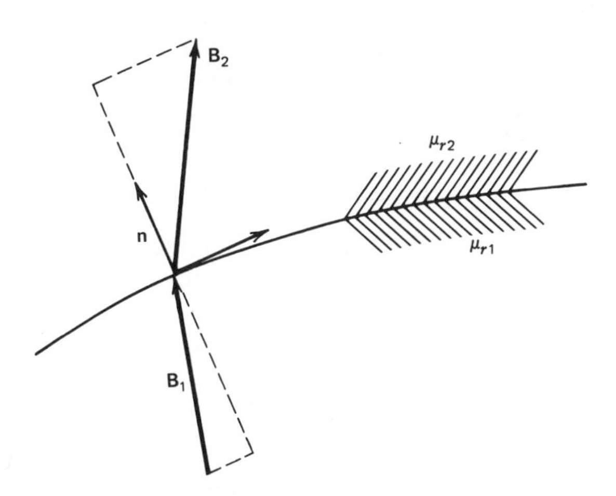

Figure 2-2 Boundary between two materials with different permeabilities showing discontinuity in magnetic field vector **B**.

magnetic fields are assumed to have different values and directions on either side of the interface. In addition, a surface charge distribution Q_s (C/m^2) and surface current distribution $\mathbf{K}$ (A/m) are assumed to exist on the interface. The various integral laws and associated boundary conditions are listed below where the notation $[f]=f_2-f_1$ is used. (See, e.g., Stratton, 1941, or Reitz and Milford, 1960 for a more complete discussion or derivation of these equations.)

$$[\mathbf{D}\cdot\mathbf{n}]=Q_s \tag{2-3.1}$$

Faraday's law of induction

$$[\mathbf{E}\times\mathbf{n}]=0 \tag{2-3.2}$$

Conservation of flux

$$[\mathbf{B}\cdot\mathbf{n}]=0 \tag{2-3.3}$$

Ampere's law

$$[\mathbf{n}\times\mathbf{H}]=\mathbf{K} \tag{2-3.4}$$

Conservation of charge

$$[\mathbf{J}\cdot\mathbf{n}]+\frac{\partial Q_s}{\partial t}=0 \tag{2-3.5}$$

2.4 ELECTROMAGNETIC CONSTITUTIVE RELATIONS

It is obvious that Eqs. (2-2.1)–(2-2.5) are not sufficient to determine all the fields since there are more unknowns than equations. Additional equations are needed which relate to material properties. These extra equations are analogous to constitutive relations in mechanics.

In a vacuum, $q=\mathbf{J}=0$ and

$$\mathbf{D}=\varepsilon_0\mathbf{E}, \qquad \mathbf{B}=\mu_0\mathbf{H} \tag{2-4.1}$$

where ε_0 and μ_0 are called the permittivity and permeability of free space, respectively. This reduces the set of equations to two:

$$\begin{aligned}\nabla\times\mathbf{B}&=\mu_0\varepsilon_0\frac{\partial\mathbf{E}}{\partial t}\\ \nabla\times\mathbf{E}&=-\frac{\partial\mathbf{B}}{\partial t}\end{aligned} \tag{2-4.2}$$

These equations can be shown to possess wavelike solutions for either $\mathbf{E}$ or $\mathbf{B}$. This leads to the conclusions that electromagnetic effects of one body on another must travel at the speed of light, $c=(1/\mu_0\varepsilon_0)^{1/2}$.

In the simplest theory of electromagnetics, material bodies are assumed to possess electric polarization **P** and magnetization **M**†. These are defined by the equations

$$\mathbf{P} = \mathbf{D} - \varepsilon_0 \mathbf{E} \tag{2-4.3}$$

$$\mathbf{M} = \frac{\mathbf{B}}{\mu_0} - \mathbf{H} \tag{2-4.4}$$

Stationary Media

For a stationary, rigid body where **E** and **B** are considered to be independent, constitutive equations of the following form must be prescribed:

$$\mathbf{P} = \mathbf{P}(\mathbf{E}, \mathbf{B})$$

$$\mathbf{M} = \mathbf{M}(\mathbf{E}, \mathbf{B})$$

and

$$\mathbf{J} = \mathbf{J}(\mathbf{E}, \mathbf{B})$$

In the classical *linear* theory of isotropic rigid, stationary, electromagnetic materials these equations take the form

$$\mathbf{P} = \varepsilon_0 \eta \mathbf{E} \quad \text{or} \quad \mathbf{D} = \varepsilon_0 (1 + \eta) \mathbf{E} \tag{2-4.5}$$

$$\mathbf{M} = \chi \mathbf{H} \quad \text{or} \quad \mathbf{B} = \mu_0 (1 + \chi) \mathbf{H} \tag{2-4.6}$$

and

$$\mathbf{J} = \sigma \mathbf{E} \tag{2-4.7}$$

The constant η is called the electric susceptibility and χ is the magnetic susceptibility. These constants as well as the electric conductivity σ can have a strong dependence on the temperature.

Additional constants μ and ε are defined by

$$\mu = \mu_0 (1 + \chi), \qquad \varepsilon = \varepsilon_0 (1 + \eta)$$

At low temperatures (between 0 and 20°K) many materials become superconducting, that is, $\sigma \rightarrow \infty$. In this state, steady closed currents can persist indefinitely. Thus voltage drops across superconducting circuits become zero for steady currents. However, for time-varying currents, an electric force is required to change the momentum of the electrons, which is proportional to **J**. Thus Ohm's law (2-4.7) is replaced by a relation of the form

$$\mathbf{E} = \mu_0 \lambda^2 \dot{\mathbf{J}} \tag{2-4.8}$$

However, this relation does not completely describe the macroscopic

†These effects can be viewed as dipole effects. One might also assume that higher multipole polarizations exist in the material.

behavior of superconductors. Of equal or greater importance are the values of critical temperature T_c, magnetic field H_c, and current density J_c, at which the normal material becomes superconducting. The set of values of T, H, and J for which the material is superconducting is bounded by a surface $J_c = f(H_c, T_c)$.

Such a transition surface is analogous to the yield surface in plasticity. Further discussion of the properties of superconducting materials is given in Chapter 6.

Moving Media

It is known from elementary physics that the motion of a conductor in a steady magnetic field can create an electric field or voltage that can induce the flow of current in the conductor. Thus the electric field in a moving frame of reference ($\mathbf{E}'$) relative to that in a stationary frame ($\mathbf{E}$) differs by a term proportional to the velocity and the magnetic field. This effect is one of the principal interactions between mechanics and electromagnetics, and is expressed mathematically by the relation

$$\mathbf{E}' = \mathbf{E} + \mathbf{v} \times \mathbf{B} \tag{2-4.9}$$

Similar relations hold for the other electromagnetic variables (see, e.g., Sommerfeld, 1952), but are not as important in magneto-solid mechanics problems as Eq. (2-4.9). These additional relations are listed below for completeness:

$$\mathbf{D}' = \mathbf{D} + \mathbf{v} \times \frac{\mathbf{H}}{c^2} \tag{2-4.10}$$

$$\mathbf{P}' = \mathbf{P} + \mathbf{v} \times \mathbf{M} \tag{2-4.11}$$

$$\mathbf{H}' = \mathbf{H} - \mathbf{v} \times \mathbf{D} \tag{2-4.12}$$

$$\mathbf{B}' = \mathbf{B} - \mathbf{v} \times \frac{\mathbf{E}}{c^2} \tag{2-4.13}$$

$$\mathbf{M}' = \mathbf{M} - \mathbf{v} \times \mathbf{P} \tag{2-4.14}$$

$$\mathbf{J}' = \mathbf{J} - q\mathbf{v} \tag{2-4.15}$$

where $c^2 = 1/(\mu_0 \varepsilon_0)$ is the square of the speed of light in vacuum. These equations are valid for velocities small compared with the speed of light.

Of importance to magneto-solid mechanics is the constitutive relation for current density; that is,

$$\mathbf{J}' = \sigma \mathbf{E}'$$

or

$$\mathbf{J} = \sigma(\mathbf{E} + \mathbf{v} \times \mathbf{B}) + q\mathbf{v} \tag{2-4.16}$$

Thus a moving conductor in a stationary magnetic field will have induced currents even if the initial electric field $\mathbf{E}=0$, or charge $q=0$.

For most problems in magneto-solid mechanics, $q \cong 0$ in good conductors, so that $\mathbf{J}'=\mathbf{J}$. Also, in time-dependent field problems where currents are induced, $|\mathbf{E}| \sim |\mathbf{v}||\mathbf{B}|$, so that for velocities less than the speed of light $\mathbf{B}' \cong \mathbf{B}$. Similarly, for nonpolarizable materials, $\mathbf{P}'=0$, and one can show that $\mathbf{M}'=\mathbf{M}$ for $v^2/c^2 \ll 1$. Thus, for most problems, the electric field will be the only variable that differs significantly between the moving material and the stationary observer.

A proper theoretical treatment of electromagnetics should be approached from the special and general theories of relativity, introducing such concepts as the energy-momentum tensor. For most engineering applications this approach is not needed and a classical, nonrelativistic treatment will be presented here. The basic paradox in electromagnetics of moving media is that the equations of electromagnetics are Lorentz-invariant,† while the classical balance equations of momentum, energy, and entropy are Galilean-invariant. This paradox manifests itself in the fact that the field variables $\mathbf{E}$ and $\mathbf{B}$ and so forth, as seen by two observers in different moving reference frames, are not the same. Thus, while relativity is not used directly, additional equations must be given for the change in $\mathbf{E}$ and $\mathbf{B}$ and so forth when one is in a moving reference along with the balance laws of mechanics and electromagnetics.

A general continuum theory of electromagnetic deformable media from the point of view of special relativity has been given by Grot and Eringen (1966), Penfield and Haus (1967), and Hutter (1973).

2.5 FERROMAGNETIC MATERIALS

One of the confusing features of electromagnetism is the existence of two measures of magnetic effects, that is, the magnetic-flux density $\mathbf{B}$ and the magnetic intensity $\mathbf{H}$. In the absence of magnetic material, $\mathbf{B}$ and $\mathbf{H}$ are identical and in MKSC units, are related by the constant $\mu_0 = 4\pi 10^{-7}$; $\mathbf{B}=\mu_0\mathbf{H}$. In a magnetic material, however, $\mathbf{B}$ and $\mu_0\mathbf{H}$ differ by the magnetization $\mathbf{M}$, that is,

$$\mathbf{B}=\mu_0(\mathbf{H}+\mathbf{M}) \tag{2-5.1}$$

$\mathbf{M}$ can be induced by external magnetic fields or in some materials can spontaneously exist in the absence of external fields. For nonferromagnetic isotropic materials we can relate $\mathbf{M}$ and $\mathbf{H}$ through a linear constitutive law

$$\mathbf{M}=\chi\mathbf{H} \tag{2-5.2}$$

For metals, such as aluminum or copper, $\chi \sim 10^{-4}$.

†See Sommerfeld (1952) or Penfeld and Haus (1967) for a discussion of these ideas.

For practical purposes, we neglect the induced magnetization in paramagnetic ($\chi > 1$) or diamagnetic ($\chi < 1$) materials. For linear ferromagnetic materials, however, we can have $\chi \sim 10^4$ or higher, and for spontaneous or "permanent" magnetization, $\mu_0 \mathbf{M} \sim 1\,\mathrm{T}$ or higher. The principal metals in this category are iron, nickel, and cobalt and their alloys.

The source of spontaneous magnetization lies in a property of atomic particles called "spin." Spin is a concept which endows atom particles with intrinsic angular momentum. When a particle moves with velocity v in a circular orbit of radius r, the angular momentum $\mathbf{h}$ is given by

$$\mathbf{h} = m\mathbf{r} \times \mathbf{v} \tag{2-5.3}$$

If the particle is an electron with charge $-e$, this circular motion produces a circulation current. Such a small circular current behaves magnetically like a magnetic dipole or magnetic moment $\mathbf{m}$, where

$$\mathbf{m} = -\frac{1}{2}\oint \mathbf{r} \times e\mathbf{v}\, ds \tag{2-5.4}$$

For a circular orbit,

$$\mathbf{m} = \frac{-e}{2m}\mathbf{h} \tag{2-5.5}$$

Thus an orbiting electron has both angular momentum as well as magnetic moment.

The relationship between angular momentum or spin and magnetic moment can be extended to isolated electrons. Thus every electron acts simultaneously as a gyroscope and a magnetic dipole. In nonferromagnetic materials the thermal motion of the electrons cancels out any net alignment of magnetic moments except in a strong magnetic field.

In ferromagnetic materials there is a tendency for electron spins to spontaneously align themselves to produce a net magnetic moment. Spontaneous magnetization takes place below a critical temperature T_c called the Curie temperature. The Curie temperature for a few metals is shown in Table B-3.

The connection between magnetization and angular momentum has been demonstrated experimentallly. A ferromagnetic cylinder suspended as a torsional pendulum can be observed to rotate when the initial magnetization along the cylinder is rotated by 180° by an external field. The change in magnetization changes the intrinsic angular momentum of the electrons and the body rotates in order to conserve angular momentum.

Intrinsic angular momentum is not a concept incorporated in classical continuum mechanics. Thus, in recent decades, so-called microcontinuum theories have been developed in which the material has intrinsic angular momentum, couple stresses, and nonsymmetric stress tensors (see, e.g., Tiersten, 1964; Eringen, 1963).

For further qualitative discussion of spin and ferromagnetism, the reader is referred to Feynman et al. (1964) (Vol. II, Chapter 37). For more advanced discussion, see Cullity (1972).

Domains

Between the atomic scale of electron spin and macroscopic magnetization lies another substructure called domains. Below the Curie temperature, electron spins of neighboring atoms will spontaneously align themselves only over a certain number of atoms, beyond which the alignment of magnetic moments will change direction as shown in Figure 2-3. Each region of aligned spins is called a *domain*. In an applied magnetic field, domains aligned with the field can grow at the expense of those that are not, thereby creating a net magnetization. Thus, while domains owe their microscopic magnetization to spin alignment, macroscopic magnetization **M** is an average of **m** over many domains.

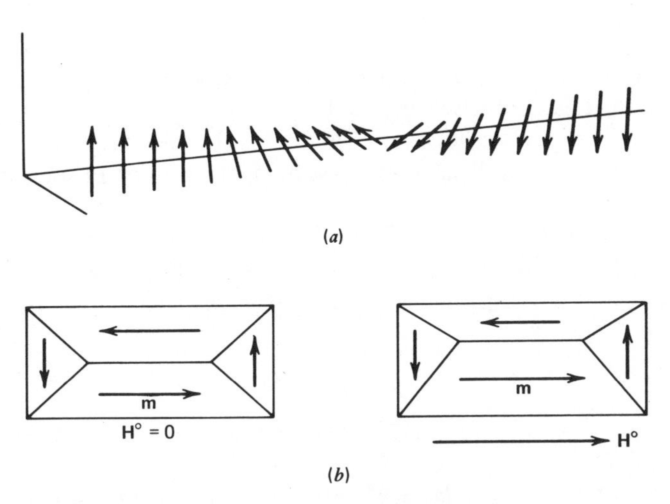

Figure 2-3 (*a*) Alignment of spins or magnetic moments in a linear array of atoms and (*b*) alignment of magnetic moments in domains.

Soft and Hard Ferromagnetism

The relation between **M** and **H** for ferromagnetic material is often written as **B**(**H**) using Eq. (2-5.1). This relation, however, is not a single-valued function and depends on the history of **H** as shown in Figure 2-4. For large |**H**|, **M** reaches a saturation value $\mathbf{M}_s$, and the **B**(**H**) curve takes the form

$$\mathbf{B} = \mu_0(\mathbf{H} + \mathbf{M}_s)$$

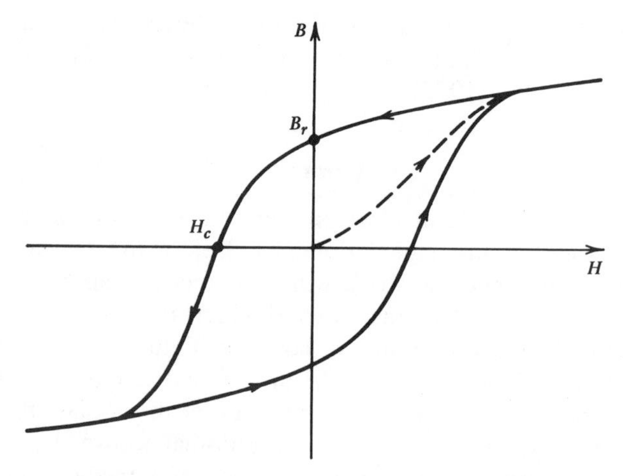

Figure 2-4 Dashed line, B–H curve for nonhysteretic material; solid lines, B–H constitutive law showing hysteresis—B_r is the remnant field and H_c is the coercive force.

For cyclic variations of **H** with zero mean, the **B(H)** relation is an open loop curve or hysteresis curve (see Fig. 2-4).

There are two idealized classes of materials. In one, the hysteresis loop is small and one can write **B** as a single-valued function of **H** such that

$$\mathbf{B}(\mathbf{H}) \to 0 \quad \text{as} \quad \mathbf{H} \to 0$$

Such materials are called *soft* magnetic materials. Transformer-grade steels are in this category. A special case is the *soft*,† *linear* ferromagnetic material where

$$\mathbf{B} = \mu_0 \mu_r \mathbf{H} \tag{2-5.6}$$

and where μ_r is called the "relative permeability." Typical values of μ_r are shown in Table B–3.

Materials for which $\mathbf{B} \to B_r$ as $\mathbf{H} \to 0$ are called *hard* magnetic materials. The so-called permanent magnetic materials are in this category. The residual magnetic-flux density B_r is called the *remnance*. If **H** is increased such that $M \to M_s$, and then reversed such that $H < 0$, the value of $-H$ such that $B \to 0$ is called the *coercivity* or *coercive force* H_c. The second quadrant of the B–H curve is called the *demagnetization curve* (see Fig. 2-5) and is important in the design of devices with permanent magnets.

To obtain a B–H curve, one takes a toroidal specimen (e.g., a short length of a hollow cylinder) and wraps two pairs of coils around the torus as shown in Figure 2-6. Through the N_1 turns of coil #1, one puts a current I and across the N_2 turns of coil #2 one measures the induced voltage $V(t)$, produced by

†Soft does not refer to its mechanical hardness.

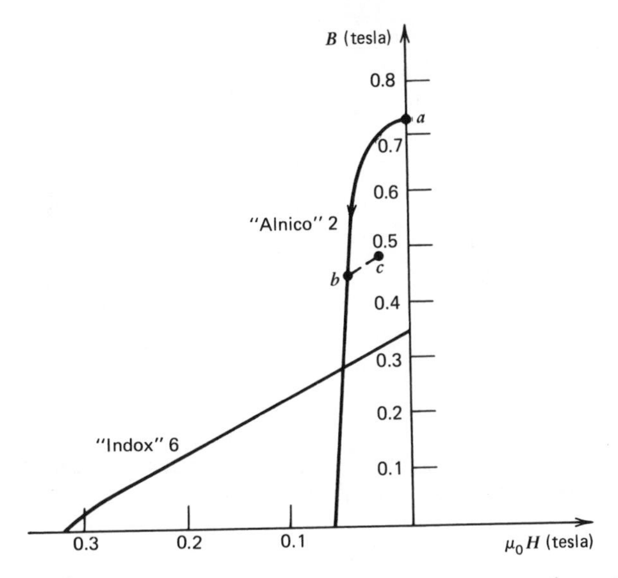

Figure 2-5 Demagnetization curves for two permanent magnetic materials; *b–c* is called the recoil curve.

the changing flux in the torus. The integral forms of Ampere's law and Faraday's law give the following relationships:

$$H = \frac{N_1 I}{2\pi R}$$
$$B = \frac{1}{N_2 A} \int_0^t V(\tau)\, d\tau \qquad (2\text{-}5.7)$$

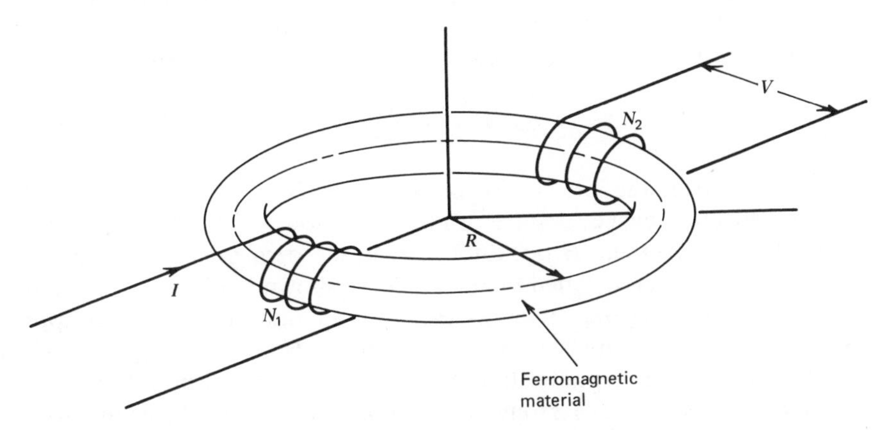

Figure 2-6 Sketch of ferromagnetic toroid for determining the *B–H* curve.

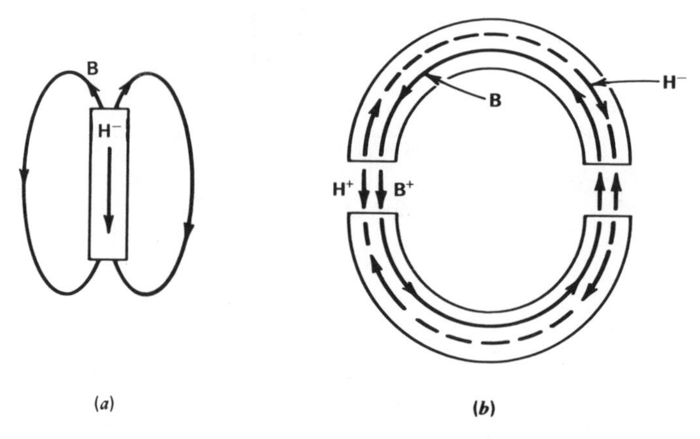

Figure 2-7 (*a*) Permanent magnetic material showing demagnetizing field H and (*b*) split permanent magnetic toroid with air gaps.

where R is the major radius of the torus, A is the cross-sectional area of the torus, and V is the voltage.

It is important to note that if the flux lines leave the material, as in the bar magnet in Figure 2-7*a*, then H inside the material cannot be zero and is negative. This follows from the fact that when $I = 0$,

$$\oint \mathbf{H} \cdot d\mathbf{l} = 0$$

Since $\mathbf{H}_{\text{out}} = \mathbf{B}_{\text{out}}/\mu_0 \neq 0$, then $\mathbf{H}_{\text{in}}$ must be a demagnetization field. Hence the importance of the demagnetization curve.

Another important feature of the B–H curve for permanent magnets is the *recoil curve* shown in Figure 2-5. Suppose our torus is split in two [see Fig. 2-7*b*], with $I = 0$ and no gap between the halves; then $\mathbf{H} = 0$ and we start at point *a* on the B–H curve in Figure 2-5. As the two halves are pulled apart by an amount u, the field $\mathbf{H}$ inside the material, $\mathbf{H}^-$, must satisfy the relation

$$\mu_0 H^- = -\frac{u}{\pi R} B(H^-) \tag{2-5.8}$$

H^- is found from the intersection of the curves $B = -\mu_0(\pi R/u)H^-$, and the demagnetization curve (point *b* in Fig. 2-5). However, if the two halves are moved back together, B does not retrace the demagnetization curve, but follows another curve called the recoil curve shown in Figure 2-5 as point *c*. In many ferromagnetic materials the slope of the recoil curve is approximately equal to the slope of the B–H curve at $H = 0$. This again shows the importance of the magnetization history of the material in determining the field B. In this respect ferromagnetic constitutive behavior is analogous to plasticity in solid mechanics.

Magnetostriction

Thus far we have not mentioned the effect of magnetization on the elastic properties of the material. As shown in Table 2-1, there is a catalog of effects of magnetization and strain in ferromagnetic solids. These phenomena and their reciprocal effects (changes in **M** due to stress or strain) are generally referred to as magnetostriction. As discussed above, the spontaneous magnetization at the domain level is due to the mutual alignment of electron spins in the $3d$ subshell in Fe, Co, and Ni. Small macroscopic change in shape due to magnetic fields is attributed to the change in the shape and size of the orbits of these $3d$ electrons due to alignment of electron spin with the applied magnetic field. There are both reversible and irreversible magnetostrictive effects, though we will only discuss the reversible effects here. (See Bozorth, 1951, or Cullity, 1972 for a more detailed discussion.)

The most elementary magnetostrictive effect is the change in length with longitudinal field. If one defines a strain measure

$$\varepsilon = \frac{\Delta L}{L}$$

where L is the original length of a specimen and ΔL the change in length, then a typical strain/magnetic field curve for an unstressed rod in a field

Table 2-1 Magnetostrictive Phenomena[a]

Phenomenon	Name
Change in length in direction of magnetic field, $\mathbf{\Delta L} \times \mathbf{H} = 0$	Joule effect
Change in **B** due to longitudinal stress	Villari–Matteucci effect
Change in dimension transverse to **H**, $\mathbf{\Delta L} \cdot \mathbf{H} = 0$	Transverse Joule effect
Change in **B** due to transverse stress	Transverse Villari effect
Change in Young's modulus due to **M**	
Bending due to a magnetic field	Guillemin effect
Change in **M** due to bending	Reciprocal Guillemin effect
Twist due to circular and longitudinal fields	Wiedemann effect
Longitudinal magnetization due to twist of a circularly magnetized rod	Inverse Wiedemann effect
Change in shear modulus due to **M**	
Change in volume due to **M**	Barrett effect
Change in **M** due to volume change	Nagaoka–Honda effect
Change in bulk modulus due to **M**	

[a]See, for example, Williams (1931).

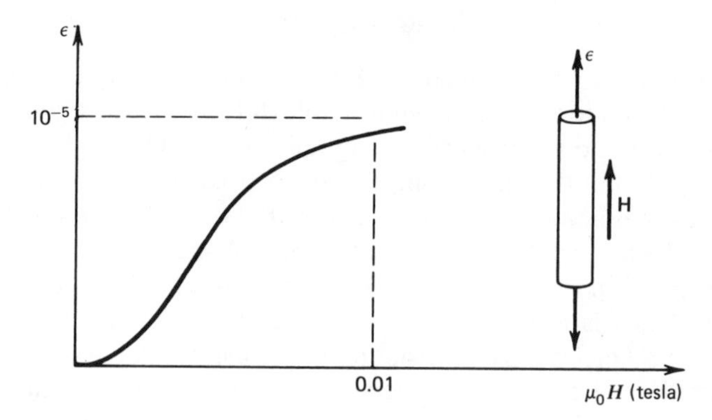

Figure 2-8 Typical low-field magnetostriction behavior for ferromagnetic materials—longitudinal strain vs. applied field.

parallel to its length is shown in Figure 2-8 for a single crystal. The first part of the curve is primarily due to domain rotation and does not change the volume of the material. Hence a change in length parallel to the applied field must be accompanied by a contraction in transverse dimensions.

For polycrystalline material such as iron, the magnetostrictive strain can reverse sign as shown in Figure 2-9. The reciprocal effect of stress on the magnetization curves for nickel is shown in Figure 2-10. It should be noted

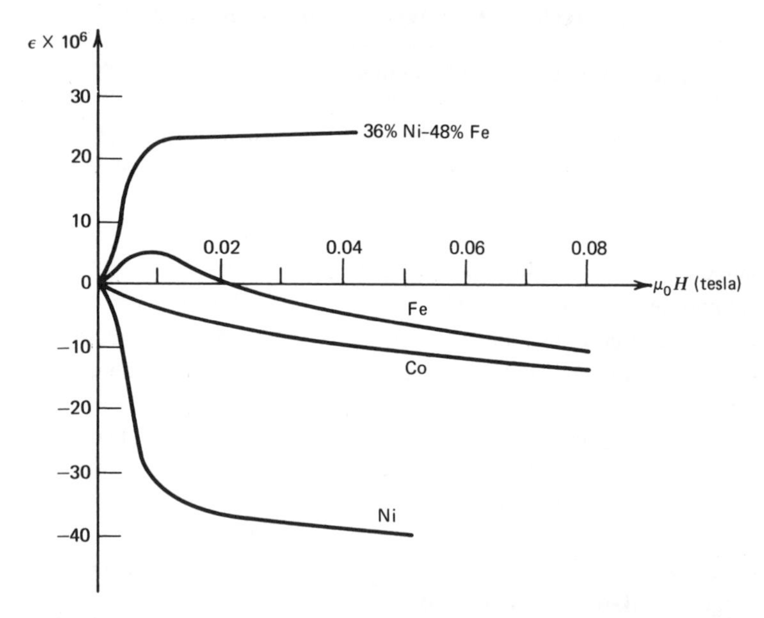

Figure 2-9 Magnetostrictive longitudinal strain vs. applied field for four ferromagnetic materials (after Cullity, 1972).

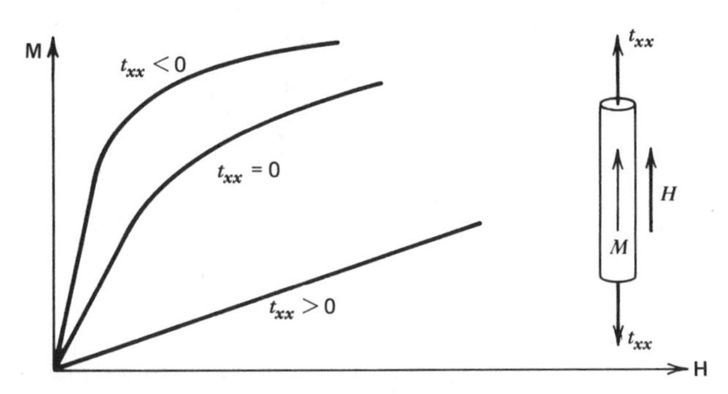

Figure 2-10 The effect of applied longitudinal stress on the induced magnetization constitutive law $M(H)$. (Reprinted with permission from C. D. Cullity, *Introduction to Magnetic Materials*, Addison-Wesley Publ. Co., copyright 1972.)

that the size of the magnetostrictive strains is very small, saturating around 10^{-5}. If the material were constrained so that $\varepsilon = 0$, the stress induced would be of the order of 200 N/cm^2 or 300 psi. Thus, for most structural ferromagnetic materials, one can often neglect magnetostrictive-related stresses. However, the magnetostrictive effects are extremely useful in transducer applications to change electromagnetic energy to sound or vibration waves in fluids or solids.

The effect of the magnetization on the elastic moduli of structural ferromagnetic materials is also small. The maximum change in Young's modulus for iron is 1% and for nickel is 6%.

Magnetostrictive effects are *not* analogous to piezoelectricity. Piezoelectric effects are changes in strain due to electric polarization in dielectric materials. The change of strain is linear in the electric field and occurs *only* in dielectrically anisotropic materials. However, magnetostrictive strains are nonlinear in the magnetic fields and can occur in isotropic materials.

Conventional stress–strain relations are of the form (repeated indices are summed over k, $l = 1, 2, 3$)

$$t_{ij} = c_{ijkl}\varepsilon_{kl} + \beta(T - T_0)\delta_{ij} \tag{2-5.9}$$

where c_{ijkl} represents the elastic constants and β represents the thermoelastic effect. Magnetostrictive effects add a term of the form

$$t'_{ij} = b_{ijkl}M_k M_l \tag{2-5.10}$$

(Here i or j represents one of three mutually orthogonal directions or planes normal to these directions; t_{ij} represents the stress on the plane with a normal in the ith direction and stress in the jth direction; t_{ii} represents a normal traction; and t_{ij} $(i \neq j)$ represents a shear or tangential traction.)

The reciprocal magnetostrictive effects (stress-induced magnetization) require specification of the B–H curve properties. For a soft, linear ferromagnetic material, these take the form

$$H_i = \chi_{ij}^{-1} M_j + b_{ijkl} \varepsilon_{kl} M_l \tag{2-5.11}$$

For isotropic materials the expression (2-5.10) takes the form

$$t'_{ij} = b_0 M^2 \delta_{ij} + b_1 M_i M_j \tag{2-5.12}$$

or

$$t'_{11} = (b_0 + b_1) M_1^2 + b_0 (M_2^2 + M_3^2)$$

$$t'_{12} = b_1 M_1 M_2$$

$$\mathrm{Tr}(\mathbf{t}') = (3 b_0 + b_1) M^2$$

In Eq. (2-5.12) we see that $b_0 + b_1$ represents the Joule effect, while b_0 represents the transverse Joule effect (see Table 2-1).

The reader should be cautioned that these relations describe an idealized model. In actual structural, ferromagnetic materials the magnetostrictive effects are nonlinear and hysteretic. What is more important perhaps is that the stresses induced by magnetostrictive effects are usually negligible compared to conventional structural stresses.

Also it should be mentioned that the constitutive relation for stress, strain, and magnetic fields must be examined in the general context of equations of motion, energy balance, and boundary conditions. A more complete discussion of these effects may be found in Brown (1966). Further discussion of these effects will be given in Chapter 3.

2.6 ELECTROMAGNETIC FORCES

The force on a stationary charge Q_1 at position $\mathbf{r}_1$, due to a charge Q_2 at $\mathbf{r}_2$, is given by an inverse-square law

$$\mathbf{F} = \frac{Q_1 Q_2 (\mathbf{r}_1 - \mathbf{r}_2)}{4\pi\varepsilon_0 |\mathbf{r}_1 - \mathbf{r}_2|^3} \tag{2-6.1}$$

It is customary to replace the effect of Q_2 by a vector

$$\mathbf{E} = \frac{Q_2 (\mathbf{r}_1 - \mathbf{r}_2)}{4\pi\varepsilon_0 |\mathbf{r}_1 - \mathbf{r}_2|^3} \tag{2-6.2}$$

or

$$\mathbf{F} = Q_1 \mathbf{E}$$

In a good conductor the net charge density is zero so that electric-field-related forces are zero in the interior of the conductor. (Electric forces may act on the surface of a good conductor where charges can accumulate.) However, moving charges in conductors (usually free electrons) experience a body force given by

$$\mathbf{f} = \mathbf{J} \times \mathbf{B}$$

For example, the force per unit length on a long filament with current I_1 in a transverse magnetic field B_0 is given by $I_1 B_0$. When the external field B_0 is created by a parallel filament with current I_2 the force per unit length is attractive if the currents have the same sense and is given by $\mu_0 I_1 I_2 / 2\pi R$. Here the force acts along the line joining the two filaments separated by a distance R.

For a closed circuit C_1, carrying steady current I_1 (see Fig. 2-11), the total force is given by the integral

$$\mathbf{F} = \oint_{C_1} I_1 \, d\mathbf{S}_1 \times \mathbf{B}$$

If $\mathbf{B}$ is created by another circuit C_2, with steady current I_2, this field is given by the expression

$$\mathbf{B} = \frac{\mu_0}{4\pi} \oint_{C_2} \frac{I_2 \, d\mathbf{S}_2 \times \mathbf{R}}{R^3} \tag{2-6.3}$$

This is a generalization of the law of Biot and Savart found in 1820 for a straight wire. Here $d\mathbf{S}_1$ and $d\mathbf{S}_2$ are differential vectors directed along the direction of currents I_1 and I_2, respectively, and $\mathbf{R}$ is a position vector from element $d\mathbf{S}_2$ to $d\mathbf{S}_1$. Thus the total force on one circuit due to current in another is given by a double integral

$$\mathbf{F} = \frac{\mu_0 I_1 I_2}{4\pi} \oint_{C_1} \oint_{C_2} \frac{d\mathbf{S}_1 \times (d\mathbf{S}_2 \times \mathbf{R})}{R^3} \tag{2-6.4}$$

This form is useful for the numerical calculation of magnetic forces between

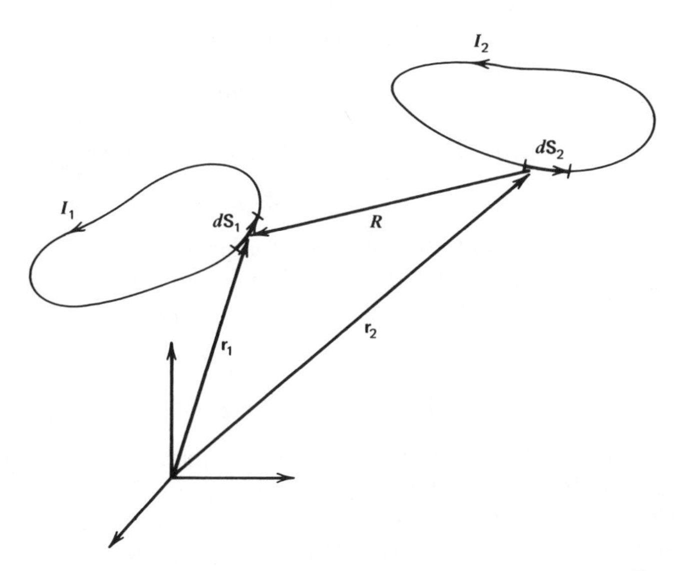

Figure 2-11 Sketch of the geometry of two interacting current filaments.

circuits. Note that the relation $|\mathbf{F}|/\mu_0 I_1 I_2$ forms a nondimensional group discussed in Chapter 1.

In a bulk solid when the body has a charge density q and current density $\mathbf{J}$, the body force per unit volume when $\mathbf{M}=\mathbf{P}=0$ is given by

$$\mathbf{f} = q\mathbf{E} + \mathbf{J} \times \mathbf{B} \tag{2-6.5}$$

When $q=\mathbf{M}=\mathbf{P}=0$, the total force on a body is then

$$\mathbf{F} = \int \mathbf{J} \times \mathbf{B}\, dv$$

This expression can be further simplified if we write $\mathbf{B}$ as the sum of two fields $\mathbf{B} = \mathbf{B}_0 + \mathbf{B}_1$; $\mathbf{B}_0$ is due to sources outside the body and $\mathbf{B}_1$ is due to the currents in the body. For low-frequency or stationary fields the self-force integral of $\mathbf{J} \times \mathbf{B}_1$ over the body is zero, so that we can write

$$\mathbf{F} = \int \mathbf{J} \times \mathbf{B}_0\, dv \tag{2-6.6}$$

When a body can be magnetized ($\mathbf{M} \neq 0$), additional forces arise. For example, the force on a small but thin magnetized needle or bar magnet can be represented by two forces acting on the poles or ends of the needle. These forces are given by $\mathbf{F}_1 = p_m \mathbf{B}_1$ and $\mathbf{F}_2 = -p_m \mathbf{B}_2$, where $\mathbf{B}_1$ and $\mathbf{B}_2$ are values of the external field at the poles and p_m is the strength of the poles. One can see that the net force on the needle is zero if the field is uniform. Thus the total force is proportional to the gradient of the external field. For a small ellipsoid of uniform magnetization $\mathbf{M}$ in a field $\mathbf{B}_0$ due to other sources outside the body this force is given by

$$\mathbf{F} = V\mathbf{M} \cdot \nabla \mathbf{B}_0 \tag{2-6.7}$$

where V is the volume of the ellipsoid. Returning to the example of the bar magnet, one can see that when $\mathbf{B}_1 = \mathbf{B}_2$ the forces on the poles form a couple. In the case of the magnetized ellipsoid, this couple is given by

$$\mathbf{C} = V\mathbf{M} \times \mathbf{B}_0 \tag{2-6.8}$$

Thus a magnetized body tends to rotate in order to align the magnetization with the applied field. Further discussion of forces and couples in magnetized materials is given in Chapters 3 and 7.

Electromagnetic Stresses

It is one of the peculiar features of magnetomechanics that the body force (2–6.5) can also be represented as a distributed surface force. These surface forces or tractions can in turn be related to a most intriguing idea, namely, electromagnetic stresses (see also Chapter 1 for a short introduction). In mechanics the concept of a stress tensor or matrix is related to tractions on the surface of a body by writing the traction vector $\boldsymbol{\tau}$ as the inner product of a 3×3 matrix $\mathbf{T}$ and the surface normal (see Fig. 2-12a),

$$\boldsymbol{\tau} = \mathbf{T} \cdot \mathbf{n} \quad \text{or} \quad \begin{Bmatrix} \tau_1 \\ \tau_2 \\ \tau_3 \end{Bmatrix} = \begin{bmatrix} T_{11} & T_{12} & T_{13} \\ T_{21} & T_{22} & T_{23} \\ T_{31} & T_{32} & T_{33} \end{bmatrix} \begin{Bmatrix} n_1 \\ n_2 \\ n_3 \end{Bmatrix} \tag{2-6.9}$$

(In Cartesian components $\tau_i = \sum t_{ij} n_j$, where the sum is over the index j.) The integral of the traction $\boldsymbol{\tau}$ over the surface of the body gives the total force

$$\mathbf{F} = \int \mathbf{T} \cdot \mathbf{n} \, da \tag{2-6.10}$$

If one assumes that the components of $\mathbf{T}$ are continuous and differentiable functions of position inside the body, then one can apply Gauss's theorem and rewrite (2-6.10) as a volume integral, that is,

$$\mathbf{F} = \int \boldsymbol{\tau} \, da = \int \boldsymbol{\nabla} \cdot \mathbf{T} \, dv \tag{2-6.11}$$

Thus $\boldsymbol{\nabla} \cdot \mathbf{T}$ can be interpreted as a body force.

The inverse procedure for electromagnetic (em) forces can be taken if we

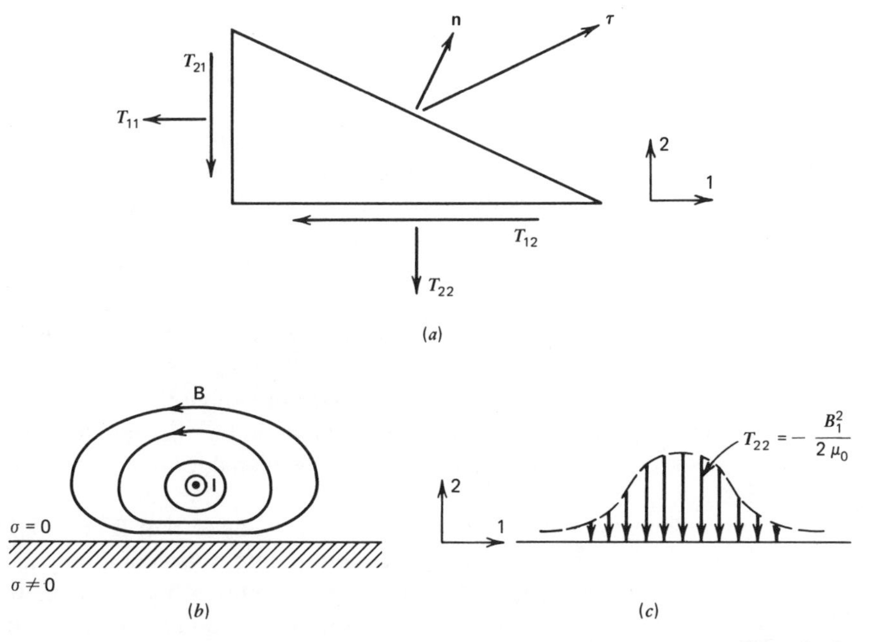

Figure 2-12 (a) Definition of stress components T_{ij} and stress vector $\boldsymbol{\tau}$, (b) exclusion of flux from a good conductor, and (c) magntic pressure distribution on the surface of the conductor in (b).

can relate the electromagnetic body forces to the divergence of a tensor $\mathbf{T}^{em}$. In this section we consider *nonpolarizable* and *nonmagnetizable* material so that the electromagnetic body force is given by Eq. (2-6.5), that is,

$$\mathbf{f} = q\mathbf{E} + \mathbf{J} \times \mathbf{B}$$

By manipulating Maxwell's equations (2-2.1)–(2-2.5), one can show

$$\mathbf{f} = \nabla \cdot \mathbf{T}^{em} - \frac{\partial}{\partial t}(\varepsilon_0 \mathbf{E} \times \mathbf{B}) \tag{2-6.12}$$

where $\mathbf{T}^{em}$ is given by

$$\mathbf{T}^{em} = \varepsilon_0\left(\mathbf{EE} - \frac{1}{2}E^2\boldsymbol{\delta}\right) + \frac{1}{\mu_0}\left(\mathbf{BB} - \frac{1}{2}B^2\boldsymbol{\delta}\right) \tag{2-6.13}$$

and where $\boldsymbol{\delta}$ is the identity tensor or matrix with 1's on the diagonal and zeros elsewhere.

The total electromagnetic force on a body carrying electric charges and currents can be written as

$$\mathbf{F} = \int \mathbf{T}^{em} \cdot \mathbf{n}\, da - \frac{d}{dt}\int \varepsilon_0 \mathbf{E} \times \mathbf{B}\, dv \tag{2-6.14}$$

If there are no other forces on the body, then Newton's law can be written in the form

$$\int \mathbf{T}^{em} \cdot \mathbf{n}\, da = \frac{d}{dt}\int (\rho\mathbf{v} + \varepsilon_0 \mathbf{E} \times \mathbf{B})\, dv \tag{2-6.15}$$

In this form, some texts interpret $\varepsilon_0 \mathbf{E} \times \mathbf{B}$ as an *electromagnetic momentum density*. This interpretation is somewhat artificial, however, since one could rewrite $\varepsilon_0 \mathbf{E} \times \mathbf{B}$ and $\mathbf{T}^{em}$ as a body force. In any case, for low frequencies or stationary fields, we can drop the $\mathbf{E} \times \mathbf{B}$ term.

For magnetic problems with currents and no charges we have

$$\int \mathbf{J} \times \mathbf{B}\, dv = \int_\Sigma \frac{1}{\mu_0}\left(\mathbf{BB} \cdot \mathbf{n} - \frac{1}{2}B^2\mathbf{n}\right) da \tag{2-6.16}$$

If we want the total force on the body, we need only know the value of the fields at the surface of the body. In fact one can extend the theorem to where Σ can be any closed surface surrounding the body which does not contain any other bodies.

In some problems the field is excluded from the body so that $\mathbf{B} \cdot \mathbf{n} = 0$. In this case the total force on the body is equivalent to integration of a *magnetic pressure* over the surface, $B^2/2\mu_0$ as illustrated in Figure 2-12 (see Chapter 1). The concept of electromagnetic stresses is useful in calculating magnetic forces when the field is known explicitly and the current is unknown. In numerical calculations it has the advantage that it replaces a volume integration by a surface integral (2-6.16).

2.7 ENERGY DISSIPATION AND POYNTING'S THEOREM

Heat Generation

In a normal, *stationary* conductor the work done by the electric field in moving charges through a resistive medium can be represented by a volume heat energy source r given by

$$r = \mathbf{J} \cdot \mathbf{E} = \frac{J^2}{\sigma} \tag{2-7.1}$$

where the right-hand equal sign holds for a linear isotropic conductor. This is sometimes called *Joule* heating. If the body is moving, with *zero charge density*, we can write $\mathbf{E}$ in terms of the field as seen in a reference moving with the body $\mathbf{E}'$, so that using Eqs. (2-4.9) and (2-4.15).

$$\mathbf{J} \cdot \mathbf{E} = \mathbf{J} \cdot \mathbf{E}' - \mathbf{J} \cdot \mathbf{v} \times \mathbf{B}$$

or

$$\mathbf{J} \cdot \mathbf{E} = \mathbf{J} \cdot \mathbf{E}' + (\mathbf{J} \times \mathbf{B}) \cdot \mathbf{v} \tag{2-7.2}$$

The second term on the right-hand side is the power produced by the body force. In the absence of other forces, this term will change the macroscopic kinetic energy of the body. Thus, for a moving body, $\mathbf{J} \cdot \mathbf{E}$ represents the work done in changing the kinetic energy as well as heat generation. For a moving body with no charges the heat generation is given by

$$r = \mathbf{J} \cdot \mathbf{E}' = \mathbf{J} \cdot (\mathbf{E} + \mathbf{v} \times \mathbf{B})$$

With charges present we replace $\mathbf{J}$ by $\mathbf{J}'$ [see Eq. (2-4.15)], that is,

$$r = \mathbf{J}' \cdot \mathbf{E}' = \frac{(J')^2}{\sigma}$$

For stationary bodies the temperature increase due to Joule heating can be calculated, if heat conduction can be neglected, by the formula

$$\theta = \int_0^t \frac{1}{C\sigma} J^2 \, d\tau \tag{2-7.3}$$

where C is the heat capacity.

Poynting's Theorem and Eddy Currents

Maxwell's equations (2-2.1)–(2-2.5) can be written in the form of an energy balance law if one takes the inner product of Ampere's law, Eq. (2-2.4), with $\mathbf{E}$ and subtracts the inner product of Faraday's law, Eq. (2-2.5), with $\mathbf{H}$. The resulting equation, discovered by Poynting and also Heaviside in 1884, can be written in the form

$$-\nabla \cdot (\mathbf{E}\times\mathbf{H}) = \mathbf{E}\cdot\mathbf{J} + \mathbf{E}\cdot\frac{\partial \mathbf{D}}{\partial t} + \mathbf{H}\cdot\frac{\partial \mathbf{B}}{\partial t} \tag{2-7.4}$$

Integrating this equation over a volume of space and applying the divergence theorem to the left-hand side we obtain

$$-\int_S (\mathbf{E}\times\mathbf{H})\cdot\mathbf{n}\,dA = \int \mathbf{E}\cdot\mathbf{J}\,dv + \frac{\partial}{\partial t}\int W\,dv \tag{2-7.5}$$

For nonhysteretic materials with constant permeability and permittivity, W takes the form

$$W = \frac{1}{2}\left(\varepsilon E^2 + \frac{B^2}{\mu}\right) \tag{2-7.6}$$

This integral has been interpreted as an energy balance with $-\mathbf{E}\times\mathbf{H}\cdot\mathbf{n}$ as the rate of energy input, $\mathbf{E}\cdot\mathbf{J}$ as the rate of energy dissipation, and the last integral as the rate of change of stored energy in the electromagnetic field.

For harmonic fields Eq. (2-7.5) assumes a form useful for the measurement of eddy-current density on the surface of a conductor. If $\mathbf{E} = \hat{\mathbf{E}}e^{i\omega t}$ and $\mathbf{H} = \hat{\mathbf{H}}e^{i\omega t}$, then

$$\mathrm{Re}\left(\int_S \hat{\mathbf{H}}\times\hat{\mathbf{E}}^*\cdot\mathbf{n}\,dA\right) = \int \frac{J^2}{\sigma}\,dv \tag{2-7.7}$$

where the asterisk denotes the complex conjugate. A sensor has been designed on this principle to measure induced currents in a conducting solid (see, e.g., Köhler, 1980).

2.8 ELECTRIC AND MAGNETIC CIRCUIT THEORY

The reader trained in mechanics is no doubt familiar with the way in which distributed kinetic energy in continuous systems can be represented by discrete mass elements and how distributed elastic energy can be represented by discrete elastic springs. The analog in electromagnetics is the localization of electric and magnetic energies in discrete circuit elements such as capacitors and inductors. This leads to a correspondence between Maxwell's partial differential equations and the ordinary differential equations of circuit theory.

Electric Circuits

For example, if one applies the conservation-of-charge law (2-2.1) in integral form (2-2.6) to a surface which encloses one plate of a capacitor (see Fig. 2-13), then one obtains a relation between the current I leaving the capacitor and the total charge; that is,

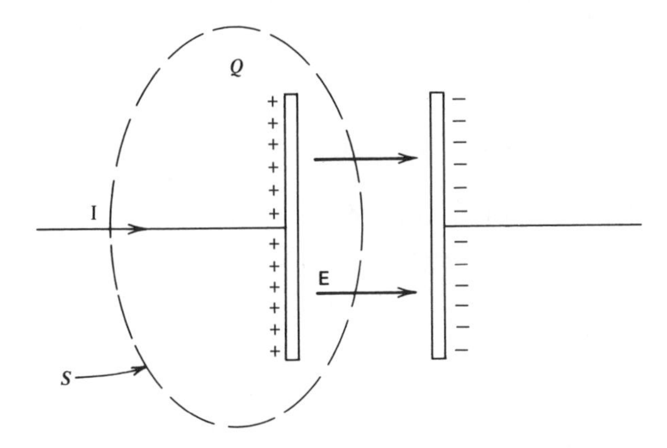

Figure 2-13 Two plates of a capacitor.

$$I = -\frac{\mathrm{d}Q}{\mathrm{d}t} \tag{2-8.1}$$

where

$$I = \int_S \mathbf{J} \cdot d\mathbf{a}, \qquad Q = \int_S q dv$$

If the current is confined to one-dimensional filaments, then at the junction of N filaments with currents $\{I_k\}$ (defined as positive out of the junction) the conservation of charge yields the relation

$$\sum_{k=1}^{N} I_k = 0 \tag{2-8.2}$$

The line integral of electric field along a path is defined as a voltage potential such that

$$V_2 - V_1 \equiv V_{21} = -\int_1^2 \mathbf{E} \cdot d\mathbf{l}$$

The surface integral of the normal component of magnetic field over an area A is defined as the magnetic flux Φ, that is,

$$\Phi = \int_A \mathbf{B} \cdot d\mathbf{a} \tag{2-8.3}$$

Then the integral form of Faraday's law (2-2.10) applied to the circuit becomes

$$V = \frac{d\Phi}{dt} \tag{2-8.4}$$

If positions around a closed circuit are labeled from 1 to N then the integral $\oint \mathbf{E} \cdot d\mathbf{l}$ will be zero and

$$\sum_{k=1}^{N} (V_k - V_{k-1}) = 0 \tag{2-8.5}$$

where $V_0 \equiv V_N$, or the sum of the voltage drops around a closed circuit is zero. Equations (2-8.2) and (2-8.5) are sometimes called Kirchhoff's circuit laws (see, e.g., Reitz and Milford, 1960).

A dissipative element or resistor is one for which

$$\int \mathbf{J} \cdot \mathbf{E} dv > 0$$

In general the relation between voltage drop and current is nonlinear, such as may occur in an arc, a vacuum tube, or solid-state diodes, that is,

$$V = f(I)$$

For a linear resistor Ohm's law (2-4.7) becomes

$$V = IR \tag{2-8.6}$$

where R is called the resistance and is in ohms. Similarly, for a linear capacitor,

$$V = CQ \tag{2-8.7}$$

where C is called the capacitance and is in farads. Expressions (2-8.6) and (2-8.7) are thus analogous to constitutive equations in mechanics.

Magnetic Circuits

In certain problems the magnetic flux is contained within well-defined flux paths, especially when ferromagnetic materials are present (see Fig. 2-14). In such cases the concept of a magnetic circuit is sometimes useful. The flux in each tube of a magnetic circuit is defined by

$$\Phi(s) = \int_A \mathbf{B} \cdot d\mathbf{a} \tag{2-8.8}$$

where s is a coordinate along the tube and $d\mathbf{a}$ is normal to the cross-sectional area of the tube. The conservation-of-magnetic-flux law (2-2.7) applied to a tube of varying cross section requires that

$$\Phi(s) = \text{constant}$$

If several flux tubes intersect, each carrying flux Φ_k, then the conservation of flux becomes

$$\sum \Phi_k = 0 \tag{2-8.9}$$

where positive flux is directed out of the junction.

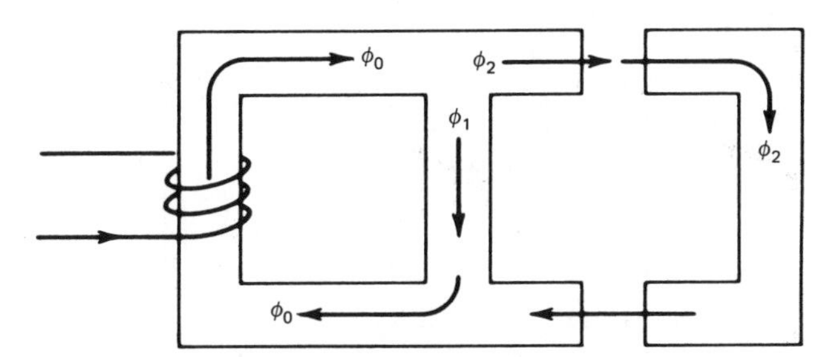

Figure 2-14 Ferromagnetic flux circuit.

In many applications the magnetic circuit is encircled by an electric circuit, as in a transformer (see Fig. 2-14). To find the relation between the current in the electric circuit and flux in the magnetic circuit, one uses the low-frequency form of Ampere's law (2-2.4). If we write $\mathbf{H}$ in the form $\mathbf{H} = \mathbf{B}/\mu$ and assume the electric current I encircles the magnetic circuit N times, then we obtain

$$NI = \Phi \oint \frac{ds}{\mu A} \tag{2-8.10}$$

The integral in Eq. (2-8.10) is called the *reluctance* $\mathcal{R}$ where

$$\mathcal{R} \equiv \oint \frac{ds}{\mu A} \tag{2-8.11}$$

and is analogous to the resistance in electric circuits.

In magnetomechanical devices such as relays and motors, the reluctance also depends on the mechanical variables such as displacements or rotations.

In linear circuits the relation between flux and current is written in the form

$$\Phi = LI \tag{2-8.12}$$

where L is called the inductance of the magnetic circuit. Thus the voltage across an electric circuit which creates a flux Φ is given by

$$V = \frac{dLI}{dt} \tag{2-8.13}$$

The interaction between discrete electromagnetic elements (circuits) and discrete mechanical systems will be discussed in Chapter 3.

In many applications permanent magnetic materials are used in a magnetic circuit with a gap as shown in Figure 2-15*a*. For a given gap volume and gap magnetic field, there is an optimum circuit reluctance $\mathcal{R}$ which will minimize the volume of permanent magnetic material required. As discussed in Section 2.5, the magnetic intensity in the permanent magnet H_m has an

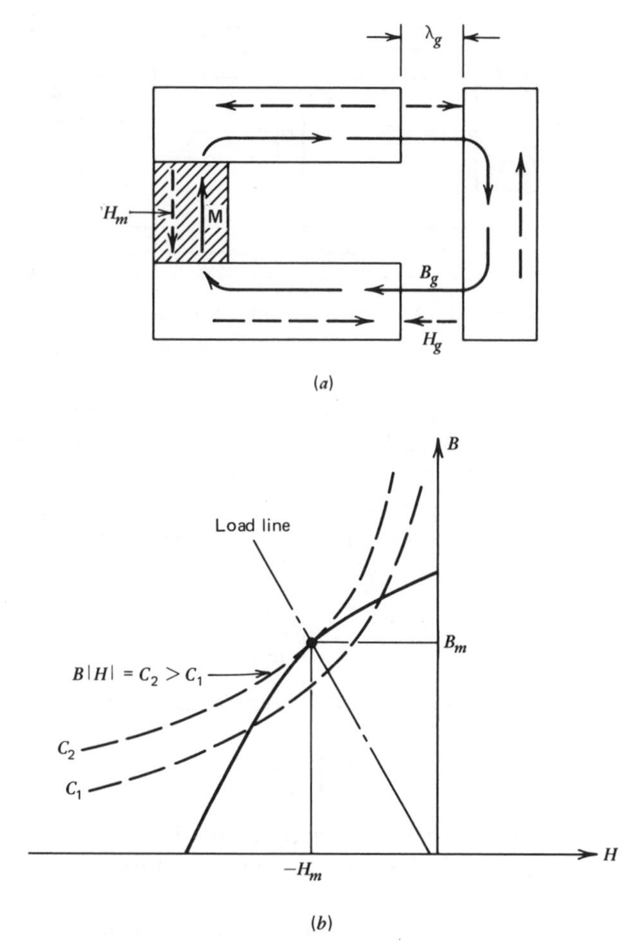

Figure 2-15 (*a*) Ferromagnetic flux circuit with permanent magnetic element (shaded area) and (*b*) demagnetization curve for permanent magnetic material and load line for flux circuit.

opposite sense to the flux in the air gap, B_g. If λ_m and λ_g are the path lengths in the magnetic material and air gap, respectively, then

$$H_m\lambda_m + H_g\lambda_g = 0$$

where H in the soft magnetic material is assumed to be approximately zero (i.e., $\mu_r \gg 1$). If we multiply this expression by the flux in the circuit, $\Phi = B_gA_g = B_mA_m$, we obtain the following expression for the volume of the permanent magnet, $V_m = A_m\lambda_m$, that is,

$$V_m = -\frac{V_gB_g^2}{\mu_0 B_m H_m} \tag{2-8.14}$$

If B_m is assumed to be positive, then H_m is negative and B_m and H_m must lie on the demagnetization curve of the B–H diagram of the permanent magnetic material (see Fig. 2-15*b*). Thus, for a minimum volume of material, the product $|B_m H_m|$ should be a maximum. This defines a unique point on the B–H curve. Since the reluctance of the circuit determines the product $B_m H_m$, $\mathcal{R}$ can be chosen to maximize $|B_m H_m|$ and minimize V_m.

3

Basic Equations of Magneto-Solid Mechanics

3.1 INTRODUCTION

In the previous chapter we saw how the mechanical variables—motion, strain, and temperature—can affect the electromagnetic behavior of a system. In summary, these magnetomechanical interactions include:

1. Induction of electric field by motion.
2. Change of magnetic field by displacement or deformation of material boundaries.
3. Change in conductivity due to temperature.
4. Change in critical current of superconductors due to strain.
5. Change in magnetization due to strain (magnetostriction).

In this chapter we examine the effect of the electromagnetic fields on the mechanical behavior. The principal effects in this class of phenomena are:

1. Change in momentum or static stresses due to electromagnetic body forces.
2. Production of angular momentum due to body couples.
3. Production of heat energy due to Joule heating or ferromagnetic hysteresis.
4. Change in strain due to magnetic fields (magnetostriction).
5. Joule-heating-induced thermoelastic strains.

These interactions may be further classified as to how they affect the balance equations of mechanics and electromagnetism, boundary conditions, or constitutive laws such as stress–strain or current–voltage relations. For large magnetic devices such as magnets or motors, which are the focus of this book, the principal interest is in electromagnetic forces.

Magnetic forces will be discussed in the next few sections. An energy method for calculating magnetic forces for discrete mechanical models will be derived. The balance laws of mechanics will be reviewed for three-dimensional solids. The resulting equations for magnetic and nonmagnetic conducting elastic solids will be derived.

3.2 ELECTROMAGNETIC FORCES

As discussed in Chapter 2, electromagnetic forces are assumed to act on distributions of charges q, current $\mathbf{J}$, electric polarization $\mathbf{P}$, and magnetization $\mathbf{M}$. In the following discussion the velocities will be assumed to be very small compared with the speed of light.

Nonmagnetizable and Nonpolarizable Material ($\mathbf{M} = \mathbf{P} = 0$)

For charge and current distributions in electric and magnetic fields, the body force per unit volume is given by the Lorentz law (2-6.5), that is,

$$\mathbf{f} = q\mathbf{E} + \mathbf{J} \times \mathbf{B} \tag{3-2.1}$$

It is instructive to examine this law for a metallic conductor. In a good conductor, the outer electrons of each metal atom are not tightly bound to the ions and are considered to form an electron "gas." Thus we may imagine a conducting solid as the interaction of two continua—an elastic ion continuum and an electron gas continuum. In general the velocity of the "free" electrons $\mathbf{v}^e$ will be different from the velocity of the solid ion continuum $\mathbf{v}^s$. Each continuum can be imagined to have corresponding charge densities Q^s and $-Q^e$. The total charge density is then

$$q = Q^s - Q^e \tag{3-2.2}$$

and the total current is given by

$$\mathbf{J} = Q^s\mathbf{v}^s - Q^e\mathbf{v}^e = Q^e(\mathbf{v}^s - \mathbf{v}^e) + q\mathbf{v}^s \tag{3-2.3}$$

If the solid is in charge equilibrium, then $Q^s = Q^e$ and

$$\mathbf{J} = Q^s(\mathbf{v}^s - \mathbf{v}^e)$$

and the current density depends on the relative velocity of the electrons to the ions. For low-frequency changes in good conductors, charge equilibrium will be the rule. However, high-frequency changes may produce a nonequilibrium state, or $Q^s \neq Q^e$ (e.g., interaction with a laser or other high-energy particles or radiation).

The body force on the solid or ion continuum is given by

$$\mathbf{f}^s = Q^s\mathbf{E} + \mathbf{J}^s \times \mathbf{B}$$

where $\mathbf{J}^s \equiv Q^s\mathbf{v}^s$ and the body force on the electrons is given by the expression

$$\mathbf{f}^e = -Q^e\mathbf{E} + \mathbf{J}^e \times \mathbf{B}$$

where $\mathbf{J}^e \equiv -Q^e\mathbf{v}^e$. The force on the electrons is transmitted to the solid or ion continuum through an interaction force. The total force on the ion–electron continuum, however, does not depend on this interaction force; that is,

$$\mathbf{f} = (Q^s - Q^e)\mathbf{E} + (\mathbf{J}^e + \mathbf{J}^s) \times \mathbf{B} \tag{3-2.4}$$

For the case of charge equilibrium, then

$$\mathbf{f} = Q^s(\mathbf{v}^s - \mathbf{v}^e) \times \mathbf{B} \equiv \mathbf{J} \times \mathbf{B}$$

For a stationary body, $\mathbf{v}^s = 0$ and $\mathbf{f} = -Q^e\mathbf{v}^e \times \mathbf{B}$.

In summary the magnetic body force $\mathbf{J} \times \mathbf{B}$ acts on the electrons which move relative to the solid. This force is transmitted from the electrons to the solid by an interaction force. If only the motion or deformation of the solid continuum is of interest, then the nature of this electron-ion interaction is not required insofar as the total force is concerned. Theories which attempt to account for the motion of both continua are known as solid-state plasma, or elastic plasma theories (see, e.g., Steele and Vural, 1969, or Moon, 1970b).

It is also instructive to calculate the work done per unit time by the body forces $\mathbf{f}^e$ and $\mathbf{f}^s$; that is,

$$\begin{aligned} P &= \mathbf{f}^s \cdot \mathbf{v}^s + \mathbf{f}^e \cdot \mathbf{v}^e \\ &= Q^s(\mathbf{E} + \mathbf{v}^s \times \mathbf{B}) \cdot \mathbf{v}^s - Q^e(\mathbf{E} + \mathbf{v}^e \times \mathbf{B}) \cdot \mathbf{v}^e \end{aligned} \tag{3-2.5}$$

We note that $\mathbf{v} \cdot \mathbf{v} \times \mathbf{B} = 0$, so that

$$P = \mathbf{J} \cdot \mathbf{E}$$

Rearranging terms one can show that

$$\mathbf{J} \cdot \mathbf{E} = \mathbf{J}' \cdot \mathbf{E}' + (\mathbf{f}^s + \mathbf{f}^e) \cdot \mathbf{v}^s \tag{3-2.6}$$

where $\mathbf{J}' \cdot \mathbf{E}' = \mathbf{f}^e \cdot (\mathbf{v}^e - \mathbf{v}^s)$. In ordinary conductors $\mathbf{J}' \cdot \mathbf{E}'$ is a heat source term. Thus $\mathbf{J} \cdot \mathbf{E}$ contains both the Joule heating and the work done by the total body force in moving with the ion or solid continuum. The expression (3-2.6) will be useful in the section on balance of energy.

3.3 FORCES ON DISCRETE CIRCUITS

When the current flow is restricted to a well-defined circuit path, the total force on the circuit can be written in the form (see Fig. 3-1)

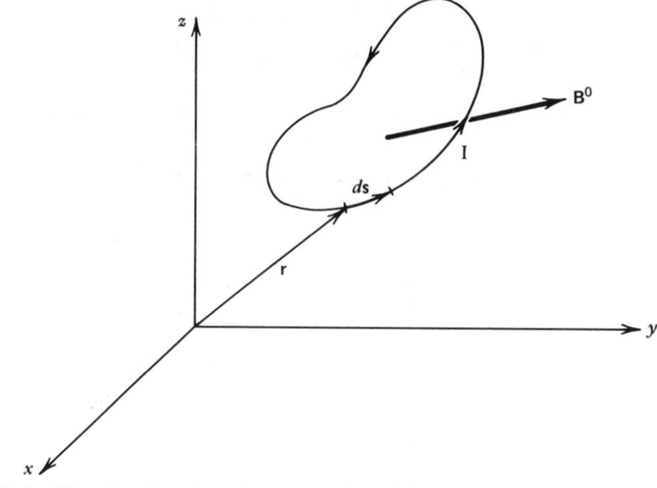

Figure 3-1 Sketch of a closed current circuit in an external magnetic field $\mathbf{B}^0$.

$$\mathbf{F} = \oint \mathbf{I} \times \mathbf{B}^0 \, ds \tag{3-3.1}$$

where $\mathbf{B}^0$ is due to all other sources of magnetic field and does not include the effect of the circuit itself. $\mathbf{I}$ is found by integrating $\mathbf{J}$ across the cross section normal to the distance element ds. If $\mathbf{B}^0$ is generated by current in another circuit, then the magnetic field is given by the Biot–Savart law (2-6.3) and the force law takes the form of Eq. (2-6.4).

One can conclude from (3-3.1) that the net force on a circuit in a uniform field is zero. However, the net moment is not necessarily zero.

When $\mathbf{r}$ is a position vector to the current element $I d\mathbf{r}$, the moment of the magnetic force on the circuit about some point is given by the formula

$$\mathbf{C} = I \oint \mathbf{r} \times (d\mathbf{r} \times \mathbf{B}^0) \tag{3-3.2}$$

When $\mathbf{B}^0$ is uniform we can reduce Eq. (3-3.2) to the form

$$\mathbf{C} = I \oint (\mathbf{B}^0 \cdot \mathbf{r}) \, d\mathbf{r}$$

Finally, using the identity

$$2(\mathbf{r} \cdot \mathbf{B}) \, d\mathbf{r} = (\mathbf{r} \times d\mathbf{r}) \times \mathbf{B} + d[\mathbf{r}(\mathbf{r} \cdot \mathbf{B})]$$

the expression for $\mathbf{C}$ takes the form

$$\mathbf{C} = \mathbf{m} \times \mathbf{B}^0 \tag{3-3.3}$$

where $\mathbf{m}$ is called the equivalent magnetic moment of the circuit and is defined by

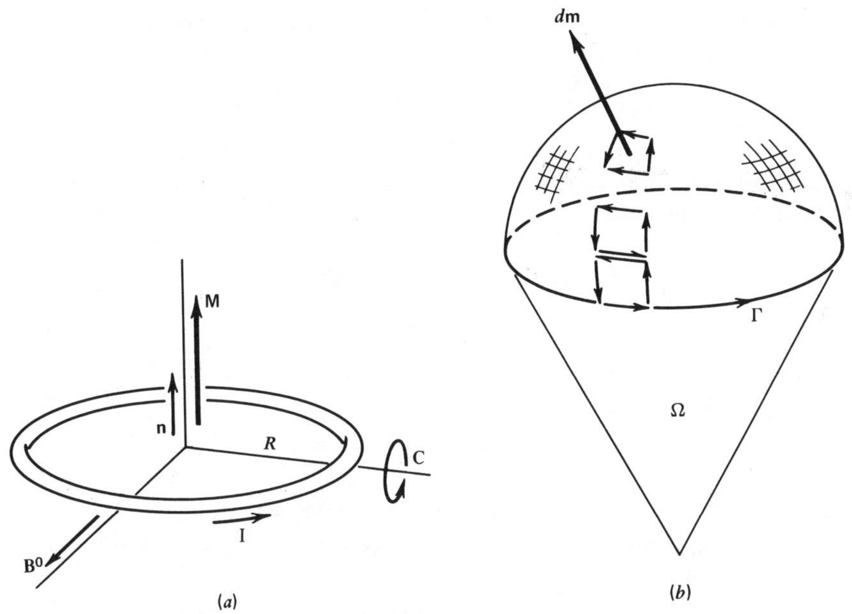

Figure 3-2 (*a*) Magnetic-moment vector for a planar circuit and (*b*) magnetic double layer or shell for a closed circuit.

$$\mathbf{m} \equiv I\frac{1}{2}\oint \mathbf{r} \times d\mathbf{r} \tag{3-3.4}$$

One can easily show that $\mathbf{m}$ is independent of the origin of $\mathbf{r}$. Thus $\mathbf{C}$ represents a true couple.

For a circular wire of radius R,

$$\mathbf{m} = I\pi R^2\mathbf{n} = IA\mathbf{n} \tag{3-3.5}$$

where $\mathbf{n}$ is normal to the plane of the circuit, defined by $\mathbf{r} \times d\mathbf{r}$, and A is the area (see Fig. 3-2*a*). The couple is given by

$$\mathbf{C} = AI\mathbf{n} \times \mathbf{B}^0 \tag{3-3.6}$$

Thus the couple acts to align the plane of the circuit normal to the magnetic field.

Circuits in Small Gradient Fields

To find the force on a rigid circuit in a magnetic field with small gradients, we can treat the circuit as a magnetic dipole. To see this consider a closed circuit with current I in the vicinity of the origin. Let $\mathbf{r}$ be the position vector to a current element $Id\mathbf{r}$. Then the total force on the circuit is given by the expression (3-3.1), and is linearly proportional to the external field $\mathbf{B}^0(\mathbf{r})$.

When $\mathbf{B}^0$ has small gradients in the neighborhood of the circuit, we expand $\mathbf{B}^0(\mathbf{r})$ in a Taylor series

$$\mathbf{B}^0(\mathbf{r}) \simeq \mathbf{B}^0(0) + \mathbf{r} \cdot \nabla \mathbf{B}^0(0) + \cdots \tag{3-3.7}$$

The force due to the constant term $\mathbf{B}^0(0)$ is zero. (The closed line integral of a perfect differential is zero.) Neglecting higher-order gradients of $\mathbf{B}^0$, the expression for the force becomes

$$\mathbf{F} = I \oint d\mathbf{r} \times (\mathbf{r} \cdot \boldsymbol{\nabla})\mathbf{B}^0(0) \tag{3-3.8}$$

We note that the dyadic $\nabla \mathbf{B}^0(0)$ is treated as a constant in the integral. With this in mind, one can rewrite Eq. (3-3.8) in the form

$$\mathbf{F} = \frac{I}{2}\left(\oint (\mathbf{r} \times d\mathbf{r}) \times \boldsymbol{\nabla}\right) \times \mathbf{B}^0 \tag{3-3.9}$$

{We have used the identity $2d\mathbf{r}(\mathbf{r} \cdot \boldsymbol{\nabla}) = (\mathbf{r} \times d\mathbf{r}) \times \boldsymbol{\nabla} + d[\mathbf{r}(\mathbf{r} \cdot \boldsymbol{\nabla})]$.}

In the previous section we defined the equivalent magnetic moment of the circuit

$$\mathbf{m} \equiv I\frac{1}{2}\oint \mathbf{r} \times d\mathbf{r}$$

Thus the total force on the circuit can be written in the form

$$\mathbf{F} = (\mathbf{m} \times \boldsymbol{\nabla}) \times \mathbf{B}^0 \tag{3-3.10}$$

If we now treat $\mathbf{m}$ as a constant with respect to the operator ∇, and use the conditions $\boldsymbol{\nabla} \cdot \mathbf{B}^0 = 0$ and $\boldsymbol{\nabla} \times \mathbf{B}^0 = 0$, then Eq. (3-3.10) can be written in the form

$$\mathbf{F} = \boldsymbol{\nabla}(\mathbf{m} \cdot \mathbf{B}^0) = (\mathbf{m} \cdot \boldsymbol{\nabla})\mathbf{B}^0 \tag{3-3.11}$$

Thus the force on the circuit is similar to that on a ferromagnetic dipole (2-6.7).

When the gradients in $\mathbf{B}^0$ are not small, we can replace the line integral over $Id\mathbf{r}$ by a surface integral over a distribution of magnetic moments; that is,

$$\mathbf{F} = \iint (d\mathbf{m} \cdot \boldsymbol{\nabla})\mathbf{B}^0$$

where $d\mathbf{m} = I\mathbf{n}dA$ [see Fig. 3-2(*b*)] and

$$\mathbf{F} = I \iint_S \mathbf{n} \cdot \nabla \mathbf{B}^0 dA \tag{3-3.12}$$

The surface must be bounded by a simple closed, unknotted circuit curve Γ, but otherwise it is arbitrary.

In replacing a circuit of current I by a surface distribution of magnetic moment, one imagines that there are positive poles on one side of the surface and negative poles on the other side. This representation has sometimes been called a *magnetic double layer* or a *magnetic shell*, especially in the early electromagnetic literature (see, e.g., Jeans, 1925).

In a like manner, the magnetic field far from a circuit can be written as if the circuit were replaced by a magnet of magnetic moment $\mathbf{m}$ given by Eq. (3-3.4); that is,

$$\mathbf{B} = \frac{\mu_0}{4\pi}\left(\frac{-\mathbf{m}}{r^3} + \frac{3(\mathbf{m}\cdot\mathbf{r})\mathbf{r}}{r^5}\right) \tag{3-3.13}$$

If we write $\mathbf{B}$ in terms of a vector potential $\mathbf{A}$, then

$$\mathbf{B} = \nabla \times \mathbf{A}$$

and

$$\mathbf{A} = \frac{\mu_0 \mathbf{m} \times \mathbf{r}}{r^3} \tag{3-3.14}$$

We may also wish to write $\mathbf{B}$ as a gradient of a scalar potential. Then we have (see, e.g., Rietz and Milford, 1960)

$$\mathbf{B} = -\mu_0 \nabla \phi$$

$$\phi = \frac{\mathbf{m}\cdot\mathbf{r}}{4\pi r^3} \tag{3-3.15}$$

3.4 FORCES ON MAGNETIZED MATERIALS ($\mathbf{J} = 0$)

The law of conservation of flux $\nabla \cdot \mathbf{B} = 0$ implies that there are no isolated magnetic poles or monopoles; that is, a source of a magnetic field line must be accompanied by a sink. It is natural then that the concept of a magnetic dipole is often used as the basic element of the continuum theory of magnetic materials since it contains its own source and sink. If a magnetic body can be thought of as a distribution of magnetic dipoles, then the magnetic field outside the body is a linear superposition of the fields produced by this dipole distribution. Thus it follows from Eq. (3-3.13) that outside the body the magnetic field is given by

$$\mathbf{B} = \frac{\mu_0}{4\pi}\int_V \left(-\frac{\mathbf{M}}{r^3} + \frac{3\mathbf{M}\cdot\mathbf{rr}}{r^5}\right) dv \tag{3-4.1}$$

In Eq. (3-4.1) $\mathbf{M}$ represents a distribution of magnetic dipoles within a volume V bounded by a surface S. If $\mathbf{B}^0$ represents the magnetic field vector due to sources outside V, then the net force and moment on the material in V are given by integrals of force and moment densities similar to Eqs. (3-3.11) and (3-3.3), respectively,

$$\mathbf{F} = \int_V (\mathbf{M}\cdot\nabla)\mathbf{B}^0 dv \tag{3-4.2}$$

and

$$\mathbf{C} = \int [\mathbf{r} \times (\mathbf{M} \cdot \nabla)\mathbf{B}^0 + \mathbf{M} \times \mathbf{B}^0] dv \tag{3-4.3}$$

where it is assumed that $\mathbf{B}^0$ is continuous and differentiable in V. The expression (3-4.2) can be transformed to a different form if we use a vector identity (see Brown, 1966)

$$\mathbf{F} = \int_V \mathbf{M} \cdot \nabla \mathbf{B}^0 dv = \int_V (-\nabla \cdot \mathbf{M})\mathbf{B}^0 dv + \int_S \mathbf{n} \cdot \mathbf{M}\mathbf{B}^0 dS \tag{3-4.4}$$

This expression has a simple interpretation if we consider a slender rod uniformly magnetized along the axis of the rod. For this case, $\nabla \cdot \mathbf{M} = 0$ in V, and we obtain [see Fig. 3-3a]

$$\mathbf{F} = MA(\mathbf{B}^0(\mathbf{r}^+) - \mathbf{B}^0(\mathbf{r}^-)) \tag{3-4.5}$$

We imagine that positive and negative magnetic monopoles or magnetic charges of intensity $p_m = \pm MA$ are concentrated at the ends of the rod and that the force on each pole is analogous to that on a charge $q\mathbf{E}$, that is,

$$\mathbf{F} = p_m \mathbf{B}_0, \qquad p_m = \mathbf{M} \cdot \mathbf{n} \tag{3-4.6}$$

With this interpretation, $\mathbf{M} \cdot \mathbf{n}$ represents a surface distribution of magnetic "charge" or poles and $-\nabla \cdot \mathbf{M}$ represents a volume distribution of magnetic poles. The right-hand expression in Eq. (3-4.4) is sometimes called the *pole model*.

Thus the representation of the magnetic body force is not unique and may even be replaced by surface tractions. This raises serious questions regarding the internal stress state in a magnetized body.

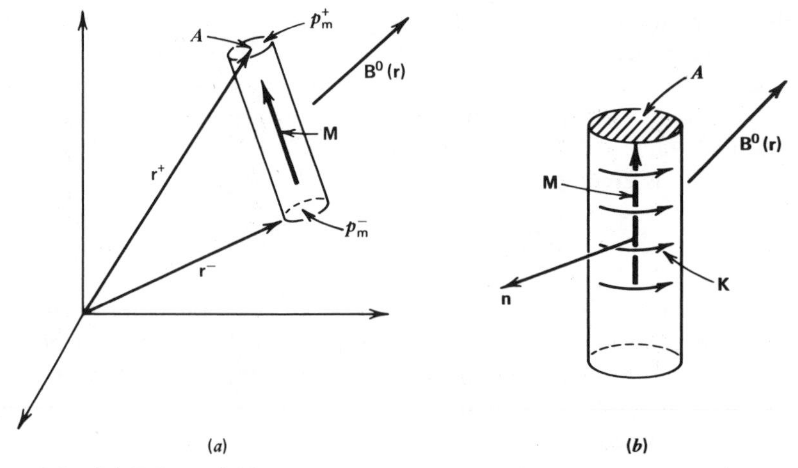

Figure 3-3 (a) Pole model for a magnetic dipole and (b) ampere-current model for a magnetic dipole.

The total force can be further transformed to a form resembling the Lorentz force (see Brown, 1966), that is,

$$\mathbf{F} = \int_V (\nabla \times \mathbf{M}) \times \mathbf{B}^0 dv + \int_S (-\mathbf{n} \times \mathbf{M}) \times \mathbf{B}^0 dS \tag{3-4.7}$$

For a uniformly axially magnetized rod, the volume integral vanishes and the force resembles that on a solenoid with surface current density $\mathbf{K} = -\mathbf{n} \times \mathbf{M}$ [see Fig. 3-3(*b*)]. This representation is called the *Ampere-current model* and $\nabla \times \mathbf{M} \equiv \mathbf{J}_m$ represents the equivalent magnetization current density and $\mathbf{J}_m \times \mathbf{B}^0$ the body force density.

In many problems the separation of the magnetic field into long-range and short-range components, that is, $\mathbf{B}^0$ and $\mathbf{B}^1$ is not known *a priori*, but only in terms of the total field. One can derive equivalent forms of Eq. (3-4.4) using the total magnetic fields $\mathbf{H}$ or $\mathbf{B}$ (see Brown, 1966, p. 57); that is,

$$\mathbf{F} = \mu_0 \int \mathbf{M} \cdot \nabla \mathbf{H} \, dv + \frac{1}{2} \mu_0 \int \mathbf{n} M_n^2 \, dS \tag{3-4.8a}$$

or

$$\mathbf{F} = \int \mathbf{M} \cdot \nabla \mathbf{B} \, dv - \frac{1}{2} \mu_0 \int \mathbf{n} M_t^2 \, dS \tag{3-4.8b}$$

where $M_t = |\mathbf{n} \times \mathbf{M}|$ on S.

There has been much written about the proper formulation to ensure a unique determination of the stresses in magnetoelasticity. However, for problems involving magnetic devices such as relays, solenoids, or electromagnets, the uncertainty in choosing the correct formulation may not be that great. The difference between using $\mathbf{M} \cdot \nabla \mathbf{B}$ or $\mathbf{M} \cdot \nabla \mathbf{B}^0$ for the body force is a hydrostatic pressure on the body of magnitude $\mu_0 M_t^2$. For the best magnetic material, $\mu_0 M_t$ is of the order of 1 T and $\mu_0 M_t^2 \sim 40$ N/cm^2 (58 psi). Since the yield stress is usually greater than 3800 N/cm^2 (20,000 psi), one can often design with this uncertainty and choose a model that is convenient to the problem.

3.5 ENERGY METHOD AND LAGRANGE'S EQUATIONS FOR DISCRETE SYSTEMS

Before considering the energy formulation for discrete magnetomechanical systems, recall first the energy form of Newton's law for a nonmagnetic system of particles, where $\mathbf{F}$ is the net force acting on the particles. The work done is equal to the change in kinetic energy $\mathcal{T}$; that is,

$$\int_{\mathbf{r}_1}^{\mathbf{r}_2} \mathbf{F} \cdot d\mathbf{r} = \int_0^t \mathbf{F} \cdot \mathbf{v} \, dt = \int_0^t d\mathcal{T} \tag{3-5.1}$$

where for a single particle $\mathcal{T} = \frac{1}{2} m v^2$.

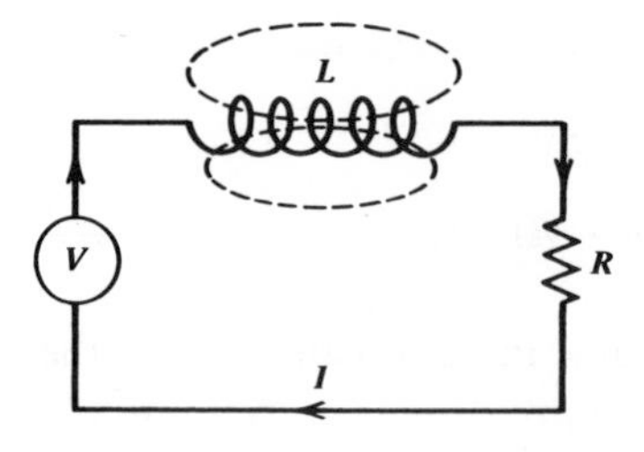

Figure 3-4 Simple circuit with inductance L, resistance R, and voltage supply.

Likewise the energy law for a *rigid circuit* with a voltage source, resistive element, and inductor (see Fig. 3-4) can be written in the form

$$\int_0^t IV dt = \int_0^t d\mathcal{W}_m + \int_0^t I^2 R dt \tag{3-5.2}$$

Here we assume that the inductor is nonhysteretic and that $\mathcal{W}_m$ is the energy stored in the magnetic field.†

Now consider a magnetomechanical system with one mechanical degree of freedom and one circuit. (It is easy to generalize to more degrees of freedom.) We assume that the inductor can be deformed, as for example, by pulling the ferromagnetic plunger out from a solenoid coil (see Fig. 3-5). We further assume that $\mathcal{W}_m$ is a single-valued function of the total flux through the inductor, Φ, and the generalized displacement s, that is, $\mathcal{W}_m \equiv \mathcal{W}_m(\Phi, s)$.

If one draws an imaginary surface around this system, we can see that the energy input consists of pumping current through the circuit using the voltage supply and the work done by the mechanical force $\mathbf{F}_{mech}$. We assume

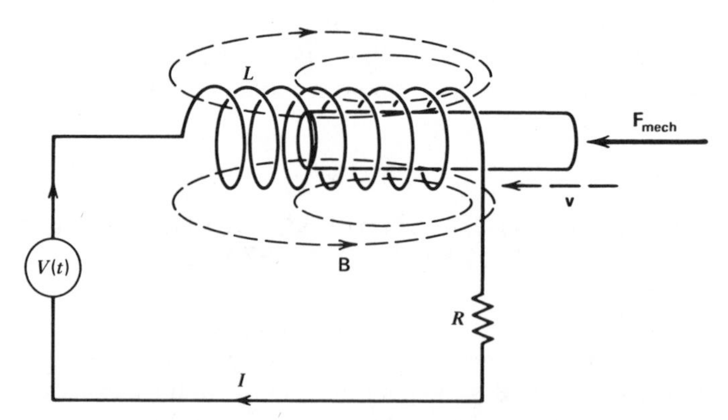

Figure 3-5 Circuit with deformable inductor and external source of mechanical work.

†In general one could also store potential energy in the electrical field in a capacitor. However, we do not consider stored charge or electrical energy in this section.

that the rate of change of electromagnetic fields is low enough to neglect radiation effects. We also assume that eddy currents are not induced in nearby conductors.

The energy principle then states that the total energy input must be equal to the sum of the energies stored in kinetic and magnetic energies plus that dissipated in the resistive element; that is,

$$\int_0^t IV dt + \int_0^t \mathbf{F}_{\text{mech}} \cdot \mathbf{v}\, dt = \int_0^t (d\mathcal{W}_{\text{m}} + d\mathcal{T}) + \int_0^t I^2 R dt \qquad (3\text{-}5.3)$$

One can see that this reduces to the two special cases of the rigid circuit ($\mathbf{v} = 0$), Eq. (3-5.2), and the nonmagnetic mass ($I = 0$, $\mathcal{W}_{\text{m}} = 0$), Eq. (3-5.1).

What is sometimes confusing when it comes to discussing energy in interdisciplinary fields is that there are two auxiliary energy laws. Thus if we isolate the deformable mechanical element, Newton's law can be written in the form of an energy principle, provided we use the total force on the particle, $\mathbf{F} = \mathbf{F}_{\text{mag}} + \mathbf{F}_{\text{mech}}$. We obtain

$$\int_0^t (\mathbf{F}_{\text{mag}} + \mathbf{F}_{\text{mech}}) \cdot \mathbf{v}\, dt = \int_0^t d\mathcal{T} \qquad (3\text{-}5.4)$$

We also have the circuit equivalent of Maxwell's equations,

$$V = IR + \frac{d\Phi}{dt} \qquad (3\text{-}5.5)$$

This too can be put into the form of an energy law by multiplying by the current I, and integrating in time,

$$\int_0^t VIdt = \int_0^t I^2 R dt + \int_0^t I d\Phi \qquad (3\text{-}5.6)$$

When the two auxiliary energy laws (3-5.4) and (3-5.6) are used in (3-5.3) to eliminate $\mathcal{T}$ and I^2R, we obtain an alternate form of the energy law for magnetomechanical systems:

$$\int_0^t I d\Phi = \int_{s_1}^{s_2} \mathbf{F}_{\text{mag}} ds + \int_0^t d\mathcal{W}_{\text{m}} \qquad (3\text{-}5.7)$$

If we assume that $\mathcal{W}_m(\Phi, s)$ is a single-valued, continuous, differentiable function of Φ and s, then one can write

$$d\mathcal{W}_{\text{m}} = \frac{\partial \mathcal{W}_m}{\partial \Phi} d\Phi + \frac{\partial \mathcal{W}_m}{\partial s} ds \qquad (3\text{-}5.8)$$

The energy law (Eq. 3-5.7) then implies the following relations:

$$I = \frac{\partial \mathcal{W}_m}{\partial \Phi}$$

and

$$F_{\text{mag}} = -\frac{\partial \mathcal{W}_m(\Phi, s)}{\partial s} \tag{3-5.9}$$

Thus we have a rule to calculate the current and the force on the circuit provided we know the magnetic energy in terms of the flux and displacement.

When the magnetic energy is known as a function of current instead of flux we can rewrite Eq. (3-5.8) by using what is known as a Legendre transformation. We define a coenergy $\mathcal{W}_m^*$ by relation

$$\mathcal{W}_m^*(I, s) = -\mathcal{W}_m(\Phi, s) + I\Phi \tag{3-5.10}$$

Using this definition we obtain

$$\Phi = \frac{\partial \mathcal{W}_m^*}{\partial I}$$

$$\text{F}_{\text{mag}} = \frac{\partial \mathcal{W}_m^*(I, s)}{\partial s} \tag{3-5.11}$$

[Note if $\mathcal{W}_m^*$ represents the area under the $\Phi(I)$ curve in Figure 3-6, then $\mathcal{W}_m$ is the area under the $I(\Phi)$ curve.] A good discussion of these ideas may be found in Crandall et al. (1968).

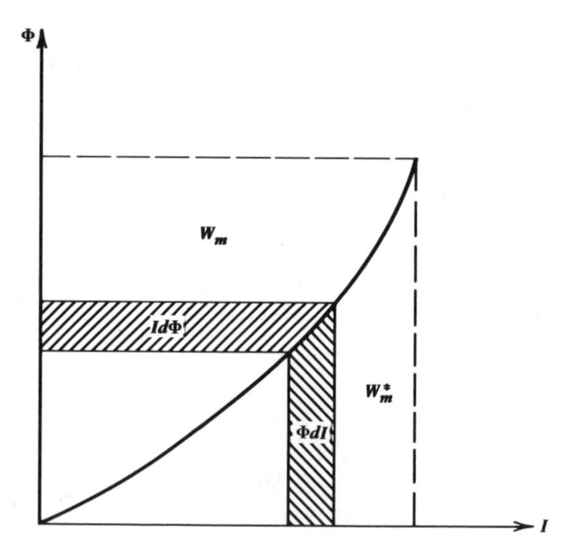

Figure 3-6 Flux-current relation for a magnetic circuit.

Lagrange's Equations

In mechanics there are two methods for deriving equations of motion. One method is based on direct application of Newton's laws and vector kinematics and the other method is based on scalar energy formulation called Lagrange's equations. There are many excellent expository texts on Lagrange's equations in mechanics (see, e.g., Goldstein, 1950) so we will only present a summary of the method. Our purpose in this section is to indicate how magnetic forces may be incorporated into Lagrange's equations and outline how the dynamical equations for coupled mechanical systems and circuits may be derived from one general principle.

Lagrange's equations for mechanical systems is based on the *principle of virtual work*. This principle starts with the idea of *virtual displacements*. For a system of particles each of whose position is described by a position vector $\mathbf{r}_i$, one imagines a set of incremental displacements $\{\delta\mathbf{r}_i\}$ which do not violate the geometric constraints on the particles. The principle of virtual work states that for a system of N particles in equilibrium under forces $\{\mathbf{F}_i\}$, the virtual work of all the forces is zero; that is,

$$\sum_{i=1}^{N} \mathbf{F}_i \cdot \delta\mathbf{r}_i = 0 \tag{3-5.12}$$

The principle of virtual work can be extended to dynamic problems as an extension of Newton's law; for example,

$$\sum_{i=1}^{N} \left(\mathbf{F}_i - \frac{d}{dt} m_i \mathbf{v}_i\right) \cdot \delta\mathbf{r}_i = 0 \tag{3-5.13}$$

where $\mathbf{v}_i = \dot{\mathbf{r}}_i$ is the velocity of each particle. The advantage of this formulation is that forces which do no work drop out in Eq. (3-5.13). Also in many problems the number of degrees of freedom is less than the number of particles and the equations of motion may be formulated in terms of an independent set of generalized coordinates $\{s_i(t)\}$, where

$$\mathbf{r}_i = \mathbf{r}_i(s_k), \qquad \mathbf{v}_i = \dot{\mathbf{r}}_i = \sum \frac{\partial \mathbf{r}_i}{\partial s_k} \dot{s}_k$$

One also defines generalized forces $\{\mathcal{F}_k\}$ which satisfy

$$\sum\sum \mathbf{F}_i \cdot \frac{\partial \mathbf{r}_i}{\partial s_k} \delta s_k = \sum \mathcal{F}_k \delta s_k \tag{3-5.14}$$

Using these definitions one can derive the following set of equations (see, e.g., Goldstein, 1950):

$$\frac{d}{dt}\frac{\partial \mathcal{T}}{\partial \dot{s}_k} - \frac{\partial \mathcal{T}}{\partial s_k} = \mathcal{F}_k \tag{3-5.15}$$

where $\mathcal{T}(s_k, \dot{s}_k)$ is the *kinetic energy* of the system of particles expressed in terms of the generalized coordinates and generalized velocities. This can sometimes be written as the sum of *conservative forces* which are related to a potential energy function $\mathcal{V}(s_k)$ and nonconservative (nc) forces; that is,

$$\mathcal{F}_k = -\frac{\partial \mathcal{V}}{\partial s_k} + \mathcal{F}_k^{\text{nc}}$$

In this case, Lagrange's equations become

$$\frac{d}{dt}\frac{\partial \mathscr{L}}{\partial \dot{s}_k} - \frac{\partial \mathscr{L}}{\partial s_k} = \mathscr{F}_k^{\text{nc}} \tag{3-5.16}$$

where $\mathscr{L} = \mathscr{T} - \mathscr{V}$ is called the *Lagrangian function.*

When electromagnetic forces are present additional terms enter Eq. (3-5.16). As an example, consider a single-charged particle moving in an electric and magnetic field. The electromagnetic force is given by

$$\mathbf{F} = Q(\mathbf{E} + \mathbf{v} \times \mathbf{B})$$

In general, the fields **E** and **B** may be related to a scalar and vector potential (see Chapter 8)

$$\mathbf{E} = -\nabla\psi - \frac{\partial \mathbf{A}}{\partial t}, \qquad \mathbf{B} = \nabla \times \mathbf{A} \tag{3-5.17}$$

It can be shown that this force may be derived from Lagrange's equations when the Lagrangian takes the form

$$\mathscr{L} = \mathscr{T} - Q\psi + Q\mathbf{A} \cdot \mathbf{v}$$

If a conservative mechanical force is also present, then $\mathscr{L}$ becomes

$$\mathscr{L} = (\mathscr{T} + Q\mathbf{A} \cdot \mathbf{v}) - (\mathscr{V} + Q\psi) \tag{3-5.18}$$

In this formulation one may think of the magnetic energy $Q\mathbf{A} \cdot \mathbf{v}$ as contributing to the kinetic energy, and the electric field energy $Q\psi$ as contributing to the potential energy.

As a second example we consider the permanently magnetized particle of magnetic current **m** in a magnetic field $\mathbf{B}^0$. The Lagrangian for this problem may be shown to be (see, e.g., Landau and Lifshitz, 1960)

$$\mathscr{L} = \mathscr{T} - \mathscr{V} + \mathbf{m} \cdot \mathbf{B}^0$$

If the particle is a soft ferromagnetic ellipsoid with a linear B–H law, in a uniform field $\mathbf{B}^0$, then the Lagrangian can be found to be

$$\mathscr{L} = \mathscr{T} - \mathscr{V} + \tfrac{1}{2}\mathbf{M} V \cdot \mathbf{B}^0$$

where **M** is the magnetization density inside the ellipsoid.

The method of Lagrange's equations may be extended to electric and magnetic circuits. Imagine a circuit path Γ which encircles a total amount of flux Φ. For a single circuit Γ, Faraday's law has the form

$$V = \frac{d\Phi}{dt} \tag{3-5.19}$$

where V is the sum of both resistive and capacitive voltage drops and sources around Γ. To calculate the virtual work for a circuit element, we must choose a proper independent variable analogous to $\delta\mathbf{r}$. One may choose either the charge flowing in the circuit Q, or the flux Φ. To illustrate the method, we choose Q as our generalized circuit coordinate and the current $I = dQ/dt$ as a circuit equivalent of the generalized velocity. For a set of *rigid circuit paths* $\{\Gamma_k\}$, the principle of virtual work becomes

$$\sum \left\{ V_k(Q_m, I_m) - \frac{d}{dt}\Phi_k(I_m) \right\} \delta Q_k = 0 \qquad \text{(3-5.20)}$$

The voltage drop across a capacitor can be related to an electric energy function

$$V = -\frac{\partial \mathcal{W}^e}{\partial Q}, \qquad \mathcal{W}^e = \frac{1}{2C}Q^2 \qquad \text{(3-5.21)}$$

where C is the capacitance. For a voltage source E_0, independent of the charge, one can write

$$V = -\frac{\partial \mathcal{W}^e}{\partial Q}, \qquad \mathcal{W}^e = -E_0 Q \qquad \text{(3-5.22)}$$

On the other hand, resistors are nonconservative elements for which an electric energy function cannot be found. For linear resistors one writes

$$V = -RI = -R\dot{Q} \qquad \text{(3-5.23)}$$

For inductive or magnetic energy storage elements we have from the first part of this section

$$\Phi = \frac{\partial \mathcal{W}^m}{\partial I} = LI \qquad \text{(3-5.24)}$$

[Here we use $\mathcal{W}^m \equiv \mathcal{W}_m^*$ used in Eq. (3-5.11) to avoid confusion with subscripts.] Thus Lagrange's equation for a single circuit path can be found in the form

$$\frac{d}{dt}\frac{\partial \mathcal{W}^m}{\partial \dot{Q}} - \frac{\partial \mathcal{W}^e}{\partial Q} = -RI \qquad \text{(3-5.25)}$$

The Lagrangian for the circuit may be written

$$\mathcal{L}^{\text{em}} = \mathcal{W}^m - \mathcal{W}^e \qquad \text{(3-5.26)}$$

since

$$\frac{\partial \mathcal{W}^e}{\partial \dot{Q}} = 0, \qquad \frac{\partial \mathcal{W}^m}{\partial Q} = 0$$

Again the magnetic energy $\mathcal{W}^m$ acts as a kinetic energy term while the electric energy $\mathcal{W}^e$ acts as a potential energy source.

For a coupled magnetomechanical system the magnetic energy function depends on the generalized displacement as well as the current. Using the results from Eq. (3-5.11) the coupled equations become

$$\frac{d}{dt}\frac{\partial \mathcal{T}}{\partial \dot{s}} - \frac{\partial(\mathcal{T}-\mathcal{V})}{\partial s} = \mathcal{F}^{\mathrm{nc}} + \frac{\partial \mathcal{W}^m(I, s)}{\partial s}$$

and (3-5.27)

$$\frac{d}{dt}\frac{\partial \mathcal{W}^m}{\partial I} - \frac{\partial \mathcal{W}^e}{\partial Q} = -RI$$

These equations can be derived from a single Lagrangian function

$$\mathcal{L}(s_1, s_2, \dot{s}_1, \dot{s}_2) = \mathcal{T}(s, \dot{s}) + \mathcal{W}^m(I, s) - [\mathcal{V}(s) + \mathcal{W}^e(Q)] \quad (3\text{-}5.28)$$

where $s_1 \equiv s$, $s_2 = Q$, and the generalized "forces" are defined by

$$\mathcal{F}_1 \delta s_1 + \mathcal{F}_2 \delta s_2 = \mathcal{F}^{\mathrm{nc}} \delta s - R\dot{Q}\delta Q$$

Equations (3-5.27) can then be derived from

$$\frac{d}{dt}\frac{\partial \mathcal{L}}{\partial \dot{s}_k} - \frac{\partial \mathcal{L}}{\partial s_k} = \mathcal{F}_k \quad (3\text{-}5.29)$$

The magnetic energy function is not always easy to determine for a magnetic system. For a linear ferromagnetic circuit it can be shown that

$$\mathcal{W}^m = \frac{1}{2}\int \mu \mathbf{H} \cdot \mathbf{H}\, dv \quad (3\text{-}5.30)$$

The integral is carried out over all material and space where $\mathbf{H} \neq 0$. One must then relate $\mathbf{H}$ to the current I through Ampere's law (2-2.4) to get the functional dependence $\mathcal{W}^m(I)$.

For a circuit with inductance L, the magnetic energy function can often be written in the form

$$\mathcal{W}^m = \tfrac{1}{2}L(s)I^2 \quad (3\text{-}5.31)$$

Formulas for the inductance of different geometrically shaped circuits may be found in Grover (1946). Examples of the use of Lagrange's equations may be found in Chapters 7 and 8 as well as in Crandall et al. (1968).

3.6 BASIC EQUATIONS FOR MAGNETOMECHANICAL CONTINUA

The balance equations of continuum mechanics may be found in dozens of references and texts (e.g., see Eringen, 1967). We will briefly summarize them for a thermoelastic body carrying electric current, but with no electric polarization or magnetization. The equations for a *linear* thermoelastic material will be given in the next section. Equations for stresses in magnetizable elastic structures are discussed in section 3.10.

Cartesian tensor notation will be implied but vector notation will be used to eliminate the need for subscripts where possible. Thus $\partial\phi/\partial x_k \equiv \phi_{,k} \equiv \nabla\phi$,

$v_{i,j} \equiv \nabla \mathbf{v}$ and $v_{i,i} \equiv \boldsymbol{\nabla} \cdot \mathbf{v}$ (repeated indices will be summed). Also $t_{ij}d_{ij} \equiv \mathbf{t}:\mathbf{d} = \mathrm{Tr}(\mathbf{t} \cdot \mathbf{d}^T)$ [Here Tr( ) indicates the trace of the tensor and $(\ )^T$ denotes the transpose].

Where necessary the identity matrix or tensor, sometimes known as the Krönecker delta $\boldsymbol{\delta}$, will be used; that is,

$$\begin{aligned} \delta_{ij} &= 0, \qquad i \neq j \\ &= 1, \qquad i = j \end{aligned}$$

The thermomechanical variables are assumed to be continuous functions of space and time. The symbols and units are defined in Appendix A. Where calculations are performed, rational MKS units will be used.

The vector $\mathbf{r}$ will denote the present spacial position of a material particle. In nonlinear continuum mechanics one also introduces material coordinates $\mathbf{R}$ to denote the position at some reference time. However, almost all problems in this book will involve either linear elasticity or linear structural theory so that material coordinates will not be needed.

Regarding time derivatives, we use the classical partial derivative $\partial/\partial t$ to indicate a time rate of change at a fixed spacial position ($\mathbf{r}$ fixed) and use a "dot" to indicate a fixed material particle ($\mathbf{R}$ fixed). These derivatives are related by

$$\dot{\phi} = \frac{\partial}{\partial t}\phi(\mathbf{r}, t) + \mathbf{v} \cdot \boldsymbol{\nabla} \phi \tag{3-6.1}$$

where $\mathbf{v} = \dot{\mathbf{r}}$ is the velocity of the particle. To calculate the time rate of change of a property ϕ inside a control volume V one must account for the flow across the boundary so that

$$\frac{d}{dt}\int \phi\, dV = \int (\dot{\phi} + \phi \boldsymbol{\nabla} \cdot \mathbf{v})dV = \int \left(\frac{\partial \phi}{\partial t} + \boldsymbol{\nabla} \cdot \phi \mathbf{v}\right)dV \tag{3-6.2}$$

Strain Measures

Each particle in a body can be described by a position vector $\mathbf{r}(x_1, x_2, x_3, t)$ (see Fig. 3-7). If the particle had a different position in some reference state, say $\mathbf{R}$, then the displacement of the particle is given as

$$\mathbf{u} = \mathbf{r} - \mathbf{R} \tag{3-6.3}$$

The concept of strain involves describing what happens to a differential line element between two particles $d\mathbf{S}$ after it undergoes deformation. The new differential line element between the same two particles is denoted by $d\mathbf{s}$. These two vectors are defined by

$$d\mathbf{S} = \mathbf{R}_2 - \mathbf{R}_1, \qquad d\mathbf{s} = \mathbf{r}_2 - \mathbf{r}_1$$

When the body is elastic we assume that the deformation is continuous and

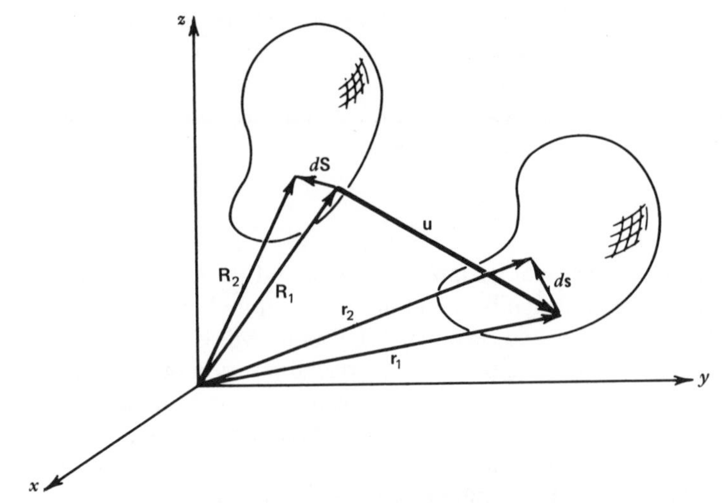

Figure 3-7 Deformable body before and after deformation.

differentiable. Thus the displacements of two neighboring points are related by

$$\mathbf{u}_2 = \mathbf{u}_1 + d\mathbf{s} \cdot \nabla\mathbf{u}_1 \tag{3-6.4}$$

It follows that

$$d\mathbf{S} = d\mathbf{s} - d\mathbf{s} \cdot \nabla\mathbf{u} = d\mathbf{s} \cdot (\boldsymbol{\delta} - \nabla\mathbf{u}) \tag{3-6.5}$$

The deformation gradient matrix $\nabla\mathbf{u}$ can be written in terms of two other tensors. In Cartesian coordinates

$$(\nabla\mathbf{u})_{ij} = \tfrac{1}{2}(u_{i,j} + u_{j,i}) + \tfrac{1}{2}(u_{i,j} - u_{j,i})$$

The first term in parentheses on the right-hand side is the infinitesimal strain tensor $\boldsymbol{\varepsilon}$, while the second term is the rotation tensor $\boldsymbol{\omega}$; that is,

$$\boldsymbol{\varepsilon} = \tfrac{1}{2}[\nabla\mathbf{u} + (\nabla\mathbf{u})^T] \tag{3-6.6}$$

and

$$\boldsymbol{\omega} = \tfrac{1}{2}[\nabla\mathbf{u} - (\nabla\mathbf{u})^T] \tag{3-6.7}$$

A measure of strain is given by the change in length of the original differential line element. However, one can show that $\boldsymbol{\omega}$ does not affect this change in length,

$$ds^2 - dS^2 = 2(d\mathbf{s} \cdot \mathbf{e}) \cdot d\mathbf{s} \tag{3-6.8}$$

where

$$\mathbf{e} = \boldsymbol{\varepsilon} - \tfrac{1}{2}(\nabla\mathbf{u})^T \cdot (\nabla\mathbf{u}) \tag{3-6.9}$$

The finite strain measure $\mathbf{e}$ can also be written in terms of the infinitesimal and rotation tensors,

$$\mathbf{e} = \boldsymbol{\varepsilon} - \tfrac{1}{2}(\boldsymbol{\varepsilon} + \boldsymbol{\omega})^T \cdot (\boldsymbol{\varepsilon} + \boldsymbol{\omega})$$

For linear elasticity problems one only uses the first term defined by (3-6.6). For certain problems involving plates and shells the strains $\boldsymbol{\varepsilon}$ can be small, while the rotations are large. In these problems the second terms may be important.

Balance of Mass

The global form of the conservation of mass is given by

$$\frac{d}{dt}\int \rho dV = 0 \tag{3-6.10}$$

The local equation of balance of mass is found to be

$$\dot{\rho} + \rho \nabla \cdot \mathbf{v} = 0 \tag{3-6.11}$$

Balance of Momentum

Newton's law states that the total linear momentum of a body can only be changed by the action of one body on another as represented by either surface tractions $\boldsymbol{\tau}$ or body forces $\mathbf{f}$. In global form this becomes

$$\frac{d}{dt}\int \rho \mathbf{v}\, dV = \int \boldsymbol{\tau} dS + \int \mathbf{f} dV \tag{3-6.12}$$

For an electromagnetic continuum with no polarization or magnetization

$$(\mathbf{P} = \mathbf{M} = 0), \qquad \mathbf{f} = \mathbf{f}^{\text{em}} + \mathbf{f}^{\text{mech}}$$

where

$$\mathbf{f}^{\text{em}} = q\mathbf{E} + \mathbf{J} \times \mathbf{B} \tag{3-6.13}$$

In local form one uses the stress tensor defined by

$$\boldsymbol{\tau} = \mathbf{t} \cdot \mathbf{n}$$

and the local balance of linear momentum is given by

$$\nabla \cdot \mathbf{t} + \mathbf{f}^{\text{mech}} + q\mathbf{E} + \mathbf{J} \times \mathbf{B} = \rho \dot{\mathbf{v}} \tag{3-6.14}$$

In Chapter 2 we showed that $\mathbf{f}^{\text{em}}$ could be written as the divergence of an electromagnetic tensor $\mathbf{T}$ [Eq. (2-6.13)]; that is,

$$\mathbf{f}^{\text{em}} = \nabla \cdot \mathbf{T} - \frac{\partial}{\partial t}(\varepsilon_0 \mathbf{E} \times \mathbf{B}) \tag{3-6.15}$$

so that Eq. (3-6.14) can be rewritten in the form

$$\nabla \cdot (\mathbf{t} + \mathbf{T} + \rho \mathbf{v}\mathbf{v}) + \mathbf{f}^{\text{mech}} = \frac{\partial}{\partial t}(\rho \mathbf{v} + \varepsilon_0 \mathbf{E} \times \mathbf{B}) \tag{3-6.16}$$

Some writers like to identify $\rho\mathbf{v} + \varepsilon_0 \mathbf{E} \times \mathbf{B}$ as the total momentum. For nonrelativistic problems this view is not very useful and the form (3-6.14) is preferred. For some static problems with $\dot{\mathbf{v}} = \mathbf{v} = 0$ we have

$$\nabla \cdot (\mathbf{t} + \mathbf{T}) = 0 \tag{3-6.17}$$

This form says that the mechanical stresses $\mathbf{t}$ must be in equilibrium with the electromagnetic stresses $\mathbf{T}$. As discussed in Chapters 1 and 2 this leads to Faraday's concepts of magnetic pressures and tensions and is sometimes useful for analyzing static stresses in magnetic structures (see Chapter 4).

In continuum mechanics one also requires a balance of angular momentum in the differential material element. In classical fluid and solid mechanics this leads to the requirement of a symmetric stress tensor, that is,

$$t_{ij} = t_{ji}$$

However, when electric polarization or magnetization is present, body couples may enter the theory, as in the dipole model [Eq. (3-4.3)] and the theory must admit a nonsymmetric stress tensor (see, e.g., Hutter and van de Ven, 1978).

Balance of Thermomechanical Energy

One can view the thermomechanical continuum as an open thermodynamic system in which the electromagnetic field provides a source of work through $\mathbf{f}^{\text{em}}$ and a source of heat through the relative Joule heating $r = \mathbf{J}' \cdot \mathbf{E}'$ (see Section 3.2). This energy principle for material in a volume V with stored energy density U, and heat flux vector $\mathbf{h}$ can be written in the form

$$\frac{d}{dt}\int_V \left(\frac{1}{2}\rho v^2 + \rho U\right) dV = \int_V (\mathbf{f}^{\text{em}} \cdot \mathbf{v} + \mathbf{f}^{\text{mech}} \cdot \mathbf{v} + r)\, dV + \int (\boldsymbol{\tau} \cdot \mathbf{v} - \mathbf{h} \cdot \mathbf{n})\, dS \tag{3-6.18}$$

If the surface traction is replaced by the stress tensor, that is, $\boldsymbol{\tau} = \mathbf{t} \cdot \mathbf{n}$, and the local momentum equation (3-6.14) is used, then the following local form of the thermomechanical energy equation can be written as

$$\rho \dot{U} + \nabla \cdot \mathbf{h} - \mathbf{t} : \nabla \mathbf{v} = \mathbf{J}' \cdot \mathbf{E}' \tag{3-6.19}$$

Balance of Thermomechanical and Electromagnetic Energy

The global energy equation for a continuous magnetomechanical system may be obtained by considering the corresponding energy equations for the electromagnetic and mechanical subsystems. We assume here that the material is *nonpolarizable* ($\mathbf{P} = 0$) and *nonmagnetizable* ($\mathbf{M} = 0$). As derived in Chapter 2, the vector equations of Maxwell can be combined into a scalar equation that looks like an energy balance statement; that is,

$$\frac{d}{dt}\int_V W dV + \int_S \mathbf{E}\times\mathbf{H}\cdot\mathbf{n}dS = -\int \mathbf{E}\cdot\mathbf{J}dV \tag{3-6.20}$$

where

$$W = \tfrac{1}{2}(\varepsilon_0 E^2 + \mu_0 H^2)$$

In a previous section we looked at an interacting mixture of electrons and ions and found that

$$\mathbf{J}\cdot\mathbf{E} = \mathbf{f}^{\mathrm{em}}\cdot\mathbf{v} + \mathbf{J}'\cdot\mathbf{E}' \tag{3-6.21}$$

$\mathbf{J}'\cdot\mathbf{E}'$ represents the work done relative to the material, such as Joule heating. (For some materials, such as a battery with chemical reactions, $\mathbf{J}'\cdot\mathbf{E}'$ can represent a source of energy.) Thus $\mathbf{J}\cdot\mathbf{E}$ represents the work done by the body force as well as the Joule heating. The expression (3-6.21) can be used to combine the electromagnetic and mechanical energy equation to obtain a single equation of balance energy:

$$\frac{d}{dt}\int_V (W + \mathcal{T} + \rho U)dV + \int_S (\mathbf{E}\times\mathbf{H}\cdot\mathbf{n} - \boldsymbol{\tau}\cdot\mathbf{v} + \mathbf{h}\cdot\mathbf{n})dS = 0 \tag{3-6.22}$$

Here $\mathcal{T}$ represents the kinetic energy density $\frac{1}{2}\rho v^2$. Thus the change in total magnetomechanical energy, as represented by the volume integral is proportional to the rate of electromagnetic energy input through S as given by $\mathbf{E}\times\mathbf{H}\cdot\mathbf{n}$, the work done by the surface tractions, and the thermal heat flux $\mathbf{h}\cdot\mathbf{n}$.

Entropy Inequality

When dissipative processes are present such as damping, heat conduction, and Joule heating, it is necessary to consider the change of entropy when formulating constitutive relations. As has been shown in many standard texts on continuum mechanics, the Clausius–Duhem entropy inequality places restrictions on the form of the constitutive equations.

We will not develop the general constitutive theory of magnetic materials in this book. The interested reader is refered to more theoretical works such as Tiersten (1964), Brown (1966), and Hutter and van de Ven (1978).

3.7 EQUATIONS FOR NONMAGNETIC ELASTIC CONDUCTORS

We present the equations for interaction of electric current with an elastic solid. Particular problems can be classified into bulk solids, which require the three-dimensional theory of elasticity, and structures such as beams, plates, and shells for which approximate theories and simpler equations may be used. In the latter, simplifying assumptions about stress distribution in the thickness direction can usually be made. In this section we present the equations for three-dimensional thermoelasticity modified for magnetic forces and heat sources. We shall neglect charge accumulation, electric polarization, and magnetization; that is, we set $q = 0$, $\mathbf{P} = 0$, and $\mathbf{M} = 0$.

Constitutive Equations

For nonferromagnetic/nonpiezoelectric materials, the stress tensor is a function of the strain and temperature. In particular, the stress in an elastic body is related to the stored internal energy. It is common to replace the energy U with the Helmholtz free energy Ψ defined by

$$\Psi = U - \theta\eta \tag{3-7.1}$$

where θ, η are the temperature and entropy respectively.

For linear elastic effects, we assume that U and Ψ depend on the infinitesimal strain tensor $\boldsymbol{\varepsilon}$ defined by Eq. (3-6.6).

The free energy is assumed to depend on the independent variables of temperature, strain, and electromagnetic fields, that is, θ, $\boldsymbol{\varepsilon}$, $\mathbf{E}$, and $\mathbf{B}$. If Hall effect and magnetization terms are neglected, then $\Psi = \Psi(\theta, \boldsymbol{\varepsilon}, \mathbf{B})$. Using Eq. (3-7.1) and the energy equation in the entropy inequality, it can be shown that the stress and entropy density must be related to the free energy by the following equations (see e.g., Eringen, 1967):

$$t_{ij} = \rho \frac{\partial \Psi}{\partial \varepsilon_{ij}}$$

and (3-7.2)

$$\eta = -\frac{\partial \Psi}{\partial \theta}$$

While general expressions for nonlinear elastic materials may be found in many references on continuum mechanics, in this book we will assume that the strains are small or that Ψ is a quadratic function of the strains.

A further assumption restricts the problems to isotropic, linear, elastic solids. For these materials we have

$$\mathbf{t} = \lambda \operatorname{Tr}(\boldsymbol{\varepsilon})\boldsymbol{\delta} + 2\mu\boldsymbol{\varepsilon} - \alpha(3\lambda + 2\mu)\theta\boldsymbol{\delta} \tag{3-7.3}$$

Here λ and μ are Lamé constants of elasticity and α is the coefficient of expansion. This expression leads to the fact that a positive change in temperature in the absence of strain results in a state of hydrostatic compression.

Current Density and Heat Flux Relations

In uncoupled electromagnetic theory and heat transfer, one usually assumes that $\mathbf{J}$ is linearly related to $\mathbf{E}$ (Ohm's law), and that $\mathbf{h}$ is linearly related to the temperature gradient $\nabla\theta$; that is,

$$\mathbf{J} = \sigma\mathbf{E} \tag{3-7.4a}$$

$$\mathbf{h} = -\kappa\nabla\theta \tag{3-7.4b}$$

These relations are for isotropic materials. For anisotropic material, the electric conductivity σ and the heat conduction coefficient κ are replaced by second-order symmetric tensors (see Section 8.2).

From the solid-state theory of electrons in solids and from the continuum theory of magnetoelasticity, one can derive a more general relationship where $\mathbf{J} = \mathbf{J}(\mathbf{E}, \mathbf{B}, \nabla\theta)$. For the moment we will consider $\mathbf{J} = \mathbf{J}(\mathbf{E}, \nabla\theta)$ and $\mathbf{h} = \mathbf{h}(\mathbf{E}, \nabla\theta)$. That $\mathbf{J}$ and $\mathbf{h}$ are not completely independent can be shown from the entropy inequality, which in the absence of motion, takes the form (Moon, 1970b; Hutter and van de Ven, 1978)

$$\mathbf{J}\cdot\mathbf{E} - \frac{\mathbf{h}\cdot\nabla\theta}{\theta} \geq 0 \tag{3-7.5}$$

Again we assume that $\mathbf{J}$ and $\mathbf{h}$ are linear functions of $\mathbf{E}$ and $\nabla\theta$, that is,

$$\begin{aligned} \mathbf{J} &= \sigma\mathbf{E} + \gamma\nabla\theta \\ \mathbf{h} &= -\kappa\nabla\theta + \beta\mathbf{E} \end{aligned} \tag{3-7.6}$$

A sufficient condition for satisfaction of Eq. (3-7.5) is to set $\beta = \gamma\theta$. This relation has also been posited on other physical grounds (see, e.g., Ziman, 1964).

The effect of thermal gradients on the current density $\mathbf{J}$ and the effect of the electric field on the heat flux vector $\mathbf{h}$ are known as *thermoelectric effects*. The induction of an electric field by a thermal gradient is called the *Seeback effect*. On the other hand, if $\nabla\theta = 0$, then a current induces heat flow (called *Peltier effect*); that is,

$$\mathbf{h} = \pi\mathbf{J} \tag{3-7.7}$$

where $\pi = \beta/\delta$ and is known as the Peltier coefficient.

These effects are related to the free-electron theory of conducting solids. In addition, it can be shown that the current-electron field relation also depends on the magnetic field. This generalized Ohm's law takes the form

$$\mathbf{J} + \sigma R\mathbf{J}\times\mathbf{B} = \sigma\mathbf{E} + \nu\nabla(\nabla\cdot\mathbf{u}) + \gamma\nabla\theta \tag{3-7.8}$$

The $\mathbf{J}\times\mathbf{B}$ term effectively adds a *magnetoresistive* term (Hall effect) to Ohm's law which for conductors such as aluminum, copper, and steel is small at room temperature and low magnetic fields. However, at low temperatures and high magnetic fields the term σRB may become important. Further discussion of the Hall effect is given in Section 8.7.

Using the constitutive relations for the stress tensor, Eq. (3-7.3), current-density vector, Eq. (3-7.4a), and heat flux vector, Eq. (3-7.4b), the momentum, heat, and magnetic field equations for linear elastic, isotropic, nonmagnetic conductors take the following forms (thermoelectric and Hall effects neglected):

Momentum equation

$$(\lambda+\mu)\boldsymbol{\nabla}(\boldsymbol{\nabla}\cdot\mathbf{u})+\mu\nabla^2\mathbf{u}-\rho\ddot{\mathbf{u}}=\alpha(3\lambda+2\mu)\nabla\theta+\frac{1}{\mu_0}\mathbf{B}\times(\boldsymbol{\nabla}\times\mathbf{B}) \tag{3-7.9}$$

Heat equation

$$\kappa\nabla^2\theta-\rho c\frac{\partial\theta}{\partial t}=\alpha\theta_0(3\lambda+2\mu)\boldsymbol{\nabla}\cdot\dot{\mathbf{u}}-\frac{1}{\sigma\mu_0^2}(\boldsymbol{\nabla}\times\mathbf{B})\cdot(\boldsymbol{\nabla}\times\mathbf{B}) \tag{3-7.10}$$

Magnetic field equation

$$\nabla^2\mathbf{B}-\mu_0\sigma\dot{\mathbf{B}}=-\mu_0\sigma\boldsymbol{\nabla}\times(\dot{\mathbf{u}}\times\mathbf{B}) \tag{3-7.11}$$

where the current density and magnetic field vectors are related by

$$\mathbf{J}=\mathbf{J}'=\sigma(\mathbf{E}+\mathbf{v}\times\mathbf{B}) \tag{3-7.12}$$

and

$$\boldsymbol{\nabla}\times\mathbf{B}=\mu_0\mathbf{J} \tag{3-7.13}$$

This constitutes a set of seven coupled, nonlinear, partial differential equations.

For some classes of problems the nonlinear term $\boldsymbol{\nabla}\times(\dot{\mathbf{u}}\times\mathbf{B})$, in the magnetic field equation (3-7.11), may be neglected in comparison to the $\dot{\mathbf{B}}$ term. The analysis behind this assumption is discussed in Chapter 8. Physically, one can say that for "stiff" problems where the particle velocities $\dot{\mathbf{u}}$ are not large, the change in flux due to $\dot{\mathbf{B}}$ is much greater than that due to convection $\dot{\mathbf{u}}$. For solids moving through a magnetic field, or thin "flexible" bodies, the convection term may be of importance (see, e.g., Chian and Moon, 1981).

When $\boldsymbol{\nabla}\times(\dot{\mathbf{u}}\times\mathbf{B})$ is neglected in Eq. (3-7.11), the equations become hierarchically coupled. Thus, $\mathbf{B}$ can be found from Eq. (3-7.11) and then θ and $\mathbf{u}$ can be solved knowing $\mathbf{B}(\mathbf{r}, t)$. In each case, the solution for $\mathbf{B}$, θ, and $\mathbf{u}$ involves linear, inhomogeneous differential equations which can be solved using Green's functions or other standard methods.

When the time variations in field are not large (see Chapter 8) one can also neglect the coupling term $\boldsymbol{\nabla}\cdot\dot{\mathbf{u}}$ in the thermoelastic equation.

It is often helpful to nondimensionalize the equations as discussed in Chapter 1. There are a number of methods, but one useful system employs the following characteristic length, time, temperature, and stress, respectively:

$$x_0=\frac{1}{\mu_0\sigma v_L} \tag{3-7.14}$$

$$t_0=\frac{1}{\mu_0\sigma v_L^2}$$

$$\theta_1 = \frac{B_0^2}{\mu_0} \frac{1}{\mu_0 \sigma \kappa}$$

and

$$t_{xx}^0 = \frac{B_0^2}{2\mu_0}$$

where v_L is the longitudinal wave speed in the solid, that is, $v_L^2 = (\lambda + 2\mu)/\rho$. In addition, two nondimensional groups enter the equations:

$$a = \frac{\sigma \mu_0 \kappa}{\rho c_v} \tag{3-7.15}$$

and

$$m = \frac{2\alpha(3\lambda + 2\mu)}{\sigma \mu_0 \kappa}$$

For one-dimensional motion in a half-space, the uncoupled magnetothermoelastic equations for $\mathbf{B} = (0, 0, B_z(x, t))$, $\mathbf{J} = (0, J_y, 0)$, $\mathbf{u} = (u, 0, 0)$, and $\eta = x/x_0$, $\tau = t/t_0$ are

$$\frac{\partial^2 \bar{B}}{\partial \eta^2} = \frac{\partial \bar{B}}{\partial \tau} \tag{3-7.16}$$

$$\frac{1}{a}\frac{\partial \bar{\theta}}{\partial \tau} - \frac{\partial^2 \bar{\theta}}{\partial \eta^2} = \left(\frac{\partial \bar{B}}{\partial \eta}\right)^2 \tag{3-7.17}$$

and

$$\frac{\partial^2 \bar{t}_{xx}}{\partial \eta^2} - \frac{\partial^2 \bar{t}_{xx}}{\partial \tau^2} = \frac{\partial^2 \bar{B}^2}{\partial \eta^2} + m\frac{\partial^2 \bar{\theta}}{\partial \tau^2} \tag{3-7.18}$$

where $\bar{B} = B_z/B_0$, $\bar{\theta} = \theta/\theta_1$, and $\bar{t}_{xx} = t_{xx}/t_{xx}^0$. The constant B_0 is proportional to the applied magnetic field on the surface of the half-space. The solution of these equations is discussed in Chapter 8.

The hierarchical nature of the coupling is clear in Eqs. (3-7.16)–(3-7.18). First, $\bar{B}$ is determined from a homogeneous diffusion equation. Next, the temperature $\bar{\theta}$ is found given $\bar{B}(\eta, \tau)$. Finally, the stress is found from an inhomogeneous wave equation. Values of x_0, t_0, θ_1, and so forth for aluminum and copper are listed in Table 8-1.

Equations for Elastic Conducting Plates

When the conductor is in the form of a conducting plate, we can integrate the magnetic forces through the thickness. The equation of motion for the lateral motion of the plate w is given by

$$D\nabla_1^4 w + \rho h \frac{\partial^2 w}{\partial t^2} = F + \mathbf{n} \cdot \boldsymbol{\nabla} \times \mathbf{c} \tag{3-7.19}$$

where for nonferromagnetic materials,

$$F = \int \mathbf{n} \cdot (\mathbf{J} \times \mathbf{B}) \, dz$$

and

$$\mathbf{c} = \int z\mathbf{n} \times (\mathbf{J} \times \mathbf{B}) \, dz$$

Here $\mathbf{n}$ is the unit vector normal to the plate surface and D is the plate stiffness constant given by

$$D = \frac{Yh^3}{12(1-\nu^2)} \tag{3-7.20}$$

and the Laplacian $\nabla_1^2 = (\partial^2/\partial x^2) + (\partial^2/\partial y^2)$. The average membrane stresses in the plate must satisfy

$$\nabla \cdot \mathbf{t} + \int (\mathbf{J} \times \mathbf{B}) \cdot (\boldsymbol{\delta} - \mathbf{nn}) \, dz = \rho h \frac{\partial^2 \mathbf{u}}{\partial t^2} \tag{3-7.21}$$

Here $\boldsymbol{\delta}$ is the identity dyadic and $\mathbf{u}$ the average in-plane displacement vector.

3.8 REDUCTION OF BODY FORCE PROBLEMS TO SURFACE TRACTION PROBLEMS

The concept of magnetic pressure can be used to replace a distributed body force problem by a more familiar surface traction analysis. This becomes useful in dynamics problems where the electric currents or fields are confined to a small layer or *skin depth* (see Chapter 2). Consider the case of a thick, electrically conducting cylinder with either an axial current or axial magnetic field (see Fig. 3-8). If the current or the field is pulsed or oscillating, the

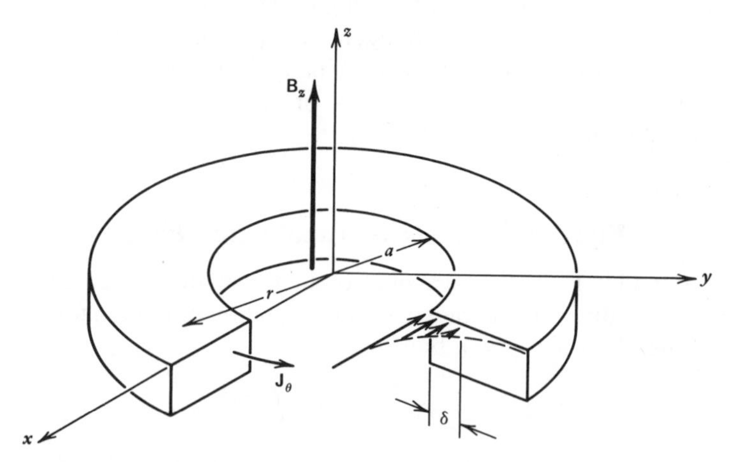

Figure 3-8 Skin effect of induced current in a cylindrical solid.

induced currents in the solid will be confined to a layer $a \le r < a + \delta$. The momentum balance equation (3-6.14) inside and outside of this region is

$$\frac{1}{r}\frac{\partial}{\partial r}(rt_{rr}) - \frac{t_{\theta\theta}}{r} - \rho\ddot{u}_r = -\frac{\partial P_m}{\partial r} \qquad (a < r < a + \delta) \tag{3-8.1a}$$

$$= 0 \qquad (r > a + \delta) \tag{3-8.1b}$$

where cylindrical symmetry is assumed for simplicity. The magnetic pressure is given by $P_m = B_z^2/2\mu_0$ [see Eq. (2-6.16)] in the axial field case shown in Fig. 3-8.

As a body force problem we have the boundary condition

$$t_{rr}(r = a) = 0$$

To obtain a surface force problem we multiply Eq. (3-8.1) by r and integrate over the skin-depth region,

$$\int_a^{a+\delta} \frac{\partial}{\partial r}[r(t_{rr} + P_m)]\,dr = \int_a^{a+\delta} (t_{\theta\theta} + P_m + \rho r\ddot{u}_r)\,dr$$

or

$$(a + \delta)t_{rr}(a + \delta) - aP_m(a) \le \delta\,|t_{\theta\theta} + P_m + \rho r\ddot{u}_r|_{\max(a<r<a+\delta)} \tag{3-8.2}$$

By decreasing the pulse time, or increasing the frequency, the skin depth can be made small (see Chapters 2 and 8). However, for fixed P_m, the stored energy in the skin-depth region tends to zero as $\delta/a \to 0$. We infer from this that $|t_{\theta\theta} + \rho r\ddot{u}_r|_{\max}$ remains bounded. Thus, in the limit as $\delta/a \to 0$,

$$t_{rr}(a + \delta) \simeq P_m(a) \tag{3-8.3}$$

Therefore to determine the stresses outside the skin-depth region, $r - a > \delta$, we may use the boundary condition

$$t_{rr}(a, t) = P_m(a, t)$$

together with the momentum equation (3-8.1b). The body force problem is thus reduced to a surface force problem. In using this approximation, one must be aware that in the actual problem there is a transition region where the normal stress goes from zero to the magnetic pressure value (see, e.g., Moon and Chattopadhyay, 1974). This problem will be treated again in Chapter 8 in a discussion of dynamic magnetoelastic stresses.

3.9 MAGNETOMECHANICAL VIRIAL THEOREM

Continuing our discussion of the mechanics of nonferromagnetic elastic conductors, we derive a theorem which relates the average stresses in a body to the stored electromagnetic energy (see also Section 6.8). This theorem is similar to one in the classical mechanics of particles called the *virial theorem* (see, e.g., Goldstein, 1950).

To derive this theorem we consider the function $G(t)$ which is the integral of the inner product of the momentum and position vector $\mathbf{r}$ over all bodies containing charge or current; that is,

$$G=\int_{\mathscr{B}} \rho\mathbf{v}\cdot\mathbf{r}\,dV \tag{3-9.1}$$

It follows that

$$\frac{dG}{dt}=2\mathscr{T}+\int_{\mathscr{B}} \rho\mathbf{r}\cdot\frac{d\mathbf{v}}{dt}\,dV \tag{3-9.2}$$

where $\mathscr{T}$ is the kinetic energy of the body or bodies. From the balance of momentum, however,

$$\rho\dot{\mathbf{v}}=\boldsymbol{\nabla}\cdot\mathbf{t}+\mathbf{f}$$

We now introduce a time average defined by

$$\langle F(t)\rangle\equiv\frac{1}{\tau}\int_0^{\tau} F(t)\,dt$$

Applying this average to Eq. (3-9.2) we find

$$\frac{1}{\tau}[G(\tau)-G(0)]=2\langle\mathscr{T}\rangle+\left\langle\int(\boldsymbol{\nabla}\cdot\mathbf{t}+\mathbf{f})\cdot\mathbf{r}\,dV\right\rangle \tag{3-9.3}$$

If the motion is bounded then the left-hand side of Eq. (3-9.3) vanishes as $\tau\to\infty$. We also obtain zero if there is no motion or if the motion is periodic. For any one of these cases, we have

$$2\langle\mathscr{T}\rangle=-\left\langle\int(\boldsymbol{\nabla}\cdot\mathbf{t}+\mathbf{J}\times\mathbf{B}+q\mathbf{E})\cdot\mathbf{r}\,dV\right\rangle \tag{3-9.4}$$

For transient problems the time average is over $0\le t<\infty$, while for periodic motions the average is over one cycle.

For static problems $\mathscr{T}=0$ and we may remove the angular brackets. For magnetic problems where $q=0$, we have

$$\int_{\mathscr{B}}(\boldsymbol{\nabla}\cdot\mathbf{t}+\mathbf{J}\times\mathbf{B})\cdot\mathbf{r}\,dV=0 \tag{3-9.5}$$

This integral may be transformed to a surface integral using the identities

$$\mathbf{r}\cdot\boldsymbol{\nabla}\cdot\mathbf{t}=\boldsymbol{\nabla}\cdot(\mathbf{r}\cdot\mathbf{t})-\mathrm{Tr}(\mathbf{t}) \tag{3-9.6}$$

and

$$\mathbf{J}\times\mathbf{B}\cdot\mathbf{r}=\boldsymbol{\nabla}\cdot(\mathbf{r}\cdot\mathbf{T}^m)+\frac{1}{2\mu_0}B^2$$

where $\mathbf{T}^m=(1/\mu_0)(\mathbf{BB}-\frac{1}{2}B^2\boldsymbol{\delta})$. We assume that all the magnetic forces are equilibrated by internal forces in the body so that the surface tractions

$\mathbf{t}\cdot\mathbf{n}=0$ on the body are zero. The virial theorem then takes the form

$$\int_{\mathcal{B}} \mathrm{Tr}(\mathbf{t})\,dV = \int_{\Sigma} \mathbf{r}\cdot\mathbf{T}^m\cdot\mathbf{n}^+\,dV + \frac{1}{2\mu_0}\int_{\mathcal{B}} B^2\,dV \tag{3-9.7}$$

where $\mathbf{n}^+$ is the outward normal to the body surface Σ. Outside the body $\mathbf{t}=0$ and the following identity holds

$$\int_{\Sigma} \mathbf{r}\cdot\mathbf{T}^m\cdot\mathbf{n}^+\,dA = \int_{V_\infty-\mathcal{B}} \frac{B^2}{2\mu_0}\,dV \tag{3-9.8}$$

This assumes that the magnetic stress vector $\mathbf{T}^m\cdot\mathbf{n}\to 0$ as $\mathbf{r}\to\infty$ and that $|\mathbf{T}^m\cdot\mathbf{n}|r^2\to 0$ as $r\to\infty$. The final form of the theorem for *static* problems becomes

$$\int_{\mathcal{B}} \mathrm{Tr}(\mathbf{t})\,dV = \int_{V_\infty} \frac{B^2}{2\mu_0}\,dV \tag{3-9.9}$$

This theorem states that a lower bound exists on the structural mass required to contain stored magnetic energy (depending on the yield or ultimate stress). Application of this concept to superconducting magnet systems is discussed in Section 6.8.

3.10 EQUATIONS FOR MAGNETIZABLE ELASTIC CONTINUA

There have been numerous attempts to formulate a field theory of magnetic interaction with elastic ferromagnetic solids. Except for the examples of buckling or vibrations of a plate in a magnetic field, few problems of interest to structural mechanics have been solved. Part of the problem is that analysts have usually dealt with a uniform magnetic field in which zero net magnetic forces result, and mechanicians have tried to come up with "exact" formulations and have avoided approximations. In this section we will examine a strength-of-materials approach to solving problems in magnetoelasticity.

The net force and couple on a magnetizable solid in an external magnetic field $\mathbf{B}^0$ were given in Section 3.4 [Eqs. (3-4.2) and (3-4.3)]. It was shown that a number of models for the magnetic force can lead to the same total force including the dipole, pole, and Amperian-current models. This has led to a plethora of general axiomatic theories for elastic magnetic materials, each using a different local force density (see, e.g., Tiersten, 1964; Brown, 1966; Pao and Yeh, 1973; Hutter and van de Ven, 1978; Alblas, 1979; Maugin and Goudjo, 1982 to mention just a few). While these works are of importance to the theoretical framework of magnetoelasticity, none has given a practical procedure for application of these general theories to machine and structural problems.

The key to the application of general magnetoelastic theories to practical problems lies in the recognition of two important facts:

1. For practical magnetic materials, the difference between the internal stress state for various magnetic force models differs by $\mu_0 M^2$ or about 40 N/cm^2 (58 psi).
2. For linear ferromagnetic materials, such as steels, iron, nickel, cobalt, and various alloys, the relative permeability is very large, μ_r or $\chi = 10^2 \rightarrow 10^5$.

These two facts form the basis for simplifying several of the general theories.

To lend some credibility to the approach, we will derive this approximate theory from one of the field theories, in particular that of Pao and Yeh (1973).

Their formation is based on the dipole model discussed above in which they take the local body force and body couple expressions to be

$$\mathbf{f}^m = \mu_0 \mathbf{M} \cdot \nabla \mathbf{H} \tag{3-10.1}$$

and

$$\mathbf{c} = \mu_0 \mathbf{M} \times \mathbf{H} = \mathbf{M} \times \mathbf{B} \tag{3-10.2}$$

We will assume a quasistatic but inhomogeneous magnetic field, and neglect any electromagnetic waves and conduction currents. A continuum theory with a body couple such as Eq. (3-10.2) generally leads to a nonsymmetric stress tensor. The body force (3-10.1) can be written as the divergence of an electromagnetic stress tensor or dyadic $\mathbf{T}^m$:

$$\mathbf{f}^m = \nabla \cdot \mathbf{T}^m \tag{3-10.3}$$

where

$$\mathbf{T}^m = \mathbf{B}\mathbf{H} - \tfrac{1}{2}\mu_0 \mathbf{H} \cdot \mathbf{H} \boldsymbol{\delta} \tag{3-10.4}$$

An important part of this theory is the boundary conditions on the stress and magnetic field. In such problems, **M**, **H**, and **B** usually have discontinuities at the surface of a material body. Thus expressions such as $\mathbf{M} \cdot \nabla \mathbf{H}$ must be defined using the limit of the derivative from the material side of surface only. The jump in **M** at the surface produces a nonvanishing force which must be accounted for in the stress boundary conditions. These become

$$\mathbf{n} \cdot [\mathbf{t} + \mathbf{T}^m] = 0$$

or

$$\mathbf{n} \cdot [\mathbf{t}] = \tfrac{1}{2}\mu_0 \mathbf{n}(\mathbf{M} \cdot \mathbf{n})^2 \tag{3-10.5}$$

The rate of work done by the magnetic field on the body can be shown to be (see Pao and Yeh, 1973)

$$\varepsilon = \mathbf{f}^m \cdot \mathbf{v} + \mu_0 \rho \mathbf{H} \cdot \frac{d(\mathbf{M}/\rho)}{dt} \tag{3-10.6}$$

where the latter term includes work due to the couple as well as the change in magnetization relative to axes that rigidly rotate with the element of the body.

The linear and angular momentum equations can be shown to be

$$\nabla \cdot \mathbf{t} + \mathbf{f}^m = \rho \frac{d\mathbf{v}}{dt} \tag{3-10.7}$$

and

$$t_{[jk]} + \mu_0 M_{[j} H_{k]} = 0 \tag{3-10.8}$$

where $A_{[ij]} = A_{ij} - A_{ji}$. The nonsymmetry of the stress tensor is a consequence of the body couple (3-10.2).

To simplify the equations of Pao and Yeh we neglect magnetostriction and piezostrictive terms in the constitutive equations between stress, strain, and magnetization and assume an isotropic linear elastic material. Under these assumptions one obtains for **t**,

$$\mathbf{t} = \lambda(\nabla \cdot \mathbf{u})\boldsymbol{\delta} + 2\mu\boldsymbol{\varepsilon} + \mu_0 \mathbf{MH} \tag{3-10.9}$$

If the magnetic material is itself linear (so called "soft" ferromagnetic material or material whose local average magnetization becomes zero when the external field is set to zero), then we can write

$$\mathbf{M} = \chi \mathbf{H}, \qquad \mathbf{B} = \mu_0 \mu_r \mathbf{H} \tag{3-10.10}$$

where $\chi = \mu_r - 1$ and for practical materials, $\mu_r \sim 0(10^2\text{–}10^4)$ (see Table B-3). For such materials the body couple (3-10.2) is zero and the stress tensor is symmetric, that is,

$$\mathbf{c} = \mu_0 \chi \mathbf{M} \times \mathbf{M} = 0$$

From Eq. (3-10.8),

$$t_{ij} = t_{ji}$$

Also,

$$\mathbf{f}^m = \mu_0 \chi \mathbf{H} \cdot \nabla \mathbf{H} = \frac{\chi}{\mu_r^2} \nabla \frac{B^2}{2\mu_0} = \frac{\chi}{\mu_r^2} \nabla P^m$$

where P^m is the magnetic pressure. Note that for $B = 1$ T, $P^m \sim 40$ N/cm^2 (58 psi). Thus, compared to the usual stress level in engineering structures, P^m is almost negligible.

The above field equations are valid when the initial rigid-body magnetization times the displacement gradients ∇u are small. Also we assume that any initial stresses caused by mechanical or magnetic forces are small.

3.11 A PLATE THEORY FOR FERROELASTIC MATERIALS

We now apply this theory to a two-dimensional plate in which we neglect stress in the x_3 direction and assume $\partial/\partial x_3 = 0$.

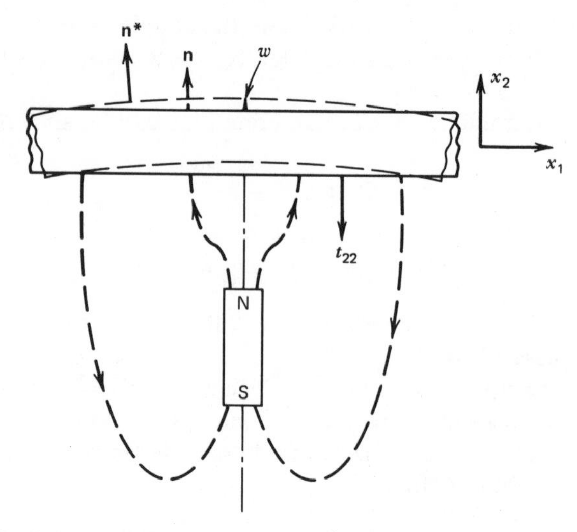

Figure 3-9 Surface normal vectors **n** and **n*** for an undeformed and bent plate.

The field equations take the form

$$\frac{\partial t_{11}}{\partial x_1}+\frac{\partial t_{21}}{\partial x_2}+\frac{\chi}{\mu_r^2}\frac{\partial P^m}{\partial x_1}=0 \tag{3-11.1}$$

$$\frac{\partial t_{12}}{\partial x_1}+\frac{\partial t_{22}}{\partial x_2}+\frac{\chi}{\mu_r^2}\frac{\partial P^m}{\partial x_2}=0$$

and

$$t_{12}=t_{21}$$

The boundary conditions (3-10.5) on the top and bottom surfaces of the plate become [$\mathbf{n}=(0, 1)$] (see Fig. 2-9)

$$t_{12}=0$$

and (3-11.2)

$$t_{22}=\frac{\chi^2}{\mu_r^2}\frac{B_2^2}{2\mu_0}$$

where only magnetic forces are assumed to be acting.

The form of Eq. (3-11.1) suggests defining new stresses $\hat{t}_{ij}$ so that

$$\hat{t}_{ji,j}=0, \qquad \hat{t}_{ij}=\hat{t}_{ji} \tag{3-11.3}$$

where

$$\hat{t}_{11}=t_{11}+\frac{\chi}{\mu_r^2}P^m, \qquad \hat{t}_{12}=t_{12}$$

and

$$\hat{t}_{22}=t_{22}+\frac{\chi}{\mu_r^2}P^m$$

Using Eq. (3-10.9) the constitutive equations for $\hat{t}_{ij}$ are given by

$$\hat{t}_{11} = \lambda(\boldsymbol{\nabla} \cdot \mathbf{u}) + 2\mu\varepsilon_{11} + \frac{\chi}{2\mu_0\mu_r^2}(B^2 + 2B_1^2) \tag{3-11.4}$$

$$\hat{t}_{22} = \lambda(\boldsymbol{\nabla} \cdot \mathbf{u}) + 2\mu\varepsilon_{22} + \frac{\chi}{2\mu_0\mu_r^2}(B^2 + 2B_2^2)$$

and

$$\hat{t}_{12} = 2\mu\varepsilon_{12} + \frac{\chi}{\mu_0\mu_r^2}B_1B_2$$

The boundary conditions on the top surface of the plate become

$$\hat{t}_{12} = 0$$

and

$$\hat{t}_{22} = \frac{\chi^2}{\mu_r^2}\frac{B_2^2}{2\mu_0}\left(1 + \frac{B^2}{B_2^2\chi}\right) \tag{3-11.5}$$

It is at this point that we make some reasonable approximations. First, we note that P_m or $B_2^2/2\mu_0$ is of the order of $10^2\ \mathrm{N/cm^2}$ compared to a yield stress of the order of $10^4\ \mathrm{N/cm^2}$, so that with $\chi \sim O(10^2)$ or larger the magnetic terms in Eq. (3-11.4) for $\hat{t}_{ij}$ are at least four orders of magnitude below the conventional stress terms $\lambda(\boldsymbol{\nabla} \cdot \mathbf{u})\boldsymbol{\delta} + 2\mu\boldsymbol{\varepsilon}$. Similarly, in the boundary conditions, we can drop the term B^2/μ_r compared to B_2^2 so long as $B_2 \neq 0$.

Thus the problem reduces to a classical elasticity problem in which the magnetic forces on the plate are represented by the jump in magnetic pressure $B_2^2/2\mu_0$ on the top and bottom of the plate.

Finally, we assume χ and μ_r are very large so that we write $\chi/\mu_r \simeq 1$.

In summary, the displacement equations become in this approximation

$$(\lambda + \mu)\nabla(\boldsymbol{\nabla} \cdot \mathbf{u}) + \mu\nabla^2\mathbf{u} = 0 \tag{3-11.6}$$

and

$$\mathbf{t} = \lambda(\boldsymbol{\nabla} \cdot \mathbf{u}) + \mu(\nabla\mathbf{u} + (\nabla\mathbf{u})^T)$$

with boundary conditions

$$\mathbf{t} \cdot \mathbf{n} = \frac{(\mathbf{B} \cdot \mathbf{n})^2}{2\mu_0}$$

and

$$\mathbf{t} \times \mathbf{n} = 0$$

Magnetic boundary conditions for the plate must account for the motion of the surface and change of surface normal; that is,

$$\mathbf{n}^* \cdot [\mathbf{B} + \mathbf{u} \cdot \nabla\mathbf{B}] = 0$$

$$\mathbf{n}^* \times [\mathbf{H} + \mathbf{u} \cdot \nabla\mathbf{H}] = 0 \tag{3-11.7}$$

where $\mathbf{n}^*$ represents the rotated normal vector which is approximately given by

$$\mathbf{n}^* = \mathbf{n} + \boldsymbol{\Omega} \times \mathbf{n}$$

(The brackets **[ ]** denote a jump in the quantity across the surface.) For a plate the rotation vector $\boldsymbol{\Omega}$ has components

$$\boldsymbol{\Omega} = \left(-\frac{\partial u_2}{\partial x_3}, \quad 0, \quad \frac{\partial u_2}{\partial x_1} \right) \tag{3-11.8}$$

The gradient terms in Eq. (3-11.7) represent Taylor-series expansions from the undeformed to the deformed surface of the plate.

These equations can be used to derive an equation for the deflection of the midsurface of a ferroelastic plate. In this approximation there is no body force. However, the transverse force in the plate is given by the jump in the magnetic pressure across the plate; that is,

$$D\nabla^4 w + \frac{[(\mathbf{B} \cdot \mathbf{n})^2]}{2\mu_0} = 0 \tag{3-11.9}$$

where w is the plate deflection, $[f] \equiv f(\text{top}) - f(\text{bottom})$, and D is the flexural stiffness of the plate per unit width [Eq. (3-7.20)].

Of course, we have not yet said anything about the determination of the magnetic field, whose value depends on the deflection of the plate through Eq. (3-11.7). Application of this theory to a particular problem will be given in Chapter 7.

4

Stress Due to Steady Currents in Structures

4.1 INTRODUCTION

In this chapter we investigate the effect of magnetic forces on structures carrying steady, electric currents. Most of the discussion will involve nonferromagnetic materials, so that magnetization-related forces will be omitted, unless noted. The subject of this chapter is also related to the design of superconducting structures; additional topics relating to magnetic forces in superconducting structures will be addressed in Chapter 6.

There are two features that distinguish magnetostructural mechanics from conventional structural problems. One is the fact that the stresses result from body forces which act in all directions, in contrast to gravity loaded structures. The second feature is the fact that electric currents are most often confined within the surface of the structure (except for arc-type problems). This constraint provides a feedback loop in which magnetic forces produce deformation of the structure, and the deformation causes a change in the magnetic forces. This interaction can lead to structural instabilities or buckling. The subject of magnetoelastic instability is introduced in more detail in Chapter 5 and also in Chapter 6.

In this chapter we study solids in the form of rods, beams, plates, and shells with steady electric current flowing through the structure. The resulting magnetic body force density is given by

$$\mathbf{f} = \mathbf{J} \times \mathbf{B} \tag{4-1.1}$$

Integration of this expression over the volume of the solid leads to the Biot–Savart force law [Eqs. (2-6.3) and (2-6.4)].

The electric currents are assumed to be maintained by either a voltage or current power supply external to the structure, or in the case of some superconducting structures, are persistent with no decay. The subject of induced currents (eddy currents), due to changes in the applied field, will be considered in Chapter 8.

It will also be assumed that the stress–strain relations are not affected by the magnetic field. Most electromagnets are wound with insulated conductors. Hence the magnetic structure has the characteristics of a *composite material*. Some features of the composite nature of magnet structures can be handled by assuming an equivalent, linear anisotropic material. However, if the coil is not "potted" in some matrix material, such as epoxy, the turns in the magnet will have relative motion due to local imperfections and initial strains in the winding process. This results in a nonlinear hysteretic behavior. An equivalent elastic material would have a low modulus for small applied loads, and then a high modulus after the initial "looseness" is taken out by the applied magnetic forces. Some discussion of this will be given in Chapter 6. The point to be emphasized is that a linear isotropic model is sometimes a very crude analytic model for actual, complex, magnet structures and a detailed numerical analysis is often required to determine the local stresses in a magnet. Nonetheless, we will use the linear elastic model as it provides some insight into the nature of stresses in magnetic force problems.

Literature Survey

The earliest works concerning the mechanics of current-carrying structures dealt with the total force and moment on the body. These include the classic works of Maxwell (1891), Jeans (1925), and Smythe (1968). Other specialized works concerning magnetic forces due to electric currents include Hague (1929), Dwight (1945), and Grover (1946). A summary of some of the known formulas for magnetic forces is given in Table B-2. (The novice in magnetomechanics should use caution in applying formulas for infinite filament problems to finite circuits since return current paths in real systems can introduce added forces not given by the formulas.)

In the design of magnetomechanical devices, it is the resulting stress distribution that is of interest. For even in self-equilibrated magnetic systems, large stresses exceeding yield stresses can be encountered. In the modern literature, there are three specialized monographs which deal with magnet design, Brechna (1973), Montgomery (1980), and Thome and Tarrh (1982). Each has a review of some of the literature concerning induced stresses in electromagnets. Two other reviews of the literature include one by Bobrov and Williams (1980) on the stresses in solenoids, and one by Moon (1978).

Our survey of current-induced stresses begins with Kapitza (1927) and Cockroft (1928) of the Rutherford Laboratory in England. If Joseph Henry is the father of the direct-current electromagnet, Kapitza might be called the father of the pulsed high-field magnet. In designing pulsed magnets with peak fields of the order of 25 T, he was able to produce stresses capable of yielding the copper and destroying the magnet. Cockroft provided tables for calculation of the magnetic forces due to $\mathbf{J} \times \mathbf{B}$. Kapitza used these tables and a method due to Cockroft to analyze the stresses in an electromagnet. The method used by Cockroft was an energy method. He reasoned that the body force on a turn of a cylindrical magnet could be derived from the mutual inductance between a single turn and the rest of the magnet (see the next section). In the actual calculation of the stresses, however, both Kapitza and Cockroft neglect the circumferential stress in the equilibrium equation,

$$\frac{\partial}{\partial r}(rt_{rr}) - t_{\theta\theta} + rf_r = 0$$

When $t_{\theta\theta}$ is dropped, one can *erroneously* calculate the radial stress for a constant current-density magnet by direct integration of the body force f_r. Although this analysis is spurious, it was unchallenged for many years.

Daniels (1953) discussed the cause of failures in high-field solenoid coils due to internal buckling of the turns. A sketch of a damaged coil, taken from his paper, is reproduced in Figure 4-1. His analysis uses an energy method

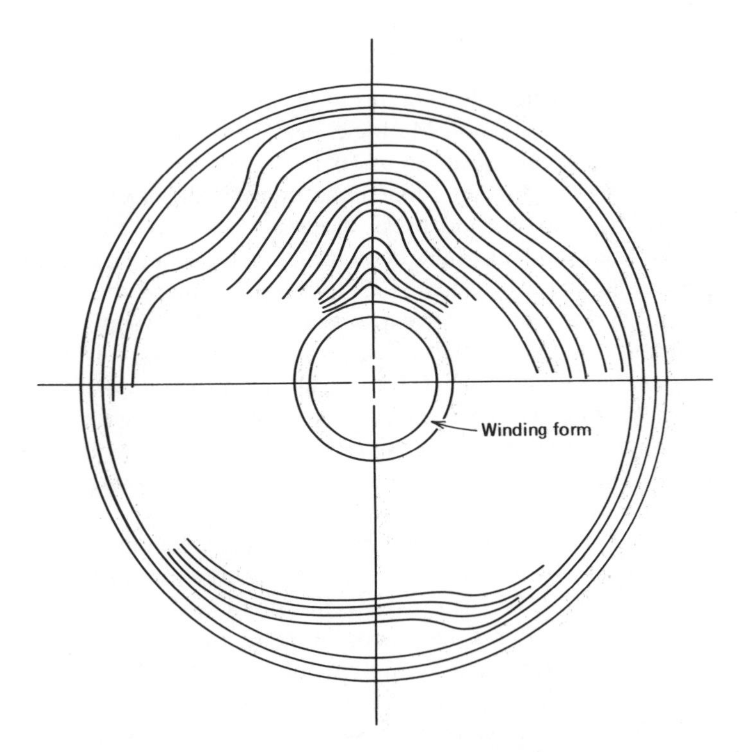

Figure 4-1 Sketch of deformed windings in a high-field solenoid magnet (after Daniels, 1953).

involving both the magnetic and elastic energies. It is one of the earliest studies of magnetoelastic stability and magnetic stiffness (see Chapter 5). However, he appears to have neglected the contribution of the compressive circumferential stress to the elastic energy.

Furth et al. (1957a, b) used the magnetic pressure concept to analyze stresses in coils. Landau and Lifshitz (1960) present a simple analysis of the stresses in a circular ring using the dependence of the inductance on the geometry of the ring. Kuznetsov (1960, 1961a) gave a two-dimensional analysis of stresses in a solenoid using a stress function approach. He includes the circumferential stress and shows the effect of dropping this term in Cockroft's analysis. He examined both the constant current-density case (1960) and the case of radial dependence of current density (1961).

Gersdorf et al. (1965) were one of the first to recognize the anisotropy of wound coils. They, in fact, measured Young's modulus ratio for the circumferential to the radial direction, that is $E_t/E_r = 20$, for copper coils with interturn insulation. They also assumed that the insulation carries no tension in the circumferential direction so they introduce a filling factor β, namely, the percentage of conductor volume to total volume (in Gersdorf's case, $\beta = 0.92$). The equilibrium equation that they use then has the form

$$\frac{\partial}{\partial r}(rt_{rr}) - \beta t_{\theta\theta} + \beta r f_r = 0 \qquad (4\text{-}1.2)$$

They also consider plastic deformations in the coil. The method of Gersdorf et al. was used by Mellville and Mattocks (1972) to analyze the failure of a pulsed cylindrical coil operating up to 16 T.

Recent studies have further refined the analysis of stresses in a wound solenoid to include winding prestress, thermal cooldown effects in superconducting magnets, and magnetic body forces (see Chapter 6). One of these is a study by Kokavec and Cesnak (1977) who calculate all three effects and provide some comparison with experimental results. Another study of superconducting solenoids is that by Arp (1977), who uses a winding-stress theory identical to Kokavec and Cesnak. He shows the cumulative effect of winding on a bobbin, bobbin removal, cooldown, and magnetic forces for both constant and variable tension winding stress. He shows that a variable tension winding scheme can lead to a more uniform total stress across the radial dimension of the coil.

Johnson et al. (1976) have treated the cylindrical solenoid as a concentric set of rings separated by an insulation layer. Each conductor ring is modeled as an anisotropic material. They show the combined effects of winding preload, thermoelastic effects due to cooling, and magnetic forces. Another analysis similar to Johnson et al. is that by Bobrov and Williams (1980), who treat the superconducting solenoid as a set of nested orthotropic rings. Their analysis includes variable winding tension. They also present stress states after winding, cooldown, and magnetic forces. This study is one of the first to examine the radial vibration of such a composite coil.

Other analyses of the stresses in a solenoid include Lontai and Marston (1965), Novitsky and Shakhtarin (1972), and Gray and Ballou (1977). The reader is referred to Bobrov and Williams (1980) for a review of these and other studies relating to stress analysis in solenoids.

Although the solenoid structure has received the most attention, almost all of the analyses treat only symmetric loading or radial deformation, and except for Bobrov and Williams (1980), do not consider dynamics. A few have considered the effect of friction between turns (Middleton and Trowbridge, 1967) and fewer still have examined the three-dimensional solenoid problem.

Fewer studies exist of stresses in nonsolenoidal conducting solids. Yuan (1972) has examined the stresses in a cylindrical rod carrying electrical current. A series of papers have discussed the elastic stability of the current-carrying rod, starting with Dolbin (1962) and ending with Chattepadhyay and Moon (1975). (See Chapter 5 for a discussion of this subject.)

Another class of problems receiving a good deal of attention are those associated with magnetic fusion reactors and magnetohydrodynamic generators. However, almost no analytic solutions are available due to the complexity of these structures. One exception is an elegant solution by File et al. (1971) in which the shape of a coil in a magnetic torus with no bending stresses is derived. This led to the design of "D"-shaped coils for tokamak fusion reactors instead of circular-shaped toroidal magnets (see Chapter 6). A large number of finite-element studies of "D"-shaped toroidal field magnets have been published which will not be reviewed here. A good source for these studies are the *Proceedings of the Symposia on Engineering Problems of Fusion Research* published by the IEEE, 1973–1983.

One of the few experimental studies of stresses in magnets is that of Pih and Gray (1976), who use mechanically loaded photoelastic models to simulate stresses in toroidal field magnets for tokamak fusion reactors.

Magnetic forces in spherical dipole and quadrapole magnets are given by Brechna (1973). Forces in an infinite helical solenoid have been derived by Kuznetsov (1961a) and by Georgievsky et al. (1974) for a helical torus (torsatron) or stellarator-type magnet for fusion.

One motivation for the interest in stress analysis in large magnets has been the failure to predict high deflections and stresses and, in some cases, failures of large magnets. In one example—the Princeton Large Torus (PLT), a normal conductor, tokamak fusion reactor—the measured deflections were from two to seven times those predicted by analysis. In a short review of this problem Bialek (1976) of the Princeton Plasma Physics Laboratory showed why a design analysis of the PLT coils, based on a monolithic structure, underpredicted the deflection, while a later model, using laminate theory, predicted the measured deflections more accurately.

In another failure, involving nested superconducting solenoids, Stevenson and Atherton (1974) attribute diamagnetic effects in the superconducting wire as the cause of increased body forces in one of the magnets.

4.2 MAGNETIC FIELDS AND FORCES

The total force and moment, as well as the internal stresses in current-carrying structures, are a consequence of the body force density [Eq. (4-1.1)]

$$\mathbf{f} = \mathbf{J} \times \mathbf{B}$$

There are two very different classes of problems involving steady currents in solids—those for which the magnetic field is independent of the induced deformation and temperature in the solid, and those which are not. Problems in which the magnetic fields and forces are dependent on the deformation of the structure are discussed briefly in Section 4.6 and in more detail in Chapters 5 and 6. The bulk of this chapter will deal with *deformation-independent* problems.

When the current density in a solid is prescribed, then the description of the body force distribution depends only on knowledge of the magnetic field. For a few cases, the magnetic field can be calculated directly using the Biot–Savart law (2-6.3) rewritten here:

$$\mathbf{B}^1(\mathbf{r}) = \frac{\mu_0}{4\pi} \int \frac{\mathbf{J} \times (\mathbf{r} - \mathbf{r}')\, dv'}{|\mathbf{r} - \mathbf{r}'|^3} \tag{4-2.1}$$

When the magnetic field has sources other than the current in the structure itself, we often write the total field in the form

$$\mathbf{B} = \mathbf{B}^0 + \mathbf{B}^1$$

where $\mathbf{B}^1$ is given by Eq. (4-2.1) and $\mathbf{B}^0$ is due to currents in other structures, magnetized material ($\mathbf{M} \neq 0$), or moving charges (e.g., plasmas).

For most practical cases of interest, the field $\mathbf{B}^1$ has to be evaluated using numerical techniques such as direct numerical integration of (Eq. 4-2.1), finite-difference or finite-element methods, and boundary integral techniques. For two-dimensional problems, such as those involving long current filaments parallel to prismatic magnetic bodies, one can sometimes employ complex variable techniques such as conformal mapping (see, e.g., Binns and Lawrenson, 1963). More often the structural analyst must rely on published tables of magnetic fields for specialized geometries or use one of a number of digital computer codes.

When discussing magnetic forces, one must distinguish between internal-force resultants and the total force and couple on the structure. The total force $\mathbf{F}$ and moment $\mathbf{C}$ with respect to some origin, on a body with circulating currents (or one with identical location of entrance and exit current leads), can only be due to sources of field outside the body, that is

$$\mathbf{F} = \int \mathbf{J} \times \mathbf{B}^0 dv \tag{4-2.2}$$

and

$$\mathbf{C} = \int \mathbf{r} \times (\mathbf{J} \times \mathbf{B}^0)\, dv \tag{4-2.3}$$

The self-field $\mathbf{B}^1$ cannot create a net force or moment on the body.

Internal-force resultants are the forces and moments between one part of a current-carrying structure and another. Although the total force on the body may be zero, the internal stresses and their resultants are never zero for finite current-carrying structures. (See Section 3.9 on force-free structures and the virial theorem.)

Using Eq. (4-2.1) one can write the magnetic body force in the form

$$\mathbf{f}(\mathbf{r}) = \mathbf{J} \times \mathbf{B}^0 + \frac{\mu_0}{4\pi} \int \frac{\mathbf{J}(\mathbf{r}) \times [\mathbf{J}(\mathbf{r}') \times (\mathbf{r} - \mathbf{r}')]}{|\mathbf{r} - \mathbf{r}'|^3} \, dv' \qquad (4\text{-}2.4)$$

From this expression one can see that the self-force, the second term on the right-hand side, scales as the square of the current density. If $\mathbf{B}^0$ is independent of $\mathbf{J}$, then the applied external force varies linearly with $\mathbf{J}$. There are, of course, examples where the magnetic field $\mathbf{B}^1$, due to $\mathbf{J}$, may magnetize another body and create a $\mathbf{B}^0$. For dynamic problems, $\mathbf{J}(\mathbf{r}, t)$, or in a moving body, $\mathbf{J}$ can create electric currents in another conductor thereby creating $\mathbf{B}^0$. In these cases the external magnetic force also scales as $\mathbf{J}^2$.

There are a number of direct and indirect methods for calculating magnetic forces. These may be grouped as follows:

1. Direct method.
2. Image method.
3. Equivalent magnetic-moment method.
4. Magnetic pressure method.
5. Cockroft's method for solenoids.
6. Energy or inductance method.

We illustrate each method with a few examples.

Direct Method—Magnetic Torus

The direct method involves the calculation of the magnetic field everywhere and direct application of the force law $\mathbf{J} \times \mathbf{B}$. There are very few examples where exact formulas exist for practical problems and the designer must often rely on numerical computer codes or tables to calculate the field. However, one exception is the case of the toroidal solenoid or torus shown in Figure 4-2. To create the torus we imagine a finite-length solenoid with circumferential currents to which we bend the axis into a circle, thereby joining the top and bottom cross sections of the solenoid. The result is a toroidal region with nonzero flux inside and zero flux outside. Application of Ampere's law (2-2.9) to a circular path inside the torus yields an expression for the resulting circumferential magnetic field

$$B_\theta = \frac{\mu_0 NI}{2\pi r}$$

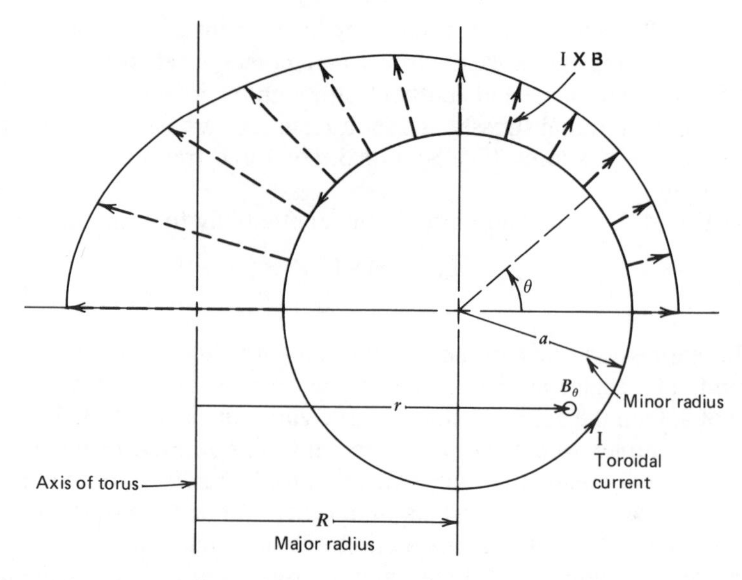

Figure 4-2 Magnetic body force distribution in a circular toroidal field coil.

where NI is the total ampere turns in the torus and r is the perpendicular distance from the toroidal axis.

In this example, we assume that the current is distributed in a thin toroidal shell in which the field decreases linearly through the thickness to a zero value outside the torus. The current in the shell thus sees an average field $\frac{1}{2}B_{\bar{\theta}}$, where $B_{\bar{\theta}}$ is evaluated on the inside surface of the shell.

The resulting body force distribution is normal to the circular cross section and is shown in Figure 4-2. If R represents the major radius of the torus, the total inward force on one turn with current I is given by the formula

$$F_x = \frac{\mu_0 I^2 N}{2\pi} \frac{a}{R} \int_0^{\pi} \frac{\cos\theta \, d\theta}{1 + (a/R)\cos\theta}$$

or

$$F_x = \frac{\mu_0 I^2 N}{2} \left\{ 1 - \left[1 - \left(\frac{a}{R} \right)^2 \right]^{-1/2} \right\} \tag{4-2.5}$$

The net vertical force on the upper half of one coil of current I is given by a similar expression

$$F_y = \frac{\mu_0 I^2 N}{4\pi} \frac{a}{R} \int_0^{\pi} \frac{\sin\theta \, d\theta}{1 + (a/R)\cos\theta}$$

or (4-2.6)

$$F_y = \frac{\mu_0 I^2 N}{4\pi} \ln \frac{1 + a/R}{1 - a/R}$$

Image Method

Consider a current filament parallel to a ferromagnetic half-space. For this case, the magnetic field outside the half-space can be shown to be equivalent to that produced by a current filament in the half-space of strength (see, e.g., Hague, 1929):

$$I_1 = \frac{\mu_r - 1}{\mu_r + 1} I_0$$

The force per unit length is then equal to that between the two current filaments I_0 and I_1, or

$$F = \frac{\mu_0 I_0^2}{4\pi h} \frac{\mu_r - 1}{\mu_r + 1}$$

where h is the height of the filament above the half-space (see Fig. 4-3a).

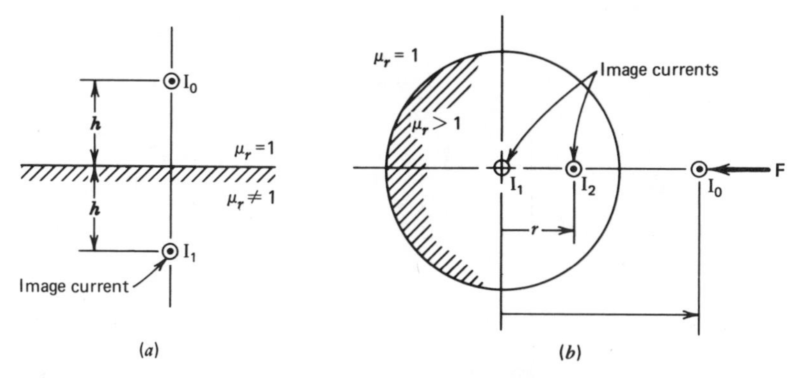

Figure 4-3 (a) Current filament and image current for an ideal soft ferromagnetic half-space and (b) current filament and images for a ferromagnetic cylinder.

Current Filament Near a Ferromagnetic Cylinder. Other examples of the image method are given in the book by Hague (1929). One interesting case is the calculation of the force between a current filament and a parallel ferromagnetic circular cylinder of radius a. For this case Hague shows that two image current filaments are required to satisfy the boundary conditions—one filament I_1 placed at the center, and a second filament I_2 placed at the inverse point $r = a^2/d$ where d is the distance of the filament I_0 from the center of the cylinder (see Fig. 4-3b). The strength of the image currents is given by

$$I_2 = -I_1 = \frac{\mu_r - 1}{\mu_r + 1} I_0 \tag{4-2.7}$$

The second image current is required since a line integral of **H** around the cylinder should produce no net current in the cylinder. The force on the real filament is given by the force between itself and the two image currents. The resulting force is attractive and is given by the expression

$$F = \frac{\mu_0 I_0^2}{2\pi d} \frac{\mu_r - 1}{\mu_r + 1} \frac{a^2}{d^2 - a^2} \tag{4-2.8}$$

Hague derives a similar expression for the case of a current filament in a ferromagnetic medium with a cylindrical hole. The reader may also try to calculate the force between a current filament and a diamagnetic cylinder with the same geometry.

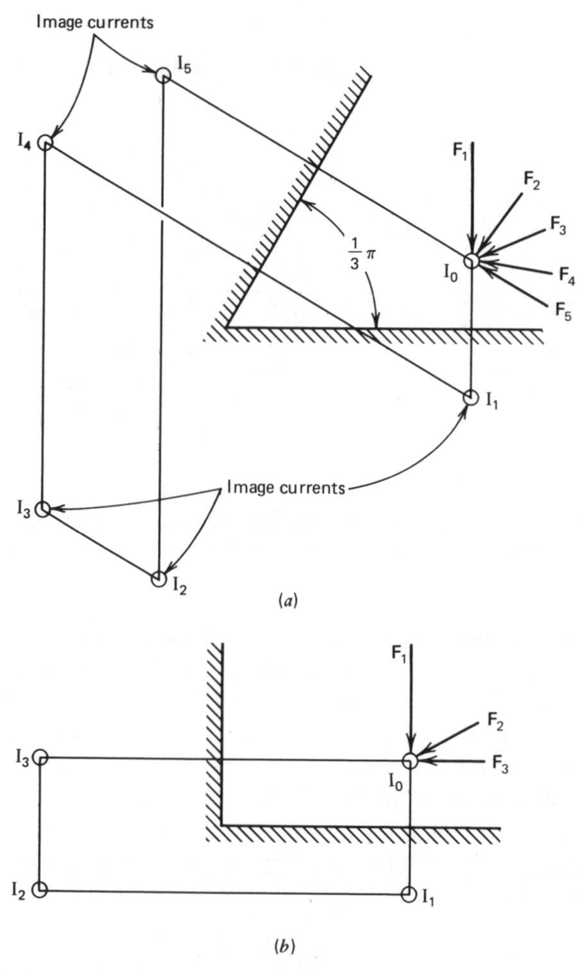

Figure 4-4 (*a*) Current filament and images for a 60° wedgelike cavity in a ferromagnetic space and (*b*) current filament and images for a 90° wedgelike cavity.

Current Filament Near a Ferromagnetic Wedge. Another example of the image method is the problem of a current filament near an ideal ferromagnetic region ($\mu_r \to \infty$) with a wedgelike corner (see Fig. 4-4). This example may be found in either Smythe (1968) or Hague (1929). When the wedge angle is an integral part of π, that is, $\theta = \pi/n$ where n is an integer, a set of image filaments may be found which will satisfy the boundary condition of zero tangential field on the outside of the wedge faces. Examples are shown in Figure 4-4 for the cases $n = 2$ and 3. Thus, for the case $\theta = \frac{1}{3}\pi$, the force on the filament is the vector sum of forces due to the five image currents.

Equivalent Magnetic-Moment Method

In Section 3.3 we discussed the case of an electric circuit in an applied magnetic field with a small gradient. For this case the external magnetic field $\mathbf{B}^0$ can be expanded in a power series in the neighborhood of the circuit. The force and couple on the circuit can be represented in terms of an equivalent magnetic moment $\mathbf{m}$ given by

$$\mathbf{m} = \frac{1}{2}\oint \mathbf{r} \times \mathbf{I}\, ds$$

$$\mathbf{F} = \mathbf{m} \cdot \nabla \mathbf{B}^0$$

and

$$\mathbf{C} = \mathbf{m} \times \mathbf{B}^0 \tag{4-2.9}$$

Coaxial Current Loops. Consider the example of two parallel coaxial circular current filaments of radii r_1 and r_2 and carrying currents I_1 and I_2, respectively, separated by a distance z (see Fig. 4-5). When $r_2 \ll r_1$ or when $r_2 \ll z$ the gradient of the field due to the circuit with I_1 will be small. The magnetic moment of circuit I_2 is given by

$$\mathbf{m}_2 = I_2 \pi r_2^2 \mathbf{e}_z$$

so that the magnetic force on I_2 due to the field of I_1 is given by

$$\mathbf{F} = I_2 \pi r_2^2 \frac{\partial \mathbf{B}^0}{\partial z}$$

In general $\mathbf{B}^0$ is given by elliptic integrals. However, on the axis, a simple expression exists for $\mathbf{B}^0$:

$$\mathbf{B}^0 = \frac{\mu_0 I_1}{2} \frac{r_1^2}{(r_1^2 + z^2)^{3/2}} \mathbf{e}_z \tag{4-2.10}$$

The calculated force on the circuit I_2 then has only an axial component given by

$$F_z = \frac{3}{2} \mu_0 I_1 I_2 \pi r_1^2 r_2^2 \frac{z}{(r_1^2 + z^2)^{5/2}} \tag{4-2.11}$$

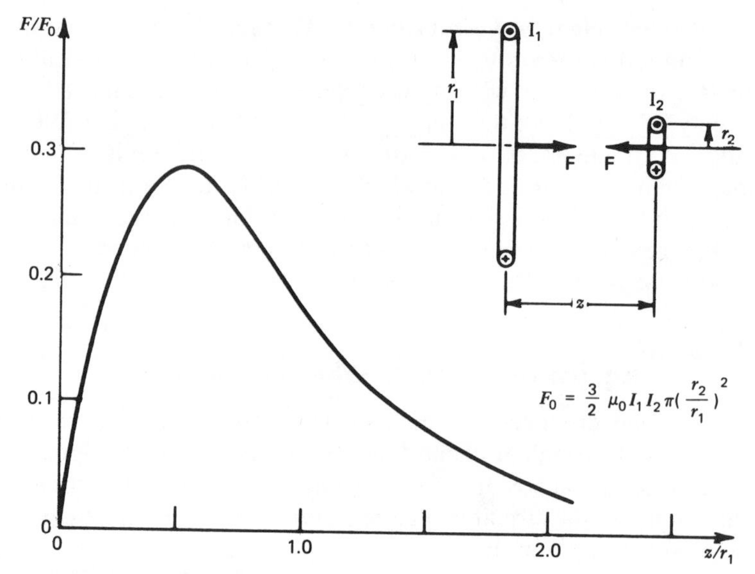

Figure 4-5 Axial magnetic force as a function of axial separation between one circular coil and another small coaxial coil [Eq. (4-2.11)].

The variation of the force with relative axial position of the two coils is shown in Figure 4-5. The force is zero when they are coplanar: is attractive if the currents are in the same direction; and reaches a maximum value when $r_1/z = 2$.

The force on circuit I_1 due to I_2 is, of course, equal, opposite, and colinear to the force on I_2. The mutual couple is zero since **m** and $\mathbf{B}^0$ are parallel. When the plane of I_2 makes an angle θ with respect to the plane of I_1, the couple on I_2 is given by

$$C = \frac{\mu_0 I_1 I_2 \pi r_1^2 r_2^2}{2(r_1^2 + z^2)^{3/2}} \sin\theta \tag{4-2.12}$$

The couple acts to turn the circuit about an axis normal to **m** and $\mathbf{B}^0$.

Magnetic Pressure Methods

The use of this method was illustrated in Chapter 1 in the example of stresses in a thin solenoid. To calculate the force on a body or part of some structure, one chooses a closed surface surrounding the body or part of the body of interest and integrates the Maxwell stress tensor (2-6.13) on this surface. To illustrate the technique, consider the example of an electrical contact between two conductors as shown in Figure 4-6— a conducting half-space and a cylindrical conductor with a hemispheric end. Under short circuit conditions, current will flow on the outside of the conductor. The resulting

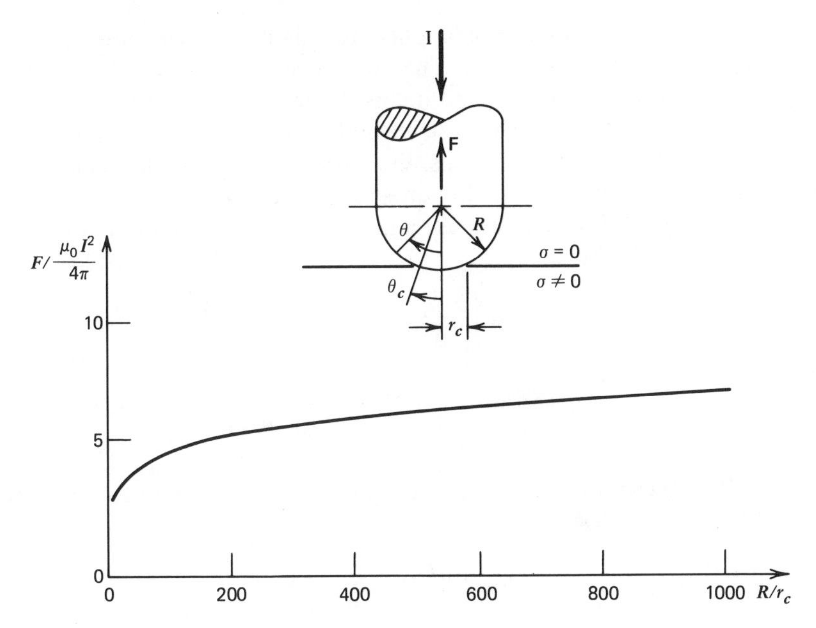

Figure 4-6 Repulsive magnetic force due to the flow of current between two conductors—a cylindrical rod with a hemispherical end and a conducting half-space [Eq. (4-2.13)].

magnetic field inside the rod conductor will be zero and that outside will be circumferential and given by

$$B_\theta = \frac{\mu_0 I}{2\pi r} = \frac{\mu_0 I}{2\pi R \sin\theta}$$

where $\theta_c < \theta \leq \frac{1}{2}\pi$ and θ_c is the angle at which the radius of the hemispheric end contacts the half-space.

The area of contact between the two bodies is assumed to be a circle of radius $r_c = R\sin\theta_c$. The current density across this area is assumed to be constant, so that the corresponding circumferential magnetic field vector is given by

$$B_\theta = \frac{\mu_0 I}{2\pi r_c^2}\, r \qquad (0 \leq r \leq r_c)$$

To calculate the force on the rod, we integrate the Maxwell stress vector on a closed area surrounding the hemispheric part of the rod at the point of contact (see Fig. 4-6).

If **n** is a unit normal to the surface, the Maxwell stress vector is given by

$$\boldsymbol{\tau} = \mathbf{n} \cdot \frac{1}{\mu_0}\left(\mathbf{BB} - \frac{1}{2}B^2\boldsymbol{\delta}\right)$$

where $\boldsymbol{\delta}$ is the identity tensor or matrix.

The Maxwell traction vector applied to this problem is equivalent to a pressure $B_\theta^2/2\mu_0$ on the surface. The surface integration is split into three terms: the area of contact A_1, the hemispheric area A_2 not in contact with the half-space, and the circular area at the top of the hemispheric part of the rod A_3. For short circuit conditions, the current in the rod flows close to the surface so that the magnetic field across A_3 is near zero.

The integrals of $B_\theta^2/2\mu_0$ over A_1 and A_2 yield axial components of force given, respectively, by

$$F_1 = \int_0^{r_c} \frac{\mu_0 I_0^2}{4\pi r_c^4} r^3\, dr$$

and

$$F_2 = 2\pi R \int_{\theta_c}^{\pi/2} \frac{B_\theta^2}{2\mu_0} \cos\theta\, d\theta$$

The result of these integrations is the total magnetic contact force given by (see, e.g., Holm, 1967)

$$F = \frac{\mu_0 I_0^2}{4\pi}\left(\ln\frac{R}{r_c} + \frac{1}{4}\right) \tag{4-2.13}$$

A graph of this force is shown in Figure 4-6. One important result of this calculation is the fact that magnetic forces tend to separate two electric contacts under large currents. This subject is discussed further in Chapter 5.

Cockroft's Method for Solenoids

For axisymmetric problems, where **J** has only a circumferential component about some axis, one can write the magnetic field **B** in terms of a potential ψ, that is,

$$\mathbf{B} = \nabla \times \psi \mathbf{e}_\theta \tag{4-2.14}$$

or

$$B_r = -\frac{\partial \psi}{\partial z}, \qquad B_z = \frac{1}{r}\frac{\partial}{\partial r}(r\psi)$$

The potential ψ is closely related to the flux Φ, through a circular loop of radius r given by the relation

$$\Phi = \int_0^r B_z 2\pi\rho\, d\rho$$

or

$$\Phi = 2\pi r \psi \tag{4-2.15}$$

On the other hand, the flux can be found in terms of the mutual inductance L between a single turn of radius r and the rest of the coil, that is, $\Phi = LI$, where I is the total current in ampere turns in the magnet. Expressions for $L(r, z)$ have been tabulated by Cockroft (1928). Using these tables for the mutual inductance, one can find the body force per unit volume from the expressions

$$f_r = \frac{1}{2\pi r} J^2 S \frac{\partial L}{\partial r}$$

and (4-2.16)

$$f_z = \frac{1}{2\pi r} J^2 S \frac{\partial L}{\partial z}$$

where S is the cross-sectional area through which the total current flows.

Energy or Inductance Method

The relation between the magnetic energy in a system of magnets and the generalized forces acting on the system was derived in (Eq. 3-5.11):

$$F = \frac{\partial \mathcal{W}(I, u)}{\partial u}$$

where u is a generalized coordinate such as an angle or a distance. In the above form we have assumed a linear magnetic system so that the magnetic energy and coenergy functions are identical, that is, $\mathcal{W} = \mathcal{W}^*$.

Suppose we consider the generalized force between one circuit and another. Then $\mathcal{W}$ may be written in terms of the mutual inductance L between the two circuits, that is,

$$\mathcal{W} = LI_1I_2$$

where I_1 and I_2 are the currents in the respective circuits. Thus

$$F = I_1 I_2 \frac{\partial L}{\partial u} \tag{4-2.17}$$

The mutual inductance has two interpretations (1) an interaction energy per unit current squared, or (2) the magnetic flux threading circuit "2" due to a unit current in circuit "1" (or vice versa). The inductance between circuits can be calculated in three ways:

$$L_{12} = \int_{A_2} \mathbf{B}_1 \cdot \mathbf{n}\, da_2 \tag{4-2.18a}$$

$$L_{12} = \oint \mathbf{A}_1 \cdot d\mathbf{s}_2 \tag{4-18b}$$

and

$$L_{12} = \frac{\mu_0}{4\pi} \oint\oint \frac{d\mathbf{s}_1 \cdot d\mathbf{s}_2}{R} \tag{4-2.18c}$$

where $\mathbf{A}_1$ is the vector potential due to the magnetic field $\mathbf{B}_1$ which is created by a unit current in circuit "1". The differential elements $d\mathbf{s}_1$ and

$d\mathbf{s}_2$ lie along the directions of the current vectors of the two circuits, while R is the distance between two circuit elements. The third formula is very useful for numerical evaluation of L_{12} since it involves only geometric properties of the circuits.

Formulas for inductance may be found in Grover (1946) or other standard reference books.

Current Filament and Circular Current Loop. As an example, consider the problem of the force between a plane circular circuit and a straight current filament lying in the same plane as shown in Figure 4-7. Since the field due to the wire is known to vary inversely with the distance from the wire, we can integrate this flux over the circle for a unit current in the filament. The resulting mutual inductance is given by

$$L = \mu_0[x - (x^2 - a^2)^{1/2}]$$

Using the relation (4-2.17), one obtains the net mutual force on the two circuits

$$F = \mu_0 I_1 I_2\left[1 - \left(1 - \frac{a^2}{x^2}\right)^{-1/2}\right] \tag{4-2.19}$$

This formula is very similar to the centering force in a torus as discussed above. To obtain the toroidal centering force we set $I_1 = NI_2$ and multiply by $\frac{1}{2}$ [see Eq. (4-2.5)].

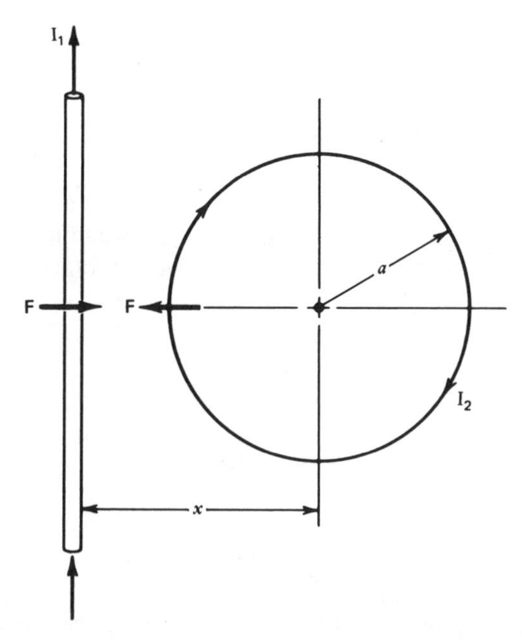

Figure 4-7 Magnetic force between a straight current filament and a circular coil.

Two Coaxial Circular Coils. As a second example, we examine the interaction between two coaxial circular circuits with currents I_1 and I_2 and radii a and b, respectively. This problem is not as straightforward as the others since the magnetic field and inductance involve elliptic integral functions. The complete problem is discussed in Smythe (1968) and here we only summarize the results.

In this example we use the vector potential $\mathbf{A}$ ($\mathbf{B} = \nabla \times \mathbf{A}$) to find the mutual inductance. In terms of cylindrical coordinates (ρ, ϕ, z), (see Fig. 4-8), the vector potential, due to a current I_1 only, has an azimuthal component given by

$$A_\phi = \frac{\mu_0 I_1}{\pi k}\left(\frac{a}{\rho}\right)^{1/2}\left[\left(1 - \frac{k^2}{2}\right)K(k) - E(k)\right] \tag{4-2.20}$$

where K and E are complete elliptic integrals of the first and second kind, respectively, with the relations

$$E = \int_0^{\pi/2} (1 - k^2 \sin^2\alpha)^{1/2}\, d\alpha$$

$$K = E - k\frac{dE}{dk}$$

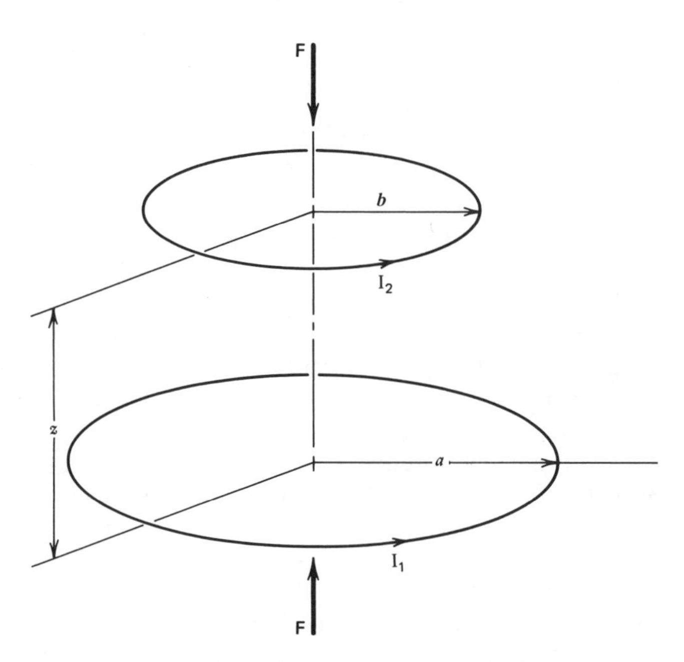

Figure 4-8 Axial magnetic force between two coaxial circular current-carrying coils.

and

$$k^2 = \frac{4a\rho}{(a+\rho)^2 + z^2}$$

Using the expression (4-2.18b) (with $\rho = b$), the mutual inductance L is given by

$$L = 2\mu_0(ab)^{1/2}[(1 - \tfrac{1}{2}k_1^2)K - E]/k_1 \qquad (4\text{-}2.21)$$

where

$$k_1^2 = \frac{4ab}{(a+b)^2 + z^2)}$$

The attractive force between the two coils is then given by

$$F_z = -\mu_0 I_1 I_2 \frac{z}{[(a+b)^2 + z^2]^{1/2}} \left(-K + \frac{a^2 + b^2 + z^2}{(a-b)^2 + z^2} E \right) \qquad (4\text{-}2.22)$$

When $b \ll a$, we can relate this expression to the problem treated in the section above on the equivalent moment method. For $k_1 \to 0$, one must use the asymptotic forms of K and E given by

$$\begin{aligned} K &= \frac{\pi}{2}\left[1 + \left(\frac{1}{2}\right)^2 k_1^2 + \left(\frac{1 \cdot 3}{2 \cdot 4}\right)^2 k_1^4 + \cdots\right] \\ E &= \frac{\pi}{2}\left[1 - \frac{1}{2^2} k_1^2 + \left(\frac{1 \cdot 3}{2 \cdot 4}\right)^2 \frac{k_1^4}{3} - \cdots\right] \end{aligned} \qquad (4\text{-}2.23)$$

The higher-order terms are required since the order 1 and order k_1^2 terms drop out in Eq. (4-2.22) for $k_1 \to 0$. The limiting force expression can be shown to be

$$F_z = \frac{3\mu_0 I_1 I_2}{2} \frac{\pi a^2 b^2 z}{(a^2 + z^2)^{5/2}} \qquad (4\text{-}2.24)$$

which is identical to that derived using the equivalent magnetic-moment method Eq. (4-2.11).

Inclined Circular Coils. As a final example of the inductance method, we consider the problem of two planar circular coils whose planes are inclined with respect to each other as shown in Figure (4-9). In this case, the analytical treatment is too complicated so we resort to a numerical solution. The mutual inductance of two inclined circular coils is found in the numerical tables given in Grover (1946) and reproduced in Figure 4-9. Qualitatively, L can be expressed in the form

$$L \simeq L_0 + \tfrac{1}{2}\lambda_1\theta^2 - \tfrac{1}{4}\lambda_2\theta^2$$

Thus the torque on the circuits would be found from the relation

$$C = I_1 I_2 \frac{\partial L}{\partial \theta} = I_1 I_2(\lambda_1\theta - \lambda_2\theta^3) \qquad (4\text{-}2.25)$$

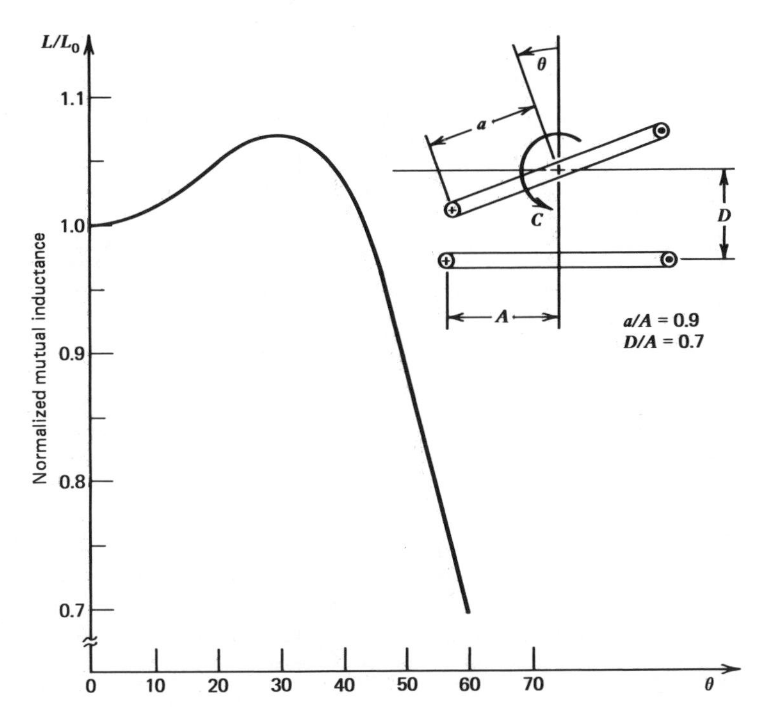

Figure 4-9 Change of mutual inductance as function of angle between axes of two inclined circular coils.

As one can see from Figure 4-9, the torque is zero for $\theta = 0$, and changes sign at $\theta = \theta^*$. In the approximate analytic model, $\theta^* = \pm(\lambda_1/\lambda_2)^{1/2}$.

4.3 STRESS IN A CIRCULAR RING WITH CURRENT

The simplest circuit is a circular filament with current. However, the exact stress distribution is not easily found because the axisymmetric magnetic field is given by elliptic integrals, Eq. 4-2.20 (see, e.g., Smythe, 1968). To obtain an approximate analysis of the stress in the ring when the cross-sectional dimension is small compared with the circuit diameter, one can use the energy method.

To understand the internal forces in the circuit, we note that the magnetic field is normal to the midplane of the circuit (Fig. 4-10). Thus radial body forces are produced given by

$$f_r = J_\theta B_z \tag{4-3.1}$$

This radial force can be shown to be outward, producing a tension in the ring given by

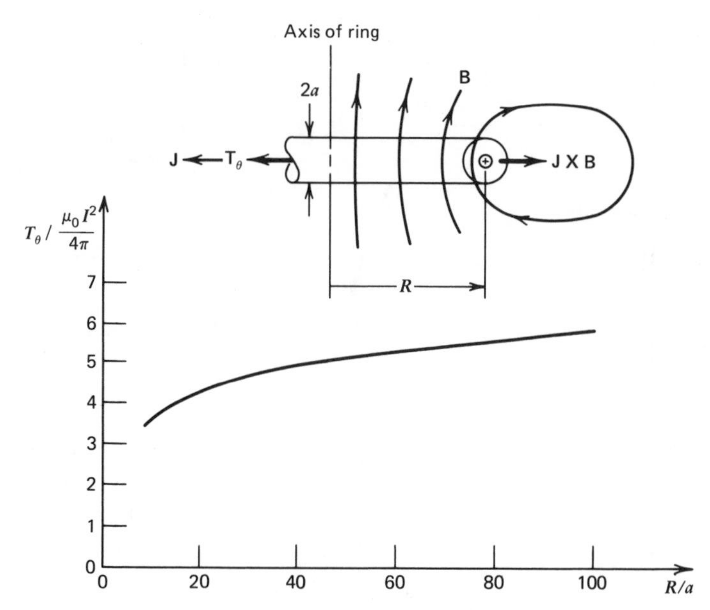

Figure 4-10 Magnetically induced circumferential force in a circular coil due to circulating current I as a function of major-to-minor radius ratio.

$$T_\theta = R \int J_\theta B_z dA = RF_r \tag{4-3.2}$$

where R is the mean radius of the ring. As noted above, when J_θ is given, B_z is not easily found in terms of simple functions. However, the self-inductance L of a circular current loop with different cross sections is known. Since L is proportional to the stored magnetic energy in the ring, we can use the energy or virtual work method of Chapter 3 to find T_θ. Restated, we equate the work done by the radial body force F_r, under a virtual displacement δR, to the change in magnetic energy when the current is held fixed; that is,

$$2\pi RF_r\delta R = \delta\left(\frac{1}{2} LI^2\right) = \frac{1}{2} I^2 \frac{\partial L}{\partial R} \delta R$$

or

$$F_r = \frac{1}{4\pi R} I^2 \frac{\partial L}{\partial R} \tag{4-3.3}$$

and

$$T_\theta = \frac{1}{4\pi} I^2 \frac{\partial L}{\partial R} \tag{4-3.4}$$

For a circular cross section of radius a,

$$L = \mu_0 R \left[\ln\left(\frac{8R}{a}\right) - \frac{7}{4}\right] \tag{4-3.5}$$

and

$$T_\theta = \frac{\mu_0 I^2}{4\pi}\left[\ln\left(\frac{8R}{a}\right) - \frac{3}{4}\right] \tag{4-3.6}$$

If we assume a uniform stress distribution $t_{\theta\theta}$ and a uniform current density J_0, then the average circumferential stress is given by

$$t_{\theta\theta} = \frac{\mu_0 J_0^2 a^2}{4}\left[\ln\left(\frac{8R}{a}\right) - \frac{3}{4}\right] \tag{4-3.7}$$

We note that if J_0 is fixed, the stress increases as R/a or 'a' increases so that stress becomes a design limitation as magnetic circuits get larger. As an example, consider a superconducting coil with $J_0 = 3 \times 10^4$ A/cm^2, $R = 1$ m, and $a = 2$ cm. For this case, the stress is about 6×10^3 N/cm^2 (8700 psi). Other examples may be obtained from Figure 4-10. Further discussion of this problem may be found in Landau and Lifshitz (1960).

4.4 STRESS ANALYSIS IN SOLENOID MAGNETS

The state of stress in a solenoid magnet was discussed briefly in Section 1.2 (see Fig. 1–3). In this section we will examine some simplified models for the stresses in solenoid magnets. The difference between theoretical models and actual solenoid magnets can be substantial. Actual solenoids are finite in length and are inherently inhomogeneous since they are wound with many turns of conductor, onto a coil form of different material, and often with insulating and structural material interleaved between the turns. In some designs, the cylindrical magnet is composed of a series of pancake wound discs assembled one on top of another. In addition to the inhomogeneity, the windings often have a pretension which leaves the cylindrical shell in a state of initial stress. If the magnet is a normal conductor, there may be holes in the conductor for cooling. If the solenoid is superconducting, the decrease in temperature from 300 to 4°K will result in further changes in the initial stress state.

Several methods of analysis have been employed including the following:

1. Thin-shell model.
2. Isolated turn model.
3. Concentric-ring model.
4. Helical ring model.
5. Homogeneous isotropic elasticity analysis.
6. Homogeneous anisotropic elasticity analysis.

Several of these methods of analysis will be discussed below. While these models may differ substantially in detail from actual solenoids, they are useful for a number of reasons. First, they give the designer or stress analyst some qualitative "feel" for the distribution of stress in these structures. Second, these models serve as "pinning points" for numerical schemes, such as finite-element codes, to provide a check on the numerical calculations.

The magnetic field and body forces for a finite-length solenoid can be calculated by numerical methods. As an example, the magnetic field for a solenoid of length equal to 4 times the inner radius, and outer-to-inner radius ratio of 3 is shown in Figure 4-11, One can see that for a constant current density, the magnetic field drops to zero almost linearly through the thickness. Also, the radial magnetic field at the quarter-plane is almost an order of magnitude smaller than the axial field.

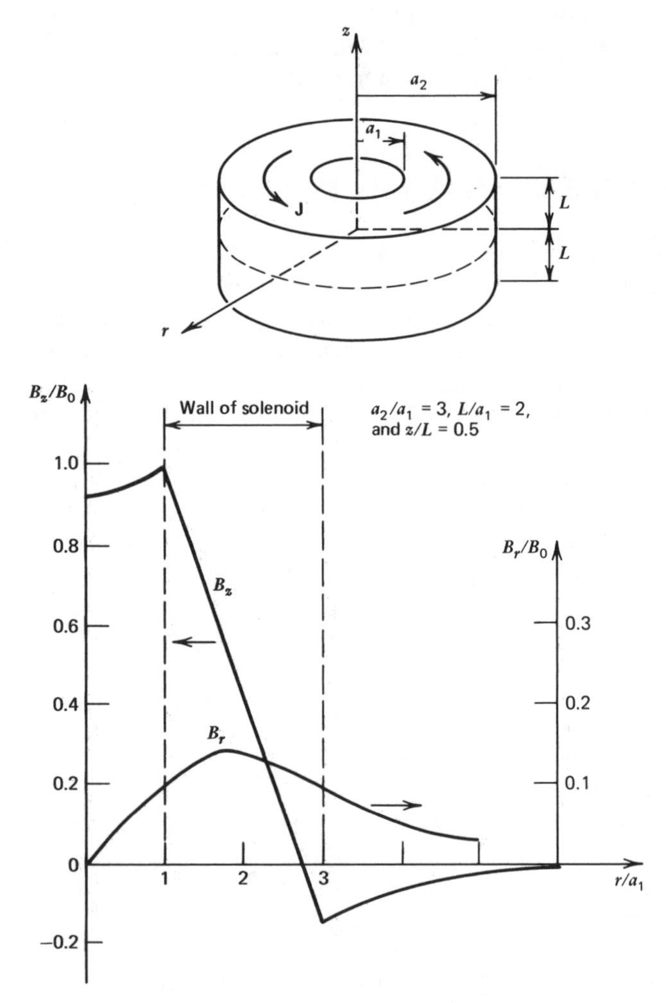

Figure 4-11 Axial and radial magnetic fields in a thick solenoid as a function of radius at one-quarter of the axial length.

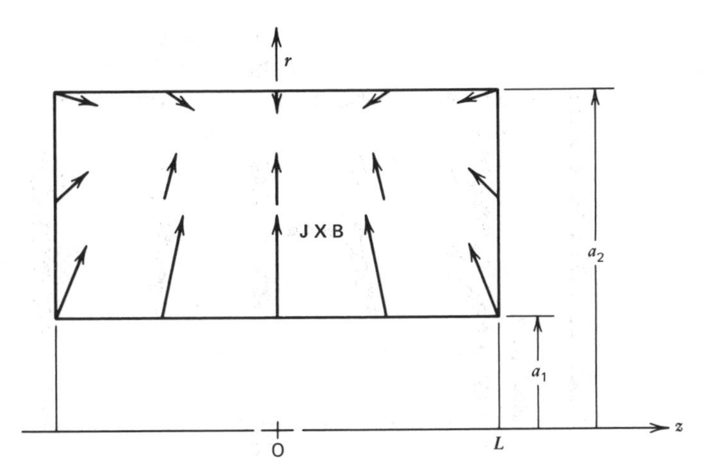

Figure 4-12 Magnetic body force distribution in a thick solenoid due to uniform circumferential density (see Fig. 4-11, $a_2/a_1 = 3$, $L/a_1 = 2$).

These properties are sometimes used to obtain an approximate model for the field in a solenoid. This model assumes that the axial field decreases linearly through the thickness and is zero at the outer radius in the central portion of the solenoid. It should be noted, however, that the small, reversed axial field near the outer radius can place the outer turns in compression. Compressive forces in the outer turns, coupled with a low radial effective elastic stiffness, can sometimes result in a buckling failure of the windings (see, e.g., Daniels, 1953). The distribution of body forces in a finite-length solenoid is shown in Figure 4-12. One can see that compressive axial stresses should result from this loading. These effects should be kept in mind when studying the various models presented below.

Long Thin Solenoid

The solenoid magnet uses circumferential currents to create an axial magnetic field. When the aspect ratio of length to diameter, $\beta = L/D$, is large, the field inside has high spacial homogeneity and the field outside, near the center section, is small. In Chapter 1 we used the idea of magnetic pressure to estimate the approximate stresses in a long thin cylindrical solenoid:

$$
\begin{aligned}
t_{\theta\theta} &= P\frac{R}{\Delta} \\
t_{zz} &= -P\frac{R}{2\Delta} \\
t_{rr} &= -P\frac{b-r}{b-a}
\end{aligned}
\tag{4-4.1}
$$

where $P = B_0^2/2\mu_0$, $R = \frac{1}{2}(a+b)$, a is the inner radius, b is the outer radius, and Δ is the thickness of the shell. B_0 is the value of the magnetic field near the center of the solenoid. The combination of circumferential tension and axial compression produces a maximum shear stress of $\tau = 3PR/4\Delta$ in a direction θ degrees from the circumferential direction given by $\tan 2\theta = 3$, or $\theta = 35.8°$.

Long, Thick, Solenoid-Isotropic Elastic Model

For a thick cylinder, the magnetic pressure model cannot be used and the actual Lorentz body forces must be employed. In order to obtain an exact solution, we make several assumptions. First, we examine only the stresses in the central section of the solenoid and assume the turns of conductor can be replaced by an equivalent homogeneous isotropic material. (The anisotropic case will be examined in the next section.) Second, we assume that the ends are constrained in some manner as to create an axial strain independent of the radius. For a long solenoid, it is further assumed that the magnetic field is zero outside the cylinder. For a uniform distribution of circumferential current $\mathbf{J} = J_0\mathbf{e}_\theta$, in a long solenoid, the magnetic field near the central section can be shown to be approximately axial; that is,

$$\mathbf{B} = \mu_0 J_0 (b - r)\mathbf{e}_z \tag{4-4.2}$$

where b and a are the outer and inner radii, respectively. The resulting body force is radial, that is,

$$\mathbf{J} \times \mathbf{B} = \mu_0 J_0^2 (b - r)\mathbf{e}_r$$

The equation of equilibrium derived from (Eq. 3-6.14) becomes

$$\frac{dt_{rr}}{dr} + \frac{t_{rr} - t_{\theta\theta}}{r} + \mu_0 J_0^2 (b - r) = 0 \tag{4-4.3}$$

An alternative method which results in the same equation is the use of the Maxwell stress tensor in the equilibrium equation (3-6.17),

$$\nabla \cdot (\mathbf{t} + \mathbf{T}) = 0$$

For this problem, $T_{rr} = T_{\theta\theta} = -B_z^2/2\mu_0$, and the equilibrium equation becomes

$$\frac{dt_{rr}}{dr} + \frac{t_{rr} - t_{\theta\theta}}{r} - \frac{B_z}{\mu_0}\frac{dB_z}{dr} = 0 \tag{4-4.4}$$

When B_z is substituted from Eq. (4-4.2) into Eq. (4-4.4), Eq. (4-4.3) results.

The equilibrium equation must be supplemented with constitutive equations and boundary conditions. For cylindrical geometry and radial body forces, the displacements are radial and axial. For an isotropic material with Young's modulus Y and Poisson's ratio ν, the constitutive equations become

$$\varepsilon_{rr} = \frac{du_r}{dr} = \frac{1}{Y}[t_{rr} - \nu(t_{\theta\theta} + t_{zz})]$$

$$\varepsilon_{\theta\theta} = \frac{u_r}{r} = \frac{1}{Y}[t_{\theta\theta} - \nu(t_{rr} + t_{zz})] \qquad (4\text{-}4.5)$$

and

$$\varepsilon_{zz} = \frac{1}{Y}[t_{zz} - \nu(t_{\theta\theta} + t_{rr})]$$

The boundary conditions at the inner and outer surfaces call for zero traction or

$$t_{rr} = 0, \qquad r = a, b$$

There are three ways to proceed from these equations. The first method eliminates the stresses using (Eq. 4-4.5) and results in an ordinary differential equation for the radial displacement u_r. The second and third methods employ a body force potential,

$$\mathbf{J} \times \mathbf{B} = -\nabla W$$

and (4-4.6)

$$W = \mu_0 J_0^2(\tfrac{1}{2}r^2 - br)$$

The second method eliminates the stresses using the Airy stress function $\mathscr{A}$, that is,

$$t_{rr} = W + \frac{1}{r}\frac{d\mathscr{A}}{dr}, \qquad t_{\theta\theta} = W + \frac{d^2\mathscr{A}}{dr^2}$$

$$\nabla^2\mathscr{A} = -\frac{1-2\nu}{1-\nu}W \qquad (4\text{-}4.7)$$

This method will not be used here, though it is useful for general two-dimensional problems. Instead a method found in Southwell (1941) will be presented which derives the stresses in the solenoid.

First, u_r may be eliminated from the constitutive equations, (4-4.5), resulting in a strain compatibility equation for the stress components. Next, we assume that

$$\frac{d\varepsilon_{zz}}{dr} = 0$$

The three equations of equilibrium, compatibility, and constant axial strain may be combined to yield the solution (see Southwell, 1941)

$$t_{rr} = A + \frac{Ba^2}{r^2} + W - \gamma\frac{1}{r^2}\int rW dr$$

and (4-4.8)

$$t_{\theta\theta} = A - \frac{Ba^2}{r^2} + (1-\gamma)W + \gamma\frac{1}{r^2}\int rW dr$$

where

$$\gamma = (1-2\nu)/(1-\nu)$$

The first two terms on the right-hand side of each equation are solutions to the homogeneous equilibrium equation. The arbitrary constant in W is absorbed in term A, which is subsequently chosen to satisfy the boundary conditions.

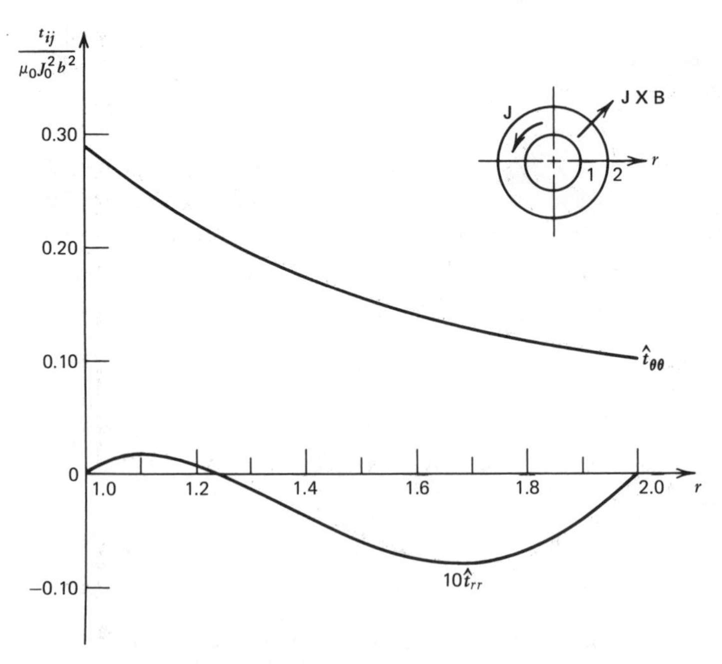

Figure 4-13 Magnetically induced radial and circumferential stresses in a long, thick-walled solenoid magnet as function of radius, $\nu = 0.35$, $b/a = 2$ [Eq. (4-4.9)].

The solution satisfying traction-free boundary conditions at $r = a$ and $r = b$ is given below. The radius r has been normalized by the inner radius a and $\alpha = a/b$:

$$t_{\theta\theta} = -\mu_0 J_0^2 b^2\left[A - B\frac{1}{r^2} - \left(1-\frac{3\gamma}{4}\right)\frac{\alpha^2 r^2}{2} + \left(1-\frac{2\gamma}{3}\right)\alpha r\right]$$

$$t_{rr} = -\mu_0 J_0^2 b^2\left[A + B\frac{1}{r^2} - \left(1-\frac{\gamma}{4}\right)\alpha^2 r^2 + \left(1-\frac{\gamma}{3}\right)\alpha r\right] \quad (4\text{-}4.9)$$

where

$$A = \frac{1}{2}\left(1-\frac{\gamma}{4}\right)(1+\alpha^2) - \left(1-\frac{\gamma}{3}\right)(\alpha^2+\alpha+1)(1+\alpha)^{-1}$$

and

$$B = -\frac{1}{2}\left(1-\frac{\gamma}{4}\right) + \left(1-\frac{\gamma}{3}\right)(1+\alpha)^{-1}$$

When $b = a + \Delta$, $\Delta/b \ll 1$, the circumferential stress can be shown to approach the limit given by the magnetic pressure method for a long thin solenoid, Eq. (4-4.1), that is, for $r = a$, $a \to b$,

$$t_{\theta\theta} \to \frac{\mu_0 J_0^2}{2} b(b-a) = P\frac{b}{b-a}$$

where the magnetic pressure P is given by

$$P = \frac{B_0^2}{2\mu_0} = \frac{\mu_0 J_0^2}{2}(b-a)^2$$

The radial and circumferential stresses are plotted in the graph in Figure 4-13. One can see that the circumferential stress has its maximum value at the inner radius and becomes much greater than t_{rr} when $a/b \to 1$.

In an actual solenoid, however, one must add to these stresses the winding stresses and thermoelastic stresses resulting from either heating, in the case of normal conductors, or cryogenic cooling, in the case of superconducting materials.

Anisotropic Solenoid Model

The analysis of the previous section can be extended to the problem of an anisotropic cylinder under radial body forces. In the case of a wound solenoid, the anisotropy results from the insulation and reinforcement between the turns of the magnet. To obtain an equivalent homogeneous anisotropic material, effective elastic moduli must be found for the radial and circumferential directions. For example, in the circumferential direction the effective Young's modulus Y_θ can be calculated from the rule of mixtures for the conductor and insulator; that is,

$$Y_\theta = f_c Y_c + f_I Y_I \tag{4-4.10}$$

where f_c and f_I are the volume fractions of conductor and insulator or reinforcement material, respectively. In the radial direction, however, a different rule should be used:

$$\frac{1}{Y_r} = \frac{f_c}{Y_c} + \frac{f_I}{Y_I} \tag{4-4.11}$$

Whenever the conducting wires are potted in epoxy, more sophisticated analyses or experiments might be necessary to obtain the effective moduli. (See, e.g., a text on composite materials, such as Christensen, 1979.)

A number of authors have analyzed the anisotropic solenoid. The outline presented here follows that of Gray and Ballou (1977). The equation of equilibrium remains the same as in the previous section, Eq. (4-4.3), except the constitutive relations (4-4.5) are changed. For the case of plane stress ($t_{zz} = 0$), these become

$$t_{rr} = \frac{Y_\theta}{n - \nu_{\theta r}^2}(\varepsilon_{rr} + \nu_{\theta r}\varepsilon_{\theta\theta})$$
$$t_{\theta\theta} = \frac{Y_\theta}{n - \nu_{\theta r}^2}(n\varepsilon_{\theta\theta} + \nu_{\theta r}\varepsilon_{rr}) \tag{4-4.12}$$

where $n = Y_\theta / Y_r$, $\nu_{\theta r}$ is similar to Poisson's ratio relating the strains in the radial and hoop directions (see, e.g., Lekhnitskii, 1963). Using the strain displacement equations (3-6.6) along with the stress–strain relation (4-4.12), the equilibrium equation can be rewritten in terms of the radial displacement u; that is,

$$\frac{d^2u}{dr^2} + \frac{1}{r}\frac{du}{dr} - n\frac{u}{r^2} = -J_\theta B_z \frac{n - \nu_{\theta r}^2}{Y_\theta} \tag{4-4.13}$$

The solution can be written as the sum of a solution to the homogeneous differential equation and a particular solution. The complementary solution can be written in the form

$$u_c = c_1 r^{n^{1/2}} + c_2 r^{-n^{1/2}} \tag{4-4.14}$$

A particular solution requires a specific $B_z(r)$ function. For a linear decrease in B_z through the thickness, that is,

$$B_z = B_i\left[1 + \beta\left(\frac{r}{a} - 1\right)\right] \tag{4-4.15}$$

the particular solution is given by

$$u_p = -\frac{J_\theta B_i a^2}{Y_\theta}(n - \nu_{\theta R}^2)\left[\frac{1-\beta}{4-n}\left(\frac{r}{a}\right)^2 + \frac{\beta}{9-n}\left(\frac{r}{a}\right)^3\right] \tag{4-4.16}$$

where the value of B_z at the inner and outer radii a and b are, respectively, B_i and B_0, and β is defined by the relation

$$\beta = \left(\frac{B_0}{B_i} - 1\right)\left(\frac{b}{a} - 1\right)^{-1}$$

If zero traction boundary conditions are imposed on the inner and outer surfaces at $r = a$ and b, then C_1 and C_2 may be determined (see Gray and Ballou, 1977 for details).

Calculation of stresses in an anisotropic cylinder are given by Gray and Ballou (1977). For the case of the moderate- to thin-wall thickness, the circumferential stress is always greater on the inside, and there is little differences in stress between isotropic and anisotropic cases. For the thick-walled solenoid, however, different anisotropies lead to stress distributions (see Fig. 4-14), which differ significantly from the isotropic case (see Fig. 4-13).

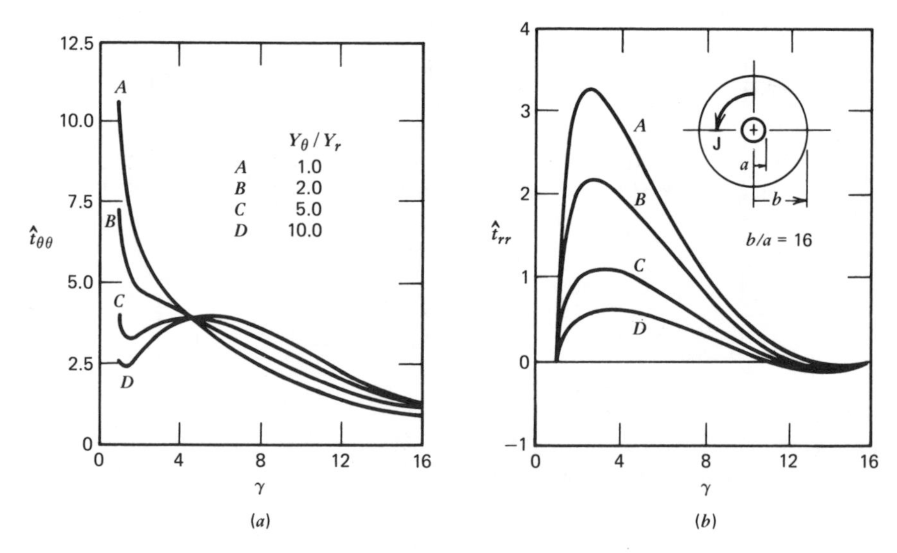

Figure 4-14 (*a*) Nondimensional circumferential stress as a function of nondimensional radius for anisotropic solenoids and (*b*) radial stress as a function of radius in anisotropic solenoids (from Gray and Ballou, 1977).

Finite Length Solenoid—A Variational Method

The stresses in a long, but finite-length, thin-walled solenoid, may be estimated using the method of virtual work (see Chapter 3). The axial and circumferential stresses can be found by replacing the $\mathbf{J} \times \mathbf{B}$ forces by a magnetic pressure inside and compressive forces N_z at the ends of the cylinder (see Fig. 4-15). (In an actual solenoid, the stresses will depend on axial position, so this is a highly simplified model.) As a generalized radial force, we choose $N_r = 2\pi RlP$, where R is the mean radius, l the length of the cylinder, and P the average magnetic pressure inside the solenoid. The wall thickness will be denoted by Δ.

If δu_r and δu_z represent changes in the radial and axial displacements, respectively, then the principle of virtual work relates the work done by N_r and N_z to the change in stored magnetic energy; that is,

$$\delta \mathcal{W} = N_r \delta u_r + N_z \delta u_z \tag{4-4.17}$$

where

$$\delta \mathcal{W}(u_r, u_z, I) = \frac{\partial \mathcal{W}}{\partial u_r} \delta u_r + \frac{\partial \mathcal{W}}{\partial u_z} \delta u_z \tag{4-4.18}$$

which leads to the relation

$$N_r = \frac{\partial \mathcal{W}}{\partial R}, \qquad N_z = \frac{\partial \mathcal{W}}{\partial l} \tag{4-4.19}$$

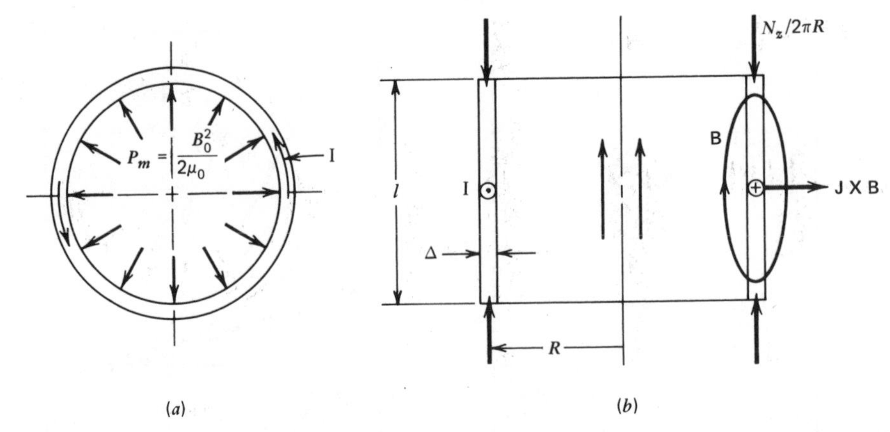

Figure 4-15 (*a*) Cross section of a thin solenoid showing magnetic pressure P_m and (*b*) equivalent axial magnetic forces on ends of a thin solenoid.

The stored magnetic energy in the solenoid can be written in terms of the current I and inductance L, that is,

$$\mathcal{W} = \tfrac{1}{2} I^2 L(R, l) \tag{4-4.20}$$

For a long solenoid,

$$L_0 = \mu_0 \pi R^2 N^2 / l \tag{4-4.21}$$

where N is the number of turns.

For a moderate-length solenoid, $L/2R > 3$, we write L in the form

$$L = k\left(\frac{l}{2R}\right) L_0 \tag{4-4.22}$$

where k represents a shape factor and has a range given by

$$0 = k(0) \leq k \leq k(\infty) = 1$$

Substituting this expression into Eq. (4-4.20) and using Eq. (4-4.19), one obtains

$$N_r = 2\pi R l P$$

$$P = \frac{\mathcal{W}}{\pi R^2 l}\left(1 - \frac{k'}{k}\frac{l}{4R}\right)$$

and

$$N_z = \frac{-\mathcal{W}}{Rl}\left(1 - \frac{k'}{k}\frac{l}{2R}\right) \tag{4-4.23}$$

where k' is the derivative of k with respect to $l/2R$. These generalized magnetic forces allow one to replace the actual magnetic force distribution

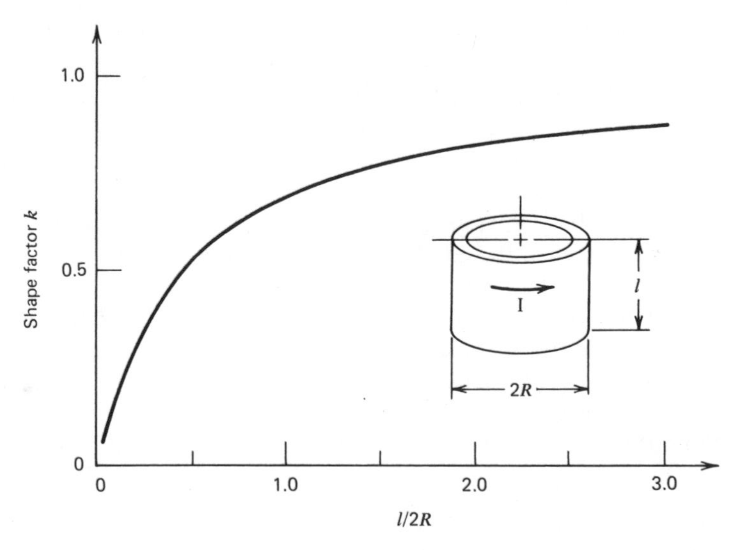

Figure 4-16 Shape factor for the inductance of an air core solenoid magnet as a function of length to diameter [see Eq. (4-4.22)] (based on table in Grover, 1946).

by an equivalent uniform pressure P on the inside of the cylinder and axial force N_z on the ends.

The shape factor can be obtained from tables (see Grover, 1946) and is a monotonically increasing function of $l/2R$ as shown in Figure 4-16. From this we can show that

$$\left(1 - \frac{k'}{k}\frac{l}{4R}\right) \geq \left(1 - \frac{k'}{k}\frac{l}{2R}\right) \geq 0$$

In the case when $l/2R \to \infty$, $W = \frac{1}{2}I^2 L_0$ and $P \to B_0^2/2\mu_0$ as expected. The above relation implies that the pressure P and the axial force always remain compressive.

This analysis has been extended to examine the elastic stability of energy storage solenoids for peak power leveling in the electric power industry (see Moon and Kim 1979).

Concentric Ring Model

In some solenoid designs, the turns are wrapped with epoxy-impregnated fiberglass tape, or insulation is interleaved between the windings. The relatively low elastic modulus of the insulation, compared with that of the conductor, results in an effective stiffness which is different in the circumferential and radial directions of the wound solenoid. One method of analysis involves using the effective anisotropic material as discussed in the previous section. An alternative method that retains some of the structural details of the insulation and conductor is the concentric-ring model. This has been employed by a number of magnet analysts. Bobrov and Williams (1980) have given a review of these methods and present a fairly complete analysis of the model. In this section, however, we outline a simplified application of the concentric ring analysis which will yield some qualitative information about stresses in solenoid magnets.

In this model, the helical winding is replaced by a series of concentric rings or cylinders. Each layer of the conductor has thickness Δ_c and will be labeled by an integer n.

Consider the axisymmetric case with radial displacement. Then the radial displacement of the outer surface of the nth layer will be designated by u_{n+1} as shown in Figure 4-17. Between each layer we assume that an elastic material exists of thickness Δ_I. We further assume that the insulation resists radial strain but has zero stiffness in the circumferential direction. Hence the insulation acts as an elastic foundation with respect to radial displacements only.

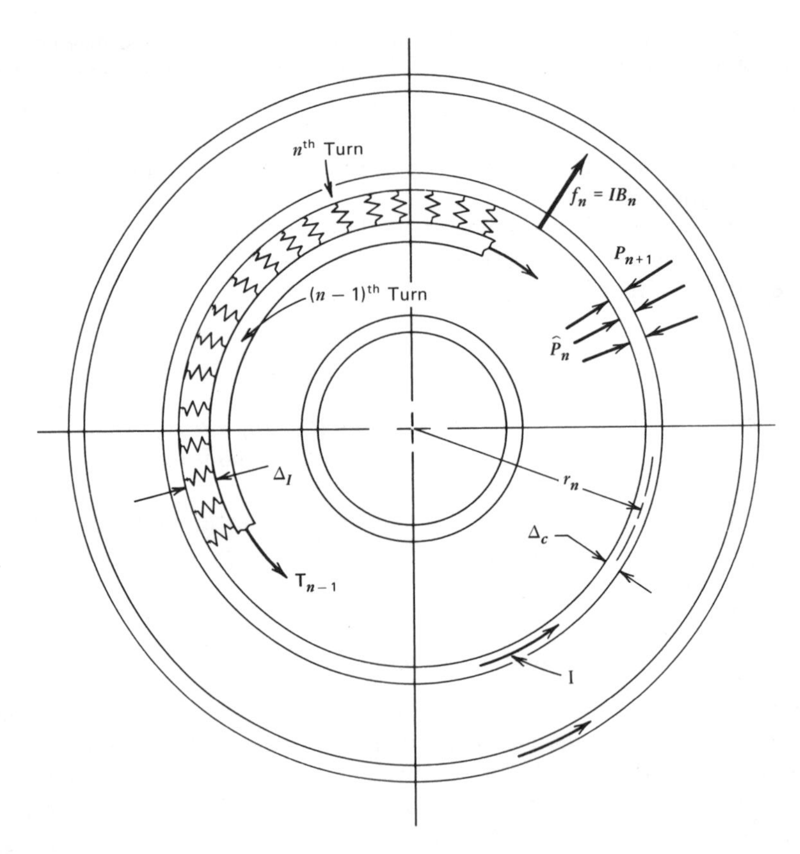

Figure 4-17 Concentric elastic ring model for calculation of stresses in a solenoid magnet.

The radial displacement of the outer surface of the nth layer of insulation (the inner surface of the nth conductor) will be denoted by $\hat{u}_n$. We approximate the displacement in each conducting layer by a linear function of the radius r, that is, in the nth layer,

$$u = \hat{u}_n + \frac{u_{n+1} - \hat{u}_n}{\Delta_c}\left(r - r_n + \frac{\Delta_c}{2}\right) \tag{4-2.24}$$

where $r_n = r_0 + n(\Delta_c + \Delta_I)$. Using this function, one can show that the average circumferential strain in the conductor is given by

$$\langle \varepsilon_{\theta\theta} \rangle = \frac{1}{\Delta_c}\int \frac{u}{r}\, dr$$

$$= \frac{u_{n+1} + \hat{u}_n}{2r_n} \equiv \varepsilon_\theta^n \tag{4-4.25}$$

where the integral is taken over the nth conductor.

A similar calculation yields the average radial strain

$$\langle \varepsilon_{rr} \rangle = \frac{u_{n+1} - \hat{u}_n}{\Delta_c} \equiv \varepsilon_r^n \tag{4-4.26}$$

Although the windings will be wound with an initial tension, u_n will only represent the displacement due to the magnetic forces.

Using a plane strain assumption for simplicity (i.e., $\varepsilon_{zz} = 0$), the tension in the nth conductor is given by

$$T_n = \Delta_c[(\lambda + 2\mu)\varepsilon_\theta^n + \lambda\varepsilon_r^n] + T_n^0 \tag{4-4.27}$$

where T_n^0 represents the initial winding tension. The stress–strain relation (3-7.3) has been used in Eq. (4-4.27) and λ and μ are the Lamé constants.

The pressure in the insulator will be assumed to follow a linear elastic law (nonlinear or plastic behavior could, of course, be included in more sophisticated analyses) given by

$$P_n = P_n^0 + S(u_n - \hat{u}_n)/\Delta_I \tag{4-4.28}$$

where S will be called the insulation modulus of elasticity. The term P_n^0 is the initial winding pressure and is assumed to be equilibrated by the initial tension T_n^0.

The equation of equilibrium for the nth conductor layer with current I, under a radial magnetic body force, is then given by

$$r_n(P_{n+1} - P_n) + \tfrac{1}{2}\Delta_c(P_{n+1} + P_n) + T_n = r_n I B_n \tag{4-4.29}$$

(We have assumed that $P_n = \hat{P}_n$.) Here B_n is the average axial magnetic field at the nth conductor. For a long solenoid with N layers,

$$B_n = B_0(1 - n/N) \tag{4-4.30}$$

Using the constitutive relations for T_n and P_n, that is, Eqs. (4-4.27) and (4-4.28), in Eq. (4-4-29) one obtains a difference equation,

$$\frac{Sr_n}{\Delta_I}(u_{n+1} - \hat{u}_{n+1} - u_n + \hat{u}_n) + \frac{\Delta_c S}{\Delta_I 2}(u_{n+1} - \hat{u}_{n+1} + u_n - \hat{u}_n)$$

$$+ \Delta_c\left((\lambda + 2\mu)\frac{u_{n+1} + \hat{u}_n}{2r_n} + \lambda\frac{u_{n+1} - \hat{u}_n}{\Delta_c}\right) = r_n IB_n \qquad (4\text{-}4.31)$$

Another equation is required. This extra equation can be supplied by assuming that the radial stress in the conductor t_{rr}^n is given by the average pressure, that is,

$$t_{rr}^n = -\tfrac{1}{2}(P_{n+1} + P_n) \qquad (4\text{-}4.32)$$

Finally, we employ one more simplifying assumption which neglects the radial strain, that is,

$$\hat{u}_n \simeq u_{n+1}$$

and (4-4.33)

$$T_n = \Delta_c(\lambda + 2\mu)\varepsilon_\theta^n$$

The resulting equation for P_n and u_n can be written as a set of two coupled first-order difference equations:

$$u_{n+1} - u_n = -\frac{\Delta_I}{S}P_n \qquad (4\text{-}4.34a)$$

and

$$r_n(P_{n+1} - P_n) + \frac{\Delta_c}{2}(P_{n+1} + P_n) + \frac{\Delta_c(\lambda + 2\mu)}{r_n}u_{n+1} = r_n IB_n$$

or

$$(1 + \alpha_n)u_{n+2} - (2 + \eta_n)u_{n+1} + (1 - \alpha_n)u_n = -\frac{\Delta_I}{S}IB_n \qquad (4\text{-}4.34b)$$

where

$$\alpha_n = \frac{\Delta_c}{2r_n}, \qquad \eta_n = \frac{\Delta_I\Delta_c(\lambda + 2\mu)}{r_n^2 S}$$

Another simple model which is embedded in this theory is the *isolated turn model* described by Montgomery (1980). It is obtained by setting $S = 0$, in Eq. (4-4.31) or letting $P_n = P_{n+1} = 0$, in Eq. (4-4.34a):

$$\frac{\Delta_c(\lambda + 2\mu)}{r_n}u_{n+1} = r_n IB_n \qquad (4\text{-}4.35)$$

This model assumes that each turn will independently support the body force acting upon it. From Eq. (4-4.34), one can see that the isolated turn theory is only valid when $\eta_n \gg 2$, or

$$\eta_n = \frac{\Delta_I\Delta_c}{r_n^2}\frac{(\lambda + 2\mu)}{S} \gg 2$$

For example, suppose for a copper conductor and some plastic insulator, $(\lambda + 2\mu)/S \sim 10^2$, $\Delta_I = 0.1\Delta_c$, and $\Delta_c/r_n = 10^{-2}$, then $\eta = 10^{-3}$; if $\Delta_c/r_n = 10^{-1}$ and $\Delta_I = \Delta_{c1}$, then $\eta = 1$. Thus, under most practical circumstances, the isolated turn model would not give a very good estimate of the stresses in the solenoid.

Equation (4-4.34) represents an inhomogeneous second-order difference equation with nonconstant coefficients. When the number of turns is not large, the difference equations may be solved using matrix techniques. For solenoids with only moderate-wall thicknesses, one may use an average α_n and η_n and analytic techniques for linear difference equations may be used.

Traction-free boundary conditions at the inner and outer surfaces for an N turn coil require that $P_1 = 0$ and $P_{N+1} = 0$, or that

$$u_1 = u_2, \qquad u_{N+1} = u_{N+2}$$

These equations coupled with the N equations for the N unknowns u_n enable one to solve for the displacements and the average circumferential stresses in each turn.

4.5 STRESS IN A THIN PLATE WITH UNIFORM CURRENT DENSITY

One of the peculiarities of magnetoelasticity is that problems for which the mechanics is straightforward are complex for determination of the magnetic field and body forces. One example is the thin plate with uniform, unidirectional current flow (see Fig. 4-18). The magnetic field near the edges has large gradients and an approximate method must be found which will avoid singularities at the edges.

The thin plate has thickness Δ and width $2a$ such that $2a/\Delta \gg 1$. We assume that the current flows along the long dimension of the plate and that the current density is uniform both in the thickness and width directions. The conservation-of-charge law (2-2.1) suggests that one employ a stream function ψ defined by the following equation:

$$\mathbf{I} = \mathbf{J}\Delta = \nabla \times (\psi \mathbf{n}) = -\mathbf{n} \times \nabla \psi \qquad (4\text{-}5.1)$$

where $\mathbf{n}$ is normal to the surface of the plate. For steady fields, Faraday's law (2-2.5) requires that

$$\Delta^2 \psi = 0 \qquad (4\text{-}5.2)$$

The current flow in the plate generates its own magnetic field $\mathbf{B}^1$ which is calculated using the Biot–Savart integral (4-2.1). An approximate technique is presented in Section 8.4 where the volume integration in (4-2.1) is replaced by a surface integral over the midplane of the thin plate (see also Moon, 1978). For one-dimensional flow in the plate, the normal component of the induced magnetic field at the plate midsurface is approximately given by

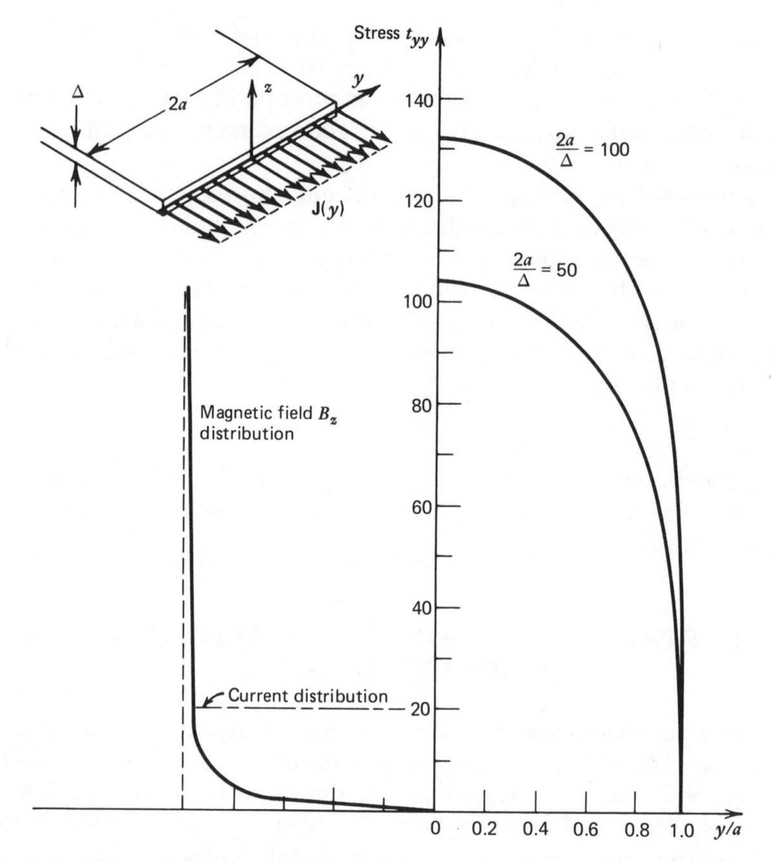

Figure 4-18 Magnetic field and compressive stress distribution in a thin plate due to a uniform axial current flow. (From Moon, 1978, reprinted with permission. Pergamon Press, Inc., copyright 1978.)

$$B_z^1(0, y, 0) = \frac{\mu_0 \psi}{\Delta} - \frac{\mu_0}{2\pi} \int_{-a}^{a} \frac{\psi(y')\,dy'}{(y' - y)^2 + \frac{1}{4}\Delta^2} \tag{4-5.3}$$

The planar component of the body force is given by

$$f_y = -B_z \nabla \psi \tag{4-5.4}$$

If we assume that $t_{xy} = 0$ and $\partial/\partial x = 0$, then the equation of equilibrium is given by

$$\frac{\partial t_{yy}}{\partial y} - (B_z^1 + B_z^0) \frac{\partial \psi}{\partial y} = 0 \tag{4-5.5}$$

where B_z^0 is the normal component of an externally generated magnetic field.

For uniform current flow, and return flow along the edges, we have

Figure 4-19 Photograph of a thin aluminum conductor before and after buckling current pulse.

$$\psi = I_0 y/2a \tag{4-5.6}$$

and the integral for B_z' can be written in the form

$$B_z^1 = \frac{\mu_0 I_0}{4a} F(\bar{y}) \tag{4-5.7}$$

and

$$F(\bar{y}) = \left(\bar{y} - \frac{1}{2\pi} \ln \frac{(A-\bar{y})^2+1}{(A+\bar{y})^2+1} - \frac{\bar{y}}{\pi}[\tan^{-1}(A-\bar{y}) + \tan^{-1}(A+\bar{y})]\right)$$

where $\bar{y} = 2y/\Delta$ and $A = 2a/\Delta$.

The stress t_{yy} produced by this distribution of magnetic field ($\mathbf{B}^0 = 0$) is compressive and its distribution across the width of the plate is shown in Figure 4-18, along with the self-magnetic field B_z^1. Such compressive forces in a thin structural member can result in elastic buckling. Buckling failures in thin sheets have been observed in lightning-strike damage to structures such as aircraft. A laboratory example of such buckling due to axial flow in thin conductors is shown in Figure 4-19. Several kiloamperes of current were passed through a sheet of aluminum foil, causing crimping or buckling of the sheet.

A related problem of interest is the stresses due to current flow around holes or cracks in thin sheets. For example, Yagawa and Horie (1982) have shown that current flow around a crack can lead to crack extension using both finite-element analysis and experiments. In a related work, Morjaria et al. (1981) have shown that current flow around cracks in thin sheets can lead to high current densities at the crack tips.

4.6 DEFORMATION-DEPENDENT FORCES AND MAGNETIC STIFFNESS

Since electric currents are almost always confined to the conductor, deformation of the conducting solid will lead to a change in the magnetic forces. In some structures, this effect may influence the elastic stability of the system (see Chapters 5 and 6). As a preliminary to a discussion of magneto-mechanical stability problems, however, one should understand the concept of magnetic stiffness. The dependence of magnetic forces on deformation is also important in predicting the load-deflection behavior of structures under both magnetic and mechanical forces.

Magnetic stiffness is analogous to elastic stiffness. Suppose the force in a particular mechanical system depends in some continuous way on a particular deformation mode described by a generalized displacement u. Then one can expand the force function $F(u)$ in a Taylor series about $u = 0$,

$$F(u) = F_0 + \left.\frac{\partial F}{\partial u}\right|_0 u + \cdots \tag{4-6.1}$$

The elastic stiffness is then defined as

$$k = -\frac{\partial F}{\partial u} \tag{4-6.2}$$

where the derivative is evaluated at $u = 0$. The negative sign indicates that if $k > 0$, the change in F will act to oppose further deformation. For a multidegree-of-freedom elastic system, one can define a set of generalized forces $\{F_\alpha\}$, given a set of generalized deformations $\{u_\alpha\}$, using the elastic energy; that is,

$$\begin{aligned} \mathcal{V} &= \tfrac{1}{2}\sum\sum k_{\alpha\beta} u_\alpha u_\beta \\ F_\alpha &= -\frac{\partial \mathcal{V}}{\partial u_\alpha} = -\sum k_{\alpha\beta} u_\beta \end{aligned} \tag{4-6.3}$$

The matrix $[k_{\alpha\beta}]$ is called appropriately the *elastic stiffness matrix*.

In an analogous manner one can define a magnetic stiffness matrix through the magnetic energy function (Chattopadhyay, 1979). Consider, for example, the two circuits shown in Figure 4-20. The magnetic energy of the system can be calculated in terms of the self-inductances L_{11} and L_{22} and the mutual inductance L_{12} through the expression

$$\mathcal{W} = \tfrac{1}{2} L_{11} I_1^2 + L_{12} I_1 I_2 + \tfrac{1}{2} L_{22} I_2^2 \tag{4-6.4}$$

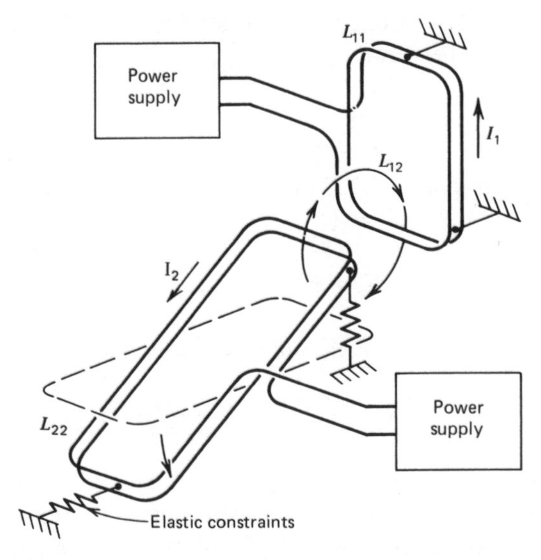

Figure 4-20 Sketch of two current-carrying circuits, one rigid and fixed and the other rigid and movable.

Since each is a continuous elastic system, there are an infinite number of deformation modes. However, for a given problem, only a finite number of modes might be important and hence we consider a finite set of deformation modes $\{u_\alpha\}$. For each deformation mode u_α we can define a generalized magnetic force (see Chapter 3)

$$F_\alpha = \frac{\partial \mathscr{W}}{\partial u_\alpha}(I_1, I_2, u_\beta) \tag{4-6.5}$$

where the partial derivative means that the currents are held fixed. The magnetic forces are thus related to the change in the inductances of the system. In the special case of rigid-body deformations, only L_{12} will change with translation or rotation of one circuit with respect to another, that is,

$$F_\alpha = I_1 I_2 \frac{\partial L_{12}}{\partial u_\alpha}$$

If we write L_{12} in a Taylor series in u_α,

$$L_{12} = L_0 + L_1 u_\alpha + L_2 u_\alpha^2 + \cdots$$

then

$$F_\alpha = I_1 I_2 (L_1 + 2L_2 u_\alpha) \tag{4-6.6}$$

and one can define a magnetic stiffness associated with the deformation u_α,

$$\kappa = -I_1 I_2 2 L_2 \tag{4-6.7}$$

For a general set of N circuits with M generalized deformations $\{u_\alpha\}$, the magnetic force takes the form

$$F_\alpha = \frac{1}{2} \sum \sum I_i I_i \frac{\partial L_{ij}}{\partial u_\alpha} \tag{4-6.8}$$

where i and j run from 1 to N. If the inductance matrix is expanded out to quadratic terms in the variables $\{u_\alpha\}$, then one obtains

$$F_\alpha = F_\alpha^0 - \sum \kappa_{\alpha\beta} u_\beta \tag{4-6.9}$$

where $[\kappa_{\alpha\beta}]$ is called the *magnetic stiffness matrix* defined by the relation

$$\kappa_{\alpha\beta} = -\frac{1}{2} \sum \sum I_i I_j \frac{\partial^2 L_{ij}}{\partial u_\alpha \partial u_\beta} \tag{4-6.10}$$

(The derivatives are evaluated at $u_\alpha = 0$.) It should be noted that the magnetic stiffness elements depend on the currents and can be positive (restoring) or negative (destabilizing). F^0 represents the generalized magnetic forces at $u_\alpha = 0$ and is defined by

$$F_\alpha^0 = \frac{1}{2} \sum \sum I_i I_j \frac{\partial L_{ij}}{\partial u_\alpha} \tag{4-6.11}$$

where the derivatives are evaluated at $u_\alpha = 0$.

Examples

1. *Magnetic stiffness for two current filaments.* For two parallel current filaments with currents I_1 and I_2 and separated by a distance d, the magnetic force on a length Λ is given by

$$F = -\frac{\mu_0 I_1 I_2 \Lambda}{2\pi d} \tag{4-6.12}$$

To find the magnetic stiffness associated with a displacement u, along the line joining the filaments and normal to each, we replace $d \to d + u$ and expand in a Taylor series. One can easily show that

$$\kappa = -\frac{\mu_0 I_1 I_2 \Lambda}{2\pi d^2} \tag{4-6.13}$$

2. *Current filament near a ferromagnetic half-space.* A current filament parallel to a soft ferromagnetic half-space will magnetize the material in the half-space. It can be shown (see Chapter 2) that for high permeability, $\mu_r \gg 1$, the induced magnetic field outside the magnetized half-space is equivalent to an image filament, of equal current and polariity of the filament outside, and at an equal distance from the boundary plane. The force of the actual filament is thus given by Eq. (4-6.12) in the above example, with $I_1 = I_2 = I$,

and $d = 2h$, where h is the distance of the filament from the half space. We note that the image also moves a distance u away from the boundary if the actual filament moves an increment u. The magnetic stiffness can be shown to be

$$\kappa = -\frac{\mu_0 I^2 \Lambda}{4\pi h^2} \tag{4-6.14}$$

for a filament of length Λ.

3. *Rotation of circular coils in a discrete coil solenoid.* One form of a magnetic fusion reactor uses a long solenoid with magnetic mirrors at each end (see Fig. 4-21). In the example shown in Figure 4-21, we want to investigate the tendency of the middle pair of coils to rotate with respect to two similar pairs on each side. The analytical solution involves elliptic integrals. It is just as easy to use a numerical method to calculate the inductance directly. In this case the generalized displacement is a rotation θ and the generalized "force" is a couple C given by

$$C = \frac{\partial W}{\partial \theta} \tag{4-6.15}$$

The magnetic stiffness with respect to rotation versus the distance between the coil pairs is shown in Figure 4-22. The data show that the stiffness can be positive for large distances and negative for distances less than the coil diameter. Thus at large distances the coils tend to align, like small dipole magnets, while at close distances they are unstable. In the latter case, proper elastic constraints are important and misalignment must be kept to a minimum.

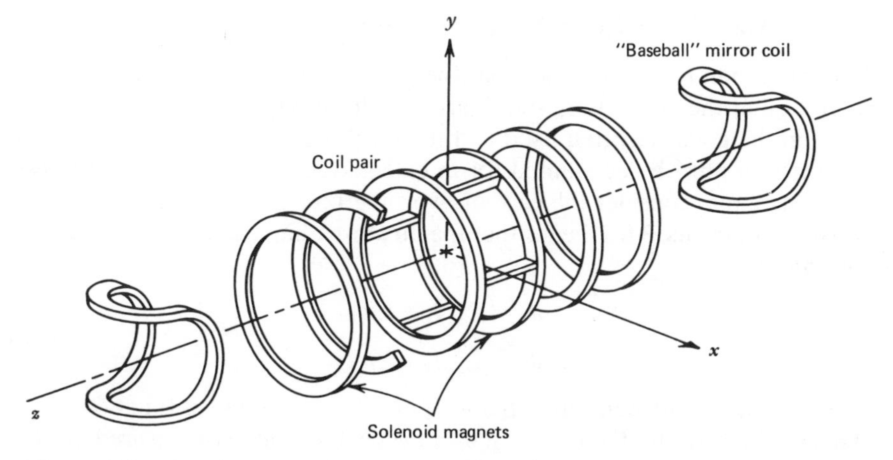

Figure 4-21 Diagram of discrete coil solenoid in a tandem mirror fusion reactor.

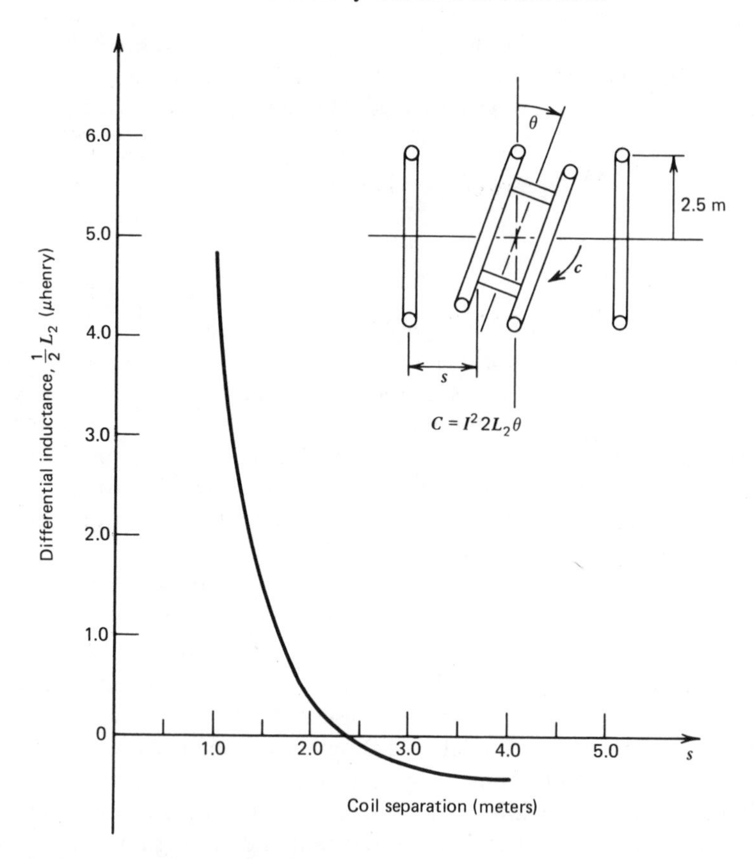

Figure 4-22 Magnetic spring constant as a function of axial separation for the rotation of two coils in a four-coil discrete coil solenoid.

Effect of Circuit Parameters on Magnetic Stiffness

The expression for the magnetic stiffness between two circuits [Eq. (4-6.2)] assumes that the currents I_1 and I_2 are held fixed. However, the voltages or fluxes in the coils are often fixed and the resulting changes in I_1 and I_2 with deformation must be accounted for in calculation of the magnetic stiffness. As an example, consider the case of two superconducting circuits each with persistent currents. This means that the flux threading each coil will remain a constant, that is,

$$\begin{aligned} \phi_1 &= L_{11}I_1 + L_{12}I_2 \\ \phi_2 &= L_{12}I_1 + L_{22}I_2 \end{aligned} \tag{4-6.16}$$

where ϕ_1 and ϕ_2 are constants. I_1 and I_2 may then be found as functions of ϕ_1 and ϕ_2. Suppose that the distance between the coils is measured by the variable u. In this case, L_{11} and L_{12} are independent of u, but the mutual inductance will change with u, so that L_{12}, I_1, and I_2 may be expanded in a Taylor series in u:

$$I_1 = I_{10} + \alpha_1 u$$

$$I_2 = I_{20} + \alpha_2 u \qquad (4\text{-}6.17)$$

$$L_{12} = L_0 + L_1 u + L_2 u^2$$

Using Eq. (4-6.16) it can be shown that $\alpha_1 = -L_1\beta_1$ and $\alpha_2 = -L_1\beta_2$, where β_1 and β_2 are linearly proportional to ϕ_1 and ϕ_2. The magnetic force is defined by

$$F = I_1 I_2 \frac{dL_{12}}{dx}$$

$$F = (I_{10} - L_1\beta_1 u)(I_{20} - L_1\beta_2 u)(L_1 + 2L_2 u)$$

The magnetic stiffness is proportional to the linear term in u, or

$$\kappa = -I_{10}I_{20}2L_2 + L_1^2(I_{10}\beta_2 + I_{20}\beta_1) \qquad (4\text{-}6.18)$$

The constant L_1 is proportional to the initial magnetic force between the circuits when $u = 0$. Thus it is clear that the magnetic stiffness under constant flux (4-6.18) differs from that under constant current (4-6.7) when there is an initial magnetic force at $u = 0$. (For an example of a levitated superconducting magnet, see Moon, 1977, p. 128.)

5

Magnetoelastic Stability of Structures

5.1 INTRODUCTION

Structural devices are often load limited by the buckling of structural elements such as beams, plates, and shells and not by the strength of the material. Also, the speeds of turbines, aircraft, and high-speed land vehicles are often limited by self-excited vibrations called "flutter". Magnetic devices and structures are also subject to static and dynamic instabilities. One distinction between magnetomechanical instabilities and conventional examples in the mechanical sciences is the fact that magnetic forces are of the body force type in comparison with surface loads such as end loads on columns and fluid pressures on a fluttering wing. When the body deforms, the magnetic forces change, introducing a feedback loop between the structure and the circuit or source of the magnetic field.

Of course, there are many uncoupled magnetoelastic problems where the deformation of the body does not measurably affect the magnitude or direction of the magnetic field (see, e.g., Chapter 4). In this chapter, however, we will examine a number of problems exhibiting mutual coupling of magnetic field and deformation which can lead to instabilities in the mechanical system or structure. We will study examples of static instabilities such as buckling or divergence.

One can also distinguish between conservative and nonconversative forces. In magnetic problems, static forces between current filaments or magnetic dipoles in most cases can be related to a force potential, namely, the magnetic energy. In this sense such forces are conservative. Magnetic forces related to induced eddy currents, such as the magnetic drag force in a linear electric motor or on a levitated vehicle, are, however, nonconservative since they cannot be derived from a potential function.

The importance of this distinction lies in the study of dynamic stability about equilibrium points. Conservative forces lead to symmetric eigenvalue problems and therefore result in either centers or saddle-point equilibrium positions in the phase plane. Thus conservative problems can only lead to buckling and not flutter, which means that a static analysis is sufficient to determine the stability of the system (see, e.g., Ziegler, 1959, Thompson, 1982). This is important since many mechanicians are aware that follower force problems can lead to flutter. However, while forces on electric filaments are follower forces in the sense that they change direction with deformation of the circuit, they can be derived from the magnetic energy and therefore can only result in buckling and not flutter.

Within the realm of static problems one must distinguish between *bifurcation-type* problems and *limit point* problems (see, e.g., Thompson and Hunt, 1973). If one considers the magnetic field or the electric current as a loading parameter and the deflection or strain in the structure as a measure of the deformation, then as the load is varied, the deflection changes continuously along a solution path in the load-deflection space (see Fig. 5-1*a*). A bifurcation problem is one in which two solution paths cross (see Fig. 5-1*a*); that is, for some value of the load there exists more than one solution for the deflection.

In limit point problems (see Fig. 5-1*b*) the loading path becomes unstable; that is, a small increase in load may result in a large deflection. In Figure 5-1*b* the solution between *A* and *B* is unstable, and when the load reaches point *A*, the deflection will jump to point *C*. Dynamically what happens is that a

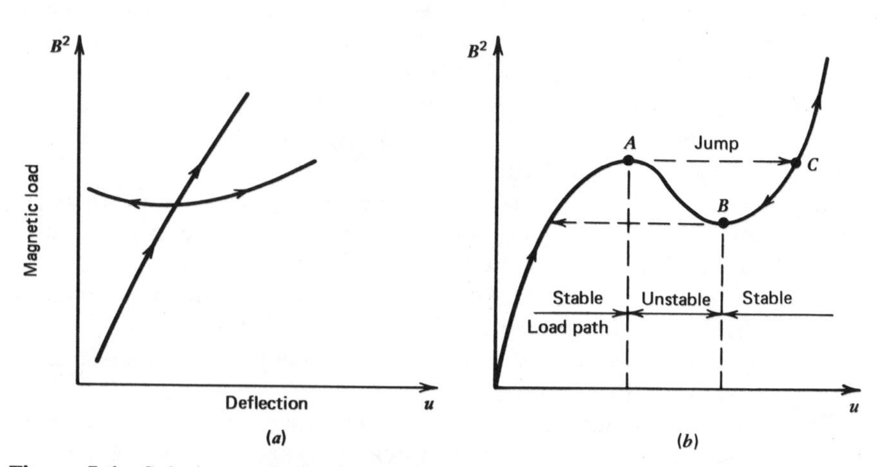

Figure 5-1 Solution paths in the magnetic force—displacement plane: (*a*) bifurcation problems and (*b*) limit point problem.

center or spiral equilibrium point becomes a saddle point and the deflection jumps to the nearest stable equilibrium point.

Symmetric Instabilities

It is instructive to look at a restricted class of static bifurcation problems with symmetric branches are shown in Figure 5-2. In this case the deflection remains either zero or constant on one of the loading paths as the current or field is increased until a critical value is reached where the two equilibrium paths cross. At the critical load parameter, the number of possible equilibrium positions changes. This is depicted in Figure 5-2*a*.

In Section 3.5, it was shown how magnetic forces could be related to the stored magnetic coenergy $\mathscr{W}(I, u) \equiv \mathscr{W}_m^*$. When conservative mechanical

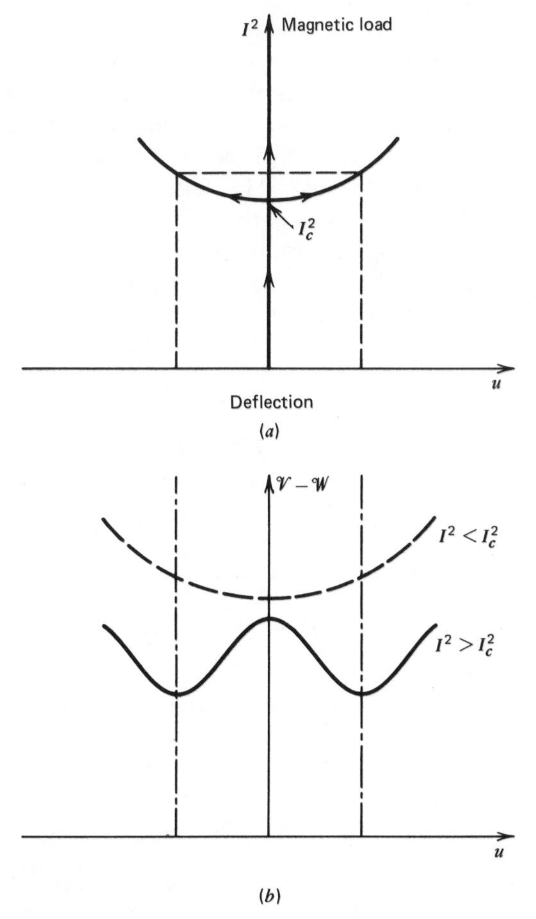

Figure 5-2 (*a*) Symmetric bifurcation problem and (*b*) energy displacement behavior for magnetic force less than and greater than the buckling value.

forces are present, the relation between elastic and magnetic energy functions for a one-degree-of-freedom system is given by

$$\frac{\partial \mathcal{V}}{\partial u} = \frac{\partial \mathcal{W}(I, u)}{\partial u}$$

Thus one can view $\mathcal{V} - \mathcal{W}$ as an *effective potential energy* for the system. When the curved equilibrium path bends up, as in Figure 5-2, the effective energy $\mathcal{V} - \mathcal{W}$ has a single trough as a function of deflection below the critical load and a double trough or three equilibrium points above the critical load.

Imperfections and Catastrophe Theory

Magnetostructural problems which exhibit bifurcation-type behavior are usually idealized mathematical models. It is impossible to load a structure along a zero deflection path without some small deflections due to small imperfections or misalignments. Thus actual load-deflection curves can only approach the bifurcation point. Beyond the buckling load, the structure may either jump to a new equilibrium position or may smoothly exhibit larger deflections as the load is increased. The latter case is still considered an instability since the zero deflection path is often the preferred state of the structure. Nonzero imperfection solutions are close to the two idealized intersecting equilibrium paths as shown by the dashed lines in Figure 5-3.

That the bifurcation point is a highly specialized condition can be seen from the mathematical representation of the effective energy of the system in the neighborhood of the bifurcation point. Along the straight path in Figure 5-3*a*, the effective potential $\mathcal{V}_1 = \mathcal{V} - \mathcal{W}$ has a minimum so that [see Fig. 5-2*b*]

$$\frac{\partial \mathcal{V}_1}{\partial u} = 0, \qquad \frac{\partial^2 \mathcal{V}_1}{\partial u^2} > 0 \tag{5-1.1}$$

Above the bifurcation point, $\mathcal{V}_1$ has a maximum or

$$\frac{\partial \mathcal{V}_1}{\partial u} = 0, \qquad \frac{\partial^2 \mathcal{V}_1}{\partial u^2} < 0 \tag{5-1.2}$$

At the bifurcation value of the loading parameter we can see that

$$\frac{\partial \mathcal{V}_1}{\partial u} = 0, \qquad \frac{\partial^2 \mathcal{V}_1}{\partial u^2} = 0 \tag{5-1.3}$$

Thus, if $\mathcal{V}_1$ has a Talor series in u, it must have the form

$$\mathcal{V}_1 = \tfrac{1}{2}(\lambda - \lambda_c)u^2 + \text{higher-order terms} \tag{5-1.4}$$

where λ_c is the critical value of the loading parameter λ.

Thus $\mathcal{V}_1$ cannot have a linear term in u for the ideal problem. A linear

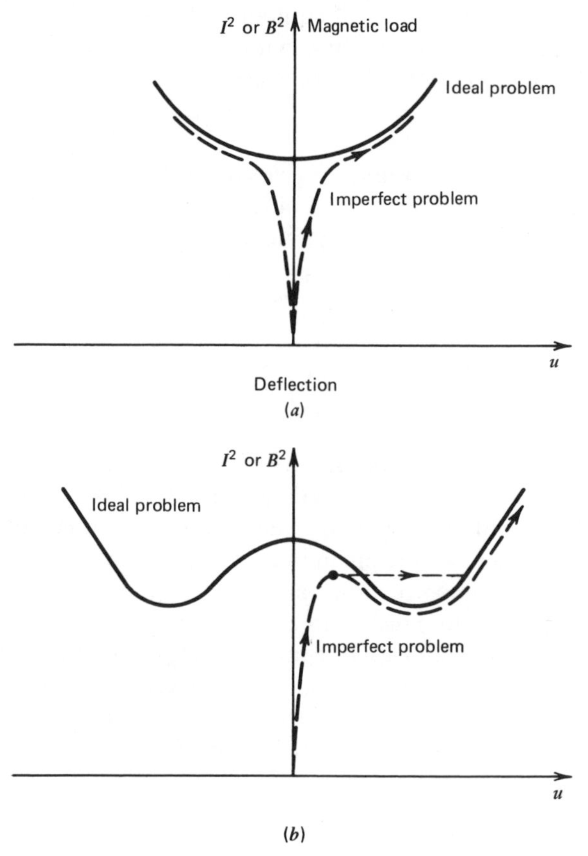

Figure 5-3 Ideal and imperfection-type load—displacement curves: (*a*) imperfection-insensitive solution and (*b*) imperfection-sensitive solution.

term in $\mathcal{V}_1$ would imply that $u \neq 0$ for $\lambda \neq \lambda_c$. Thus the linear term in $\mathcal{V}_1$ represents an *imperfection* or *misalignment parameter.*

Along the ideal path, $\mathcal{V}_1$ is said to have a singularity since both first and second derivatives vanish. A potential for a nonideal structure, that is, one with a linear term in u, is said to be an *unfolding* of the singularity.

The study of mathematical singularities under changes in parameters is known as *catastrophe theory* (see, e.g., Poston and Stewart, 1978). One considers a potential $\mathcal{V}$ as a function of generalized coordinates $\{q_i\}$ and a set of parameters $\{\lambda_k\}$, which in our case are the loads, currents, and geometric imperfections. The equilibrium points of this function are given by

$$\frac{\partial \mathcal{V}}{\partial q_i}(q_j; \lambda_k) = 0 \qquad (5\text{-}1.5)$$

Catastrophe theory is concerned with the locus of points in the space of $\{\lambda_k\}$

parameters at which the number of equilibrium points change. For example, consider the buckling of a structure with symmetric bifurcation potential when imperfection and nonlinear forces are present:

$$\mathscr{V} = \varepsilon u + \tfrac{1}{2}(\lambda - \lambda_c)u^2 + \tfrac{1}{4}\alpha u^4 \tag{5-1.6}$$

The equilibrium condition $\partial \mathscr{V}/\partial u = 0$ gives one the relation between u and λ. When $\varepsilon \neq 0$, the structure loses its stability at a limit point. The set of points at which the number of equilibrium points changes is given by

$$\frac{\partial^2 \mathscr{V}}{\partial u^2} = 0$$

This leads to a relation between λ and ε, that is,

$$4(\lambda_c - \lambda)^3 = 27\alpha\varepsilon^2 \tag{5-1.7}$$

As seen in Figure 5-4, this curve has a cusp at $\varepsilon = 0$ and $\lambda = \lambda_c$, and shows that imperfections decrease the buckling load λ. This is known as a *cusp catastrophe*. When other degrees of freedom are admitted, the number of load and imperfection parameters increases and the catastrophe set (i.e., the values of load and imperfections at which the number of equilibrium positions changes) becomes two-, three- or higher-dimensional. For example, for two degrees of freedom with one load parameter λ and two imperfection parameters ε_1 and ε_2, the catastrophe set is a set of surfaces in $(\lambda, \varepsilon_1, \varepsilon_2)$ space. For further discussion of these ideas the reader is referred to Poston and Stewart (1978) for a readable presentation of the subject in the context of problems in the physical sciences.

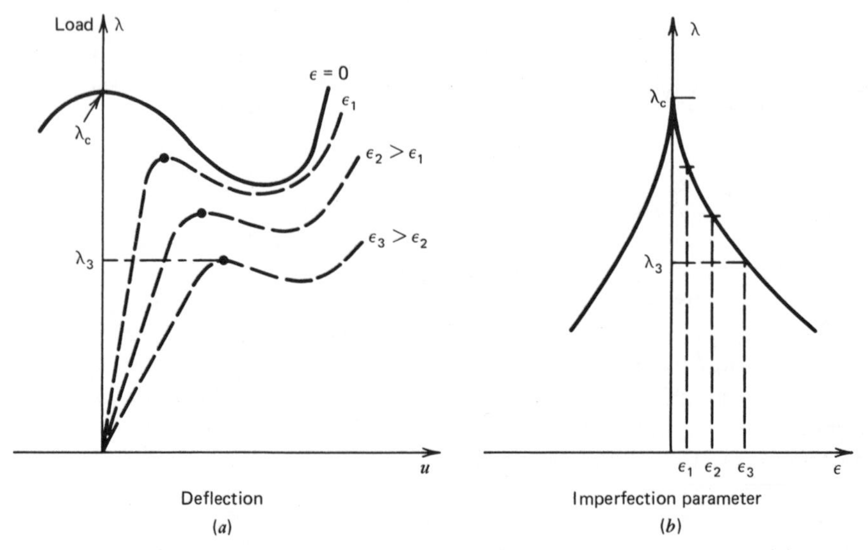

Figure 5-4 (*a*) Load-displacement curves for different imperfections and (*b*) critical or limit load vs. imperfection parameter.

5.2 EARNSHAW'S THEOREM

Central to a discussion of static magnetic stability problems is Earnshaw's theorem. Stated simply this theorem says: "A body with steady charges, magnetization or currents placed in a steady electric or magnetic field cannot rest in stable equilibrium under the action of electric and magnetic forces alone." (See, e.g. Jeans, 1925.) This theorem is based on the fact that steady electric and magnetic fields satisfy Laplace's equation. Continuous, bounded, solutions of Laplace's equation, or potential functions, cannot have any maxima or minima in a closed region except at the boundary of the region. Thus one can extend the theorem to say it is impossible to find a stable equilibrium configuration of a body under any force potential satisfying Laplace's equation, including gravity. This theorem has not prevented untold members of inventors to patent levitation devices based on permanent magnets even though the theorem was known as far back as 1839.

The Reverend Samuel Earnshaw (1805–1888) was a mathematical physicist who discovered this theorem while studying the properties of the electromagnetic ether. Early in the history of electromagnetism some hope emerged that all phenomena in the physical world would be explained by gravitational and electrostatic forces between charged particles. These hopes were extinguished, however, when Earnshaw concluded that inverse-square force laws between particles alone cannot account for the stability of a collection of particles. He presented this theorem in a lecture in 1839 which was later published (Earnshaw, 1842). However, while most texts in this field often refer to this theorem, no citation is given for the work. A brief note by Scott (1959) gives the proper reference for the work.

A practical result of the theorem is that it is impossible to suspend a ferromagnetic body below a permanent magnet without some feedback forces. Although an equilibrium position can be found where the magnetic force balances the force of gravity, the object either slams into the magnet or falls to the floor for small departures from equilibrium. There are exceptions to Earnshaw's theorem, however, the most dramatic being the suspension of multiton vehicles with superconducting magnets. Earnshaw's theorem is always true when all the sources of electromagnetism are held fixed in time and space as the test object is moved from equilibrium. However, magnetizable or superconducting bodies automatically adjust the magnetization or circulatory currents as the external field changes, in order to satisfy certain boundary conditions. One can show, however, that for flux attractors, that is, paramagnetic or ferromagnetic bodies, the theorem still holds; but for flux excluders, such as diamagnetic or superconducting bodies, the theorem may be violated and stable suspension under magnetic forces alone can be found.

A classic counterexample to Earnshaw's theorem is the stable levitation of a permanent bar magnet in a superconducting bowl (see Fig. 5-5*a*). The induced superconducting currents in the bowl act to exclude the total flux from the superconductor. These currents act as repelling magnets and

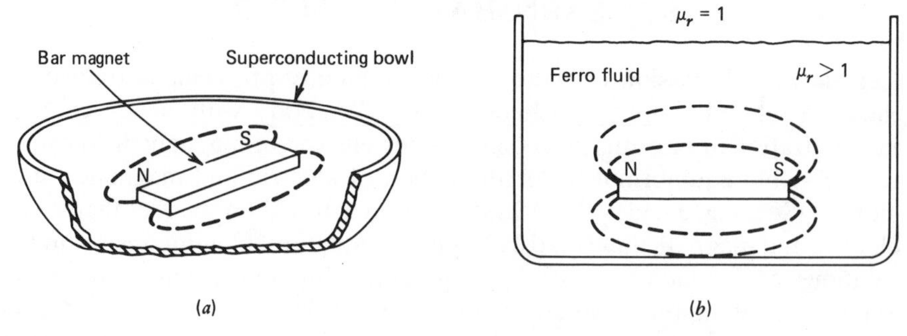

Figure 5-5 Two exceptions to Earnshaw's theorem: (*a*) levitation of a permanent bar magnet in a superconducting bowl and (*b*) suspension of a bar magnet in ferrofluid.

automatically adjust their strength and position as the bar magnet moves.

One can, of course, use active feedback to thwart Earnshaw's maxim, as is done in electromagnetic levitation of trains. An effect analogous to the superconducting bowl can be obtained by moving a magnet in a conducting trough (see, e.g., Fig. 1-8b). The induced eddy currents will screen out the total flux and stable levitation can be obtained.

Another counterexample of Earnshaw's theorem involves magnetic fluids (see Fig. 5-5*b*). If one places a magnet in a medium of high permeability, such as a magnetic fluid, the flux will again be screened out at the boundaries of the fluid and stable levitation can be obtained (Rosensweig, 1966; Berkovsky and Rosensweig, 1979).

We offer here a proof of Earnshaw's theorem for constant current circuits (Moon, 1980a). For simplicity, consider two circuits carrying currents I_1 and I_2 (see Fig. 5-6). Assume that the I_2 circuit is fixed and we ask whether it is possible to find a stable equilibrium position for circuit I_1 under the action of mutual magnetic forces alone. The magnetic energy for this system is given by the expression

$$\mathcal{W} = \tfrac{1}{2}L_{11}I_1^2 + L_{12}I_1I_2 + \tfrac{1}{2}L_{22}I_2^2 \tag{5-2.1}$$

The generalized force on circuit I_1, due to a generalized displacement s_i, is given by (see Section 3.5)

$$F_i = \frac{\partial \mathcal{W}}{\partial s_i} = I_1 I_2 \frac{\partial L_{12}}{\partial s_i} \tag{5-2.2}$$

For equilibrium, we must have $\partial \mathcal{W}/\partial s_i = 0$.

Near the equilibrium position we can expand the potential $\mathcal{W}$ in a Taylor series up to quadratic terms; that is,

$$\mathcal{W} = \frac{1}{2}\sum\sum W_{ij}(s_i - s_i^0)(s_j - s_j^0) + W_0 \tag{5-2.3}$$

For stability, $\mathcal{W}$ must be a relative *maximum*.

The matrix $[\mathcal{W}_{ij}]$ is real and symmetric and as such we can find a new set of generalized coordinates $\{x_i\}$ in which the matrix $[\mathcal{W}_{ij}]$ is diagonal. The coordinates $\{x_i\}$ are related to the set $\{s_i - s_i^0\}$ by an orthogonal transformation. In this new set of coordinates, $\mathcal{W} - W_0$ has the form

$$2(\mathcal{W} - W_0) = \lambda_1 x_1^2 + \lambda_2 x_2^2 + \cdots + \lambda_n^2 x_n^2$$

If W has a maximum then it can be shown that

$$\lambda_k < 0, \qquad k = 1, 2, \ldots, n$$

The trace of $\mathcal{W}$ is defined by $\sum_{i=1}^n W_{ii} \equiv \mathrm{Tr}(W_{ij})$. The trace is invariant under an orthogonal change of coordinates. Thus

$$\mathrm{Tr}(W_{ij}) = \sum \lambda_i < 0 \tag{5-2.4}$$

for a maximum. If $\{s_i\}$ are the Cartesian components of the position vector $\mathbf{r}$, to some point on the I_1 circuit, then the condition (5-2.4) reduces to

$$\mathrm{Tr}(W_{ij}) = \nabla^2 \mathcal{W} < 0. \tag{5-2.5}$$

For two circuits with constant currents, however, one can show that this condition can never be met. If $\mathbf{r} + \boldsymbol{\rho}_1$ is a position vector to any point in the I_1 circuit and $\boldsymbol{\rho}_2$ is its counterpart for the I_2 circuit (see Fig. 5-6), then we have

$$L_{12} = \frac{\mu_0}{4\pi} \int\int \frac{d\mathbf{S}_1 \cdot d\mathbf{S}_2}{|\mathbf{r} + \boldsymbol{\rho}_1 - \boldsymbol{\rho}_2|} \tag{5-2.6}$$

where $d\mathbf{S}_1$ and $d\mathbf{S}_2$ are differential vectors along the respective current

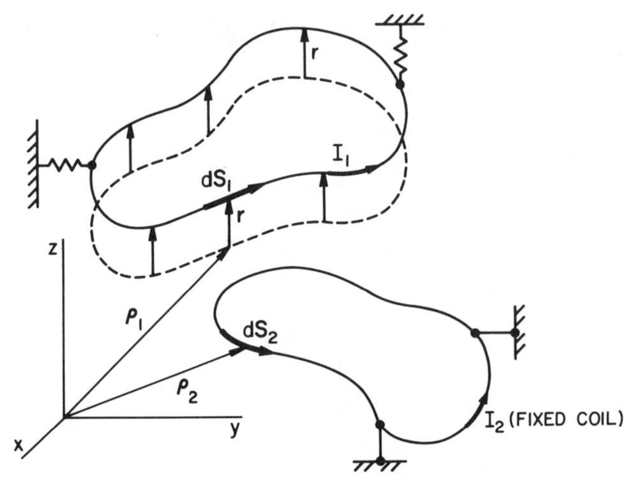

Figure 5-6 Interaction of two rigid circuits, one fixed and the other movable.

filaments in each circuit. We assume that the orientation of $d\mathbf{S}_1$ does not change as $\mathbf{r}$ is varied. Then, if $\mathbf{r} \equiv (x_1, x_2, x_3)$,

$$\frac{\partial^2 \mathcal{W}}{\partial x_i \partial x_j} = \frac{\mu_0}{4\pi} I_1 I_2 \int\int \frac{\partial^2}{\partial x_i \partial x_j} \frac{d\mathbf{S}_1 \cdot d\mathbf{S}_2}{|\mathbf{r} + \boldsymbol{\rho}_1 - \boldsymbol{\rho}_2|} \tag{5-2.7}$$

Assuming that the smallest distance between circuits is not zero, one can show that

$$\nabla^2 \mathcal{W} = 0 \tag{5-2.8}$$

and that the necessary condition for stability, Eq. (5-2.4), cannot be met.

It should be noted, however, that for superconducting circuits with persistent currents, I_1 and I_2 can change with $\mathbf{r}$ to maintain flux in one or both of the circuits (see, e.g., Chapter 6), and it may be possible for condition (5-2.4) to be met (Thornton, 1973).

One can conclude from this discussion that for a system of coils with constant currents, there will always be at least one mode of static instability.

5.3 THE TWO-WIRE INSTABILITY

An elementary example of a magnetoelastic instability is the case of two wires carrying currents in the same direction separated by a linear elastic constraint (see Fig. 5-7). For low-enough currents the elastic spring is capable of holding the mutually attracting filaments apart; but at a critical value of $I_1 I_2$ the wires will snap together. This one-degree-of-freedom problem is an example of what is known in structures as a *limit point instability*. It is also an example of what is called a *fold catastrophe*. These statements will be clarified below.

For convenience we assume one wire is fixed and call the horizontal displacement of the other wire u. The equilibrium condition equating elastic and magnetic forces is given by

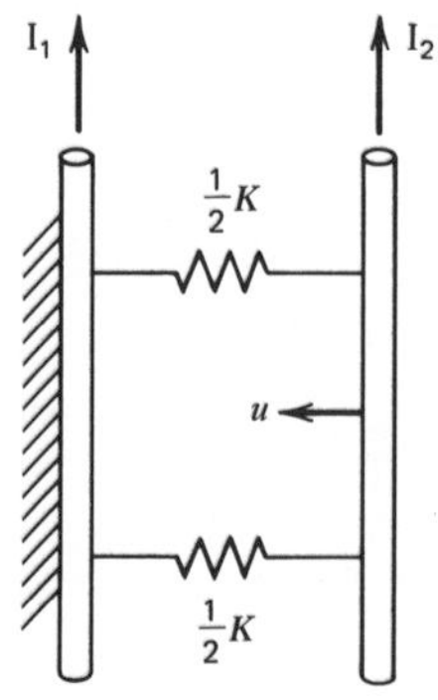

Figure 5-7 Parallel current filaments with elastic constraints.

$$Ku = \frac{\mu_0 I_1 I_2}{2\pi(d_0 - u)} \tag{5-3.1}$$

where K is the linear spring constant and d_0 is the initial separation of the wires when either $I_1 = 0$ or $I_2 = 0$.

The equilibrium values of u may be found by plotting each force as a function of u as in Figure 5-8. The intersection of the elastic and magnetic

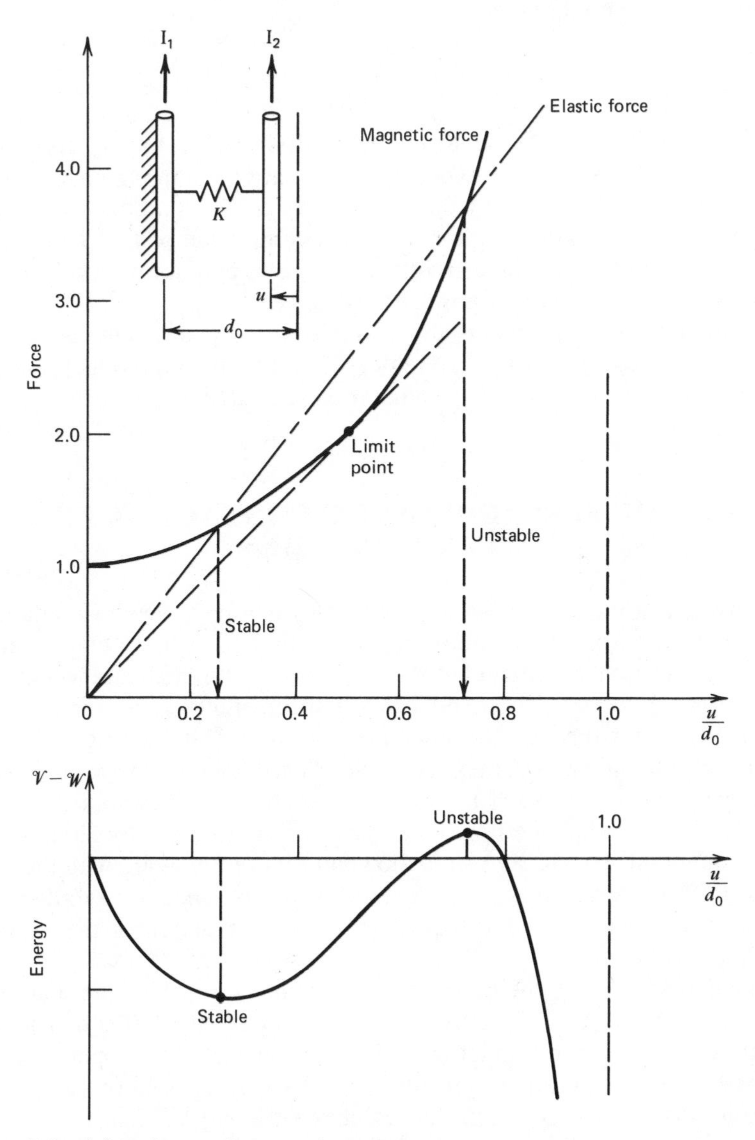

Figure 5-8 (a) Balance of magnetic and elastic forces as function of displacement and (b) effective potential energy vs. displacement.

functions of u shows that there are either two, one, or no equilibrium positions depending on the value of $\mu_0 I_1 I_2$. Since the single equilibrium position occurs for only a special value of $\mu_0 I_1 I_2$, this value is called the critical value or buckling load. The critical value of $\mu_0 I_1 I_2$ is given by $(\mu_0 I_1 I_2)_c = \frac{1}{2}\pi K d_0^2$. When $\mu_0 I_1 I_2$ is below this value there are two equilibrium positions. However, it can be shown that the smaller value of u is a stable equilibrium position, while the other value is unstable. This can be shown by investigating the effective energy $\mathcal{V}_1 \equiv \mathcal{V} - \mathcal{W}$:

$$\mathcal{V}_1 = \frac{1}{2} K u^2 - \frac{\mu_0 I_1 I_2}{2\pi} \ln(d_0 - u) \qquad (5\text{-}3.2)$$

As shown in Figure 5-8 the lower equilibrium value for u represents a minimum for $\mathcal{V}_1$. As $\mu_0 I_1 I_2 \to (\mu_0 I_1 I_2)_c$ the relative minimum coalesces with the maximum. In the load-deflection plane one can see that beyond $(\mu_0 I_1 I_2)_c$ there are no equilibrium positions. The buckling value of $\mu_0 I_1 I_2$ is called a limit point. A potential which suffers a change of equilibrium position from two to zero is said to exhibit a fold catastrophe.

In actual physical problems the wire would snap to a new equilibrium position when $\mu_0 I_1 I_2$ is slightly greater than the buckling value. This new value would be determined by nonlinear elastic forces.

5.4 THE ELECTROELASTIC CONTACT—A FOLD CATASTROPHE

The passage of electricity between two conducting bodies in contact is a classic but little-known instability. If one of the bodies has a finite radius of curvature, the area of contact between the two will be small and will increase with the force between the conductors. Current flowing between the conductors must pass through the small area of contact, increasing the current density and creating a high magnetic field. These high magnetic fields act on the flowing current in such a way as to force the conductors apart. The magnetic forces at an electric contact have been discussed by Holm (1967). A diagram of an electric contact is shown in Figure 5-9 along with the forces acting on the conductors. The instability arises since the contact area decreases with the magnetic force, which further increases the magnetic separation force.

To calculate the magnetic force on the contact consider the diagram in Figure 5-9 where r_c is the radius of the contact circle and R is the radius of curvature of the upper contact (see also Section 4.2). The upper contact is assumed to be isolated mechanically from the main source of current by a spring as shown in Figure 5-9. To calculate the net vertical force on the upper contact, we will use the concept of magnetic pressure or the Maxwell

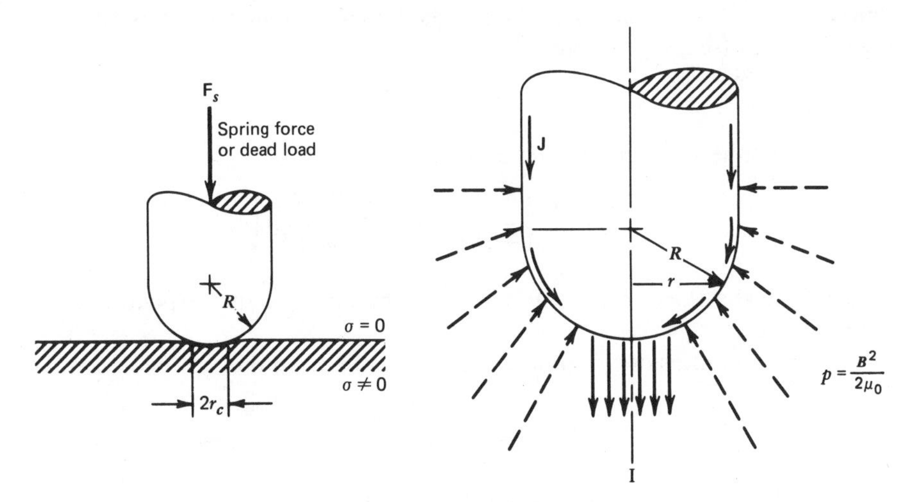

Figure 5-9 Current flow through elastic contacts.

stress tensor. Since the current flow is axial and radial, the magnetic field **B** outside the conductor will be circumferential. The magnetic pressure equivalent to the **J × B** body force will be $p = B^2/2\mu_0$ (see Sections 2.6 and 4.2). With the aid of the diagram in Figure 5-9 one can show that the vertical magnetic force on the contact is given by a term due to magnetic forces on the outside surface of the contact and a force directly under the contact area itself:

$$F_m = \int_{r_c}^{R} p2\pi r dr + \int_{0}^{r_c} p2\pi r dr \tag{5-4.1}$$

Ampere's law relates B and I outside the contact area, that is,

$$2\pi r B = \mu_0 I \tag{5-4.2}$$

Inside the contact area we assume uniform current density so that $B = \mu_0 I r/2\pi r_c^2$. Thus the magnetic force is proportional to the square of the current. Using Eq. (2-6.16) one can show that (Eq. 4-2.13)

$$F_m = \frac{\mu_0 I^2}{4\pi}\left(\ln\frac{R}{r_c} + \frac{1}{4}\right) \tag{5-4.3}$$

To establish the initial contact area, the spring force F_s is applied to the top of the contact and is balanced by a resisting contact force F_c at the bottom. This contact force is known to depend on the elastic properties of the conducting bodies and is proportional to the relative approach of the two contacting bodies, α. This force, known as the *Hertz contact force*, is found to have the form (see, e.g., Love, 1922)

$$F_c = k\alpha^{3/2} \tag{5-4.4}$$

where for a spherical surface and an elastic half-space,

$$k = \frac{4}{3} R^{1/2}\left[\frac{1}{(1-\nu_1^2)/Y_1 + (1-\nu_2^2)/Y_2}\right]$$

ν and Y are the Poisson ratio and Young's modulus of the respective contacting bodies.

Further, the radius of the area of contact depends on the approach α, that is,

$$r_c = (R\alpha)^{1/2} \tag{5-4.5}$$

The force balance for static equilibrium becomes

$$F_s = F_m + F_c \tag{5-4.6}$$

or

$$F_s(\alpha) = \frac{\mu_0 I^2}{4\pi} \ln \frac{R}{b\alpha^{1/2}} + k\alpha^{3/2} \tag{5-4.7}$$

where $4 \ln(1/b) = 1$.

The applied force F_s could depend on the approach α, as in the case of a spring. When the applied force is a constant such as a dead-weight load, the equilibrium values of α are plotted in Figure 5-10. When the current is absent, there is *one* equilibrium value of α where the applied force F_s equals the Hertzian contact force. However, when current flows through the contact, there are *two* values of α for equilibrium as shown in Figure 5-10. In this figure I^2 is held fixed, α_0 depends on I^2 through Eq. (5-4.10), and the dead-weight force F_s is considered as a variable.

Above the critical load there are one stable and one unstable equilibrium point. As the applied load is decreased, these two points coalesce into one point called the *limit point*. Below there are no equilibrium values for α and the contact will accelerate under the magnetic force. This behavior is characteristic of the *fold catastrophe* (see, e.g., Poston and Stewart, 1978). The most elementary potential near the limit point for a fold is the form

$$\mathcal{V} = \tfrac{1}{3}x^3 + ax \tag{5-4.8}$$

This gives two solutions for $a < 0$ and none for $a > 0$ (i.e., set $\partial V/\partial x = 0$). The catastrophe set consists of a single point $a = 0$, found from $\partial^2 V/\partial x^2 = 0$. The corresponding value of I, regarded here as a control variable, is found from the relation

$$\frac{\partial F_s}{\partial \alpha} - \frac{\partial F_c}{\partial \alpha} - \frac{\partial F_m}{\partial \alpha} = 0 \tag{5-4.9}$$

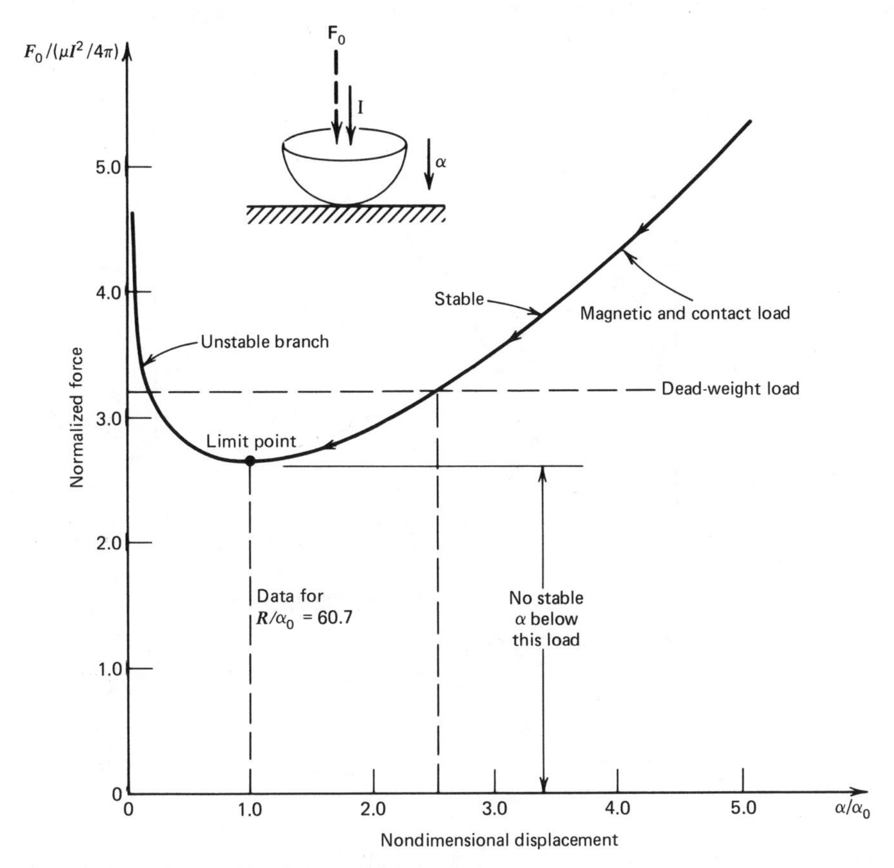

Figure 5-10 Sum of magnetic and Hertz contact forces as function of displacement.

When F_s is relatively insensitive to α and equal to a constant, this equation defines the single solution α_0:

$$k\alpha_0^{3/2} = \frac{1}{3}\frac{\mu_0 I^2}{4\pi} \tag{5-4.10}$$

When substituted into Eq. (5-4.7) we obtain an equation for the limit point value of I_0. Figure 5-11 shows typical critical current levels for copper contacts with a radius of curvature of $R = 1$ cm.

One should note that when the contact radius becomes very small, the current density is high enough to heat the solid to near melting. It is obvious that the elastic model used here will break down. Furthermore, in some problems arcing may occur across the contact surfaces. The critical value of current calculated above must then be thought of as an upper bound on the permissible current for a stable contact.

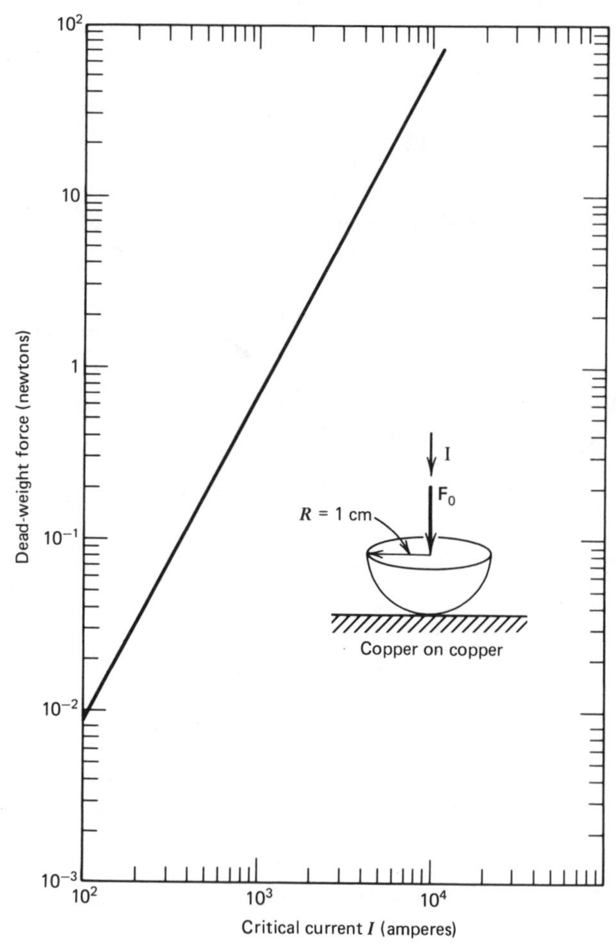

Figure 5-11 Critical applied contact force to avoid liftoff during current pulse.

5.5 CURRENT FILAMENT NEAR A FERROMAGNETIC PLATE

Another example of a limit point instability is that of a rigid current filament near an elastic ferromagnetic plate. We assume a wire runs parallel to the bending axis of a thin, soft ferromagnetic plate of permeability $\mu_r \gg 1$, as shown in Figure 5-12.

We make certain simplifying assumptions which result in an analytical model that retains the essential phenomenon. First, we consider only bending in the plane normal to the filament axis so that the plate acts as a beam. The beam-plate is assumed to be simply supported between two points separated by a length L.

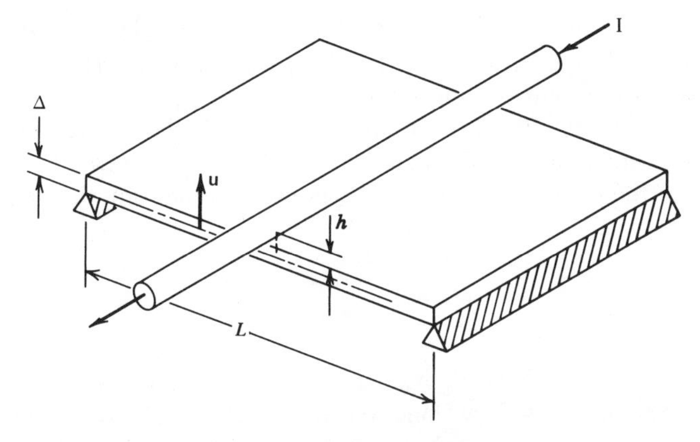

Figure 5-12 Rigid current filament near an elastic ferromagnetic plate.

The simplest case is that for which the wire is located at the midspan position. The magnetic field of the wire magnetizes the plate creating a magnetic attractive force which deflects the plate closer to the wire, further increasing the force. A current level is reached where a very small increment in current results in a large change in deflection and the plate slams up against the wire. This is a case of a *limit point instability*.

To model the magnetic force we use the *image method* (see Section 4.2). If the permeability is large enough, the combined magnetic field lines due to the wire and the plate will exit the plate almost normal to the surface. If the distance of the wire to the plate is much less than the span L, and we neglect the curvature, the magnetic field pattern will be that of a wire near a half-space with infinite permeability (see Fig. 5-13). In this case, the field outside the half-space is identical to that of the filament and an image

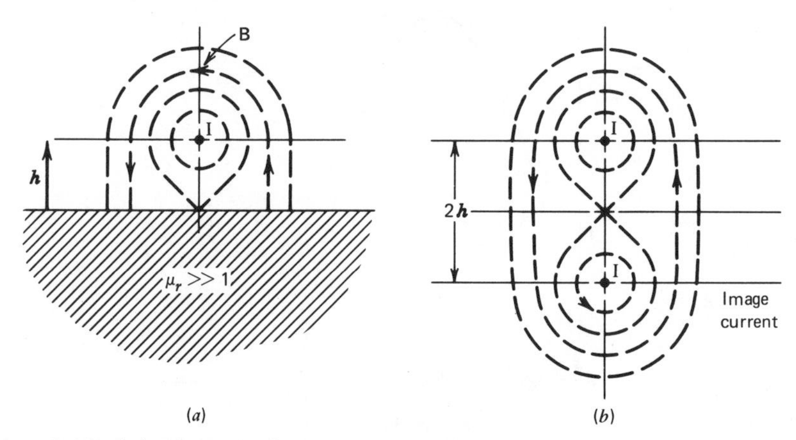

Figure 5-13 (*a*) Magnetic field lines of a current filament near a ferromagnetic half-space and (*b*) current filament and image for an ideal ferromagnetic half-space.

filament carrying the same current as the actual wire. The force between the wire and the half-space, or in this case the wire and the plate, is simply the force between the filament and its image.

If h is the initial distance between wire and plate, after bending the distance between filament and image will be $2(h-u)$, where u is the midspan deflection of the plate. The vertical magnetic force per unit filament length is then

$$F = \frac{\mu_0 I_0^2}{4\pi(h-u)} \tag{5-5.1}$$

Our final assumption is to consider the magnetic force as concentrated at the midspan point. This allows one to use the known solution for the midspan force and deflection of a simply supported plate

$$u = \frac{FL^3}{48D} \tag{5-5.2}$$

where $D = Y\Delta^3/[12(1-\nu^2)]$. Here Y is Young's modulus, ν is Poisson's ratio, and Δ is the plate thickness.

Substituting the magnetic force into Eq. (5-5.2) we have

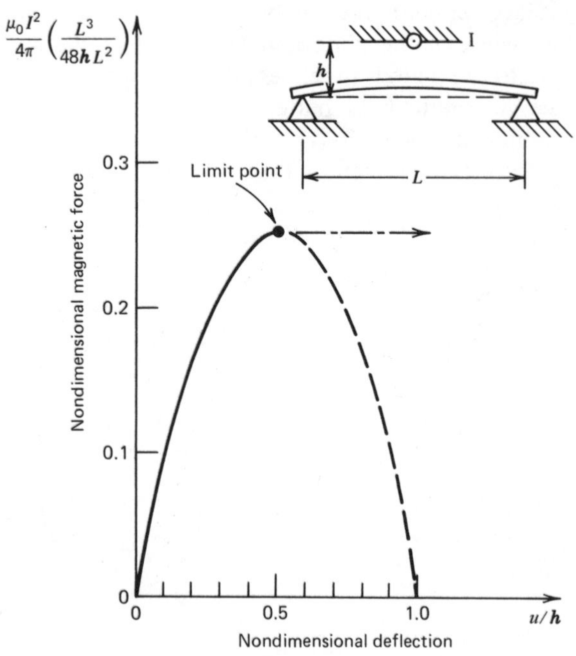

Figure 5-14 Plate displacement vs. magnetic force parameter for a current filament near an elastic plate [Eq. (5-5.4)].

$$\frac{\mu_0 I^2}{4\pi} = \frac{48D}{L^3}(h-u)u \qquad (5\text{-}5.3)$$

This relation is shown plotted in Figure 5-14. The limit point occurs where $dI^2/du = 0$, or at $u = \frac{1}{2}h$. This condition gives the maximum value of I, that is,

$$\left(\frac{\mu_0 I^2}{4\pi}\right)_{\max} = \frac{12Dh^2}{L^3} \qquad (5\text{-}5.4)$$

Typical values of $I_{\max}$ for a ferromagnetic plate are given in Figure 5-15.

It should be noted here that this model assumes that all the flux is trapped in the plate, which enabled us to neglect magnetic forces on the top of the plate. However, in actual materials the plate will saturate and some flux will exit the top of the plate. In this case one should use the Maxwell stress tensor to calculate the net force on top and bottom of the plate (see Section 7.7).

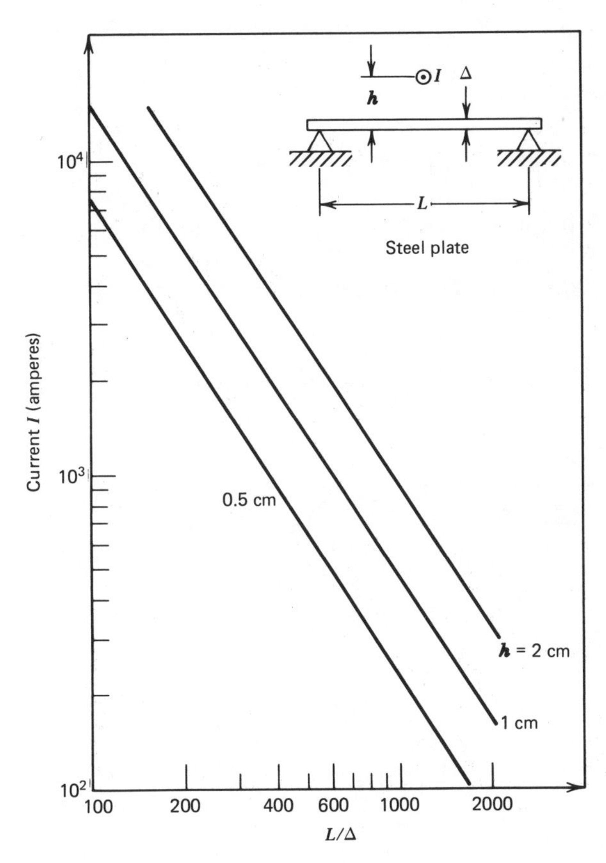

Figure 5-15 Buckling current vs. length-to-thickness ratio for a current filament near an elastic plate.

5.6 ELECTROMAGNET NEAR AN ELASTIC KEEPER

Our next example is similar to the previous problem and illustrates the *energy method* for magnetoelastic stability problems. We imagine a "U"-shaped magnetic circuit in which the plate acts as a movable keeper (see Fig. 5-16). N turns of electric current loop around the "U"-shaped, soft magnetic core and create the magnetic flux Φ. The flux is assumed to be confined to the ferromagnetic material, except where it jumps the gap between the electromagnet and the platelike keeper (see Fig. 5-16). We assume that the keeper is either attached to an elastic spring or is part of a beamlike plate.

In this example it will be convenient to work with the flux Φ. Thus if $\mathscr{W}$ is the stored magnetic energy in the electromagnet, the attractive force between the plate keeper and electromagnet is given by [see Eq. (3-5.9)]

$$F_m = -\frac{\partial \mathscr{W}(\Phi, z)}{\partial z} \tag{5-6.1}$$

where z is the gap separation. (Note the change of sign when $\mathscr{W}$ is a function of flux and not current.) The stored magnetic energy is given by the integral (see Section 2.8)

$$\mathscr{W} = \int_0^\Phi NI d\Phi \tag{5-6.2}$$

where NI is the number of ampere turns encircling the electromagnet.

Our object here is to find the relation between either Φ or I and the equilibrium gap distance. The critical current will occur where $dI^2/dz = 0$.

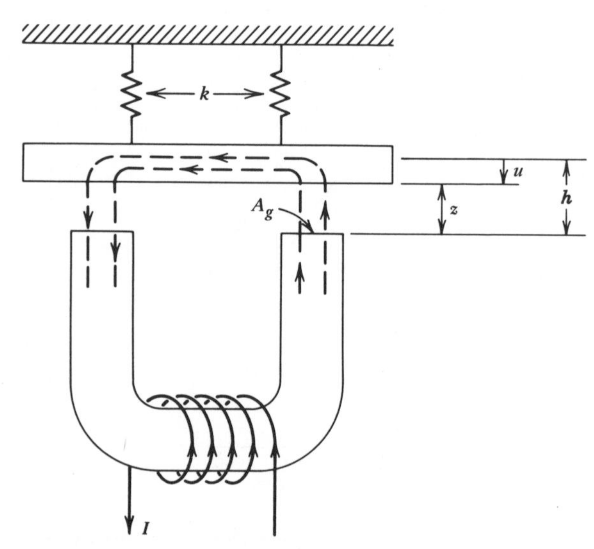

Figure 5-16 Electromagnet with movable keeper.

In order to carry out this program we need a relation between NI and Φ. This is found by using Ampere's law in integral form for a path around the magnetic circuit. The integral form of Ampere's law (2-2.9) for the magnetic circuit shown in Figure 5-16 is

$$NI = \int_{\text{fe}} \mathbf{H} \cdot d\mathbf{l} + 2H_g z$$

or

$$NI = \Phi[\mathcal{R}_{\text{fe}}(\Phi) + \mathcal{R}_g(z)] \tag{5-6.3}$$

where

$$\mathcal{R}_{\text{fe}} = \int_{\text{fe}} \frac{dl}{\mu_0 \mu_r A}$$

$$\mathcal{R}_g = \frac{2z}{\mu_0 A_g}$$

The expression $\mathcal{R}_{\text{fe}} + \mathcal{R}_g$ is called the *reluctance* [see Eq. (2-8.11)] and is analogous to the resistance in an electric circuit. We have implicitly used the approximation that the flux density is uniform across the cross-sectional area A of the ferromagnetic circuit and across the area of the gap A_g; that is,

$$H = \frac{B}{\mu_0 \mu_r} = \Phi \frac{1}{\mu_0 \mu_r A}$$

This analysis is not restricted to linear ferromagnetic material since we can let μ_r depend on B or Φ. Using Eqs. (5-6.1) and (5-6.2) one obtains the formula for the magnetic force:

$$F_m = -\frac{\Phi^2}{2} \frac{d\mathcal{R}_g}{dz} = -\frac{\Phi^2}{\mu_0 A_g} \tag{5-6.4}$$

If the plate spring force is linear with displacement, $u = h - z$, where h is the zero current gap distance, the equilibrium condition is given by

$$\frac{\Phi^2}{\mu_0 A_g} = k(h - z) \tag{5-6.5}$$

If one can control the flux there is one value of z for a given Φ, and there is no instability. However, if we replace Φ by the current I, we have

$$\mu_0 (NI)^2 = \frac{ku[\mu_0 A_g \mathcal{R}_{fe} + 2(h - u)]^2}{A_g}. \tag{5-6.6}$$

In general $\mathcal{R}_{\text{fe}}$ will be an implicit function of I if μ_r depends on the flux density. For the special case where μ_r is a constant, $(NI)^2$ will have a maximum value and hence a limit point for the displacement

$$u^* = \frac{\mu_0 A_g \mathcal{R}_{\text{fe}}}{6} + \frac{h}{3} \tag{5-6.7}$$

For most ferromagnetic materials, μ_r is very large $O(10^3)$–$O(10^5)$ so that $\mathcal{R}_{\text{fe}}$ is a very small number. In this case the keeper will snap toward the electromagnet at a small distance beyond $u = \frac{1}{3}h$.

This type of problem is of importance in the design of relay switches and solenoid devices. One large-scale application involves the levitation or suspension of bodies or vehicles by electromagnets. As this example shows a constant current device will be unstable and for stable levitation at all gap distances the flux must be controlled through feedback devices (see Section 7.4).

5.7 THE THREE-WIRE PROBLEM

One of the simplest examples of magnetoelastic bifurcation instability is the case of a straight current filament between two parallel current filaments as shown in Figure 5-17. If all three carry current in the same direction, the center filament will be attracted to both the left and right filaments, and will be unstable unless there is some elastic constraint between the wires. This example is a simple analog of the more complex problem of the stability of an array of mutually attracting magnets, as in fusion reactors and other applications (e.g., see Section 6.7).

Consider the arrangement in Figure 5-17 where a long filament with current I_0 is placed parallel and in the same plane between two fixed filaments with currents I_1 and I_2. We assume that nonlinear elastic constraints are placed between the wires and that the deflection of the center filament u is uniform along its length. The equilibrium positions are determined from the balance equation between elastic and magnetic forces; that is,

$$Ku(1 + \beta u^2) = \frac{\mu_0 I_0}{2\pi}\left(\frac{I_2}{D_2 - u} - \frac{I_1}{D_1 + u}\right) \tag{5-7.1}$$

The linear spring constant per unit length is K and the nonlinear spring

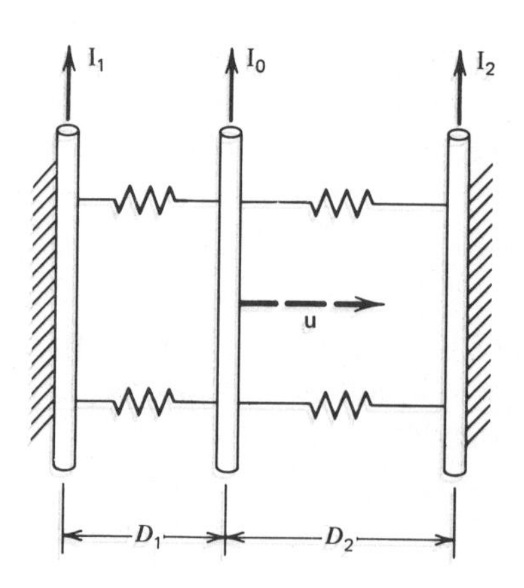

Figure 5-17 The elastic three-wire problem.

constant is $K\beta$. For $u = 0$ to be a solution we require that I_1 and I_2 be related; that is,

$$\frac{\mu_0 I_0 I_1}{2\pi D_1} = \frac{\mu_0 I_0 I_2}{2\pi D_2} \equiv F \tag{5-7.2}$$

so that F will be considered as a load parameter. Besides $u = 0$, the relation (5-7.1) admits another solution given by

$$F = (1 + \beta u^2)\left(1 - \frac{u}{D_2}\right)\left(1 + \frac{u}{D_1}\right)\frac{KD_1D_2}{D_1 + D_2} \tag{5-7.3}$$

In general this will result in an unsymmetric bifurcation problem where $dF/du \neq 0$ for $u = 0$, as shown in Figure 5-18. As a special case we assume that the problem is symmetric, which requires that $D_1 = D_2 \equiv D$ and $I_2 = I_1$. Then Eq. (5-7.3) takes the form

$$F = F_c\left[1 - (1 - D^2\beta)\left(\frac{u}{D}\right)^2 - D^2\beta\left(\frac{u}{D}\right)^4\right] \tag{5-7.4}$$

The term F_c is the critical buckling load value $\frac{1}{2}KD$ (or $\mu_0 I_0 I_1 = \pi D^2 K$) and the first two terms on the right-hand side of Eq. (5-7.4) represent a classic "pitchfork" bifurcation. If $D^2\beta > 1$, that is, a large elastic nonlinearity, then the buckling is *imperfection-insensitive* and the deflection follows a smooth curve for $F > F_c$ as shown by the dotted line in Figure 4-19(a). If $D^2\beta < 1$, that is, a weak elastic nonlinearity, the bifurcation is *imperfection-sensitive*. This means that for an actual problem the deflection will follow the dashed line in Figure 5-19(b), which results in a limit point or snap instability at a load value less than F_c.

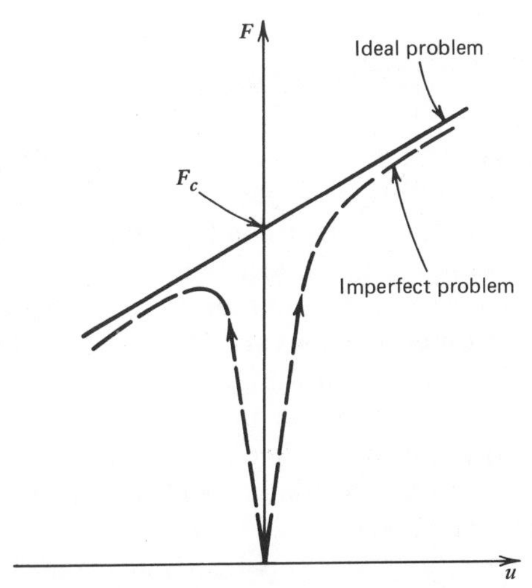

Figure 5-18 Unsymmetric bifurcation problem.

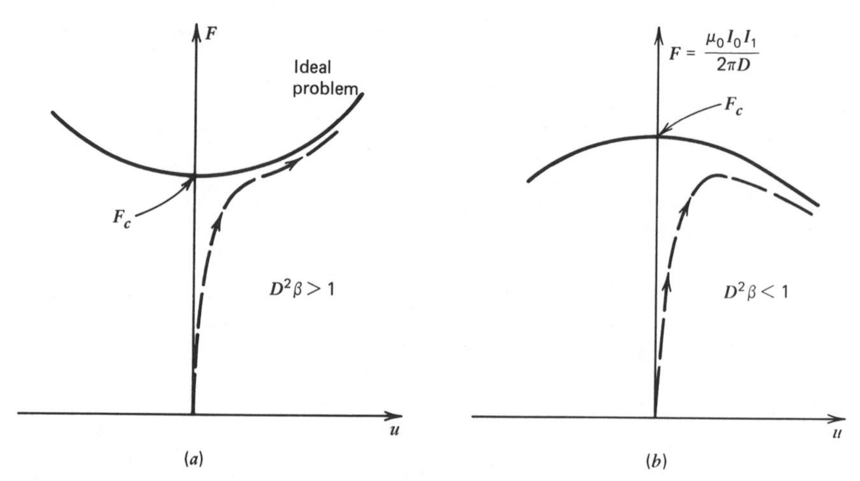

Figure 5-19 (*a*) Imperfection-insensitive load-displacement curve for the symmetric three-wire problem and (*b*) imperfection-sensitive load-deflection curve for the three-wire problem.

One interesting property of this example is the secondary buckling that occurs in the first case where $D^2\beta > 1$ for large u. Here, when $F > F_c$, one can have a limit point instability at $F = F_{c2}$ due to the quartic term in Eq. (5-7.4).

This example is known more generally as a *cusp catastrophe*. In order to understand this description one must examine the three-wire problem under nonideal conditions as outlined in Section 5.1.

5.8 THE *N*-WIRE TORUS

The three-wire problem can be extended to the case of N wires arranged in a circular array with an axial return wire (see Figure 5-20). This problem is of importance in the design of magnetic fusion reactors (tokamaks) since it is a model of a toroidal array of N magnets (see, e.g., Moon and Swanson, 1976; Moon, 1978). A coaxial array of filaments creates a circular magnetic field B_θ in the region between the outer filaments and inner wire. In three dimensions, radial current paths (neglected here) would confine the magnetic field to a toroidal region. The array is also similar to a coaxial cable.

In the symmetric array shown in Figure 5-20 there is only an outward radial magnetic force or pressure on the outer filaments which is assumed to be equilibrated by mechanical restraints. However, since the outer wires are all attracting one another, any tangential displacements that destroy the symmetry lead to destablizing circumferential magnetic forces on the wires.

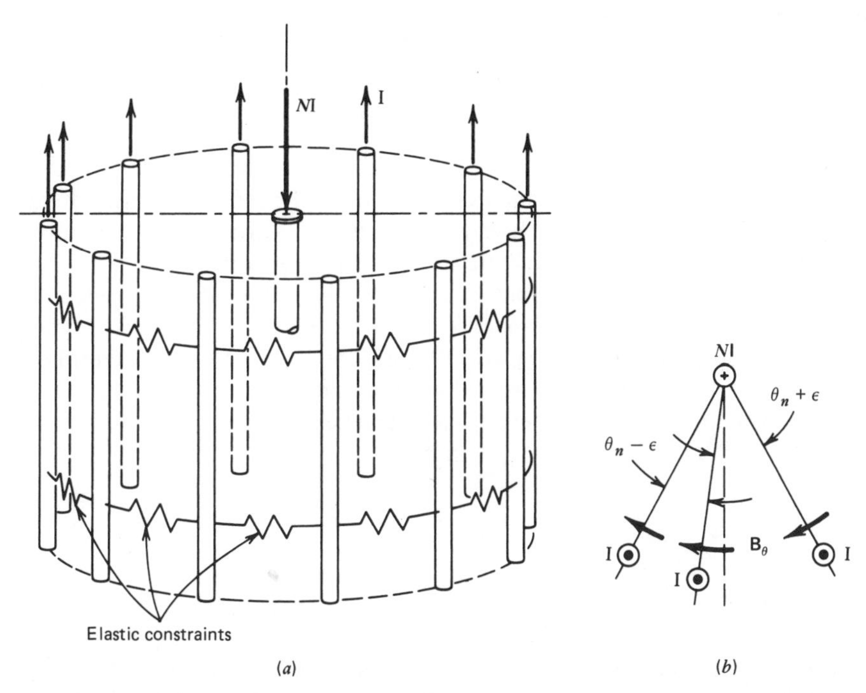

Figure 5-20 (a) Circular array of parallel current filaments with elastic constraints and (b) top view of displaced current filament.

If there are elastic constraints between the wires, the array will be stable up to a critical value of the current.

To calculate the critical buckling current, we use the example of the three-wire problem. Assume that the unperturbed angle between one wire and its nth neighboring pair of wires is $\theta_n = 2\pi n/N$, and that the center filament moves an angle ε closer to the left wire and away from the right wire. We note that the tangential magnetic force is proportional to the radial field produced by the two side wires and that the axial current NI only produces a tangential field and hence does not contribute to the tangential force on the center wire.

The radial magnetic field, as seen by the perturbed wire, can be shown to be

$$B_r = \frac{\mu_0 I}{4\pi R}\left(\cot \tfrac{1}{2}(\theta_n - \varepsilon) - \cot \tfrac{1}{2}(\theta_n + \varepsilon)\right) \tag{5-8.1}$$

For $\varepsilon \ll \theta_n$, this expression becomes

$$B_r = \frac{\mu_0 I}{4\pi R} \frac{\varepsilon}{\sin^2 \frac{1}{2}\theta_n}$$

The total force on the filament from all pairs is given by

$$F_\theta = \frac{\mu_0 I^2}{4\pi R}\varepsilon \sum_{n=1} \frac{1}{\sin^2(n\pi/N)} \tag{5-8.2}$$

For $n\pi/N \ll 1$, we see that the force from successive pairs falls off as $1/n^2$, so that only the first and second nearest neighbors effects the perturbed wire. Expression (5-8.2) can also be derived from the variation of the mutual inductance between the center wire and the two side wires.

If all the wires are perturbed, each will see forces due to successive neighboring pairs. The equations of equilibrium can be written for a typical wire with displacement $\varepsilon_m \equiv u_m/R$. We assume that the wires have elastic constraints between them and an external structure, and that the forces are proportional to the displacements. The balance between elastic forces and magnetic perturbation forces is given by

$$\begin{aligned} ku_m &= \frac{\mu_0 I^2}{8\pi R}\left(\frac{1}{\sin^2(\pi/N)}(2u_m - u_{m-1} - u_{m+1}) \right. \\ &\left. + \frac{1}{\sin^2(2\pi/N)}(2u_m - u_{m-2} - u_{m+2})\right) \end{aligned} \tag{5-8.3}$$

where k is the elastic spring (only first and second nearest neighbor forces are included).

This equation will be satisfied for discrete values of I^2. The lowest value can be shown to be given by the solution $u_m = -u_{m+1}$, which leads to a critical current given by (see "Periodic Array of Magnets" in Section 6.7)

$$\mu_0 I^2 = 2\pi Rk \sin^2(\pi/N) \tag{5-8.4}$$

In terms of the mean magnetic pressure at the wire, $P \equiv B^2/2\mu_0$ where $B = \mu_0 NI/2\pi R$, the critical pressure is given by

$$P_c = \frac{\pi k}{4R}\left(\frac{\sin(\pi/N)}{\pi/N}\right)^2 \tag{5-8.5}$$

We see that for a large number of coils, the *buckling magnetic pressure* in the torus is independent of N and decreases with the radius R.

Other problems of this type relating to superconducting fusion reactors and energy storage devices are discussed in Section 6.7.

5.9 CURRENT FILAMENT BETWEEN FERROMAGNETIC PLATES

The magnetic field of a current filament placed near ferromagnetic material will magnetize the material and induce an attractive force between the filament and the ferromagnetic body. When a current-carrying filament is

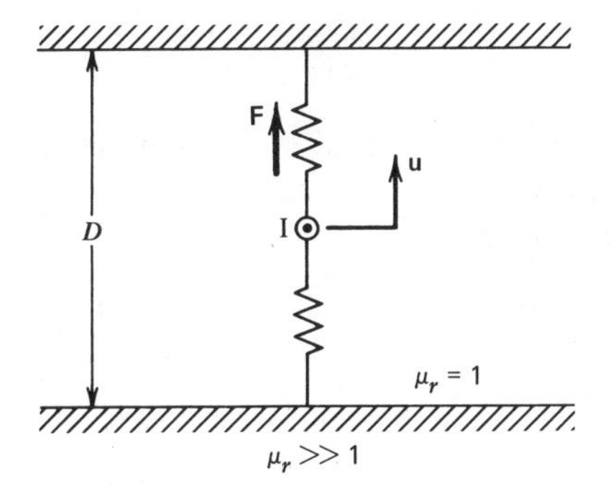

Figure 5-21 Current filament between two ferromagnetic half-spaces.

placed midway between two ferromagnetic walls (see Fig. 5-21) the competing attractive forces on the wire from each wall can lead to a bifurcation-type buckling problem.

This problem is more than academic since current filaments are placed in slots in ferromagnetic rotors in rotating machinery. To calculate the force on the wire one can use the *image method* as used in Section 5.5 (see also Section 4.2). However, because of the two walls, the image filaments due to one wall create image filaments behind the other wall which results in an infinite array of filaments. This problem has been worked out by Hague (1929). If D is the distance between the walls (see Fig. 5-21) and u the vertical displacement of the wire from the midplane, the vertical force per unit length on the wire is given by

$$F = \mu_0 I^2 \frac{\pi}{D} \tan \frac{\pi u}{D} \tag{5-9.1}$$

If the displacement u is constant along the length of wire, with an elastic stiffness per unit length κ and a nonlinear symmetric stiffness coefficient $\alpha\kappa$, the balance of elastic and magnetic forces leads to ($|\pi u/D| \ll 1$)

$$u = 0$$

or

$$\frac{\mu_0 I^2 \pi^2}{D^2 \kappa} = \frac{1 + \alpha u^2}{1 + \pi^2 u^2/3D^2} \tag{5-9.2}$$

This classic bifurcation or "pitchfork instability" is imperfection-sensitive (-insensitive) if $3\alpha D^2 - \pi^2 < 0$ (>0).

A more interesting problem results if we allow the filament to bend along its length under the action of attractive forces from the ferromagnetic walls. If the ratio of gap to span is small, that is, $D/L \ll 1$, then we may assume that locally the attractive force between the wire and the wall is given by Eq. (5-9.1).

The balance of elastic and magnetic forces on a differential length of the filament results in a differential equation for bending of the wire,

$$Y\mathscr{I}u'''' - \frac{\pi^2}{D^2}\mu_0 I^2 u = 0 \tag{5-9.3}$$

where $\mathscr{I}$ is the area moment of the cross section. Thus the ferromagnetic walls act as an elastic foundation with negative spring constant.

Again we note that $u = 0$ is a solution as well as $u = A\sin(n\pi/L)x$. The lowest current nontrivial periodic solution is given by

$$\mu_0 I^2 = \frac{\pi^2 Y\mathscr{I}}{L^2}\left(\frac{D}{L}\right)^2 \tag{5-9.4}$$

5.10 FERROELASTIC BEAM IN A UNIFORM MAGNETIC FIELD

Aside from practical applications, stability and vibration problems are important because they allow one to test the theoretical model against experiments. This is because they lead to eigenvalue problems for some parameter in the theoretical model, which are easy to observe in experiments.

In the case of ferroelastic buckling, two classes of stability problems have received attention in the last decade. One class deals with hard ferromagnetic materials in the form of slender bodies in magnetic fields; the other class involves soft ferroelastic beams and plates in magnetic fields.

In the first class of problems, the stability of saturated ferromagnetic materials is considered. Here the magnetization vector **M** has a constant magnitude, but its orientation may vary. **M** initially lies along the direction of an applied magnetic field $\mathbf{B}_0$, but as the field is lowered the magnetization **M** switches to a different direction. The critical field for this switching is called the *nucleation field* (Brown, 1966). Solutions for the case of the rigid saturated ferromagnetic problem have been given by Frei et al. (1957), Aharoni (1963), and Brown (1966) for ellipsoids. Three distinct modes have been found for the direction cosines of **M** called *curling*, *coherent rotation*, and *buckling*, though in the latter there is no physical deformation of the body.

Elastic buckling of ferromagnetic whiskers has been observed by DeBlois (1967) as shown in Figure 5-22. If a pole model is used (see Chapter 3) the ends of the elastica can be imagined to have poles of strength ρ_m and forces $\rho_m\mathbf{B}_0$, which provide compressive stresses on the initially straight whisker. The postbuckling shape of the whisker shows the classic shape of the elastica as shown in Figure 5-22.

One of the early papers on magnetoelastic buckling of soft magnetic materials was by Mozniker (1959). In his experiments he placed a nonferromagnetic cantilevered beam, with a ferromagnetic solid on the end, between the poles of an electromagnet (see Fig. 5-23*a*). The equation of motion for the lowest transverse mode was given by

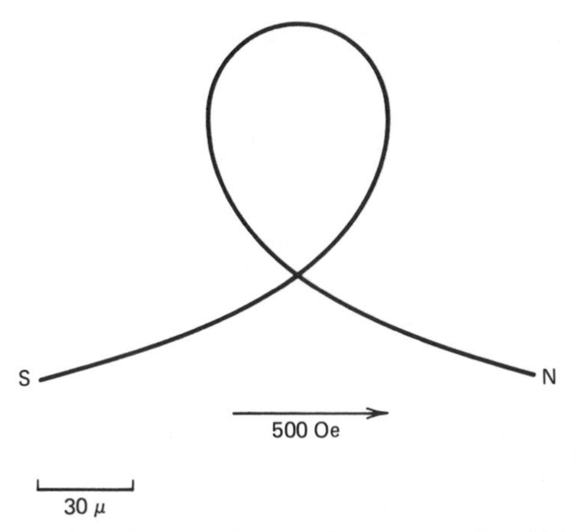

Figure 5-22 Photograph of the buckling of a ferromagnetic whisker in a longitudinal magnetic field. (Personal communication from R. W. DeBlois.)

$$\frac{d^2u}{dx^2} + \omega_0^2 u - KB_0^2 \frac{Gu}{(G^2 - u^2)^2} = 0 \tag{5-10.1}$$

where G is the gap between the magnet and the beam and B_0 is the magnetic field produced by the electromagnet. The linearized equation has a vibratory solution whose frequency decreases with current, that is

$$\omega^2 = \omega_0^2\left(1 - \frac{KB_0^2}{G^3\omega_0^2}\right) \tag{5-10.2}$$

Mozniker measured this decrease in natural frequency with magnetic field.

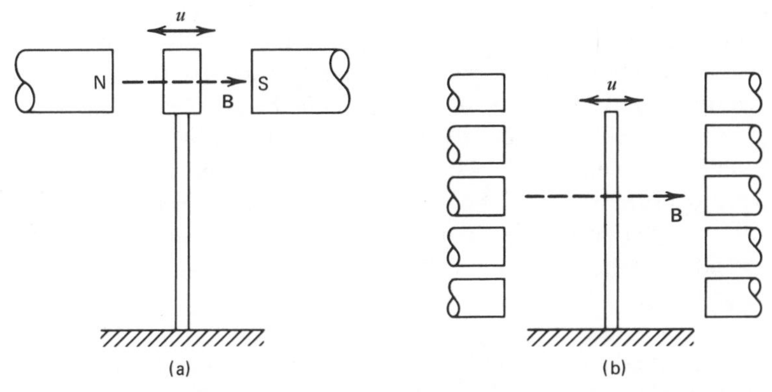

Figure 5-23 (*a*) Vibration of a ferromagnetic mass between poles of an electromagnet and (*b*) vibrations of a ferromagnetic beam between a series of magnetic poles. (From Moon, 1978, reprinted with permission. Pergamon Press, Inc., copyright 1978.)

This behavior is characteristic of such solutions as one approaches the bifurcation or buckling field. The buckling field is given by $\omega = 0$, or

$$B_{0c}^2 = \frac{G^3 \omega_0^2}{K} \tag{5-10.3}$$

Panovko and Gubanova (1965) extended the theoretical problem for a beam placed between a set of poles (see Fig. 5-23*b*) in which they postulated a negative magnetic spring force distribution along the beam leading to the equation

$$Y\mathcal{I} \frac{d^4 u}{dx^4} = \frac{K B_0^2}{G_0^3} u \tag{5-10.4}$$

For sinusoidal deformation of the form $u = A \sin(\pi x / L)$, this theory leads to a buckling magnetic field value of the form

$$B_c \sim \left(\frac{h}{L}\right)^2$$

where h is proportional to the beam or plate thickness. However, no experiments were reported by Panovko and Gubanova and subsequent experiments have shown that

$$B_c \sim \left(\frac{h}{L}\right)^{3/2} \tag{5-10.5}$$

which calls into question the validity of their assumptions.

Many analyses of this problem have been carried out in the last 12 years. Magnetoelastic buckling has been used as a way to test macroscopic theories of interaction between electromagnetic fields and deformable media.

Axiomatic theories of soft ferroelastic solids by Kaliski (1962) and Dunkin and Eringen (1963) when applied to a plate in a transverse magnetic field did not predict any static instability. Moon and Pao (1968, 1969) carried out early experiments on this problem, and used an *ad hoc* theory employing a dipole model for the magnetic forces on a beam-plate, which correctly predicts the critical field-length dependence, Eq. (5-10.5). Later, Kaliski (1969), Pao and Yeh (1973), Hutter and van der Ven (1978), and Maugin and Goudjo (1982) showed how axiomatic formulations of magnetic field interactions with elastic solids could correctly predict the observed buckling phenomenon. Ambartsumyan (1982) has given a review of similar work in the Soviet Union.

While continuum theories for electromagnetic solids must correctly predict magnetoelastic buckling, the phenomenon has been explained by more than one formulation in the high permeability limit. Dipole, Ampere stress, and pole model formulations have all been shown to lead to identical values for the critical buckling field in the limits as $\mu_r \to \infty$ in the linearized theory. In one sense it is encouraging that different logical approaches will yield equivalent results for a specific linearized problem. However, engineers are still faced with a welter of axiomatic theories if they want to solve more general linear and nonlinear problems in magnetoelasticity.

In the next section we present one analysis of the magnetoelastic plate buckling using the *pole model* for magnetic forces and linear elasticity.

Elastic Plate in a Transverse Magnetic Field

We consider an elastic soft ferromagnetic plate, bounded by planes $y = \pm h$, through which a constant external magnetic field $\mathbf{B}^0$ penetrates normal to the undeformed surface (see Fig. 5-24). We note that if the plate suffers a rigid-body rotation of angle Ω, a very large component of $\mathbf{M}$ will appear along the plate and transverse to $\mathbf{B}^0$. For a rotated rigid plate, this magnetization is given by

$$\mathbf{M} = \frac{\chi}{\mu_0} \mathbf{B}_0 \left(\frac{\cos \Omega}{\chi + 1} \mathbf{n} + \sin \Omega \mathbf{s} \right) \tag{5-10.6}$$

where χ is the magnetic susceptibility of the plate and $\mathbf{n}$ and $\mathbf{s}$ are unit vectors normal and tangential to the plate surface, respectively. Since the external field is uniform, that is, $\nabla \mathbf{B}^0 = 0$, there is no net force on the plate. From the dipole model, however, one can calculate a body couple $\mathbf{M} \times \mathbf{B}^0$ given by

$$\mathbf{M} \times \mathbf{B}^0 = \frac{B^2}{2\mu_0} \frac{\chi^2}{\chi + 1} \sin 2\Omega (\mathbf{s} \times \mathbf{n}) \tag{5-10.7}$$

This couple is destabilizing in that it tends to align the plate parallel to the external magnetic field. Mechanical constraints are thus necessary to hold the plate normal to the magnetic field. If the plate is elastic, however, there will exist a critical magnetic field at which the plate will attempt to rotate locally about the points of constraint. If a pole model is used, then one should also expect to obtain a net couple from the magnetic stresses on the surfaces of the plates.

The simplest problem one can address analytically is the case of periodic simple supports at $x_n = \pm nL$, where the displacement $\mathbf{u}$ is restricted to plane

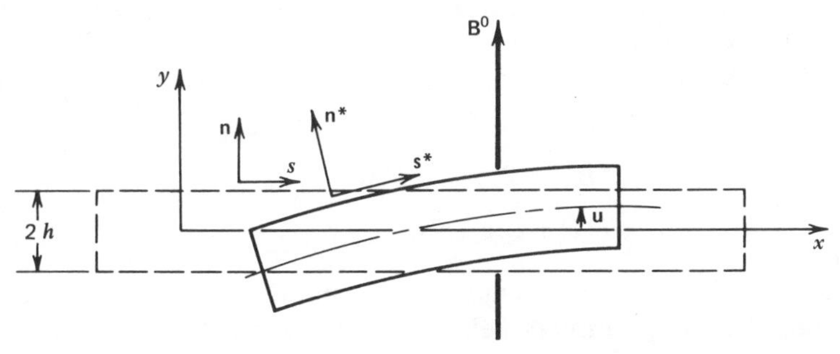

Figure 5-24 Change of surface normal and tangential vectors on a deformed plate.

strain and u_x is antisymmetric and u_y symmetric in the y direction; that is,

$$u_x(x, y) = -u_x(x, -y) \quad \text{and} \quad u_y(x, y) = u_y(x, -y) \tag{5-10.8}$$

(There is another mode of buckling called the "sausage" mode in which u_y is antisymmetric in y. This will not be considered here, but may be found in Moon, 1967.)

Guided by our observations on the rigid plate, we must somehow connect the local rotation of the plate surfaces to the change in induced magnetization. To accomplish this we borrow a trick from linearized wing theory in aerodynamics, and consider the magnetic boundary conditions applied to the *rotated*, but undisplaced, surfaces $y = \pm h$.

The magnetic field is written in terms of perturbed quantities:

$$\mathbf{B} = B_0\mathbf{e}_y + \mathbf{B}^1, \qquad \mathbf{H} = H_0\mathbf{e}_y + \mathbf{H}^1 \tag{5-10.9}$$

$$\mathbf{H}^1 = \nabla\phi$$

and

$$\nabla^2\phi = 0$$

The rotation of the top surface of the plate is given by

$$\boldsymbol{\omega} = \frac{\partial u_y}{\partial x}\mathbf{e}_x \times \mathbf{e}_y \tag{5-10.10}$$

and the rotated normal to the plate surface can be expressed in terms of $\boldsymbol{\omega}$ for small rotations; that is,

$$\mathbf{n}^* = \mathbf{n} + \boldsymbol{\omega} \times \boldsymbol{n} \tag{5-10.11}$$

where $\mathbf{n} = \mathbf{e}_y$.

When a gradient exists in the external field, the magnetic field has to be written as a Taylor series about the undisturbed surface position $\mathbf{r}_s$; for example,

$$\mathbf{B}^0(\mathbf{r}_s + \mathbf{u}) = \mathbf{B}^0(\mathbf{r}_s) + \mathbf{u} \cdot \nabla\mathbf{B}^0$$

This is not necessary in the uniform field problem, but in experiments the neglect of the field gradient term $\mathbf{u} \cdot \nabla\mathbf{B}^0$ may lead to differences between theory and experiment.

For the uniform field case, the magnetic boundary conditions on the rotated surfaces lead to

$$\begin{aligned} [B_y^1] &= 0, \qquad y = \pm h \\ [H_x^1 + \omega H_0] &= 0, \qquad y = \pm h \end{aligned} \tag{5-10.12}$$

For a sinusoidal deformation along the beam, these conditions lead to an expression for the perturbed field potential ϕ inside the plate. For a displacement field,

$$\begin{bmatrix} u_x \\ u_y \end{bmatrix} = \begin{bmatrix} U(y)\sin kx \\ V(y)\cos kx \end{bmatrix} \tag{5-10.13}$$

and

$$\phi^- - \frac{VB_0}{\mu_0\Delta}\cosh ky \cos kx$$

where $\Delta = (\chi+1)\sinh kh + \cosh kh$ and $k = n\pi/L$.

From ϕ^- we can obtain the magnetization in the plate and from $\mathbf{M}$ we can determine the forces on the plate when it bends. As we saw in Chapter 3, however, there are many ways to express the magnetic forces. In the axiomatic method developed by Pao and Yeh (1973) they used what amounted to a pole model (see Section 3.10). In the limit of high permeability, this theory is equivalent to classical elasticity with magnetic stresses on the boundaries representing forces between surface magnetic poles $\mathbf{M}\cdot\mathbf{n}$ and the magnetic field.

In this problem we neglect the self-forces between perturbed magnetized elements in different parts of the plate since they are nonlinear in the deformation. Thus we assume traction boundary conditions between *surface magnetic poles* and the *external magnetic field* $\mathbf{B}^0$. Following Eqs. (2-4.6), and (3-4.4), the stress vector on the plate surfaces is given by

$$\mathbf{t}\cdot\mathbf{n}^* = (\mathbf{M}\cdot\mathbf{n}^*)\mathbf{B}^0$$

where $\mathbf{n}^*$ is given by Eq. (5-10.11). To first order in the strains or displacements this condition takes the form (neglecting the Maxwell stresses in the undeformed plate)

$$t_{yy} = \chi B_0 \frac{\partial \phi^-}{\partial y}$$

and

$$t_{xy} = 0 \tag{5-10.14}$$

on the top and bottom surfaces of the plate, respectively.

We remind the reader that magnetic bódy forces (3-4.4) for a linear ferromagnetic body are identically zero; that is,

$$\mathbf{f} = (\nabla\cdot\mathbf{M})\mathbf{B}^0 = 0$$

since $\nabla\cdot\mathbf{M} = \chi\nabla^2\phi = 0$. Thus only magnetic surface tractions enter the solution and the displacement $\mathbf{u}$ satisfies the classical equilibrium equation of linear elasticity,

$$(\lambda+\mu)\nabla(\nabla\cdot\mathbf{u}) + \mu\nabla^2\mathbf{u} = 0 \tag{5-10.15}$$

where λ and μ are elastic constants [see Eq. (3-11.6)]. Substituting the displacement vector (5-10.13) into (5-10.15), one obtains expressions for $U(y)$ and $V(y)$:

$$\begin{bmatrix} U \\ V \end{bmatrix} = b_1 \begin{bmatrix} -\sinh ky \\ \cosh ky \end{bmatrix} + b_2 \begin{bmatrix} (3-4\nu)\sinh ky + ky \cosh ky \\ -ky \sinh ky \end{bmatrix} \tag{5-10.16}$$

From the stress–strain law for isotropic elastic materials, one can find the stresses; that is,

$$t_{yy} = \frac{Y}{(1+\nu)(1-2\nu)} \cos ky[(1-\nu)V'(y) + \nu kU(y)]$$
$$t_{xy} = G \sin ky[U'(y) - kV(y)] \tag{5-10.17}$$

where G is the shear modulus and ν is Poisson's ratio.

The boundary conditions (5-10.14) lead to two homogeneous equations for b_1 and b_2. The vanishing of the determinant of the coefficients of this set of homogeneous equations then yields a condition on $B_0^2/2u$, for a sinusoidal deformation to exist.

Alternatively, we can use $t_{xy} = 0$ to solve for b_2 in terms of b_1 and obtain an expression for t_{yy}. In the thin-plate limit, $2h/L \ll 1$, we can expand the

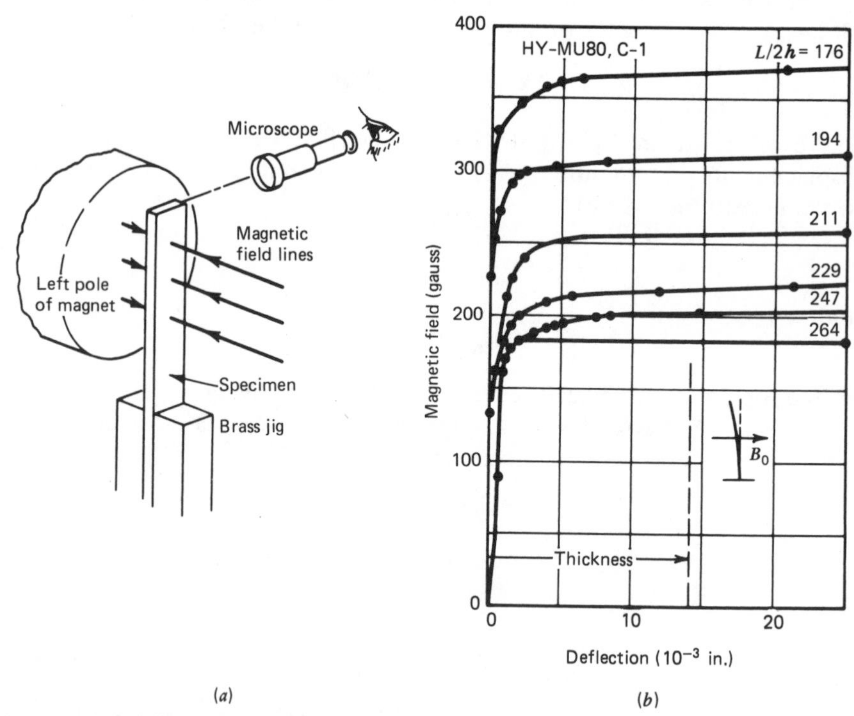

Figure 5-25 (a) Ferromagnetic plate in a transverse magnetic field and (b) experimental buckling curves for a ferroelastic plate in a transverse magnetic field (from Moon and Pao, 1969).

cosh kh and sinh kh in the parameter kh. In this limit the normal stress becomes

$$t_{yy} \simeq \frac{Y}{3(1-\nu^2)} b_1 k(kh)^3 \cos kx \tag{5-10.18}$$

Equating t_{yy} in Eq. (5-10.17) to the magnetic stress boundary condition (5-10.14) results in the critical value for B_0; that is,

$$\frac{B_0^2}{2\mu_0} = \frac{Y}{6(1-\nu^2)}(kh)^2 \frac{1+(\chi+1)kh}{\chi+1} \tag{5-10.19}$$

In experiments where $\mu_r = \chi + 1 = O(10^3)$ or larger, we can have $(\chi+1)kh \gg 1$. In this limit we find the *critical buckling field* to be

$$\frac{B_0^2}{2\mu_0} \simeq \frac{Y}{6(1-\nu^2)}(kh)^3 = \frac{Y\pi^3}{6(1-\nu^2)}\left(\frac{h}{L}\right)^3 \tag{5-10.20}$$

Typical experimental buckling curves of field versus deformation are shown in Figure 5-25 for a cantilevered beam-plate. Critical field versus $2h/L$ for theory and experiment are shown in Figure 5-26. A number of

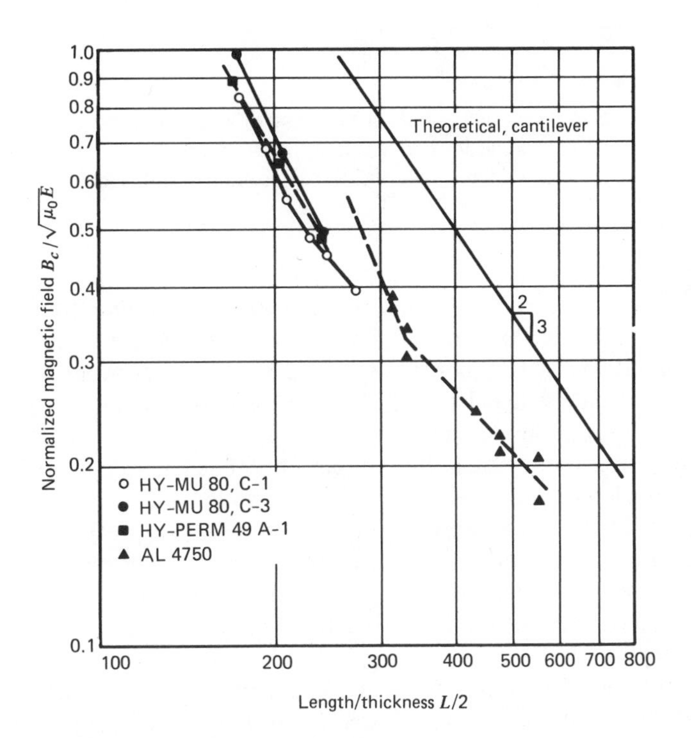

Figure 5-26 Critical magnetic field for buckling of high-permeability ferromagnetic beam-plates vs. length-to-thickness ratio (from Moon and Pao, 1969).

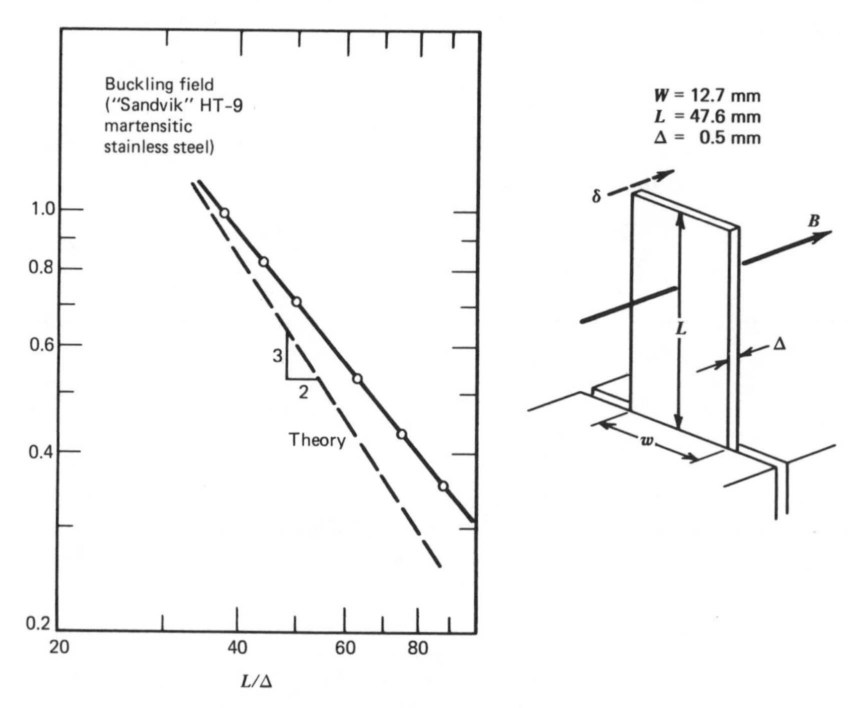

Figure 5-27 Critical magnetic field for buckling of a low-permeability stainless-steel plate in a transverse magnetic field vs. length-to-thickness ratio. (From Moon and Hara, 1982, reprinted with permission. North-Holland Publ. Co., copyright 1982.)

mechanicians have reported this phenomenon, including Moon and Pao (1968), Yeh (1971), Popelar and Bast (1972), Dalrymple et al. (1974), and Miya et al. (1978). All studies show the 3/2-power law between B_0 and $2h/L$. However, all data fall below the theoretical value by up to a factor of 2. Different explanations for the difference between theory and experiment have been attempted, including imperfection-sensitivity (Popelar, 1972), edge effects due to finite width (Wallerstein and Peach, 1972), electromagnet pole effects (Yeh, 1971), and magnetic hysteresis (Miya et al. 1978).

Recently, experiments on low μ_r materials ($\mu_r < 100$), such as martensitic stainless steel by Moon and Hara (1982), have shown good agreement between the theoretical buckling field [Eq. (5-10.19)] and experimental values (see Fig. 5-27).

The problem of a circular plate in a transverse field has been solved by the author (1970a). In this solution it was found that the critical field is sensitive to small departures of the field from the normal direction.

5.11 CIRCULAR FERROELASTIC ROD IN A TRANSVERSE MAGNETIC FIELD

The plate in a magnetic field is similar to a circular rod in a uniform transverse magnetic field (see Moon, 1978). In this problem, however, we avoid the problem of edge effects, which could not be handled in the plate solution. But, in the case of the rod we can have bending deformation parallel or transverse to the magnetic field (see Fig. 5-28). It has been observed experimentally that motion parallel to the field suffers a decrease in natural frequency, while vibration transverse to the field appears to increase the frequency of the lowest bending mode of a cantilevered rod (see Fig. 5-29).

For a theoretical analysis of motion *parallel* to $\mathbf{B}^0$, we use beam theory with a magnetic body couple per unit length given by

$$\mathbf{C} = C\mathbf{e}_x = \mu_0 \int_0^a \int_0^{2\pi} \mathbf{M} \times \mathbf{H}_0 \, r dr d\theta \tag{5-11.1}$$

The displacement of the axis of the rod is assumed to be of the form

$$\mathbf{u} = u\mathbf{e}_y = u_0 e^{ikz}(\cos\theta\mathbf{e}_\theta + \sin\theta\mathbf{e}_r)$$

and assumed to satisfy the classical Bernouli–Euler equation with added body couple; that is,

$$Y\mathcal{I}\frac{\partial^4 u}{\partial z^4} - \frac{\partial C}{\partial z} + m\frac{\partial^2 u}{\partial t^2} = 0 \tag{5-11.2}$$

where m is the mass per unit length and $Y\mathcal{I}$ is the flexural stiffness. The perturbed magnetic field is again assumed to be expressed in terms of a magnetic potential ϕ. The boundary conditions on ϕ, applied on the deformed surface, are related to the undeformed surface, $r = a$, by expanding the fields in a Taylor series.

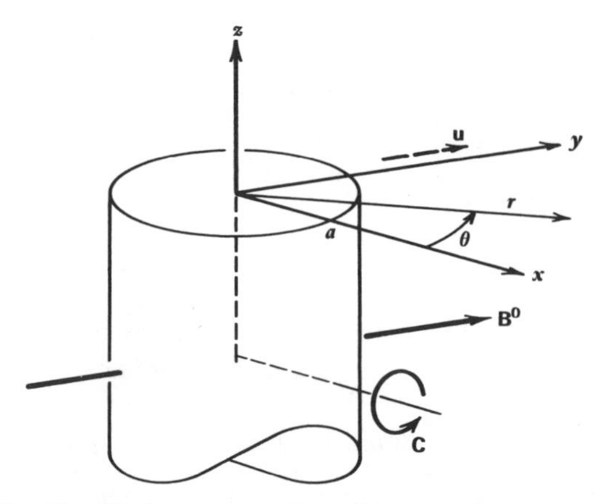

Figure 5-28 Circular ferromagnetic rod in a transverse magnetic field.

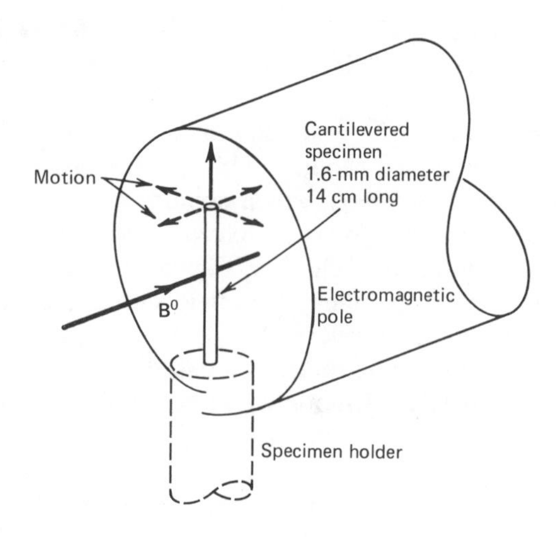

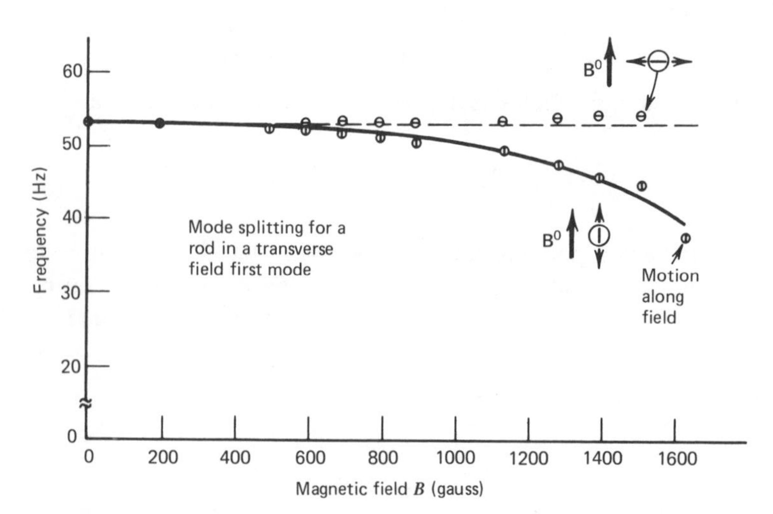

Figure 5-29 Natural frequency of a ferroelastic rod of circular cross section in a transverse magnetic field. (From Moon, 1978, reprinted with permission. Pergamon Press, Inc., copyright 1978.)

While time dependence is retained in the equation of motion, the magnetic problem is considered to be quasistatic and induced eddy currents are neglected. The magnetic potential inside the rod can be shown to be expressed in terms of the modified Bessel functions $I_0(kr)$ and $I_2(kr)$:

$$\phi = A_1 I_0(kr) + A_2 I_2(kr) \cos 2\theta \qquad (5\text{-}11.3)$$

This leads to an expression for the couple C,

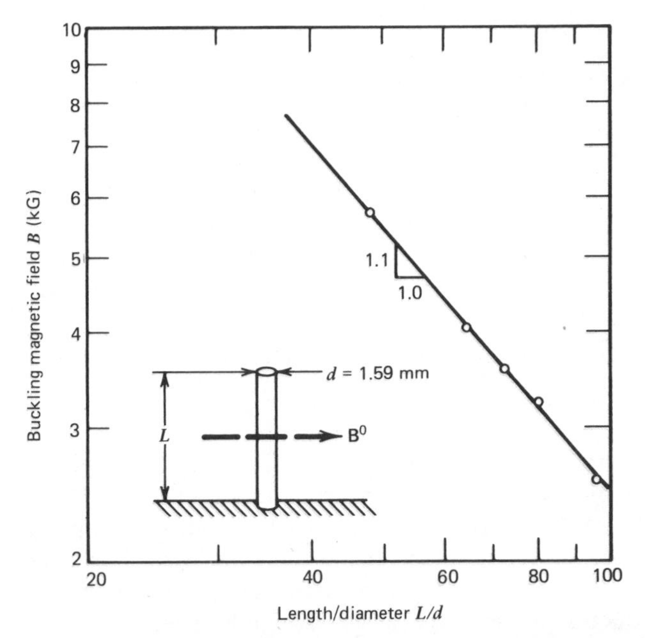

Figure 5-30 Critical buckling field vs. length-to-diameter ratio for a cold rolled steel rod in a transverse magnetic field. (From Moon, 1978, reprinted with permission. Pergamon Press, Inc., copyright 1978.)

$$C = -2\pi a\chi B_0 e^{ikz} A_1 i I_1(ka) \tag{5-11.5}$$

A_1 is determined from the magnetic boundary conditions which, in turn, depend on the displacement of the rod. For a periodically pinned rod, with $k = \pi/L$, the natural frequency has the form

$$\omega^2 = \omega_0^2\left(1 - \frac{B^2}{B_c^2}\right)$$

where the buckling magnetic field is given, for $ka \ll 1$, by the expression

$$\frac{B_c^2}{2\mu_0 Y} = \frac{\mu_r + 1}{\chi^2}\frac{(ka)^2}{2}[1 + \mu_r(ka)^2|\ln(ka)|]$$

Experimental values of B_c versus $L/2a$ are shown in Figure 5-30 for a cantilevered, cold rolled steel rod.

5.12 STABILITY OF CURRENT-CARRYING RODS

A classic instability in plasma physics occurs when an initially straight high current arc suffers a "kink" or bend above a critical value of the current (see

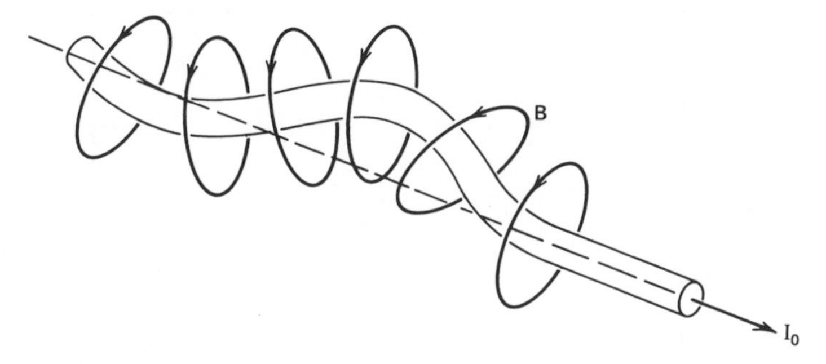

Figure 5-31 Kink instability in a flexible current-carrying conductor.

Fig. 5-31). One is naturally led to ask if a similar instability can occur for an elastic conductor. Of course, the restraining forces in a solid conductor are much stronger than those in a gaseous conductor so that much higher currents would be required to buckle the solid wire. However, for superconducting wire or transient short circuit conditions, it is possible for an elastic conductor to buckle for large-enough currents.

To estimate the current necessary to deform an otherwise straight conductor under its own self-field, one must first determine the destabilizing magnetic forces as a function of the deformation of the filament. Next, these forces must be equilibrated with the elastic constraining forces under the given geometric constraints or support conditions.

An elementary calculation of the lateral forces on an initially straight, current-carrying filament of cross-section radius a, when it is displaced laterally in a sinusoidal displacement $w(x) = w_0 \sin kx$, gives a *force proportional to the curvature* of the deformed filament (see, e.g., Thompson, 1962):

$$F \sim \frac{\mu_0 I^2}{4\pi} \frac{\partial^2 w}{\partial x^2} \ln(ka) + O((ka)^2) \tag{5-12.1}$$

One of the first analyses of this problem for an elastic conductor was given by Leontovich and Shafronov (1961), who looked at the combined effect of the self-field of the current as well as of a longitudinal magnetic field. Dolbin (1962) extended this analysis to the case of an elastic cylinder with wavelike deformations. The particular case of bending deformation in a current-carrying rod was treated by Dolbin and Morozov (1966). They concluded that a periodically supported rod could buckle under high-enough static current when no tension forces or other constraints were placed on the rod.

More recently, Chattopadhyay and Moon (1975) reexamined this problem and established experimental evidence for this instability. With superconductor applications in mind, they assumed a uniform distribution of longitudinal current across the cross section of the rod. Elastic vibrations were included in the analysis, but the magnetic field and forces were

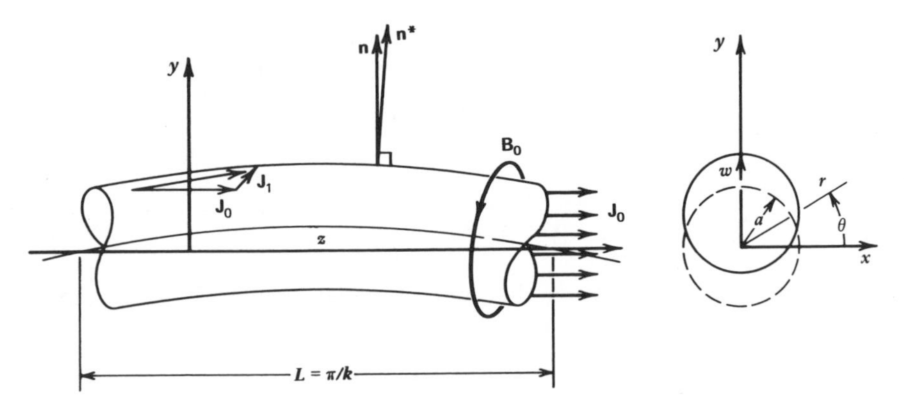

Figure 5-32 Element of a bent circular wire carrying electric current.

calculated using magnetostatic theory, which neglects radiation and eddy-current damping.

A sketch of the geometry of this problem is shown in Figure 5-32. The boundary conditions used required the current to flow tangential to the surface of the deformed rod. Thus the inclusion of the rotation of the surface normal is an important key to solving the problem. The bending vibrations of the rod must satisfy the balance-of-momentum equation for a differential element of the rod

$$\rho A \frac{\partial^2 w}{\partial t^2} + Y\mathscr{I} \frac{\partial^4 w}{\partial x^4} - T \frac{\partial^2 w}{\partial x^2} = F_y + \frac{\partial C_x}{\partial z} \tag{5-12.2}$$

Here F_y is the lateral magnetic force on the rod per unit length, C_x is the magnetic body couple per unit length, and T is either a prescribed tension in the rod or tension induced by magnetic forces when the conductor is part of a closed circuit.

The transverse body force is calculated from

$$\mathbf{F} = \iint \mathbf{J} \times \mathbf{B}\, da \tag{5-12.3}$$

The body couple is calculated from the integral

$$\mathbf{C} = \iint \mathbf{r} \times (\mathbf{J} \times \mathbf{B})\, da \tag{5-12.4}$$

where $\mathbf{r}$ is a vector in the plane of the circular cross section with origin at the center.

Without reproducing the original paper, we will sketch the method of solution as used by Chattopadhyay and Moon. The following is a little technical. Readers only interested in the solution should skip to Eq. (5-12.18).

If $\mathbf{n}^*$ is the outward normal to the deformed rod, the boundary condition on the surface of the rod becomes

$$\mathbf{J}\cdot\mathbf{n}^* = 0, \qquad r = a \tag{5-12.5}$$

where

$$\mathbf{n}^* = \mathbf{e}_r - \sin\theta \frac{\partial w}{\partial z}\mathbf{e}_z \tag{5-12.6}$$

The total current density is the sum of initial current $\mathbf{J}_0 = J_0\mathbf{e}_z$ and a perturbed current $\mathbf{J}_1 = \sigma\mathbf{E}$ where

$$\mathbf{E} = \nabla\phi, \qquad \nabla^2\phi = 0 \tag{5-12.7}$$

The boundary condition (5-12.5) then takes the form

$$\frac{\partial\phi^-}{\partial r} = \frac{J_0}{\sigma}\frac{\partial w}{\partial z}\sin\theta \tag{5-12.8}$$

where ϕ^- is the potential in the interior of the rod.

The other boundary condition requires $\mathbf{E}\times\mathbf{n}^*$ to be continuous across the rod. A solution which satisfies both Laplace's equation and the boundary conditions for a sinusoidal deformation $w = w_0 e^{ikz}$, when the rod has initial uniform current density J_0, is

$$\phi^- = \frac{iJ_0 w_0}{\sigma I_1'(ka)} I_1(kr)(\sin\theta)e^{ikz}, \qquad r < a \tag{5-12.9}$$

where I_1 is a modified Bessel function of the first kind and the prime I_1' indicates a derivative.

Once the perturbed current density is known, the perturbed magnetic field $\mathbf{B}$ and the forces (5-12.3) and (5-12.4) can be calculated. The magnetic field can be found using a vector potential $\mathbf{B} = \nabla\times\mathbf{A}$ which guarantees that $\nabla\cdot\mathbf{B} = 0$.

The vector potential, in turn, is directly related to $\mathbf{J}$ via Maxwell's equations; that is,

$$\nabla^2\mathbf{A} = -\mu_0\mathbf{J} = -\mu_0\sigma\nabla\phi^-, \qquad r < a \tag{5-12.10}$$

and

$$\nabla^2\mathbf{A} = 0, \qquad r > a$$

along the gauge condition $\nabla\cdot\mathbf{A} = 0$. When ϕ^- is given, a particular solution for $\mathbf{A}$ can be found which is added to the solutions of the homogeneous equation $\nabla^2\mathbf{A} = 0$. Thus we need boundary conditions on $\mathbf{B}$ at the surface. In the absence of surface currents, $\mathbf{B}$ must be continuous. However, this condition must be applied at the *deformed* surface using a Taylor-series expansion for $\mathbf{B}$, that is,

$$\mathbf{B}(\mathbf{r}_s + \mathbf{w}) \simeq \mathbf{B}(\mathbf{r}_s) + (\mathbf{w}\cdot\nabla)\mathbf{B}(\mathbf{r}_s) \tag{5-12.11}$$

This is applied to the solution for inside the rod, $\mathbf{B}^-$ and outside the rod, $\mathbf{B}^+$. The boundary conditions for the perturbed magnetic field $\mathbf{B}_1$ become (to terms linear in w)

$$[B_{1r}] = 0, \qquad [B_{1z}] = 0 \tag{5-12.12}$$

and

$$[B_{1\theta}] = \mu_0 J_0 w \sin\theta$$

where

$$[B] = B^+(a) - B^-(a)$$

(See Moon and Chattopadhyay (1975), for further details.)

The calculation of the perturbed force on the conductor due to its own field has some interesting subtleties and it is instructive to consider this calculation in some detail. First, we note that when bending is absent an initial radially compressive body force $\mathbf{J}_0 \times \mathbf{B}_0$ acts on the rod producing zero net force. When the rod is bent, the perturbed force is found by integrating $\mathbf{J} \times \mathbf{B}$ over the *deformed volume* element.

To transform an integration over a deformed element to one over the undeformed element, we relate the spacial position coordinates x,y,z to the material coordinates X, Y, Z by the equations

$$x = X, \qquad y = w + Y, \qquad z = Z - Y\frac{\partial w}{\partial Z} \tag{5-12.13}$$

The differential element length dz is related to the undeformed length dZ by the curvature, which for small slope is given by

$$dZ = \left(1 + Y\frac{\partial^2 w}{\partial Z^2}\right) dz \tag{5-12.14}$$

Writing the current density and magnetic field in terms of initial and perturbed variables, the body force has the following form (to terms linear in w):

$$\begin{aligned} &\mathbf{f} = \mathbf{f}_0 + \mathbf{f}_1 \\ &\mathbf{f}_0 = \mathbf{J}_0 \times \mathbf{B}_0, \qquad \mathbf{f}_1 = \mathbf{J}_1 \times \mathbf{B}_0 + \mathbf{J}_0 \times \mathbf{B}_1 \end{aligned} \tag{5-12.15}$$

For linear terms only, we integrate $\mathbf{f}_1$ over the undeformed volume element. However, we must integrate $\mathbf{f}_0$ over the deformed element by expanding $\mathbf{f}_0$ in the material coordinates; that is,

$$\int_{\text{deformed}} f_{0y}\, dz\, da = \int_{A_0} \left(f_{0y}(Y) + w\frac{\partial f_{0y}}{\partial Y}\right)\left(1 - \frac{\partial^2 w}{\partial Z^2}\, Y\right) dA\, dZ \tag{5-12.16}$$

Integration of f_{0y} over the undeformed area is zero since

$$f_{0y} = -\frac{\mu_0 I_0^2}{2\pi^2 a^4}\, Y$$

This leads to the expression

$$\int_{\text{deformed}} f_{0y} da = \frac{\mu_0 I^2}{2\pi a^2}\left(\frac{a^2}{4}\frac{\partial^2 w}{\partial Z^2} - w\right) \tag{5-12.17}$$

The integration of f_{1y} involves modified Bessel functions, where sinusoidal deformation is assumed of wavelength $\lambda = 2\pi/k$ and $ka \ll 1$. Under these assumptions, one can show that

$$F_y = \frac{\mu_0 I^2}{4\pi} wk^2\left(\ln\frac{2}{ka} - \Gamma - \frac{1}{4}\right) \tag{5-12.18}$$

where $\Gamma = 0.577$ is Euler's constant. This is similar in form to Eq. (5-12.1) used in plasma physics.

Under identical assumptions, the couple per unit length due to the moment of the body force distribution about the centroid of the cross section can be shown to be

$$C_x \simeq \frac{\mu_0 I^2}{8\pi} ikw \tag{5-12.19}$$

This couple is stabilizing since it acts to decrease the slope $\partial w/\partial z = ikw$.

The expressions for body force and couple per unit length, (5-12.18) and (5-12.19), respectively, when substituted into the equation of motion (5-12.2) yield a dispersion relation of the form

$$\left(\frac{\omega}{\omega_0}\right)^2 = 1 - \frac{I^2}{I_c^2} \tag{5-12.20}$$

where

$$\mu_0 I_c^2 = 4\pi(ka)^2 \frac{Y\mathcal{I}}{a^2}\left(\ln\frac{2}{ka} - 1.327\right)^{-1} \tag{5-12.21}$$

which has the same form as found in Dolbin and Morozov (1966) except for the constant in the denominator. The relation (5-12.21) neglects any magnetically induced tension in the conductor which might stabilize the rod [see Eq. (5-12.2)].

Example: A Magnetic Levitation Coil

Vehicles have been designed using superconducting magnets to levitate above a conducting guideway (see, e.g., Thornton, 1973). The conductors in these magnets must be cooled to liquid helium. Hence the number of support connections from the cold conductor to the room-temperature magnet casing must be minimized to avoid excessive helium-gas boil-off. Fewer supports, however, might lead to buckling.

Consider, for example, a long rectangular coil carrying 5×10^5 A. Neglecting the effect of the other three sides, we investigate the minimum support length to avoid a buckling instability for a cross-section radius of

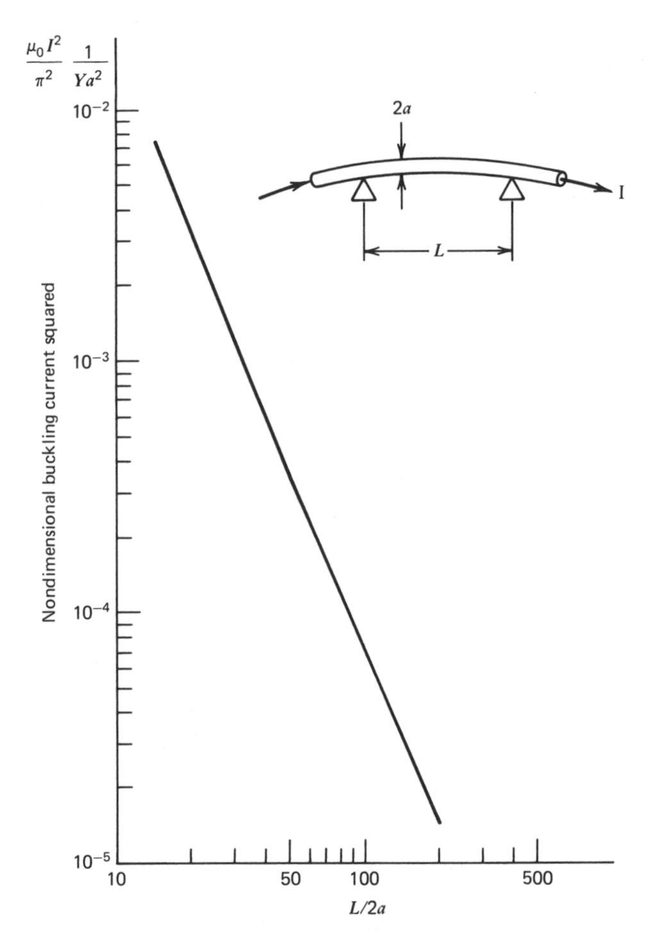

Figure 5-33 Buckling current vs. length-to-diameter ratio for a circular current-carrying rod [see Eq. (5-12.21)].

$a = 2$ cm and modulus $Y = 10^{11}$ N/m^2. These conditions when substituted into the expression (5-12.21) give a minimum length of 1.4 m between supports. A graph of buckling current versus support length is shown in Figure 5-33.

Experimental confirmation of the decrease of natural frequency with current due to destabilizing self-magnetic forces has been obtained by Chattopadhyay and Moon (1975). Actual dynamic buckling of a filament under short circuit currents of 20 kA has been photographed by Abramova et al. (1966). However, it is highly probable that thermoelastic effects due to Joule heating were present in these experiments.

6

Mechanics of Superconducting Solids and Structures

6.1 INTRODUCTION

An increase by orders of magnitude in the ability to store or transmit energy or information has often been a prelude to a new technological age. Joseph Henry's discovery of the multiturn electromagnet increased the ability to lift objects with magnetic forces from ounces to tons and helped open a new era in the use of magnetic forces to replace human, animal, and water power. The discovery of superconductivity by Kamerlingh Onnes in 1911 promised another such revolution in magnetomechanics. But this discovery did not lead to its full potential until the 1960s when superconducting materials were discovered which could sustain high current densities and magnetic fields without becoming resistive or *normal*.

Although over 25 superconducting elements and hundreds of superconducting compounds have been discovered, most can carry only small current densities before becoming resistive. However, a class of intermetallic compounds with a crystal structure called A-15 (see, e.g., Brechna, 1973) was discovered in the early 1960s to have the property of very high current densities in high magnetic fields. The most common commercially available A-15 compound is Nb_3S_n.

This revolution in magnetic energy storage capacity can be measured by comparing the power required to sustain a given amount of stored energy by a resistive, copper-wound magnet with one made of superconducting wire. For example, a 1.25-m-long copper-wound solenoid with a 0.18-m diameter, built for a plasma physics experiment in the 1960s, stored 41 kJ of magnetic energy with a power input of 1.5 MW, or about 37 kW per kilojoule stored. In the 1980s superconducting magnets have been built which store over 10^2 MJ with a cryogenic cooling loss of 10^2 kW at room temperature, or less than 10^{-3} kW per kilojoule stored energy. This remarkable improvement in magnetic energy storage has led to the development of a number of superconducting devices and applications, including electric-utility energy storage devices; magnetohydrodynamics (MHD) magnets; superconducting transmission lines; large magnets for fusion reactors, superconducting generators, and motors; high-speed levitated vehicles using superconducting magnets; and nuclear-magnetic-resonance (NMR) whole body magnets for medical applications.

This new technology brings with it a new set of engineering problems in both the electrical and mechanical sciences. Many of these problems are of an interdisciplinary nature. Some involve the design and manufacture of the superconducting material itself, while others involve magnet construction or over-all system problems. In this chapter we will try to review a number of engineering-related problems involving the principles of mechanics.

First, we present a brief review of superconducting phenomena and physics and describe some of the physical properties of commercially available superconducting material. The application of composite materials theory to superconducting solids will be reviewed. A more detailed discussion will be given on the subject of superconducting structures, including stress analysis, structural stability, and dynamics. Finally, we will discuss the relation of the virial theorem to large-scale superconducting magnets.

It should be noted that there are several very good books available on the subject of superconductivity. A book by Williams (1970) and another by Tinkham (1975) deal with the physics. Books related to the engineering aspects of superconducting magnets include those by Brechna (1973), Montgomery (1980), and Thome and Tarrh (1982).

6.2 THE PHENOMENA OF SUPERCONDUCTIVITY

Below a critical cryogenic temperature (less than 22°K), some pure metals, alloys, and compounds exhibit a marked change in their physical properties; chief among these changes is the loss of resistance to the flow of electrical current. Above this temperature, the metal is called *normal*, and below this temperature it is called *superconductive* or *superconducting*, though for some materials further distinctions in the superconducting state can be made. In the superconducting state, currents in closed superconducting circuits can flow indefinitely without applied voltage. While zero resistance (or infinite conductivity) is the most notable property of superconducting materials, it is by no means the only important property with regard to magnetomechanical devices.

A list of the major macroscopic properties of superconductors includes the following:

1. Zero resistance.
2. Flux exclusion—diamagnetism.
3. Flux quantization.
4. Type-I and -II materials—vortex structure.
5. Change in thermal properties.
6. Strain dependence of critical current.

Zero Resistance

The salient properties of superconductivity only occur below certain critical temperatures T_c depending on the particular material. However, the state can become normal if the value of the transverse magnetic field becomes too high. Further, transport current can create a transverse field. Thus a limiting value of J exists, for a given temperature and field, above which the material becomes normal. These properties are illustrated in Figure 6-1.

For a superconducting wirelike conductor carrying current in a transverse magnetic field, the material will become normal if the values of T, B, and J

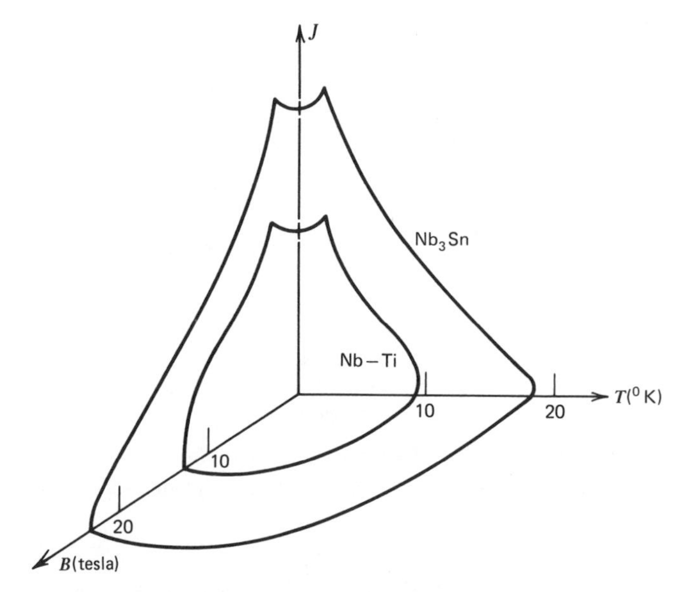

Figure 6-1 Superconducting–normal-conducting transition surface for Nb_3Sn and Nb-Ti.

do not lie in a corner of the space† of (T, B, J) where $T > 0$. If any of the three variables put the state out of this corner, the material will become resistive. Typical values of T, B, and J on the critical surface are shown in Table B-4 for a number of materials. This critical surface is similar to the yield surface in plasticity in solid mechanics.

When the transport current is small, the dependence of the critical magnetic field on the temperature is given by the relation

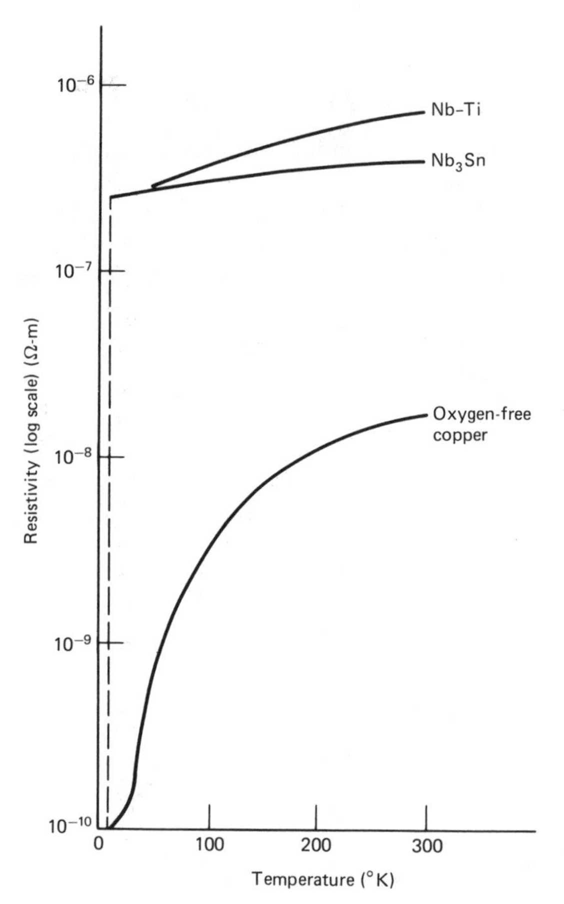

Figure 6-2 Electrical resistivity of superconducting wires and oxygen-free copper. (Adapted from Brechna, 1973, with permission of Springer-Verlag, Heidelberg, copyright 1973.)

†For the more mathematically inclined, we note that the triad (T, B, J) do not form a proper vector space, since $T < 0$ is excluded. Also **B** and **J** are vectors that are related through Maxwell's equations. Nonetheless, 6-1 is useful for describing the limitations in the superconducting state.

$$B_c = B_{0c}\left(1 - \frac{T^2}{T_c^2}\right) \tag{6-2.1}$$

For example, for Nb-Ti at $T = 4.2°K$ (liquid helium), $B_c / B_{0c} = 0.82$. If conductor motion in the magnet produces friction and heating which raises the temperature by 2°K, then $B_c / B_{0c} = 0.62$. If this value were below the design field in the magnet, the conductor would become locally normal.

One of the paradoxes of superconductivity is the fact that these materials are generally more resistive in their normal state than normal conductors, such as copper or aluminum, as illustrated in Figure 6-2.

Flux Exclusion

One of the principal properties of superconductors is their ability to screen out magnetic flux from the interior of the conductor. This is illustrated in Figure 6-3 for a cylindrical conductor in a transverse magnetic field. Above the critical temperature T_c, the field penetrates the boundary of the conductor. Below T_c, and for small values of the magnetic field, the flux will be excluded from all but a thin layer near the conductor surface. For low-enough values of the magnetic field, this exclusion is complete and is called the *Meissner effect*. The screening is accomplished by persistent currents that

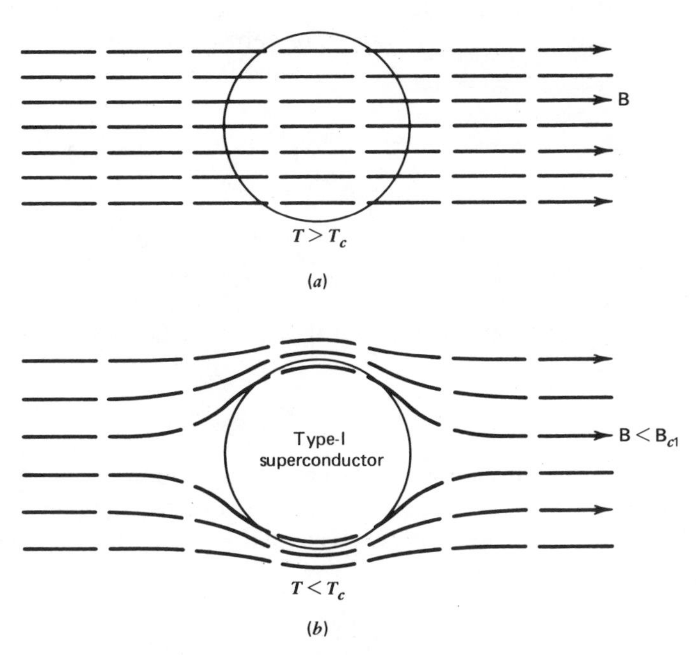

Figure 6-3 Meissner effect: (a) superconductor in magnetic field above critical temperature and (b) superconductor in magnetic field below critical temperature.

circulate near the boundary of the superconductor. Exclusion of flux from a material is often called *diamagnetism*, as contrasted with paramagnetism or ferromagnetism in which flux lines are attracted to the material. In effect, we can think of the material as having a negative magnetic susceptibility. (Suppose $\mathbf{B} = \mu_0(\mathbf{H} + \mathbf{M}) = 0$ and $\mathbf{M} = \chi\mathbf{H}$, then $\chi = -1$.) In an external field $\mathbf{B}^0$ a diamagnetic body can experience a force and couple given by

$$\mathbf{F} = \mathbf{m} \cdot \nabla\mathbf{B}^0$$

and (6-2.2)

$$\mathbf{C} = \mathbf{m} \times \boldsymbol{B}^0$$

where for a uniformly magnetized body $\mathbf{m} = \mathbf{M}V$ and V is the volume of the diamagnetic body. Physically, the equivalent force on a superconducting body results from the Lorentz force $\mathbf{J} \times \mathbf{B}^0$ on the circulating screening currents.

One explanation for the exclusion of flux from a superconductor is based on the semiclassical theory of electrical conduction in solids which uses a Newtonian model for a localized electron carrying charge $-e$, and possessing mass m; that is,

$$m\frac{d\mathbf{v}}{dt} + b\mathbf{v} = -e(\mathbf{E} + \mathbf{v} \times \mathbf{B}) \tag{6-2.3}$$

where the current density is defined by $\mathbf{J} = -ne\mathbf{v}$ and n is the number of conduction electrons per unit volume. For zero resistance, $b = 0$, and for a small magnetic effect, Eq. (6-2.3) assumes the form

$$\frac{d\mathbf{J}}{dt} = \frac{ne^2}{m}\mathbf{E} \tag{6-2.4}$$

This represents a constitutive law for $\mathbf{J}$ in superconducting materials and replaces Ohm's law $\mathbf{J} = \sigma\mathbf{E}$, for normal conductors. When Eq. (6-2.4) is combined with Maxwell's equations (2-2.1)–(2-2.4) (where displacement currents are neglected), an equation for the magnetic field in a superconductor is obtained:

$$\nabla^2\dot{\mathbf{B}} = \lambda^{-2}\dot{\mathbf{B}} \tag{6-2.5}$$

where $\lambda^2 = m/ne^2\mu_0$. Equations (6-2.4) and (6-2.5) form the basis of one of the early theories of superconductivity proposed by the London brothers in 1935. They tried to explain the Meissner effect using a one-dimensional solution of Eq. (6-2.5), that is,

$$B = B_0 e^{-x/\lambda}$$

which predicts that flux will be confined to a thin layer near the surface of a superconductor.

Flux Quantization

Another of the fundamental properties of the superconducting state is the fact that the flux contained in a closed circuit of current is an integral number of flux quanta. The smallest flux that can penetrate a circuit is $\Phi_0 = 2.07 \times 10^{-15}$ Wb. This value is given by the ratio of Planck's constant and the electron charge, that is, $\Phi_0 = h/2e$. This property is a macroscopic manifestation of quantum mechanics. One of the consequences of flux quanta is that magnetic flux can only penetrate a superconductor in discrete flux bundles called *fluxoids*. One important application of this property is a very sensitive magnetic-flux measuring device called a "SQUID," an acronym for "superconducting quantum interference device" (see Section 9-6).

Derivation of flux quantization can be found in any number of reference texts on superconductivity. However, a heuristic mechanics-based derivation is given below. In this model, the circulating current is treated as a collection of moving charges in an electromagnetic field. The classical equation of motion for one such charge is given by

$$q(\mathbf{E} + \mathbf{v} \times \mathbf{B}) = m\frac{d\mathbf{v}}{dt} \tag{6-2.6}$$

where q and m are the charge and mass of the electron, respectively, and $\mathbf{v}$ is the velocity. In Section 8.3 we show that $\mathbf{E}$ and $\mathbf{B}$ are related to scalar and vector potentials ψ and $\mathbf{A}$, respectively, by

$$\mathbf{B} = \nabla \times \mathbf{A}, \qquad \mathbf{E} + \frac{\partial \mathbf{A}}{\partial t} = -\nabla\psi \tag{6-2.7}$$

The equation of motion in terms of ψ and $\mathbf{A}$ is given by

$$q\left(-\nabla\psi + \nabla(\mathbf{v} \cdot \mathbf{A}) - \frac{d\mathbf{A}}{dt}\right) = m\frac{d\mathbf{v}}{dt} \tag{6-2.8}$$

This equation can be written in a form resembling Lagrange's equation (see Chapter 3), provided we define a Lagrangian by

$$\mathcal{L} = \mathcal{T} - q\psi + q(\mathbf{v} \cdot \mathbf{A}) \tag{6-2.9}$$

where $\mathcal{T}$ is the kinetic energy of the charge. The generalized momentum is then given by

$$\mathbf{p} = \frac{\partial \mathcal{L}}{\partial \mathbf{v}} = m\mathbf{v} + q\mathbf{A} \tag{6-2.10}$$

where the added term $q\mathbf{A}$ accounts for the energy stored in the magnetic field.

Now imagine a closed superconducting circuit with current density $\mathbf{J} = nq\mathbf{v}$. The flux threading a circular circuit is given by

$$\Phi = \int \mathbf{A} \cdot d\mathbf{l} = 2\pi r A_0 \tag{6-2.11}$$

Thus the momentum takes the form

$$p_\theta = \frac{mJ}{qn} + q\frac{\Phi}{2\pi r} \tag{6-2.12}$$

From elementary quantum mechanics the momentum is related to a probability wave function which has wavelength λ, so that $p_\theta = h/\lambda$ where h is Planck's constant. For a closed orbit, λ must be a fraction of the orbit circumference, that is, $\lambda = 2\pi r/N$ where N is an integer. To simplify the problem, we assume that all the current flows on the surface of the closed circuit. Thus, inside $J = 0$, and

$$\frac{hN}{2\pi r} = q\frac{\Phi}{2\pi r}$$

or

$$\Phi = \frac{Nh}{q} \tag{6-2.13}$$

It has been observed that superconduction electrons travel in pairs, so that $q = -2e$, or $\Phi = Nh/2e$. Thus the flux can only take on integral values of the quantity $\Phi_0 = h/2e$, as mentioned earlier.

Type-I and -II Superconductors—Vortex Structure

Superconducting materials are classified as either Type I or Type II (sometimes *soft* or *hard* is used). The Type-I materials are often the pure metals and have low values of the critical magnetic field and critical current. Within the superconducting state, they exhibit the property of complete flux exclusion.

Type-II materials are generally alloys or compounds, such as Nb-Ti or Nb_3S_n, which are able to carry very high current density in high transverse magnetic fields without becoming normal. At low magnetic fields, Type-II materials will exhibit perfect diamagnetism. However, another state exists in which the flux penetrates the material in clusters of flux lines. A schematic of this effect is shown in Figure 6-4 for a thin-film superconducting material. In this case the screening currents circulate around each flux bundle like small vortices. The center of each vortex is normal, while the region of zero to low field is superconducting.

There exists then two values of the critical magnetic field **H**. Below H_{c_1} the material behaves as a Type-I material with complete flux exclusion. Above H_{c_1} and below H_{c_2} the flux can penetrate the material creating normal and superconducting regions. This is illustrated in Figure 6-5, and values of H_{c_1} and H_{c_2} are tabulated in Table B-4 for a few materials. For example, the lower critical field for Nb_3S_n is 0.023 T (230 G), while the upper critical field is 23 T. Thus, for practical applications, flux penetrates a Type-II superconductor in bundles of flux lines. This means that for a Type-II material,

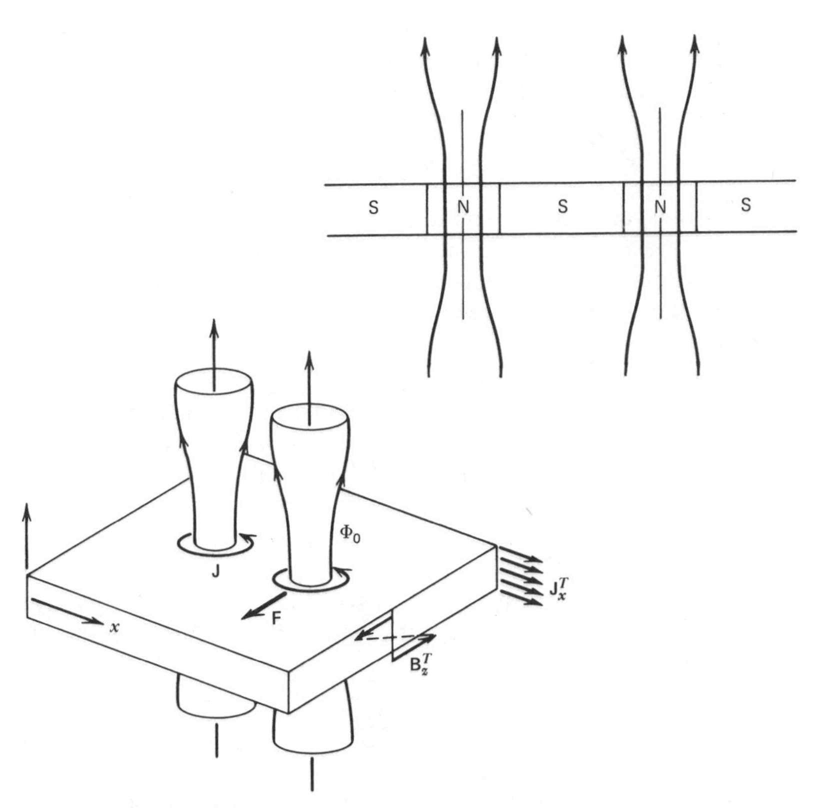

Figure 6-4 Flux bundles and current vortices in a Type-II superconductor.

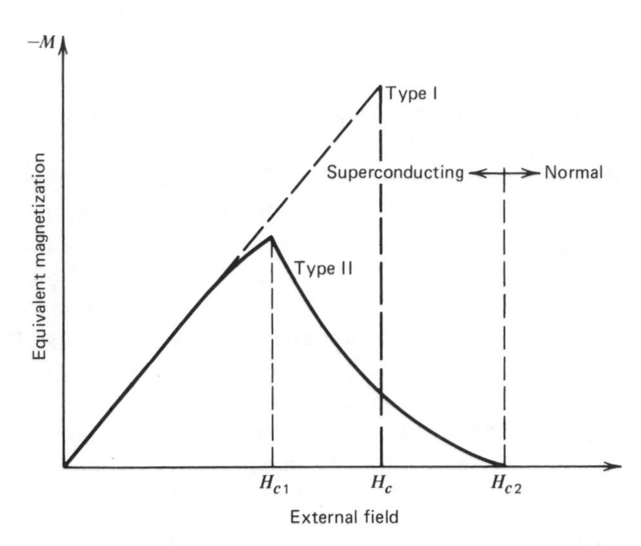

Figure 6-5 Equivalent diamagnetism vs. magnetic field for Type-I and Type-II superconducting materials.

above H_{c_1}, the diamagnetic force (6-2.3) is greatly reduced. However, Stevenson and Atherton (1974) claim that diamagnetic forces were responsible for the failure of a Type-II-material wound magnet.

Critical Current

The critical transport current for a Type-II material is governed by the interaction of the current with the magnetic fluxoid lattice. One of the important concepts in the understanding of these effects is the force on the fluxoid due to the transport current.

To calculate the force on the circulating current surrounding a fluxoid, we assume the flux is confined to a normal cylindrical region as shown in Figure 6-4. The circulating current acts as a magnetic dipole with pole strength Φ_0/μ_0, where Φ_0 is the total flux through the cylinder. If we denote the dipole strength per unit length by $\mathcal{M}$, the force on the dipole due to an external field $\mathbf{B}^T$ is given by

$$\mathbf{F} = \mathcal{M} \cdot \nabla \mathbf{B}^T \tag{6-2.14}$$

Suppose a transport current flows transverse to the fluxoid axis with a density J_x^T. Using the axes shown in Figure 6-10, the magnetic field associated with J_T must satisfy

$$\mu_0 J_x^T = \frac{\partial B_z^T}{\partial y} \tag{6-2.15}$$

For uniform current density J_x^T, B_z^T is a linear function of y, and the nonzero component of the force is given by

$$F_z = \frac{\Phi_0}{\mu_0} \frac{\partial B_z^T}{\partial y} = \Phi_0 J_T \tag{6-2.16}$$

If B_0 denotes the average fluxoid flux per unit area, then the force on the fluxoid per unit area is $J_T B_0$.

Thus, in the absence of resistive forces, flux lines in Type-II superconductors would move freely through the conductor, leading to dissipation and eventually driving the material into the normal state. In practical materials, however, the flux-line motion is resisted by lattice defects. These defects act to "pin" the fluxoid at various points along the vortex resulting in so-called *pinning forces*. The pinning forces depend on the magnetic field as well as temperature and the nature of the defects. (See, e.g., Kramer, 1975; Huebener, 1979 for a review of pinning forces in hard superconductors.)

In one theory for the critical transport current advanced by Bean (1972), it is assumed that all flux lines move to maximize the pinning force. For a one-dimensional problem, with the transport current J orthogonal to B_0, one has

$$JB = \alpha_c(B) \tag{6-2.17}$$

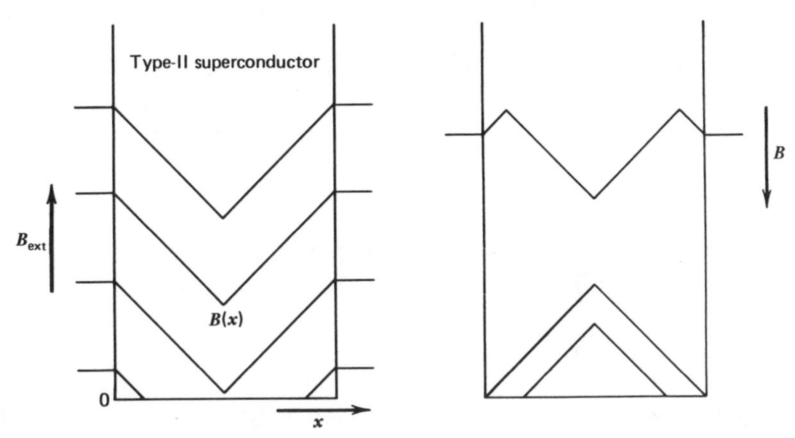

Figure 6-6 Penetration of flux into a Type-II superconductor—Bean critical-state model.

where α_c is the maximum value that a pinning force can attain. This is called the *critical-state model*. Bean, for example, assumes that $\alpha_c \sim B$ so that J has a constant value whenever it penetrates the superconductor. (Critical current behavior in commercial superconductors is more complicated; see, e.g., Fig. 6-11). The Bean model leads to a constant field gradient in the Type-II superconductor as illustrated in Figure 6-6. This figure illustrates the hysteretic behavior of flux penetration in hard superconductors.

Strain Dependence of Critical Current

It has been discovered that strain in superconducting materials will affect the value of the critical current (see, e.g., Koch and Easton, 1977). This is especially true for Nb_3S_n as shown in Figure 6-7. For most commercial

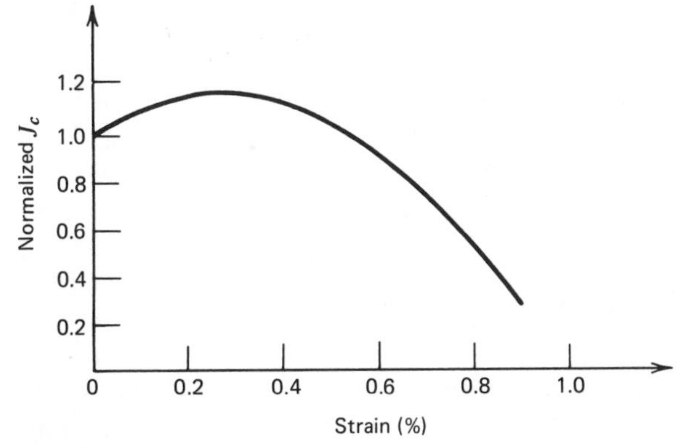

Figure 6-7 Critical current vs. longitudinal strain for Nb_3Sn.

superconducting wire, the critical current is observed to increase initially with tensile strain and then decrease with increasing tension. In one theory, the stress at the maximum J_c is thought to cancel out initial compressive stress produced by manufacturing and thermoelastic strains at cryogenic temperatures. (See Section 6.3 for a further discussion of this matter.)

6.3 PROPERTIES OF COMMERCIAL TYPE-II SUPERCONDUCTING SOLIDS

Superconducting materials that can be manufactured in large quantities and used in practical devices include alloys of niobium, such as Nb-Ti or Nb-Zr, or compounds of niobium and vanadium, such as Nb_3S_n, Nb_3Ge, and V_3Ga. These materials are made in long lengths for winding into magnets. The cross-sectional geometry has a variety of topologies depending on the material and the application. Several topologies are illustrated in Figure 6-8.

An isolated superconducting filament can suffer a thermal instability. A small rise in temperature produces a normal resistive zone that grows under Joule heating until the whole magnet becomes normal. This problem has been mitigated in practical materials by placing a good conductor, such as copper or aluminum, in parallel with the superconducting conductor. This has led to two basic types of conductors—the multifilament and the layered or flexible-tape superconducting composite conductor. The term "composite" is used to denote the fact that two or more materials are used. Thus the properties of the wire depend on those of the constituents. Elastic constants of multifilament Nb-Ti/Cu potted in an epoxy matrix are given in Table 6-1.

Table 6-1 Elastic Stiffness Constants for a Composite Nb-Ti/Cu Epoxy-Potted Coil[a] (Units in 10^{10} N/m^2)

T (°K)	c_{11}	c_{12}	c_{13}	c_{33}	c_{44}	c_{66}
0	5.1	1.9	2.4	10.6	1.1	1.6
100	5.0	1.9	2.3	10.3	1.1	1.5
300	3.8	1.4	1.8	9.4	0.74	1.2

Source: Weston (1975).

[a] These constants are dependent on ν_{13} and were calculated from the static value which was taken as independent of temperature.

In the continuous-filament-type conductor, such as Nb-Ti, the normal conducting matrix is often copper. The wire is made by stacking up arrays of copper and Nb-Ti rods inside a copper cylinder. This billet is then drawn

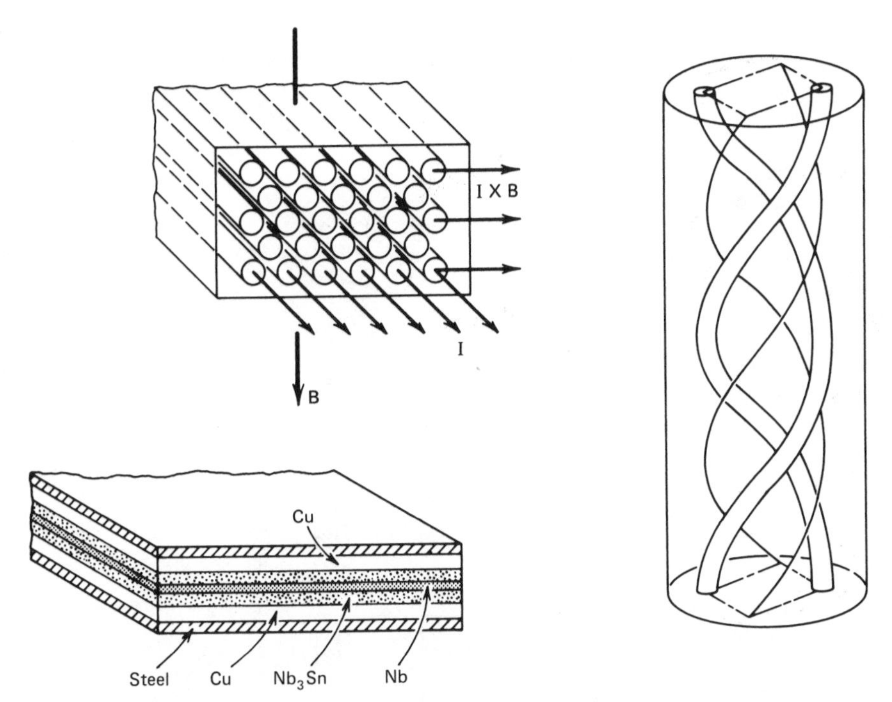

Figure 6-8 Different topologies for superconducting composites.

down by passing the billet through successive dies until a certain cross-sectional shape and size are achieved (see Fig. 6-9). This process obviously involves large strains and can only be used for ductile materials. Nb-Ti has sufficient ductility to be drawn down to a 10-μm size without fracture, but Nb_3Sn is very brittle and requires different methods of manufacture.

In the layered composite, a niobium layer and a zinc or bronze layer are thermally reacted so that the niobium and zinc diffuse toward the interface to make Nb_3Sn. Additional copper and stainless-steel layers can be bonded to this composite to provide thermal stability and mechanical strength (see Fig. 6-8.)

One method of manufacturing a multifilament Nb_3Sn composite is to array niobium and bronze rods in a billet, draw it down to wire form, and then thermally react the wire to form Nb_3Sn (see Fig. 6-10). A cylindrical tantalum barrier is often used to prevent diffusion into the copper (see, e.g., Hoard, 1980).

Another macrostructure involves aligned, but random, arrays of superconducting fibers of finite lengths in a normal conductor matrix (see, e.g., Flükiger, 1980). These materials are similar to chopped fiber composites in structural mechanics.

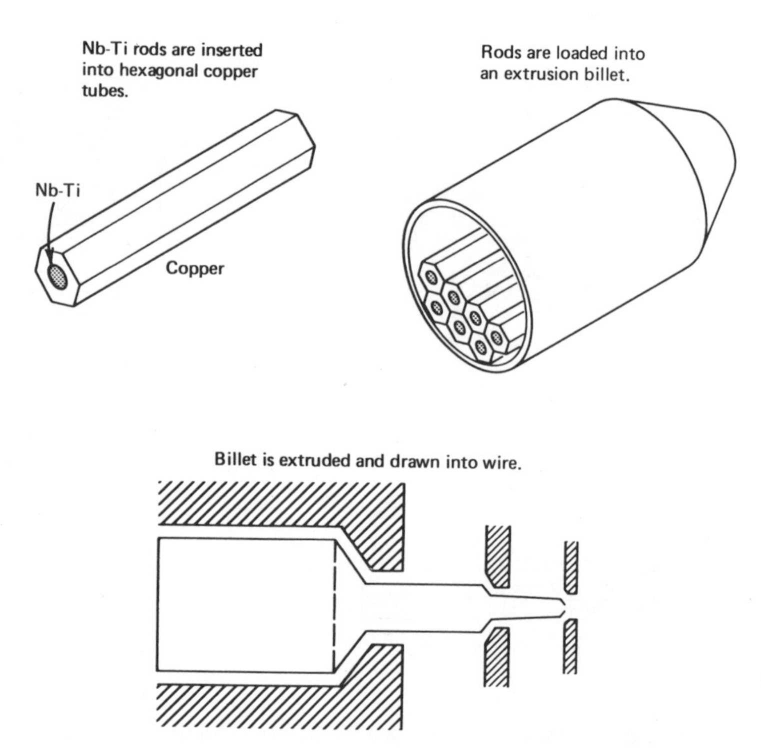

Figure 6-9 Steps in the manufacture of superconducting composites (from Scanlan, 1979).

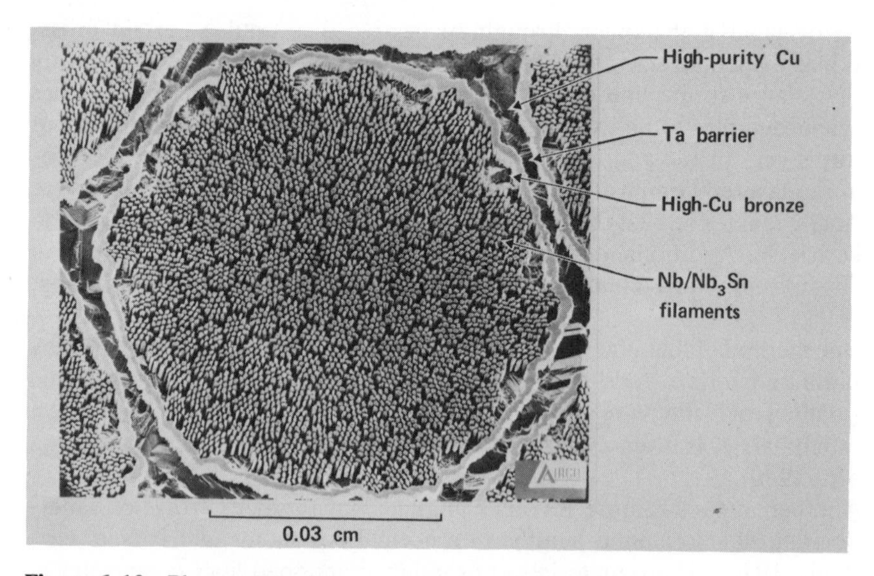

Figure 6-10 Photograph of the cross section of a multifilamentary superconducting composite. (from Hoard, 1980)

Superconducting Properties

The properties of various Type-II superconductors are given in Table B-4. The exact value of the critical current depends not only on the magnetic field, but on the degree of thermal stability required and the heat transfer properties. The maximum applied current density versus applied transverse magnetic field is shown in Figure 6-11 for a number of superconductors. Commercially available conductors of Nb-Ti range from 0.5-mm-diam. wire, with 400 filaments which can carry 200 A at low fields (~1.0 T), to a $1.2 \times 1.2\ cm^2$ square conductor designed for 6000 A for the mirror fusion yin–yang magnets used at Livermore, California (see Fig. 1-6). For applications which call for fields below 9 T, Nb-Ti/Cu composites are presently employed. For higher-field environments, Nb_3Sn composites are currently used. As discussed below, the critical current, critical temperature T_c and field H_{c2} in these conductors are strain-sensitive.

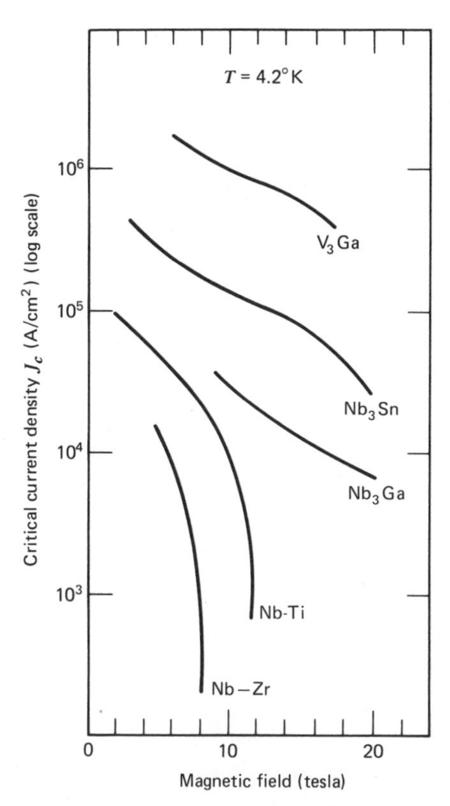

Figure 6-11 Critical current density vs. magnetic field for several superconducting materials. (From Hein, 1974, reprinted with permission. American Association for the Advancement of Science, copyright 1974.)

Inelastic Properties

Because superconducting wire and cable sometimes have two, three, or four constituents, the over-all stress–strain behavior has different regimes governed by the different yield stress or elastic limit stress properties of the constituents. Consider, for example, the properties of a Nb-Ti copper matrix composite wire as shown in Figure 6-12. Both the copper matrix and the Nb-Ti filaments are elastic up to a stress of a little over 1.2×10^8 N/m^2 (18×10^3 psi) after which the copper yields. Between 1.2 and 4.4×10^8 N/m^2 the filaments carry most of the additional stress. Beyond a strain of 10^{-2}, the filaments begin to fracture and the whole composite yields. These effects lead to hysteretic stress–strain behavior.

Another inelastic effect is a serrated stress–strain curve for superconducting filaments. A stress–strain curve for Nb-Ti wires at 4.2°K is shown in Figure 6-13 (Koch and Easton, 1977). The effect has also been observed for Nb-Zr. Associated with these serrations are acoustic noise or acoustic emission signals. These serrations have been observed for both loading and unloading. One theory, advanced by Koch and Easton, is based on a

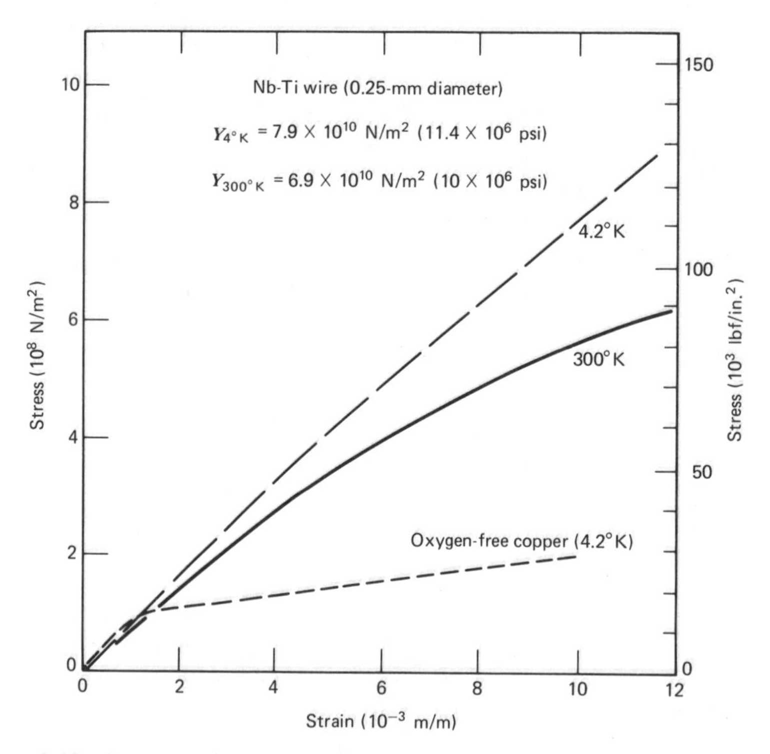

Figure 6-12 Stress–strain curves for Nb-Ti wire at 300 and 4°K. (Personal communication from D. Cornish, Lawrence Livermore Laboratory, Livermore, California.)

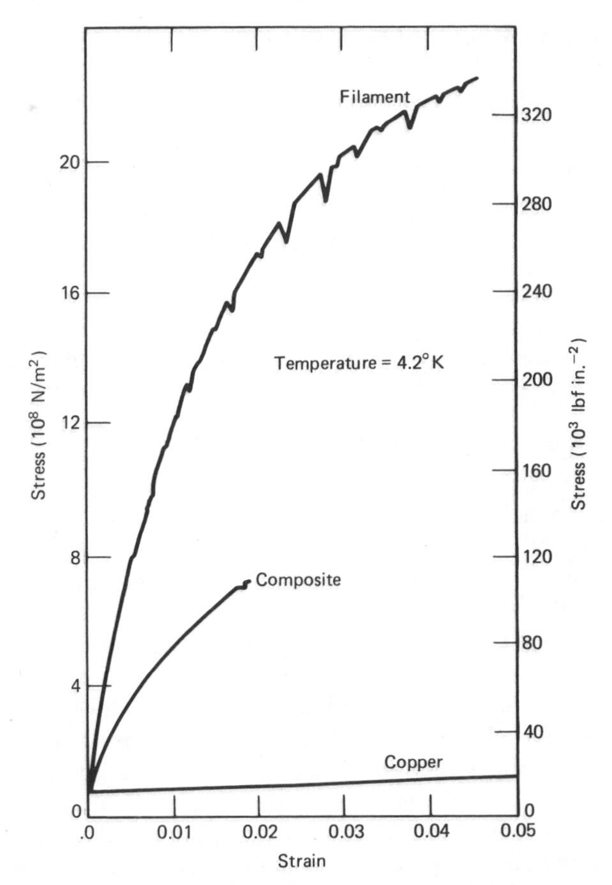

Figure 6-13 Stress–strain curves for Nb-Ti filament, copper matrix, and superconducting composite. (From Koch and Easton, 1977, reprinted with permission. Butterworth Scientific Ltd., copyright 1977.)

reversible stress-induced phase transformation or stress-induced twinning.

The thermoelastic behavior of composite superconductors is another important property. The integrated thermal contraction strain from 293 to 4.2°K for Nb-Ti cable is of the order of 200×10^{-5}. This contraction is greater than copper and 304 stainless steel (Brechna, 1973).

Strain Effects on Critical Current and Temperature

The high current capacity of Type-II commercial superconducting materials is a consequence of pinning the magnetic-flux vortex lattice by crystal defects in the metal. It should not be surprising then that the maximum sustainable current density is strain-dependent in these materials. The effect of strain on critical current has been reviewed by a number of researchers, including an extensive paper by Koch and Easton (1977) and a doctoral thesis by Hoard (1980).

Welch (1980), in a review paper, summarizes the effects of the three-dimensional strain tensor $\boldsymbol{\varepsilon}$ on the critical temperature T_c and the upper critical field H_{c2}. His analysis is based on a constitutive law for $T_c(\boldsymbol{\varepsilon})$ for a single crystal

$$T_c(\boldsymbol{\varepsilon}) = T_c(0) + \boldsymbol{\Gamma} \cdot \boldsymbol{\varepsilon} + \tfrac{1}{2}\boldsymbol{\varepsilon} \cdot \boldsymbol{\Delta} \cdot \boldsymbol{\varepsilon} \tag{6-3.1}$$

where $\boldsymbol{\Gamma}$ and $\boldsymbol{\Delta}$ are tensors, and $\boldsymbol{\Delta}$ has the same symmetry as the elastic modulii. He concludes that the linear term in Eq. (6-3.1) is negligible for multifilament composite superconductors. He cites evidence that the strain dependence of J_c may be derived from the strain dependence of T_c and H_{c2} for multifilament Nb_3Sn composite superconductors at high fields.

In a Nb-Ti/Cu composite, Ekin et al. (1977) observed a decrease in J_c (at 4.2°K and 8 T) of about 15–20% at an axial tensile stress that was $\frac{3}{4}$ of the fracture stress. Much of the effect was reversible. In general, the higher the transverse magnetic field the greater the effect. In experiments of this kind, the critical current is defined as the current at which the sample achieved a specific resistivity, sometimes chosen to be 10^{-10} or 10^{-11} Ω cm. Experiments by Ekin and Clark (1976) on Nb-Ti/Al composite showed a similar effect with a noticeable decrease in J_c for axial strains of the order of 0.5–1.5%. Again the effect was almost completely reversible. However, Fisher et al. (1975) performed cyclic strain tests on a Nb-Ti/Cu composite of from 200 to 1300 cycles and notices accumulated degradation of J_c of up to 8% at axial tensile strains below 0.5% and compressive strain less than 0.1%.

In Nb_3Sn composite conductors, more-pronounced effects of strain on J_c have been measured. A typical set of data is shown in Figures 6-7 and 6-14. What is often observed is an initial increase in J_c, with axial tensile stress, to a maximum J_c and then a decrease in J_c with further increases in stress (see Fig. 6-7). The present interpretation of this behavior is based on different thermal contractions of the constituents. The difference between the thermal contraction of the Nb_3Sn layer and the bronze matrix places the Nb_3Sn in a state of axial compression. The applied axial strain ε_m at which J_c is maximum is thought to be the zero intrinsic strain of the superconductor. Indeed the data show a symmetry about $\varepsilon = \varepsilon_m$, perhaps indicating that intrinsic axial tensile or compressive strains have equivalent degradation effects on J_c.

Hoard (1980), in a doctoral dissertation, uses an axially symmetric thermoelastic model for two types of Nb_3Sn composite (see Fig. 6-15). His analysis employs the equilibrium equation for stress without body forces, and a stress–strain constitutive law with thermal expansion; that is,

$$\frac{d}{dr} t_{rr} + \frac{1}{r}(t_{rr} - t_{\theta\theta}) = 0 \tag{6-3.2}$$

and

$$t_{rr} = (\lambda + 2\mu)\varepsilon_{rr} + \lambda(\varepsilon_{\theta\theta} + \varepsilon_{zz}) + \alpha \Delta T \tag{6-3.3}$$

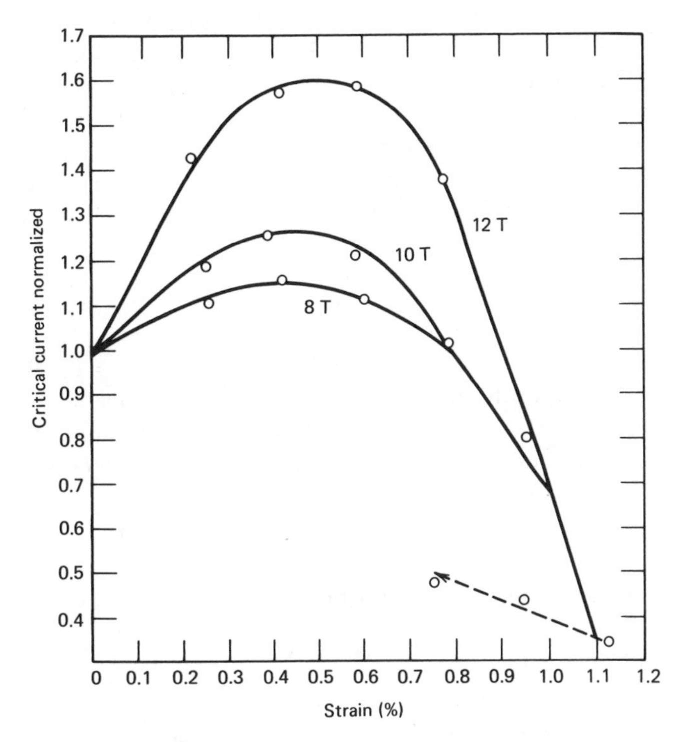

Figure 6-14 Effect of longitudinal strain on critical current density for Nb_3Sn composite superconductor (from Hoard, 1980).

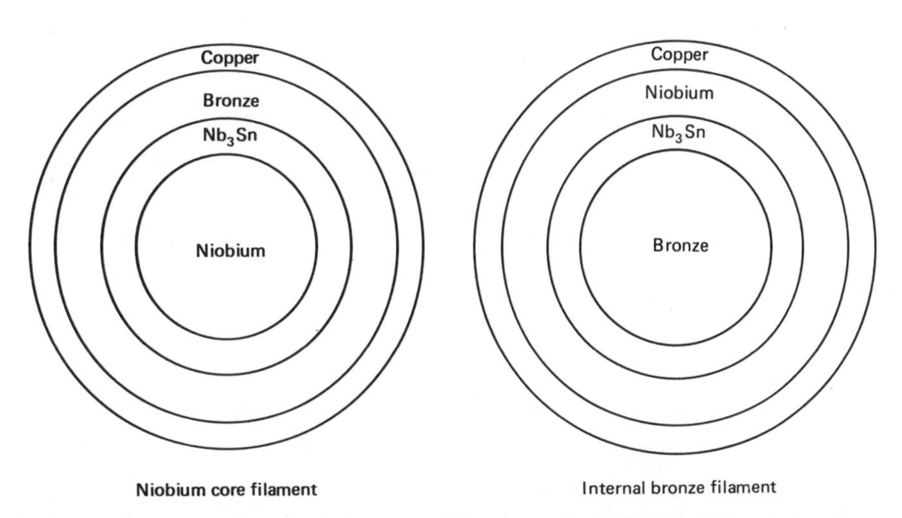

Figure 6-15 Thermoelastic models for analysis of residual strains in superconducting composites (from Hoard, 1980).

with similar expressions for $t_{\theta\theta}$ and t_{zz}. The reference temperature for $\Delta T = T - T_0$ was chosen as the Nb_3Sn diffusion reaction temperature $T_0 \sim 10^3$ K. The constants λ, μ, and α will differ for each layer or constituent and continuous displacements and stress t_{rr} were assumed to exist at each interface. Zero radial stress at the outer cylindrical surface was also assumed. Using this model, Hoard analyzed two problems (1) the state of internal stress/strain with temperature change, but zero total axial force, that is, across the circular face

$$\int_0^R t_{zz} r dr = 0 \tag{6-3.4}$$

and (2) the effect of an applied axial strain with $\Delta T = 0$. Hoard also attempted to incorporate the effects of plasticity in his model using a modified Poisson's ratio and Young's modulus. As a measure of the strain energy of distortion, he calculates an effective stress $\bar{\sigma}$ given by

$$\bar{\sigma} = \frac{1}{\sqrt{2}}[(t_{rr} - t_{\theta\theta})^2 + (t_{\theta\theta} - t_{zz})^2 + (t_{zz} - t_{rr})^2]^{1/2} \tag{6-3.5}$$

His analysis shows that at the maximum J_c ($\varepsilon = \varepsilon_m$), the state of stress in

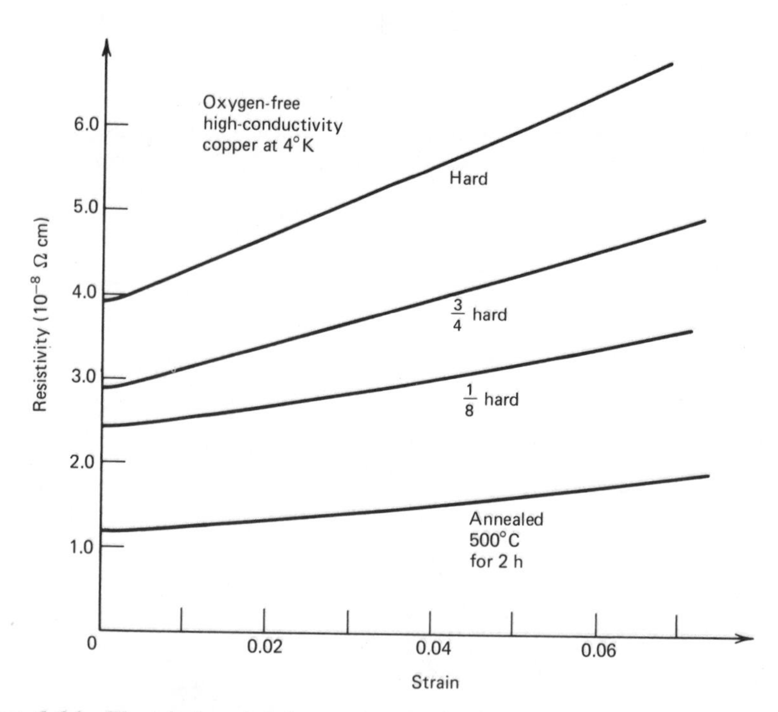

Figure 6-16 Electrical resistivity vs. longitudinal strain in oxygen-free copper at 4.2°K for different hardness. (Personal communication from D. Cornish, Lawrence Livermore Laboratory, Livermore, California.)

the superconducting layer is not zero, but that $\bar{\sigma}$ is a minimum. Welch (1980) also discusses the three-dimensional strain in a filamentary superconducting composite.

The effect of strain on the electrical properties of the composite, especially the matrix material, is of great importance since the Joule heating and thermal stability requirements depend on the resistivity. For oxygen-free copper, at 4.2°K, cold working raises the yield, but at the expense of increased resistivity, as shown in Figure 6-16. For a given hardness of copper, additional strain leads to an increase in resistivity.

6.4 COMPOSITE STRUCTURE OF SUPERCONDUCTING MATERIALS

Commercial superconductors are by nature multiconstituent materials. The constituents are chosen to maximize current-carrying capacity, heat transfer,

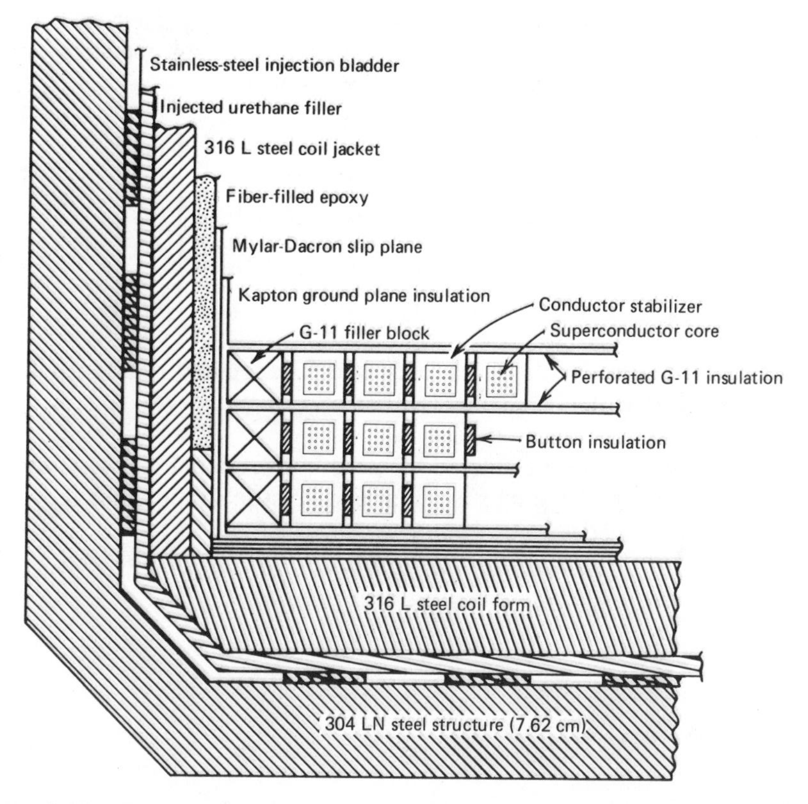

Figure 6-17 Cross section of the mirror fusion test facility (MFTF) yin–yang magnet for fusion research. (From Horvath, 1980, with permission of ASME, copyright 1980.)

strength, and elastic stiffness. There are a variety of geometric arrangements, as illustrated in Figure 6-8, including composite plates, fiber matrix, chopped fiber composite, helical fiber matrix, and nested cylindrical fiber-matrix systems. It should be noted that there are two scales of composite behavior. There is the material or conductor level such as a Nb-Ti fiber/copper matrix composite where the fiber diameters are on the scale of 1–50 μm. Then there is the magnet winding structure shown in Figure 6-17 which consists of composite conductor, electrical insulation, and sometimes reinforcement material, where the distance between repeating elements or cells is of the order of centimeters. Each scale requires its own analysis.

Elastic Properties of Fiber Composites

One of the systems which has received a great deal of attention in the mechanics literature is the fiber-reinforced composite (see, e.g., Christensen (1979); Tsai and Hahn, 1980). The main commercial superconducting materials in this category are the Nb-Ti composites with either a copper or aluminum matrix. For this system, the bulk elastic properties can be considered to be transversely isotropic. In matrix notation, the stress–strain law for orthotropic material takes the following form:

$$\begin{bmatrix} t_{11} \\ t_{22} \\ t_{33} \\ t_{13} \\ t_{23} \\ t_{12} \end{bmatrix} = \begin{bmatrix} c_{11} & c_{12} & c_{13} & 0 & 0 & 0 \\ c_{12} & c_{22} & c_{23} & 0 & 0 & 0 \\ c_{13} & c_{23} & c_{33} & 0 & 0 & 0 \\ 0 & 0 & 0 & c_{44} & & 0 \\ 0 & 0 & 0 & 0 & c_{55} & 0 \\ 0 & 0 & 0 & 0 & 0 & c_{66} \end{bmatrix} \begin{bmatrix} \varepsilon_{11} \\ \varepsilon_{22} \\ \varepsilon_{33} \\ \varepsilon_{13} \\ \varepsilon_{23} \\ \varepsilon_{12} \end{bmatrix} \quad (6\text{-}4.1)$$

For transverse isotropy, there are only five independent constants. If the 3-axis is aligned with the filaments, then it can be shown that $c_{11} = c_{22}$, $c_{13} = c_{23}$, $c_{66} = \frac{1}{2}(c_{11} - c_{12})$, and $c_{44} = c_{55}$.

The constants c_{ij} can be related to more conventional elastic constants such as the anisotropic Young's modulus and Poisson's ratio defined by (see, e.g., Christensen, 1979)

$$Y_{11} = \frac{t_{11}}{\varepsilon_{11}} \quad (6\text{-}4.2)$$

$$\nu_{12} = -\frac{\varepsilon_{22}}{\varepsilon_{11}}$$

and

$$\nu_{13} = -\frac{\varepsilon_{33}}{\varepsilon_{11}}$$

The composite mechanics problem that results then is to find the five constants c_{11}, c_{33}, c_{44}, c_{12}, and c_{13} in terms of the properties of the elastic fiber and matrix materials.

When uniaxial tension is analyzed, for the case when fiber and matrix have identical Poisson's ratios, the simple rule of mixture results for the Young's modulus in the fiber direction, that is,

$$Y_{11} = \eta Y_f + (1 - \eta) Y_m \tag{6-4.3}$$

where η is the volume fraction of the fiber constituent. When $\nu_f \neq \nu_m$ then a more complex relationship results (see, e.g., Chirstensen, 1979). The equivalent elastic modulii for a fiber-matrix composite can be obtained from an elasticity analysis of an infinite medium with a periodic array of fibers.

Measurements of the elastic properties of composite superconducting materials are reviewed by Koch and Easton (1977). A comparison of experimentally determined values of elastic constants with predictions from various composite mechanics theories was made by Sun and Gray (1975). The material used was Nb-Ti/Cu composite. The fiber-to-matrix Young's modulus ratio for this composite is around 0.70. The longitudinal Young's modulus and shear modulus agreed very closely with the theoretical predictions. However, the transverse Young's modulus and shear modulus did not. The fiber volume fraction was low (i.e., 0.14) so that the comparison was not a severe test for the composite theories. However, there is a paucity of data in the literature of the complete anisotropic elastic constant for superconductors. Koch and Easton (1977) review a number of papers reporting the longitudinal Young's modulus for different superconductor systems. There seems to be general agreement that the simple rule-of-mixtures formula works well for the longitudinal modulus. Attempts have been made to use the rule of mixtures for predicting tensile strength by Reed et al. (1977); but the experimental values were about 25% higher than theoretical values.

The mechanical behavior of superconducting composites must be studied in three different regimes:

1. Both fiber and matrix and elastic.
2. The fiber is elastic and the matrix has yielded.
3. Both fiber and matrix are in the inelastic region.

Another complexity concerns the state of initial stress in the composite. Unlike many structural composite systems such as boron/epoxy or graphite/epoxy, the fiber and matrix structure is drawn down to size with numerous cold working and annealing cycles. Further, the material is cooled from 300 to 4.2°K, which generates additional internal stresses due to different thermal expansion properties of fiber and matrix (see end of previous section).

Filamentary composite superconductors often are twisted during drawing in order to prevent flux jumps and associated thermomagnetic instabilities. Heim (1974) has studied the effect of helical filaments on the strength and stress distribution in a Nb-Ti composite. However, to date no thorough study has been done on the effect on twisting on the complete elastic behavior. One would expect, for example, that a helical fiber structure would couple the axial strain and the torsional twisting along the conductor axis. However, this effect has not been reported at the time of this writing.

Structural Scale Composite Behavior

We have seen how the multifilament superconductor can be considered as an equivalent anisotropic material using the mechanics of composite materials. This conductor is usually wound on a coil form with insulation between the turns and sometimes structural-reinforcement material. The resulting multi-turn system of conductor, insulation, and reinforcement can also be considered as a composite material with equivalent elastic properties. An example of the turn structure is shown in Figure 6-17. Modeling of this turn structure is important to determine how the magnetic forces, which are applied to the conductor, are transferred to the structural case.

There are two types of magnet winding which should be distinguished here. In smaller magnets, the conductor turns are sometimes potted in a polymer matrix, such as epoxy. In such magnets, the application of composite theory works well in both the circumferential and transverse directions. However, in large magnets, voids must be present to permit the flow of helium and the turns are held together by compressive stresses due to winding pretension, thermoelastic cooldown stresses, and magnetic body forces. The application of composite theory to this type of turn structure does not work well, especially in the transverse direction. In fact it has been found that significant *nonlinear* elastic behavior is present to question the notion or usefulness of an equivalent linear elastic model.

The composite elastic constants of the epoxy-*potted* multiturn coil have been reported by Weston (1975). In this study he used an ultrasonic resonant piezoelectric oscillator method to obtain the equivalent elastic constants. The turn structure consisted of copper-stabilized Nb-Ti rectangular-cross-sectional wire (0.56 mm × 0.72 mm) with fiberglass cloth (0.1 mm thick) between the wire layers. The copper-to-superconductor ratio was 1.8 : 1 and the wire-to-epoxy cross-sectional-area fraction was about 3 to 1. The density of the coil composite was 6.0 g/cm^3. Weston measured these constants as a function of temperature and fitted curves to the data. The dynamically measured values were within 5–10% of static measurements. The elastic stiffness constants at 0 and 300°K are shown in Table 6.1. He assumed the material was transverse isotropic so that $c_{66} = \frac{1}{2}(c_{11} - c_{12})$, with the 3-axis along the wires. The anisotropic nature of the coil composite is large, with $c_{33}/c_{11} = 2.1$.

An example of *nonlinear* elastic phenomenon in unpotted magnet windings is shown in Figure 6-18. These data were obtained by Horvath (1980) for a yin–yang magnet for the mirror fusion machine at Livermore, California. At low pressures there is an extremely soft, elastic response, whereas at

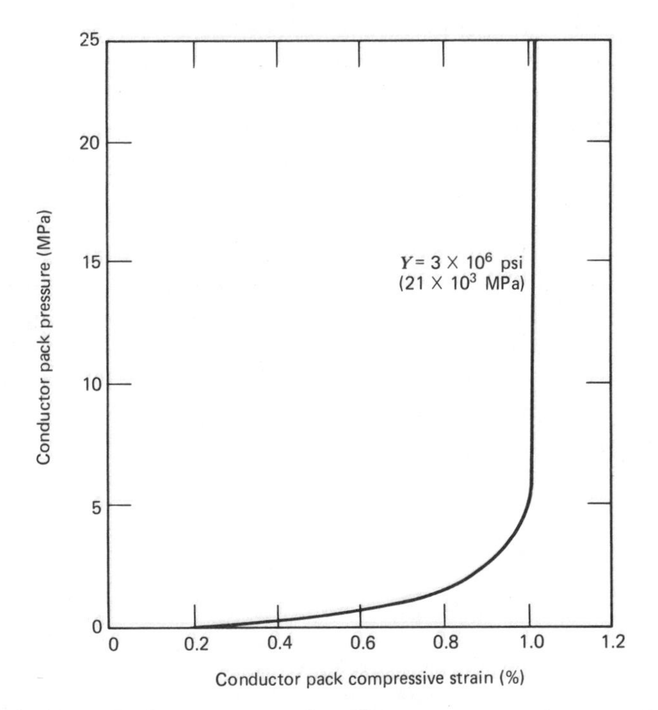

Figure 6-18 Compressive response of multilayer superconductor and insulation material for the MFTF yin-yang magnet in Figure 6-17 in the interlayer direction. (From Horvath, 1980, with permission of ASME, copyright 1980.)

higher pressures, the tangent modulus approaches the value given by composite theory. This behavior is analogous to that of certain geological material, such as granite or dolomite, in which microcracks and internal pores are thought to be closed with pressure. In the case of a layered or wound structure with no matrix bonding, this nonlinear behavior might be due to the closing or collapse of surface irregularities. As the pressure between turns is increased, the number of Hertz contacts increases until finally the entire areas of the two layers are in contact.

6.5 BENDING-FREE MAGNETS

Although some analysts have proposed force-free or stress-free structures for magnets (e.g., Kuznetzov, 1961b), the virial theorem (see Sections 3.9, 6.8) establishes that a structure must have a minimum amount of mass proportional to the stored magnetic energy. The same theorem says that structural mass associated with bending stresses does not contribute to this minimum-mass requirement. Though design constraints may require structures with bending stresses, the most mass-efficient designs are those that use truss or cable structures. One application where this has been exploited has been in the design of "bending-free" toroidal field magnets for fusion reactors (see Fig. 1-5). In this system the symmetric placement of a set of coils around a torus leads to cancellation of mutual attraction forces, and creates an inward centering force on the small radius parts of the torus, and outward body forces on the outer arcs of the coils. The near-total confinement of magnetic flux inside the coils can be pictured as a magnetic pressure inside the torus. Calculation of the total inward force on each magnet was given in Chapter 4.

Because of the variation of the toroidal magnetic field with radius, the outward body force $\mathbf{I} \times \mathbf{B}$ varies over the arc. In general, a variable body force on a curved ringlike structure requires shear and bending stresses for equilibrium. This can be shown by examining the equations of equilibrium for an arc under planar body forces (see Fig. 6-19):

$$\begin{aligned} \frac{dN}{ds} + T\kappa - IB &= 0 \\ \frac{dT}{ds} - N\kappa &= 0 \qquad (6\text{-}5.1) \\ \frac{dG}{ds} + N &= 0 \end{aligned}$$

where N is the internal shear force, T is the internal tension, and G is the

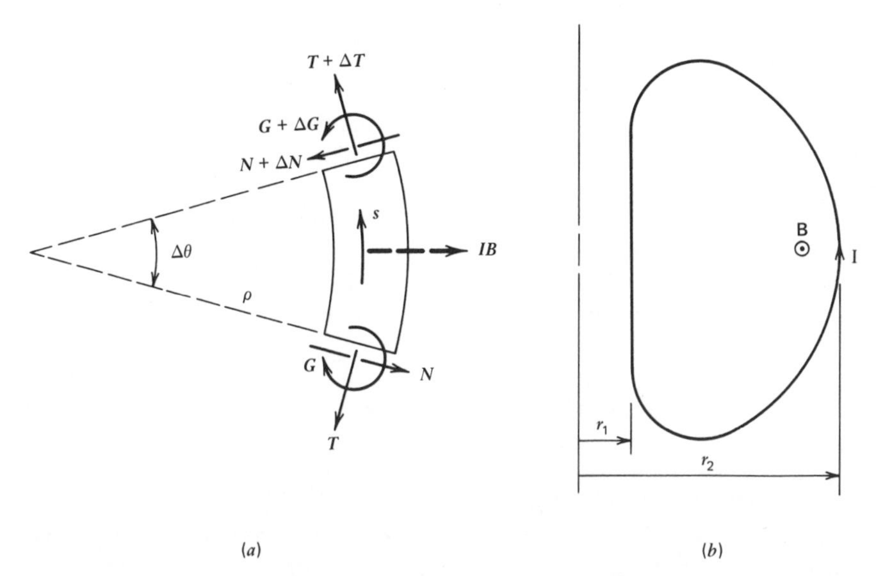

Figure 6-19 (*a*) Element of planar coil and (*b*) "D"-shaped pure tension coil for a toroidal magnetic fusion reactor.

bending moment. The radius of curvature ρ is the inverse of the curvature κ.

One can see from Eqs. (6-5.1) that for the shear and bending moment to vanish, the curvature must be proportional to the body force IB, and the tension T is a constant along the arc length s; that is,

$$T = IB\rho = \text{constant} \tag{6-5.2}$$

According to Ampere's law, the toroidal magnetic field is inversely proportional to the radial distance from the axis of the torus; that is,

$$B = \frac{\mu_0 NI}{2\pi r} \tag{6-5.3}$$

where N is the number of coils in the torus. Thus ρ must be proportional to r or $\rho = kr$, where k is a constant. In Cartesian coordinates this condition takes the form ($r \equiv x$)

$$\frac{d^2y}{dx^2}\left[1+\left(\frac{dy}{dx}\right)^2\right]^{-3/2} = \frac{1}{kx} \tag{6-5.4}$$

This condition was first given by File et al. (1971). They also were able to integrate Eq. (6-5.4) to obtain an explicit rule for the shape of the arc

$$x = \int_1^y \frac{\pm \ln y \, dy}{(k^2 - \ln^2 y)^{1/2}} \tag{6-5.5}$$

The constant is related to the inner and outer radii of the magnetic arc, r_1 and r_2, respectively, by the relation

$$k = \frac{1}{2}\ln\frac{r_2}{r_1}$$

and the tension is

$$T = \frac{\mu_0 I^2}{4\pi} N \ln\frac{r_2}{r_1} \tag{6-5.6}$$

The resulting magnet shape is a "D"-shaped coil with a straight leg at $r = r_1$ and the bending-free arc from $r_1 \leq r \leq r_2$.

In another study, however, Gralnick and Tenney (1976) (see also Thome and Tarrh, 1982) were able to evaluate the integral (6-5.5) by writing Eq. (6-5.5) in terms of the tangent angle of the arc θ; that is,

$$y(k, \theta) = k(r_1 r_2)^{1/2} \int_0^\theta (\sin\phi) e^{k \sin\phi} \, d\phi \tag{6-5.7}$$

where $r = (r_1 r_2)^{1/2} e^{k \sin\theta}$. The solution found by Gralnick and Tenney, shown in Figure 6-20, can be written in terms of modified Bessel functions of the first kind, $I_n(k)$; that is,

$$y(k, \theta) = k\frac{\partial \mathcal{S}}{\partial k}(r_1 r_2)^{1/2} \tag{6-5.8}$$

where

$$\mathcal{S}(k, \theta) = \theta I_0(k) + \sum_{n=1}^{\infty} \frac{iI_n(k)}{n}(e^{-in\theta} - 1)(1 + e^{in(\theta+\pi)})e^{in\pi/2}$$

With these curves, a designer can choose different constant tension coil shapes as shown in Figure 6-20. Using this expression, they are able to obtain formulas for the geometric properties of the coil, including the height of the straight segment or support cylinder h, the coil area A, the arc length S_0, and the inductance of the coil in the torus. For a "D"-shaped coil with vertical tangents at the straight section one finds

$$\begin{aligned} h &= 2\pi r_0 k I_1(k) \\ S_0 &= 2\pi r_0 k[I_0(k) + I_1(k)] \\ A &= 2\pi r_0^2 k[I_1(2k) - e^{-k} I_1(k)] \\ L &= 2\pi r_0 k^2[I_0(k) + 2I_1(k) + I_2(k)] \end{aligned} \tag{6-5.9}$$

where $r_0^2 = r_1 r_2$.

The inductance represents the magnetic flux through the coil in the full torus due to a unit current in all the coils. Expressions for other shaped coils are given in the paper by Gralnick and Tenney.

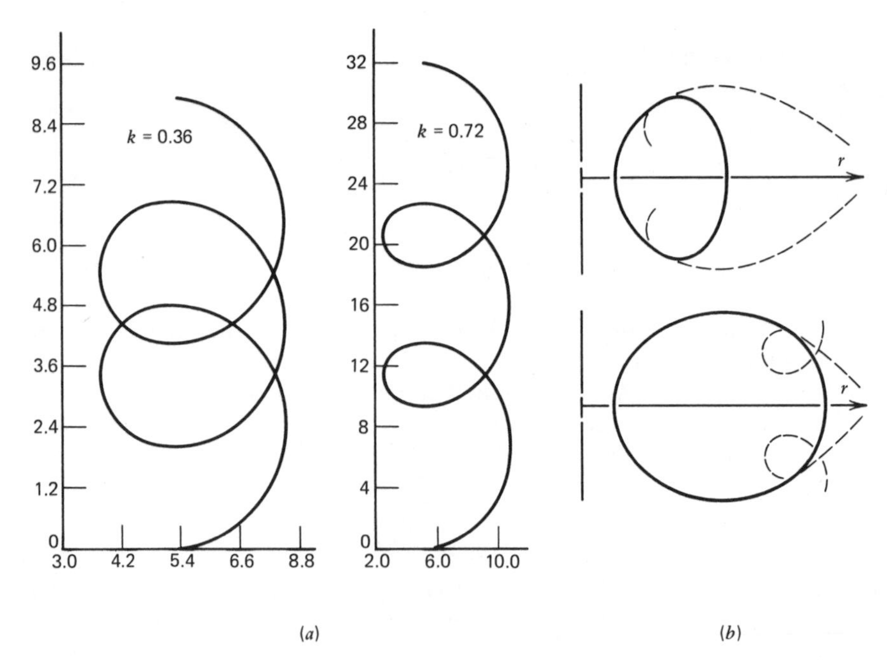

Figure 6-20 (a) Solution curves for a pure tension filament in a toroidal magnetic field and (b) compound coils made from pure tension solution curves (from Gralnick and Tenney, 1976).

One error implicit in the analysis of the zero bending coil for a discrete coil torus is that the magnetic field (6-5.3) used to derive Eq. (6-5.4) is for a continuous shell of toroidal current. Moses and Young (1975) have obtained approximate analytic corrections for the forces on coils in a discrete coil or sectored torus. Moses and Young derive an expression for the Lorentz body force per unit arc length $\mathbf{I} \times \mathbf{B}$ for an N coil sectored torus (see the diagram in Figure 6-19; note $dr = ds \sin \phi$):

$$f = \frac{\mu_0 N I^2}{4\pi r}\left[1 + \frac{1}{N}\left(\cos\phi + \frac{r}{\rho}\ln\frac{\alpha r}{Na}\right)\right] \tag{6-5.10}$$

where a is a linear dimension of the cross section of the coil.

For a coil with circular cross section and radius a, $\alpha = 1.284$, while for a square cross section of width a, $\alpha = 1.284/0.573$. Using this approximate, but more accurate, representation of the magnetic body force in a discrete coil torus, the condition for constant tension, zero bending along the arc of the magnet becomes

$$\rho = r\frac{[4\pi T/\mu_0 I^2 - \ln(\alpha r/Na)]}{N + \cos\phi} \tag{6-5.11}$$

When N is large this expression approaches the condition derived by File et al. (1971). Moses and Young used this condition to calculate more accurate shapes of zero bending TF (toroidal field) coils in a discrete coil torus.

The pure tension coil is not without its critics. Although the lack of bending should lead to a smaller coil cross section, these designs generally result in a greater toroidal volume or arc length than say a circular coil torus. Thus the pure tension coil may lead to higher stored energy in the torus than is needed for the plasma application.

Other work on pure tension coils includes the addition of truss of cable elements of a coil made up of constant tension segments (Bonanos, 1981). Bobrov and Marston (1982) have discussed momentless structures in the design of superconducting MHD magnets.

6.6 STRESSES IN SUPERCONDUCTING MAGNETS

As superconducting devices get larger, the importance of stress analysis assumes a more central role in the design of these structures. Stress analysis due to magnetic forces was discussed in Chapter 4. However, in superconducting magnets, there are at least two other sources of stress—winding pretension and thermal stresses due to cooling from 273 to 4.2°K. Other effects include potting the coil in a polymer matrix, the effects of the winding coil form or bobbin, and the interaction between the current-carrying windings and the structural case.

There are two classes of problems. In the first class, the magnetic body forces $\mathbf{J} \times \mathbf{B}$ are equilibrated by stresses in the composite conductor itself. In the second case, additional structural material is added to keep the stresses below a safe value. In this class there is a further division between those magnets in which the structural material is added throughout the windings (*endostructural reinforcement*), as in the Argonne bubble chamber magnet (see, e.g., Brechna, 1973), and those designs in which the forces are transferred from the conductor to a structural case (*exostructural reinforcement*), as in the Livermore yin–yang magnets for mirror fusion. These two schemes are illustrated in Figure 6-21.

When internal reinforcement is used, or when the composite conductor itself carries the load, the analysis can proceed using an equivalent anisotropic elastic model. This approach works well when the coil is potted with a rigid filler such as epoxy. In unpotted magnets, the effective modulus normal to the windings is usually very low and often nonlinear (see Section 6.4).

Several studies have analyzed stresses in solenoid magnets due to winding pretension, thermal cooldown, and magnetic loading. Therse include Johnson et al. (1976), Arp (1977), Kokavec and Cesnak (1977), and Bobrov and Williams (1980).

Winding Stresses

The analysis of winding stress can be found in the early mechanics literature of this century, such as in Case (1938). While the technical process involves nonaxisymmetric loading in a helically wound tape, most analyses assume a symmetric state of stress and imagine the coil as constructed from a series of concentric rings placed one on top of the other. The radial stress between the

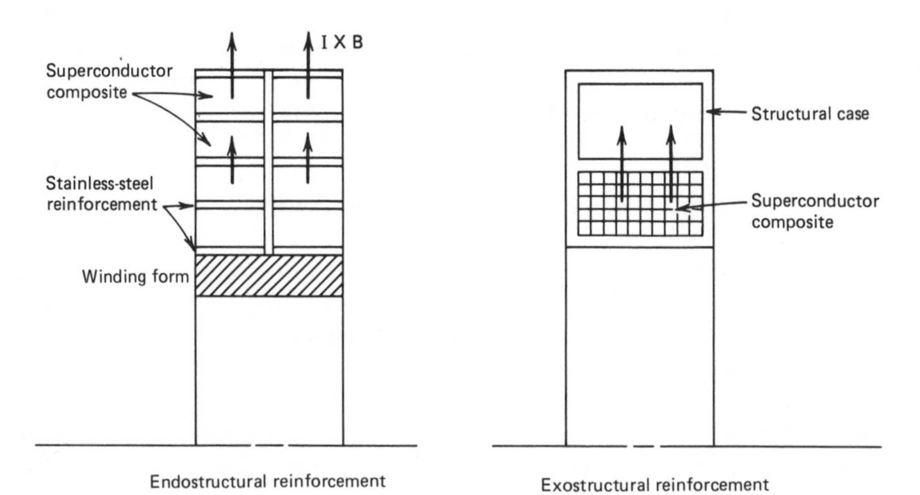

Figure 6-21 Two reinforcement schemes for superconducting coil structures.

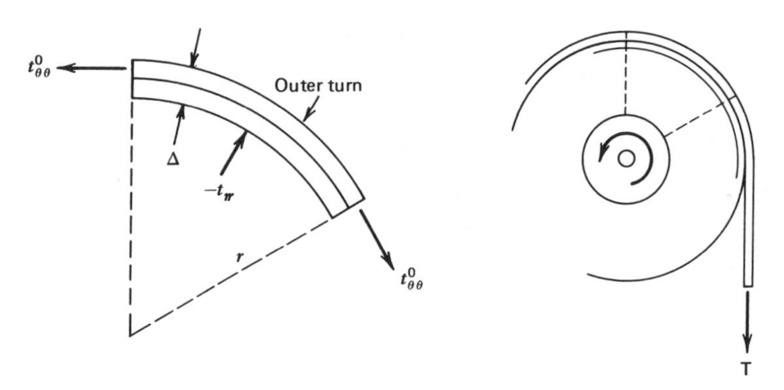

Figure 6-22 Element of the outer turn of multiturn coil showing winding stresses.

last ring and the other nested rings is related to the winding stress $t^0_{\theta\theta}$ by the relation (see Fig. 6-22)

$$t_{rr} = -\frac{t^0_{\theta\theta}\Delta}{r} \tag{6-6.1}$$

where Δ is the radial thickness of the winding and r is the radius of the last ring. Arp in his analysis takes the limit where $\Delta \equiv dr \to 0$ and $t_{rr} \to dt_{rr}$ and arrives at a boundary condition at the outer winding radius

$$\frac{dt_{rr}}{dr} = -\frac{t^0_{\theta\theta}}{r} \tag{6-6.2}$$

Typical of the results of these analyses is that successive windings add an incremental compression to the inner turns, which leads to circumferential compression in the inner turns.

Thermoelastic Effects

The winding process creates an anisotropy in the effective thermal expansion coefficients in the radial and circumferential directions which leads to thermal stresses in the magnet even when thermal gradients are absent. For example, Bobrov and Williams use a thermal strain of $\varepsilon_r = -3.66 \times 10^{-3}$ in the radial direction and $\varepsilon_\theta = -2.73 \times 10^{-3}$ in the circumferential direction for a small Nb-Ti wound magnet due to cooldown from 273 to 4.2°K. Arp has shown that the thermal stresses in a solenoid magnet are proportional to $Y_\theta \Delta\alpha \Delta T$, where $\Delta\alpha = \alpha_r - \alpha_\theta$ is the difference between radial and circumferential thermal expansion coefficients, and ΔT is the change in temperature.

Combined Effects

The effects of combined winding, cooldown, and magnetic loads are illustrated for two magnets using the data of Arp (1977) and Bobrov and Williams (1980) (see Figs. 6-23–6-25). In both cases, the magnet is wound on a bobbin or winding form. In Arp's analysis he removes the form before magnetic loading whereas in Bobrov and Williams, the coil separates from the form under magnetic loading.

In Arp's paper the ratio of outer-to-inner radii is $R_0/R_i = 1.26$. From Figure 6-23 one can see that after the bobbin is removed some part of the winding is under compressive circumferential stress. Thermal stresses shift the tensile stress from the outside to the inside. The magnetic forces then put all the windings under circumferential tension with the highest stresses at the inner radius. Arp uses a modulus ratio of $Y_0/Y_r = 2$ at 4.2°K, but uses different modulii for the winding and cooldown phases.

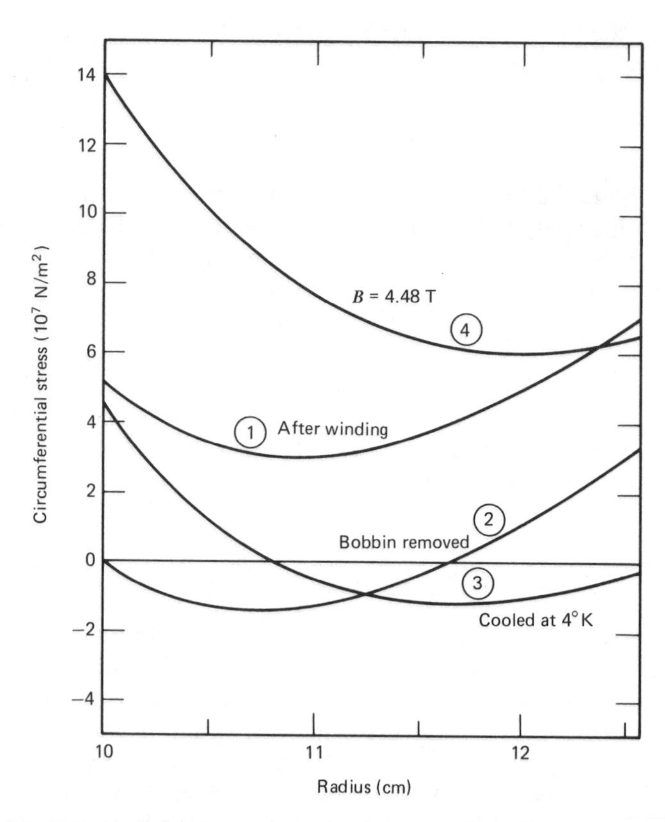

Figure 6-23 Calculation of cumulative circumferential stresses vs. radius due to winding, cooldown, and magnetic forces for a solenoid magnet wound at constant tension. (From Arp, 1977, reprinted with permission. American Institute of Physics, copyright 1977.)

For the magnet analyzed by Bobrov and Williams (1980) (Figs. 6-24 and 6-25), they use a ratio $R_0/R_i = 1.26$ and $Y_\theta/Y_r = 2.4$ at 4.2°K. This magnet was built and used for a nuclear-magnetic-resonance spectrometer. The Nb-Ti composite tape was wound on an aluminum cylinder and the outer four turns were beryllium copper. The winding tension was not constant. Under magnetic loads the coil separates from the coil form as shown by the zero radial stress at the inner radius in Figure 6-24. In this small magnet, we can see from Figure 6-25 for the circumferential stress, that the winding and thermal stresses dominate the stress due to magnetic forces alone.

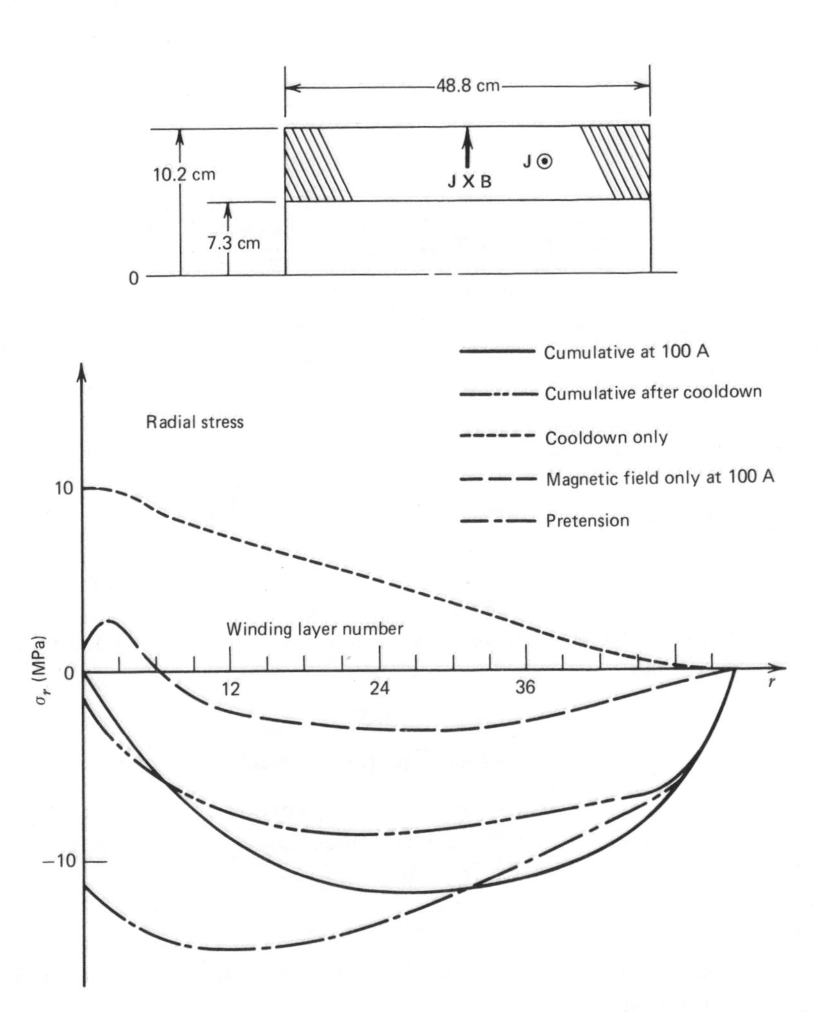

Figure 6-24 Calculation of cumulative radial stress vs. radius due to winding, cooldown, and magnetic forces in a multiturn, solenoid magnet (from Bobrov and Williams, 1980).

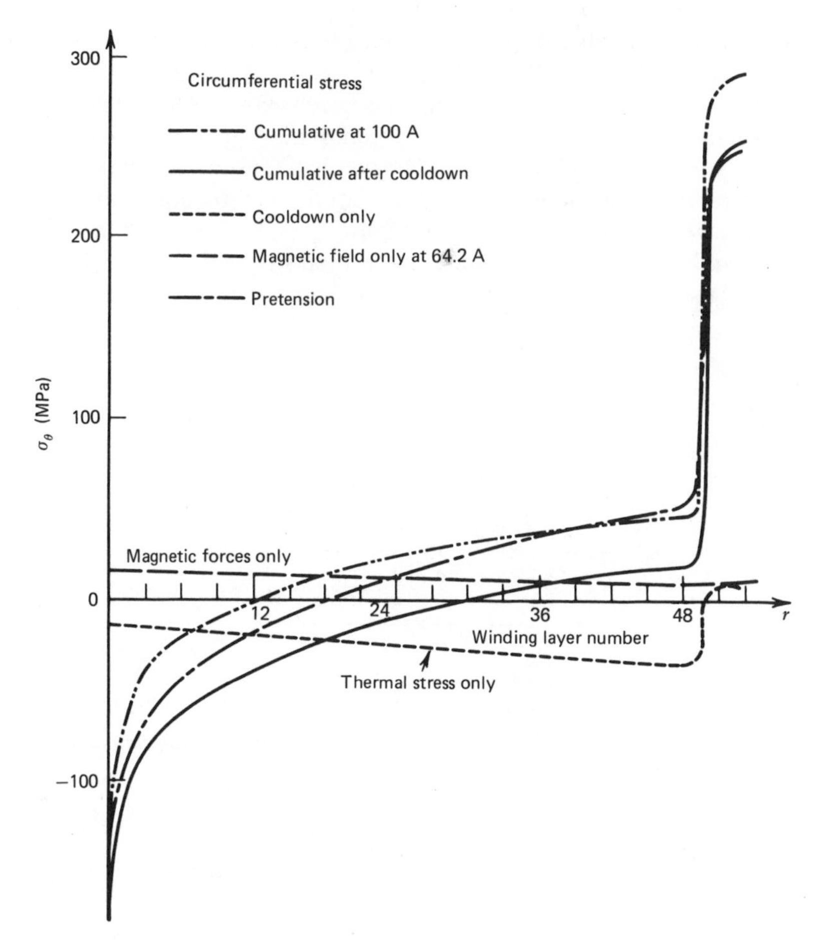

Figure 6-25 Calculation of cumulative circumferential stress vs. radius due to winding, cooldown, and magnetic forces in a multiturn solenoid magnet (from Bobrov and Williams, 1980).

Other Stress-Producing Effects

There are other strain- or force-producing effects that the magnet designer must provide for in the structure. Often these loads are related to fault or off-design conditions. This list includes the following:

1. Seismic loads.
2. Unbalanced magnetic forces due to misalignment or one or more neighboring coils out.
3. Dynamic forces due to eddy currents.
4. Thermal gradients due to cooling failure.

5. Construction-related forces.
6. Internal pressure due to helium boil-off.
7. Swelling due to radiation.
8. Fatigue crack propagation due to cyclic loading.

While most large superconducting magnet designs employ paramagnetic stainless steels for the structure, cold working of the material or ferrite in the weld material can leave some part of the structure ferromagnetic (Dalder, 1981). Although ferromagnetic forces are usually small in comparison with $\mathbf{J} \times \mathbf{B}$ forces, they might act in a direction where forces were not anticipated. When ferromagnetic material is employed, additional body forces and body couples must be added to the analysis (see Chapter 3 or 7; also Moon and Hara, 1982).

6.7 MAGNETOELASTIC STABILITY OF SUPERCONDUCTING STRUCTURES

As a consequence of Earnshaw's theorem (Chapter 5), a magnetostatic system of current-carrying structures has at least one unstable mode of deformation in the absence of mechanical constraints and feedback control currents. With elastic constraints, such a system can suffer magnetoelastic buckling at a critical value of the current. Superconducting structures are more susceptible to such instabilities for at least three reasons. First, they generally have a higher aspect ratio (global length dimension, such as coil diameter, divided by the square root of the cross-sectional area) than normal coils as a consequence of their higher current-carrying capacity. Second, the need to maximize the volume of high magnetic field space results in what is known in the industry as "stay-out zones," spaces between the magnets where intercoil support structures cannot intrude. Finally, the third reason for the sensitivity of superconducting structures to instabilities is the desire to minimize thermal leaks into the cryogenic magnet through structural connections from the warm to the cold environment.

In Chapter 5 we illustrated some simple, but somewhat academic, examples of buckling of current-carrying structures, such as the three-wire problem. In this section we present examples that are closer to those in superconducting magnetic devices and discuss methods for obtaining the critical buckling currents.

There are four types of interaction that might lead to buckling of current-carrying coils:

1. Perturbed magnetic fields due to the coil's own deformation—self-field effects. The kink instability found in plasma physics is in this class. (See Section 5.12.)

2. Interaction of the coil with magnetic fields from nearby coils.
3. Magnetic forces on the coil due to magnetic fields induced by the coil in nearby ferromagnetic material.
4. Initial stress due to magnetic forces in the *undeformed* coil.

There is no clearer example of these effects than the case of a closed superconducting circuit, such as a circular filament. For most practical applications, such as "*D*"-shaped toroidal field magnets or nonplanar mirror magnets (e.g., the yin–yang magnet), one must analyze the effects of initial stress on elastic stiffness using numerical methods. However, in the case of the circular circuit structure some analytical solutions can be obtained.

A problem which has appeared in the literature is the case of a circular coil in an axisymmetric poloidal magnetic field created by other coils or sources of magnetic field. Large superconducting rings suspended or levitated by magnetic fields have been used in plasma-physics experiments. The stability of such coils has been examined by Tenney (1969). Three floating superconducting rings were built at Princeton, New Jersey, Livermore, California and Culham, England (see, e.g., Cornish, 1981) for use in multipole magnetic fusion experiments.

The elastic stability of a straight conductor under magnetic forces resulting from bending of a straight wire was examined in Section 5.12. The self-magnetic force resulting from a "kink" in the conductor was found to be proportional to the bending curvature [Eqs. (5-12.1) and (5-12.18)]. When the conductor is bent in a circle, however, a self-induced tension is created in the conductor which often dominates the "kink" effect. In addition, magnetic forces produced by other turns in the magnet or neighboring magnets must be considered in determining the magnetoelastic stability.

Elastic Stability of a Circular Coil

We consider the static and dynamic deformation of an elastic circular ring of radius R and carrying current I (see Fig. 6-26). The magnetic forces arise from two sources. The first is due to the self-magnetic field due to the current I. In a circular ring this results in a circumferential tension. For an undeformed circuit of a circular cross section of radius a, the tension T is given by the expression (4-3.6),

$$T = \frac{\mu_0 I^2}{4\pi}\left[\ln\left(\frac{8R}{a}\right) - \frac{3}{4}\right] \tag{6-7.1}$$

The other contribution to the magnetic force is due to sources external to the structure.

To study the deformation of the ring, we examine a differential element with internal-force and moment resultants acting on it (see Fig. 6-26). As a simplification, we consider only the displacement of the centroid of the cross

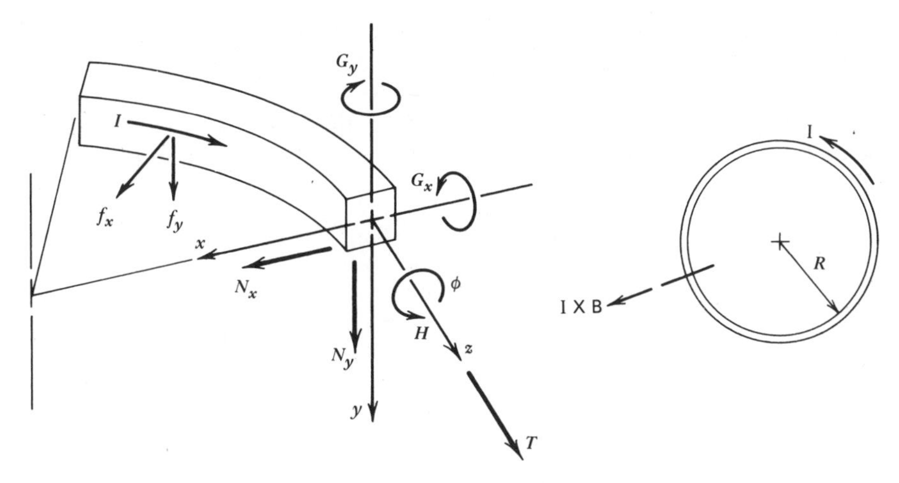

Figure 6-26 Element of a circular ring showing bending stress resultants.

section, (u_x, u_y, u_z), and twist about the centroidal axis, ϕ, as a function of the position around the circumference and time, where the z axis is tangent to the circumference and the x axis is directed *toward* the center of curvature.

In addition to the displacement and twist, $\mathbf{u}(z, t)$ and $\phi(z, t)$, respectively, one must introduce the radii of curvature, or their inverse κ_1 and κ_2, and a measure of the twisting called the torsion τ. The radius of curvature of the central filament of the ring in the $x-z$ plane is denoted by $\rho_1 = 1/\kappa_1$, and the corresponding radius in the $y-z$ plane is represented by $\rho_2 = 1/\kappa_2$. For a circular ring, the initial curvatures are $\kappa_1 = 1/R$ and $\kappa_2 = 0$. When the ring is bent and twisted, the curvatures and torsion are related to the displacement and twist by the relations (Love, 1922):

$$\kappa_1 = \frac{1}{R} + \frac{u_x}{R^2} + \frac{\partial^2 u_x}{\partial z^2}$$

$$\kappa_2 = \frac{\phi}{R} - \frac{\partial^2 u_y}{\partial z^2} \tag{6-7.2}$$

$$\tau = \frac{d\phi}{dz} + \frac{1}{R}\frac{\partial u_y}{\partial z}$$

On the cross-sectional area normal to the centroidal axis, the stresses can produce net transverse shear forces N_x and N_y, as well as a tension T. The moments of the stresses about the centroid produce bending couples G_x and G_y about the x and y axes, respectively, and a twisting couple H, about the centroidal or z axis. (This notation is similar to that found in Love, 1922.) Using this notation, there are equations of motion for the change of linear momentum (Newton's laws) and three equations representing the change of angular momentum (Euler's equations). For deformations with circumferential wavelengths that are long compared to the small radius a, we neglect the angular momentum of the differential element about a point on the centroid axis. With this assumption, the equations of motion become

$$\begin{aligned} \frac{\partial N_x}{\partial z} - N_y\tau + T\kappa_1 + f_x &= m\frac{\partial^2 u_x}{\partial t^2} \\ \frac{\partial N_y}{\partial z} - T\kappa_2 + N_x\tau + f_y &= m\frac{\partial^2 u_y}{\partial t^2} \\ \frac{\partial T}{\partial z} - N_x\kappa_1 + N_y\kappa_2 + f_z &= m\frac{\partial^2 u_z}{\partial t^2} \end{aligned} \tag{6-7.3}$$

$$\begin{aligned} \frac{\partial G_x}{\partial z} - G_y\tau + H\kappa_1 - N_y + c_x &= 0 \\ \frac{\partial G_y}{\partial z} - H\kappa_2 + G_x\tau + N_x + c_y &= 0 \\ \frac{\partial H}{\partial z} - G_x\kappa_1 + G_y\kappa_2 + c_z &= 0 \end{aligned} \tag{6-7.4}$$

In the above equations, $\mathbf{c}$ represents an applied body couple and $\mathbf{f}$ the magnetic body force. Diamagnetic effects and twisting of the filaments in the superconductor can lead to a nonzero body couple, but are neglected here. Thus we set $\mathbf{c} = 0$.

The body force $\mathbf{f}$ is the integral of the Lorentz force on the cross-sectional area, that is,

$$\mathbf{f} = \int \mathbf{J} \times (\mathbf{B}^0 + \mathbf{B}^1)\, dA \tag{6-7.5}$$

where $\mathbf{B}^0$ is due to external sources and $\mathbf{B}^1$ is the self-field.

To complete this set of equations, one requires a relation between the bending and twisting moments G_x, G_y, and H and the deformation. In the classical theory of elastic rings these relations are given by

$$\begin{aligned} G_x &= A\kappa_2 \\ G_y &= D\left(\kappa_1 - \frac{1}{R}\right) \\ H &= C\tau \end{aligned} \tag{6-7.6}$$

where A and D represent bending stiffnesses and C represents the torsional stiffness for the cross section.

Circular Coil in a Transverse Magnetic Field

A problem of practical importance is the buckling of the outer turns of a solenoid or ring magnet (see Fig. 6-27). This problem was analyzed by Daniels (1953), who observed such a failure in a normal conductor solenoid. (See also Fig. 4-1.)

To examine the elastic stability we split the body force into initial and deformation-dependent terms and linearize the equilibrium equations. The initial stress state in the circular flat ring is circumferential tension T_0 given by

$$T_0 = f_0 R \tag{6-7.7}$$

where f_0 is the Lorentz force (6-7.5) in the circular unbent configuration. For a coil with circular cross section, T_0 is given by

$$T_0 = \frac{\mu_0 I_0^2}{4\pi}\left[\ln\left(\frac{8R}{a}\right) - \frac{3}{4}\right] + I_0 B_y^0 R \tag{6-7.8}$$

Let us compare the two terms in Eq. (6-7.8) by forming the nondimensional group

$$\Gamma = \frac{\mu_0 I_0}{4\pi B_y^0} \frac{\ln(8R/a)}{R}$$

First, we note that as $R \to \infty$, $\Gamma \to 0$. Also, for large magnets where $I_0 = 10^3$ A, $B_y^0 = 10^{-1}$ T, $R = 1$ m, and $a = 10^{-2}$ m, we have $\Gamma = 10^{-3} \ln 800$. Thus

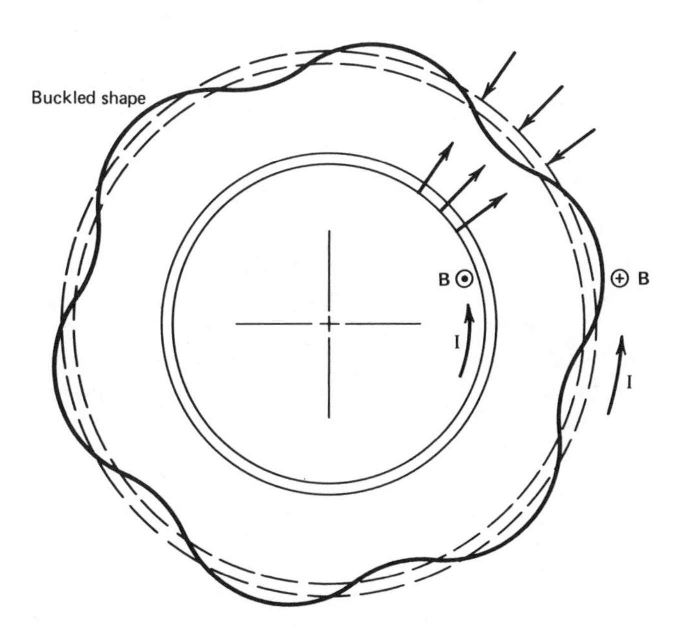

Figure 6-27 Buckling of outer turns of a solenoid.

we *neglect* the self-field forces in the following analysis, that is, we assume that $T_0 = I_0 B_y^0 R$.

There is, of course, an associated radial displacement with this tension; however, we reserve the symbol u_x for the radial deflection due to in-plane bending. The linearized equations of motion then take the following form:

$$\frac{\partial N_x}{\partial z} + T_0\left(\kappa_1 - \frac{1}{R}\right) + \frac{T_1}{R} + f_x^1 = m\frac{\partial^2 u_x}{\partial t^2}$$

$$\frac{\partial N_y}{\partial z} - T_0 \kappa_2 + f_y^1 = m\frac{\partial^2 u_y}{\partial t^2} \tag{6-7.9}$$

$$\frac{\partial T_1}{\partial z} - N_x \frac{1}{R} + f_z^1 = m\frac{\partial^2 u_z}{\partial t^2}$$

$$\frac{\partial G_x}{\partial z} + H\frac{1}{R} - N_y = 0$$

$$\frac{\partial G_y}{\partial z} + N_x = 0 \tag{6-7.10}$$

$$\frac{\partial H}{\partial z} - G_x \frac{1}{R} = 0$$

In these equations, T_1 is the added tension due to the bending and $\mathbf{f}^1$ is the perturbed magnetic body force. The terms $N_y\tau$ and $N_x\tau$ which appear in Eq. (6-7.3) are not retained in Eq. (6-7.9) because they are nonlinear in the displacements. We consider two special cases—plane and out-of-plane deformations. When the cross section is symmetric, f_x^1 depends only on u_x and f_y^1 depends only on u_y and ϕ, and we can uncouple the two cases.

In-Plane Deformation

For this case, we set $u_y = \phi = 0$. Also, it follows that $N_y = 0$ and $G_x = H = 0$. To simplify the analysis, consider the static case, that is, $\ddot{u}_x = 0$. To determine if buckling is possible, we look for a solution near the initially stressed state. Using Eqs. (6-7.2) and (6-7.6) it is easily shown that Eqs. (6-7.9) and (6-7.10) can be reduced to the following equations for the internal shear force $N_x \equiv N$ and the inward radial displacement $u_x \equiv u$ (using primes to indicate differentiation with respect to the original circumferential distance z):

$$N'' + T_0\left(u''' + \frac{u'}{R^2}\right) + \frac{N}{R^2} + \left(f_x' - \frac{f_z}{R}\right) = 0 \tag{6-7.11}$$

and

$$D\left(u''' + \frac{u'}{R^2}\right) + N = 0 \tag{6-7.12}$$

Here f_z and f_x represent *perturbed* magnetic force components. The force $\mathbf{f}$ is a consequence of two effects. First, if $u' \neq 0$, the direction of $\mathbf{I} \times \mathbf{B}$ is no longer radial. The inward, radial component of $\mathbf{I}$ produces a circumferential force component $f_z = I_0 B_0 u'$.

The second perturbed force results from the fact that when $u \neq 0$ the actual length of the differential element becomes $R(1 - u/R)d\theta$, instead of $Rd\theta = dz$, to terms first order in u. Thus the difference between the actual and initial radial force is $f_x = I_0 B_0 u/R$. This term is required in order for the buckling behavior to be independent of rigid-body translation in the plane of the ring. With these assumptions we have

$$\mathbf{f} = \frac{I_0 B_0 u}{R} \mathbf{e}_x + (I_0 B_0 u')\mathbf{e}_z \tag{6-7.13}$$

and

$$\frac{\partial f_x}{\partial z} - \frac{f_z}{R} = 0$$

In this analysis we have assumed that the applied field B_0 is uniform. If B_0 has a nonzero gradient, then an additional perturbed magnetic force will appear in Eq. (6-7.13). Thus the perturbed magnetic forces do not affect the buckling directly. Actually, their effect is represented in the term $T_0(\kappa - 1/R)$ which accounts for changes of direction or curvature of the current filament $\mathbf{I}_0$.

A solution which satisfies Eqs. (6-7.11) and (6-7.12) has the form

$$u = U_0 \cos \frac{nz}{R}$$

and

$$N = N_0 \sin \frac{nz}{R} \tag{6-7.14}$$

where n represents the number of wavelengths around the circumference. The resulting solution can be shown to be

$$U_0 = N_0 = 0 \qquad (I_0 B_0 \text{ arbitrary})$$

or

$$N_0 = 0, \qquad n = 1 \qquad [I_0 B_0 \text{ and } U_0 \text{ arbitrary (rigid-body case)}]$$

or

$$I_0 B_0 = -\frac{D}{R^3}(n^2 - 1) \qquad (U_0 \text{ arbitrary}) \tag{6-7.15}$$

where n is an integer greater than unity for a complete ring with no constraints. Thus the lowest buckling mode is $n = 2$.

We see that for buckling, the magnetic body force must produce a *compressive* initial stress in the ring and that $f^0 = |I_0 B_0|$ must attain a minimum value

$$f_c^0 = \frac{3D}{R^3} \tag{6-7.16}$$

For a ring with circular cross section of radius a, this value is given by

$$f_c^0 = \frac{3Ya}{4}\left(\frac{a}{R}\right)^3 \tag{6-7.17}$$

At this value it can be shown that the noncircular ring shape $u \neq 0$ is of lower energy than the circular shape $u = 0$. The linear theory, however, cannot determine the amplitude of u and a nonlinear analysis must be carried out (see Chapter 5).

Out-of-Plane Deformation

In this case we assume that $u_x = f_x = f_z = 0$. Also, it can be shown that $N_x = G_y = T_1 = 0$. When the center of current and the centroid coincide, the perturbed magnetic force f_y due to the external field can be shown to be proportional to ϕ. This is because the xyz coordinate system rotates with the element of the rod. However, the magnetic body force $I_0 B_y$ remains in the plane of the undisturbed coil thereby producing a component $I_0 B_y \phi$ along the rotated y axis.

It is instructive to include vibrating inertia terms in the analysis, so we assume that all variables vary as $e^{i\omega t}$ and we use the notation

$$u_y = ue^{i\omega t}, \qquad \phi \equiv \phi e^{i\omega t}$$

Using primes for differentiation with respect to z, the equations of motion can be reduced to two coupled equations for u and ϕ:

$$A\left(u'' - \frac{\phi}{R}\right)'' - \frac{C(\phi + u/R)''}{R} - I_0 B_0 R u'' = m_0 \omega^2 u$$

and

$$A\left(u'' - \frac{\phi}{R}\right) + CR\left(\phi + \frac{u}{R}\right)'' = 0 \tag{6-7.18}$$

As with the in-plane case we seek a solution in the form

$$\phi = \Phi_0 \cos\frac{nz}{R}$$

and (6-7.19)

$$u = U_0 \cos\frac{nz}{R}$$

When $I_0 B_0$ is arbitrary, Eqs. (6-7.18), along with the solutions (6-7.19), determine the natural frequency through the relation

$$m_0 \omega^2 = \frac{I_0 B_0 n^2}{R} + \frac{AC(n^2 - 1)^2 n^2}{R^4 (A + Cn^2)} \tag{6-7.20}$$

Thus we conclude that magnetically produced circumferential tension increases natural frequencies. Conversely, a compressive magnetic force decreases the out-of-plane stiffness of the coil and decreases the frequency. This may be expressed in the nondimensional form

$$\left(\frac{\omega}{\omega_0}\right)^2 = \left(1 - \frac{f}{f_c}\right) \tag{6-7.21}$$

where $f = -I_0B_0$ and ω_0 is the frequency when the force is absent. A critical value of the magnetic force results when $f = f_c$ and is given by

$$f_c = \frac{AC(n^2-1)^2}{R^3(A+Cn^2)} \tag{6-7.22}$$

This corresponds to *out-of-plane buckling*.

With this expression we can compare out-of-plane and in-plane buckling forces

$$\frac{(I_0B_0)_{\text{out}}}{(I_0B_0)_{\text{in}}} \equiv \eta = \frac{A}{D}\frac{(n^2-1)}{(n^2+A/C)} \tag{6-7.23}$$

From this formula we can conclude the following:

1. If $A \leq D$, out-of-plane buckling occurs before in-plane buckling for $I_0B_0 < 0$.
2. For $n = 2$, in-plane buckling can only occur if

$$A < \frac{4D}{3(1-D/3C)}, \qquad D < 3C$$

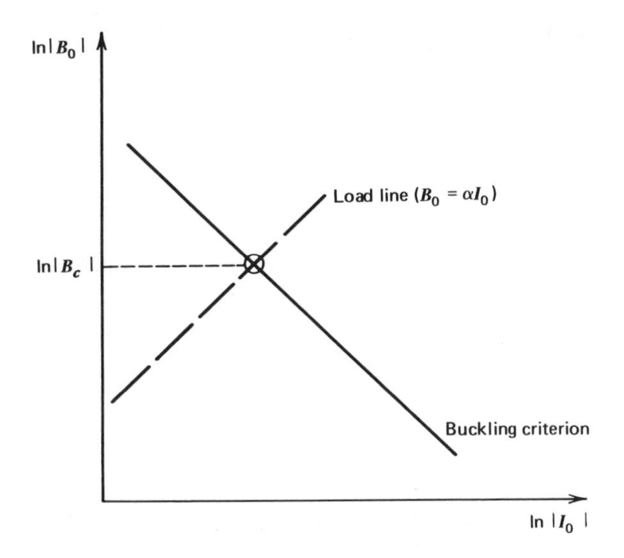

Figure 6-28 Planar buckling condition of a circular coil in a transverse magnetic field.

For a circular cross section, the buckling condition for out-of-plane deformation is ($n = 2$):

$$|I_0 B_0| = \frac{\pi}{2}\left(\frac{a}{R}\right)^3 aY \tag{6-7.24}$$

This is sketched on a log-log scale in Figure 6-28. If the conductor is part of a solenoid coil then B_0 is proportional to I, that is, $B_0 = \alpha I_0$ (which is similar to a load line also shown in Fig. 6-28), which determines the critical current.

Circular Coil in a Toroidal Magnetic Field

Another important example of magnetoelastic buckling of circular planar coils is the problem of a current-carrying coil in a toroidal magnetic field. This problem occurs in magnetic fusion tokamak reactors (see Fig. 1-5). Large circular coils of up to 10–20-m diameter, carrying 7×10^6 At of current, have been proposed. These coaxial parallel coils are used to induce the plasma current and stabilize the plasma. They create poloidal magnetic fields and are sometimes called PF coils in contrast to the toroidal field or TF coils. In some designs these coils are placed inside the TF coils in order to get closer to the plasma region and minimize the coil currents. This places the coils in a large toroidal magnetic field which is parallel to the coils. If the PF coils are precisely aligned with respect to the TF coils, the magnetic force on a PF coil element is zero,† that is, $\mathbf{I}_p \times \mathbf{B}_T = 0$. However, for small deformations or misalignments of the coils, perturbation forces will occur which tend to destabilize the PF coils.

In actual designs, both the coil and the supports around the circumference will have some flexibility or stiffness. However, to simplify the analysis we will examine two cases:

1. Elastic PF coil on rigid supports.
2. Rigid PF coil on elastic supports.

The first case was analyzed by the author in a U.S. Department of Energy contractors' report (Moon, 1976). A summary of the analysis is given here. The second case has been reported in the literature and experimental evidence for this type of buckling instability has been given (Moon, 1979a).

Case 1: elastic PF coil on rigid supports. The magnetic forces have two sources—one from the magnetic field of the ring itself and the other from the

†This is only an approximation for a toroidal field with no field ripple. However, in a discrete torus, the discrete nature of the coils will result in some radial magnetic field. It can be shown that this will lead to a torque between each of the PF coils and the TF coils. For a complete set of TF coils the net torque on the PF coil will vanish; however, periodic forces around the circumference will be present.

interaction with the toroidal magnetic field. As in the previous analysis of a ring in a transverse field, we neglect the effect of the self-field on both the initial tension in the coil and on the perturbed magnetic forces. When the circular magnet is undeformed, the interaction with the toroidal field is nil. However, when the ring is deformed (see Fig. 6-29), perturbation forces arise which are proportional to the deformation, that is,

$$f_x = I_0 B_0 \frac{\partial u_y}{\partial z}$$
$$f_y = -I_0 B_0 \frac{\partial u_x}{\partial z} \tag{6-7.25}$$

The perturbed forces in Eq. (6-7.25) couple the in-plane and out-of-plane motions and are derived from the Lorentz expression for magnetic force

$$\mathbf{f} = \mathbf{I} \times \mathbf{B}_0 = (I_0 \mathbf{e}_z + \boldsymbol{\Omega} \times I \mathbf{e}_z) \times B_0 \mathbf{e}_z \tag{6-7.26}$$

where the rotation vector of the center of current is given by

$$\boldsymbol{\Omega} = -\frac{\partial u_y}{\partial z} \mathbf{e}_x + \frac{\partial u_x}{\partial z} \mathbf{e}_y \tag{6-7.27}$$

To establish the existence of an instability, we seek values of $I_0 B_0$ which permit a nonzero deformation field of arbitrary amplitude. A buckling solution is a static deformation field which satisfies the equations of motion (6-7.9) and (6-7.10) with the dynamic terms dropped. These equations, along

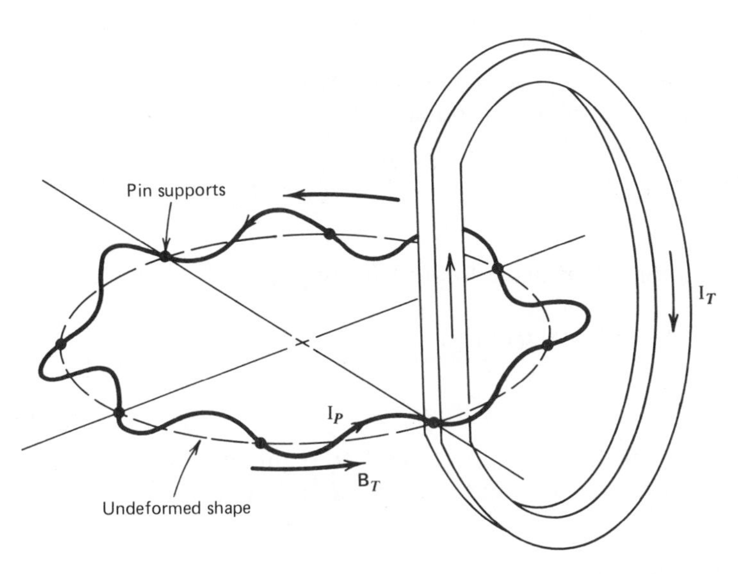

Figure 6-29 Helical buckling of an elastic poloidal coil in a toroidal magnetic field.

with the constitutive equations (6-7.6), take the form (where a prime denotes a derivative with respect to the arc length z)

$$\left[\frac{Cu_y}{R^2} - Au_y''\right]'' + \frac{A+C}{R}\phi'' - I_0 B_0 u_x' = 0$$

$$D\left[\frac{u_x}{R^2} + u_x''\right]'' - \frac{T}{R} - I_0 B_0 u_y' = 0$$

$$\frac{A+C}{R}u_y'' + C\phi'' - \frac{A}{R^2}\phi = 0 \tag{6-7.28}$$

$$T'' + \frac{T}{R^2} + \frac{I_0 B_0}{R}u_y' = 0$$

If the number of pin supports is M, we will seek solutions of the form

$$\begin{bmatrix} u_x \\ u_y \end{bmatrix} = \begin{bmatrix} a_1 \\ a_2 \end{bmatrix} e^{iMz/R} \tag{6-7.29}$$

A solution can be found which will satisfy both the equilibrium equations (6-7.29) and the pin constraints at

$$z_n = \frac{2\pi R}{M}n, \qquad n = 0, 1, \ldots, M-1 \tag{6-7.30}$$

$$u_x = U_0\left(1 - \cos\frac{Mz}{R}\right)$$

$$u_y = \alpha U_0 \sin\frac{Mz}{R} \tag{6-7.31}$$

This deformation is in the form of circular helix (see Fig. 6-29). The values of $I_0 B_0$ for which this solution is possible are

$$I_0 B_0 = \frac{D^{1/2}}{R^3}(M^2 - 1)\left(\frac{(C + M^2 A)(A + M^2 C) - M^2(A + C)^2}{A + M^2 C}\right)^{1/2} \tag{6-7.32}$$

and the ratio of out-of-plane to in-plane amplitudes is

$$\alpha = -\frac{D^{1/2}(M^2 - 1)}{M}\left(\frac{(C + M^2 A)(A + M^2 C) - (A + C)^2 M^2}{A + M^2 C}\right)^{-1/2} \tag{6-7.33}$$

To get an idea of the magnitude of $I_0 B_0$ required for buckling, consider a circular cross section of homogeneous material. For this case

$$A = D = \tfrac{1}{2}C = \tfrac{1}{4}Y\pi r_0^4$$

For this case, Eq. (6-7.32) becomes

$$I_0 B_0 = \frac{\pi}{4} YR\left(\frac{r_0}{R}\right)^4 (M^2-1)\left(\frac{(2+M^2)(1+2M^2)-9M^2}{1+2M^2}\right)^{1/2} \tag{6-7.34}$$

Thus associated with the PF coil in a toroidal magnetic field is a nondimensional number expressing the ratio of magnetic buckling force to elastic restoring force

$$\text{M.E.} \equiv \frac{I_0 B_0}{Y\pi R}\left(\frac{R}{r_0}\right)^4 \tag{6-7.35}$$

For a given number of constraints M, this magnetoelastic number (M.E.) must satisfy the *stability condition*.

$$\text{M.E.} < G(M)$$

where

$$G(M) \equiv \frac{M^2-1}{4}\left(\frac{(2+M^2)(1+2M^2)-9M^2}{1+2M^2}\right)^{1/2} \tag{6-7.36}$$

The value of $G(M)$ for different numbers of pin constraints is shown in Table 6-2.

Table 6-2 Values of Magnetoelastic Number [Eq. 6-7.36] for Different Numbers of Pin Constraints

M	$G(M)^a$	$\frac{1}{4}M^3$
1	0	
2	1.061	2
4	13.85	16
6		54
8	123.5	128
16		1024
24	3443	3456

[a] The value of $\frac{1}{4}M^3$ is a good approximation for G when the number of constraints $M > 8$.

Example: Application of this theory to a PF coil in the design of a tokamak reactor UWMAK-III by the University of Wisconsin fusion group can readily be made using the following parameters (see Fig. 6-30):

D–6 "diverter coil"

$R = 19.0\text{ m}$

$I_0 = 6.8 \times 10^6\text{ A}$

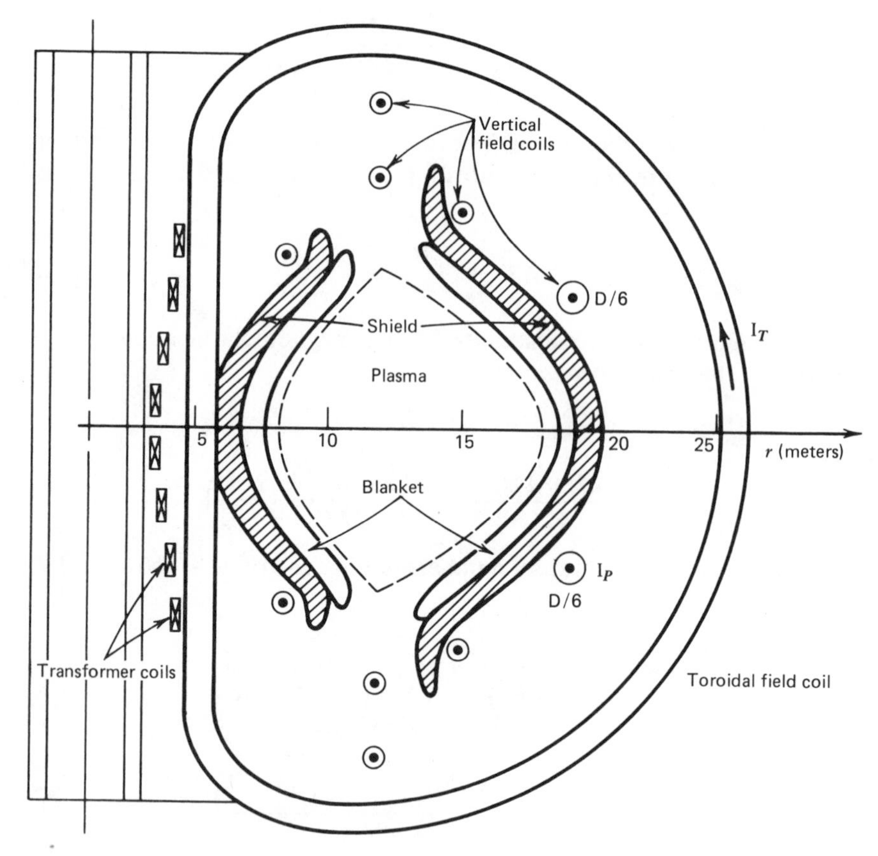

Figure 6-30 Cross section of a toroidal magnetic fusion reactor (tokamak) showing position of veretical (poloidal) field coils (from University of Wisconsin design UWMAK-III).

$B_0 = 2.5\ \text{Wb/m}^2$
$r_0 = 0.23\ \text{m}$
$Y = 2 \times 10^{11}\ \text{N/m}^2$
based on $J = 4100\ \text{A/cm}^2$ in the composite superconductor

For these parameters

$$\text{M.E.} = \frac{I_0 B_0 R^3}{Y \pi r_0^4} = 35.7$$

This implies that the coil will need at least 6–8 supports to avoid helical buckling in the toroidal magnetic field.

Case 2: rigid PF Coil on Elastic Supports. For a rigid circular coil in a toroidal magnetic field, there will be six degrees of freedom—three translation and three rotation modes.

Because of the symmetries present both in the coil and the circumferential magnetic field, one can show that a simple translation alone or pure pitch will not produce a net magnetic force or moment on the ring. It can be shown, however, that a combination of translation in the plane and pitch about an axis parallel to the translation will produce both a magnetic force and moment.

The total force on the displaced ring can be calculated by integrating the Lorentz force $\mathbf{I}_P \times \mathbf{B}_T$ around the ring (see, e.g., Moon, 1979a). The magnetic couple can be found in a similar way. One can also use an equivalent method based on the magnetic interaction energy between the poloidal coil and the toroidal field. In this method the change in magnetic energy of the system, $\mathscr{W}$, with the generalized displacements q_i is related to changes in the elastic energy function $\mathscr{V}$ (see Section 3.5), that is,

$$\frac{\partial \mathscr{W}}{\partial q_i} = \frac{\partial \mathscr{V}}{\partial q_i} \tag{6-7.37}$$

If U and Ω represent generalized coordinates, the magnetic force and couple can be derived from a magnetic potential energy $\mathscr{W}$ which is derived below.

In this analysis we use cylindrical coordinates with the origin placed at the center of the original position and the z axis normal to the original plane of the coil. The position of each element of the displaced coil is given by (see Fig. 6-31)

$$\mathbf{r} = (R + x)\mathbf{e}_x + z\mathbf{e}_z$$

Then imagine a radial $x - z$ plane at an angle θ. As this plane sweeps around the torus, the locus of points $(x(\theta), z(\theta))$ traces out a closed oval-shaped curve of area A on this plane. Since the toroidal magnetic field is normal to the radial plane, the toroidal flux linkage with the poloidal coil is

$$\Phi_T = \int_A B_T \, dA \tag{6-7.38}$$

Thus the magnetic interaction energy between the toroidal field and the poloidal coil of current I_P is

$$\mathscr{W} = -\Phi_T I_P \tag{6-7.39}$$

If N_T is the number of coils in the torus and I_T the current in each coil, then the flux density of the toroidal field is given by†

$$B_T = \frac{\mu_0 N_T I_T}{2\pi(R + x)} \tag{6-7.40}$$

†This expression neglects ripple in the field due to the discrete coil torus.

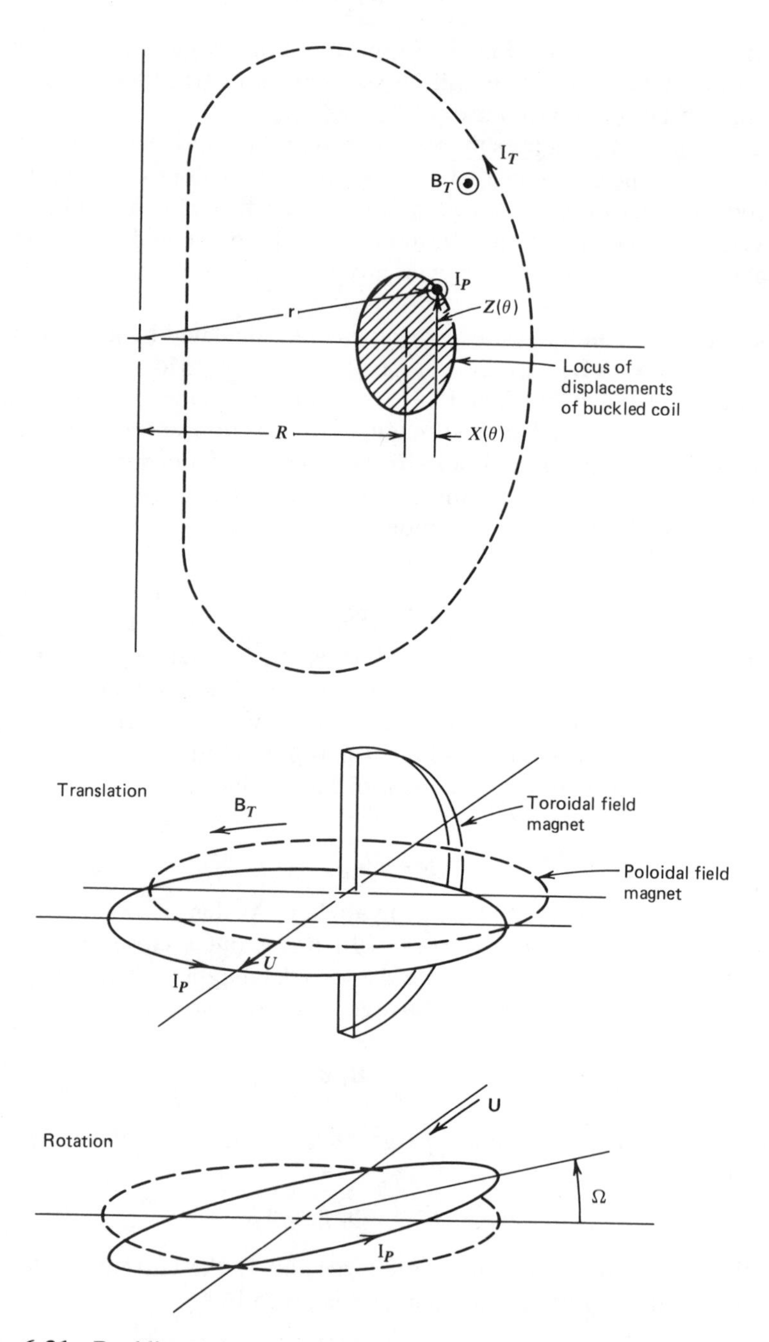

Figure 6-31 Buckling deformations of a rigid poloidal coil on elastic supports in a toroidal magnetic field.

To integrate the flux through the oval area, we note that B_T is independent of $z(\theta)$ and is symmetric about the $z = 0$ plane so that Φ_T becomes

$$\Phi_T = \frac{\mu_0 N_T I_T}{\pi} \int_A \frac{dA}{R+x} = 2B_0 \int_0^{\pi} \frac{z(\theta)}{1 + x(\theta)/R} \frac{dx}{d\theta} d\theta \tag{6-7.41}$$

where $B_0 \equiv \mu_0 N_T I_T / 2\pi R$ is the value of a toroidal field at the original position of the poloidal coil.

Thus, when $z(\theta)$ and $x(\theta)$ are known functions, the flux can be found. For example, in the linear approximation

$$x = U \cos\theta, \qquad z = V \sin\theta \tag{6-7.42}$$

where $V \equiv \Omega R$. One can easily show that the curve $z(x)$ is an ellipse. In this case the flux is given by the simple expression $\Phi_T = B_0 \pi UV$, and the magnetic energy $\mathscr{W}$ [Eq. (6-7.39)] becomes

$$\mathscr{W} = -\pi UVI_p B_0 \tag{6-7.43}$$

This is sufficient for the determination of the buckling current field product $I_P B_0$.

The elastic energy of the coil attached to spring supports is assumed to be of the form

$$\mathscr{V} = \tfrac{1}{2} K_1 U^2 + \tfrac{1}{2} K_2 V^2 \tag{6-7.44}$$

The extension of Eq. (6-7.37) to the dynamic case involves constructing a Lagrangian of the form (see Section 3.5):

$$\mathscr{L} = \tfrac{1}{2} m \dot{U}^2 + \tfrac{1}{2} \mathscr{J} \dot{\Omega}^2 - \mathscr{V} + \mathscr{W} \tag{6-7.45}$$

where m is the mass of the ring and $\mathscr{J}$ is the moment of inertia. One obtains two coupled equations of motion for the displacement U and the pitch rotation $\Omega = V/R$,

$$m\ddot{U} + K_1 U = -I_P B_0 \pi R \Omega \tag{6-7.46}$$

$$\mathscr{J}\ddot{\Omega} + K_2 R^2 \Omega = -I_P B_0 \pi R U$$

One can show that this leads to two modes—in one case the natural frequency increases with $I_P B_0$, while in the other case the frequency decreases with $I_P B_0$. For the case when the zero current frequencies are equal, that is, $K_1/m = K_2 R^2/\mathscr{J}$, the nonzero current frequencies become

$$\omega^2 = \omega_0^2 \pm \frac{\pi I_P B_0 R}{(m\mathscr{J})^{1/2}} \tag{6-7.47}$$

A buckling instability occurs when one of the frequencies goes to zero, or when

$$I_P B_0 = \frac{\omega_0^2 (m\mathscr{J})^{1/2}}{\pi R} \tag{6-7.48}$$

When the frequencies are not equal the condition for buckling becomes

$$I_P B_0 = \frac{(K_1 K_2)^{1/2}}{\pi} \tag{6-7.49}$$

This instability has been observed experimentally by the author (Moon, 1979a). The toroidal magnetic field was created by a 16-coil, 30-cm-diam superconducting torus with a 0.4-T toroidal magnetic field. A circular 16-turn superconducting coil was threaded around the torus inside the toroidal coils and *connected in series* with the 16 toroidal magnets. This arrangement allowed the use of one power supply. Since the magnetic perturbation force on the coil is proportional to $I_P B_T$, and B_T is proportional to I_P by virtue of the series connection of all the coils, the magnetic forces were proportional to I_P^2. The elastic restoring forces were provided by four stainless-steel springs attached between the circular poloidal coil and the plastic support structure.

The frequencies of the lateral and pitching modes showed a decrease with current. The zero frequency intercept for the lower mode showed a critical buckling current of 90.0 At or about 15% above the value determined from the static displacement tests. The deflection of the coil showed a large increase near 80 At (see Fig. 6-32*a*). A plot of deflection δ versus δ/I^2 (called a Southwell plot) showed a slope of 79.1 At for the buckling current.

The theoretical prediction gave a value of $I_0 = 68.2$ At as compared to the experimental value of 90.0 At from the vibration tests and 79.1 At from the deflection measurements.

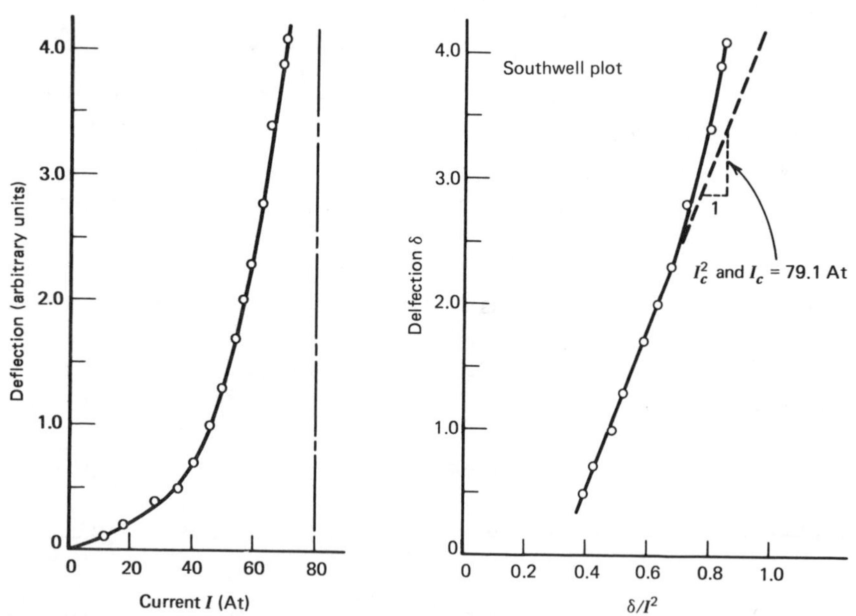

Figure 6-32 Experimental buckling of a PF coil in a toroidal magnetic field: (*a*) deflection vs. current and (*b*) Southwell plot—deflection δ vs. δ/I^2.

Periodic Array of Magnets

When a magnetic field is required over an extended volume, such as in a magnetic torus, stellarator, or high-energy particle experiment, magnets are often employed in a periodic array as shown in Figure 6-33. In this case each magnet has the identical geometry and is in the same magnetic field environment as every other magnet. (In an open system, such as a discrete coil solenoid, this may hold for only the magnets in the central section since the end magnets will experience magnetic forces that are different from the central magnets.) The vibrations and elastic stability of each magnet are coupled to those of its neighboring magnets. One system that has been studied quite extensively is the magnetic fusion TF coil array (Moon and Swanson, 1976; Swanson and Moon, 1977; Moon, 1979b, Miya et al., 1980; Miya and Uesaka, 1982).

The analysis of such systems is similar to that of a periodic chain of masses and springs (see, e.g., Brillouin, 1953). We now discuss the procedure to

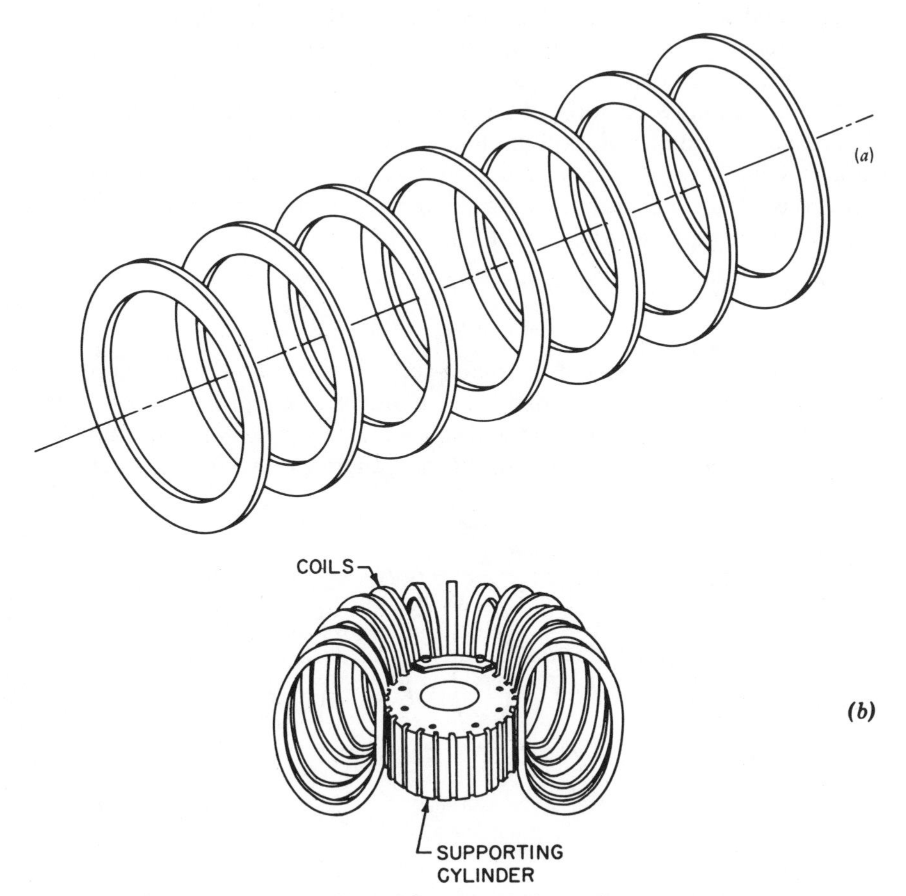

Figure 6-33 Periodic arrays of magnets: (*a*) linear discrete coil solenoid and (*b*) discrete coil torus.

determine the vibration frequencies and modes of the set of magnets, as well as possible buckling modes.

Central to the analysis of a periodic array of objects is the concept of a *cell*. The cell is the smallest group of magnets that repeat. While two- and three-dimensional cell structures could be examined, the linear array will be discussed here. Each cell will be identified by an integer n (see Fig. 6-34).

Further, though each magnet theoretically has an infinite number of degrees of freedom, we consider only a finite set of deformation modes to be identified by a nodal amplitude $u_{n\alpha}$. For example, the out-of-plane vibrations of a circular ring may be expressd in terms of harmonic functions of the circumferential coordinate z,

$$v(z, t) = \sum_{\alpha=1}^{\infty} u_\alpha(t)\cos\frac{\alpha\pi z}{L} \tag{6-7.50}$$

where v is the out-of-plane displacement and L is the length of the circumference. Thus $u_{n\alpha}(t)$ is the α deformation mode amplitude in the nth magnet and is not identified with the deformation of a particular point, but is treated as a *generalized coordinate*. The set $\{u_{n\alpha}\}$ can also include the rigid-body modes of the magnets in each cell.

The equations of motion of such a set of magnets can be derived from Lagrange's equations of motion. To do this one obtains expressions for the kinetic energy $\mathcal{T}$, elastic potential energy $\mathcal{V}$, and magnetic energy $\mathcal{W}$ as functions of the generalized coordinates $u_{n\alpha}(t)$ and generalized velocities $\dot{u}_{n\alpha}(t)$. These equations take the form (see Section 3.5):

$$\frac{d}{dt}\frac{\partial\mathcal{T}}{\partial\dot{u}_{n\alpha}} - \frac{\partial\mathcal{T}}{\partial u_{n\alpha}} + \frac{\partial\mathcal{V}}{\partial u_{n\alpha}} = \frac{\partial\mathcal{W}}{\partial u_{n\alpha}} \tag{6-7.51}$$

Other forces acting on the magnets, such as those producing damping or time-dependent forces, are neglected here. In many problems, the dependence of the kinetic energy on the displacements can be neglected and $\mathcal{T}$ can be written as a quadratic function of the generalized velocities; that is,

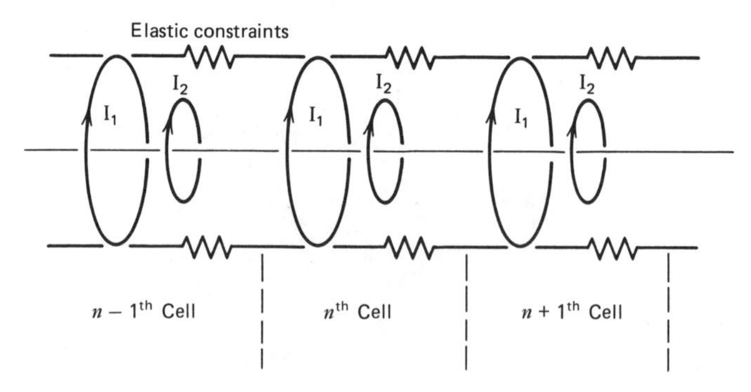

Figure 6-34 Definition of cells in a periodic array of magnets.

$$\mathscr{T}=\frac{1}{2}\sum_{\alpha}\sum_{\beta} m_{\alpha\beta}\dot{u}_{n\alpha}\dot{u}_{n\beta} \tag{6-7.52}$$

where $\mathscr{T}$ is the kinetic energy of the nth cell and $m_{\alpha\beta}$ is called the mass matrix.

The elastic potential energy has two parts, that is, $\mathscr{V}=\mathscr{V}_1+\mathscr{V}_2$. One part involves only the displacements in the nth cell and is written as

$$\mathscr{V}_1=\frac{1}{2}\sum_{\alpha}\sum_{\beta} k_{\alpha\beta}u_{n\alpha}u_{n\beta} \tag{6-7.53}$$

where $k_{\alpha\beta}$ is called the elastic stiffness matrix. This includes elastic constraints between magnets in the same cell and between magnets and the outside environment. The other part of $\mathscr{V}$ includes the interaction between the α mode in the nth cell and the β mode in the $(n+m)$th cell. Similarly, one can split $\mathscr{W}$ into a part that involves magnetic forces between modes or magnets within one cell and forces between modes in different cells. From Section 4.6, we have

$$\frac{\partial \mathscr{W}}{\partial u_{n\alpha}}=F_{\alpha}^{0}-\sum_{\beta}\kappa_{\alpha\beta}u_{n\beta} \tag{6-7.54}$$

where $\kappa_{\alpha\beta}$ is called the *magnetic stiffness matrix*. We further assume, for simplicity, that the initial magnetic forces F_{α}^{0} are equilibrated. The equations of motion can be shown to take the form

$$\sum_{\beta} m_{\alpha\beta}\ddot{u}_{n\beta}+\sum_{\beta}(k_{\alpha\beta}+\kappa_{\alpha\beta})u_{n\beta}=\sum_{m}(f_{nm}^{(\alpha)}+F_{nm}^{(\alpha)}) \tag{6-7.55}$$

where $f_{nm}^{(\alpha)}$ and $F_{nm}^{(\alpha)}$ represent the interaction between the elastic and magnetic forces in different cells.

To illustrate the method, we consider a problem with one degree of freedom per cell. Further, we assume that the forces between cells only depend on the difference between modal displacements u_n-u_{n+m}, with only nearest neighbor interaction between the elastic constraints. With these simplifying assumptions, the equation of motion for the nth cell becomes

$$m\ddot{u}_n+(k_0+\kappa_0)u_n+k_1(2u_n-u_{n-1}-u_{n+1})=\sum_{m}F_{nm} \tag{6-7.56}$$

If I is the current in the magnets, then we can write F_{nm} in the form

$$F_{nm}=I^2\lambda_m(2u_n-u_{n-m}-u_{n+m}) \tag{6-7.57}$$

The magnetic stiffness between cells is $\kappa_m=-I^2\lambda_m$. If the λ_m are positive the magnetic forces are destabilizing.

Equation (6-7.56) is called a *difference differential equation*. It can be shown that a vibrating solution has the form

$$u_n=Ue^{i\omega t}e^{iqn} \tag{6-7.58}$$

where q is the phase shift between cells. We note that $2\pi/q$ represents a nondimensional wavelength. The equation of motion (6-7.56), along with Eq. (6-7.57), establishes a relation between the frequency ω and the phase shift q which is given by

$$m\omega^2 = k_0 + \kappa_0 + 4k_1 \sin^2 \frac{q}{2} - 4I^2 \sum_{m=1}^{M} \lambda_m \sin^2 \frac{mq}{2} \tag{6-7.59}$$

where M is the number of symmetric pairs of magnets near the nth cell.

Before interpreting this result, we note that κ_0 depends on the current I and can be positive or negative. If $\kappa_0 = 0$, then the magnetic forces will decrease the natural frequencies if $\lambda_m > 0$ for all m. In practice, only nearest and next nearest neighbors are important since the mutual magnetic forces decrease quite rapidly as the distance between cells increases.

So far, Eq. (6-7.59) does not determine a unique relation for the frequency. For this, one requires boundary conditions. Consider the example of a cyclic set of magnets as in a tokamak or stellarator. If N is the number of magnets, then

$$\begin{aligned} M &= \tfrac{1}{2}(N-1) \qquad (N \text{ odd}) \\ &= \tfrac{1}{2}N \qquad\qquad (N \text{ even}) \end{aligned}$$

and the phase number must satisfy the relation

$$e^{iqn} = e^{iq(n+N)}$$

This leads to the condition

$$q = \frac{2\pi\alpha}{N} \tag{6-7.60}$$

where α is an integer. The motion is then a linear combination of modes given by

$$u_n(t) = \sum_{\alpha=0}^{N-1} U_\alpha e^{i\omega_\alpha t} e^{i2\pi n\alpha/N} \tag{6-7.61a}$$

where

$$m_0\omega_\alpha^2 = k_0 + \kappa_0 + 4k_1 \sin^2 \frac{\pi\alpha}{N} - 4I^2 \sum_{m=1}^{M} \lambda_m \sin^2 \frac{\pi m\alpha}{N} \tag{6-7.61b}$$

As a specific example, consider the case where $\kappa_0 = 0$ and $M = 1$. When $\lambda_1 > 0$, the different frequencies ω_α decrease with increase in current. Buckling occurs when the total stiffness is zero or when one of the frequencies $\omega_\alpha \to 0$. For N even, this occurs for the mode $\alpha = \frac{1}{2}N$ and the critical value of the current for buckling is given by

$$I_0^2 = \frac{k_0 + 4k_1}{\lambda_1} \tag{6-7.62}$$

Since all the coils are coupled magnetically, all the coils buckle in a specific mode. For the case under discussion, the phase difference between cells during buckling is given by $q = 2\pi\alpha/N = \pi$. Thus $u_{n+1} = -u_n$ or the set

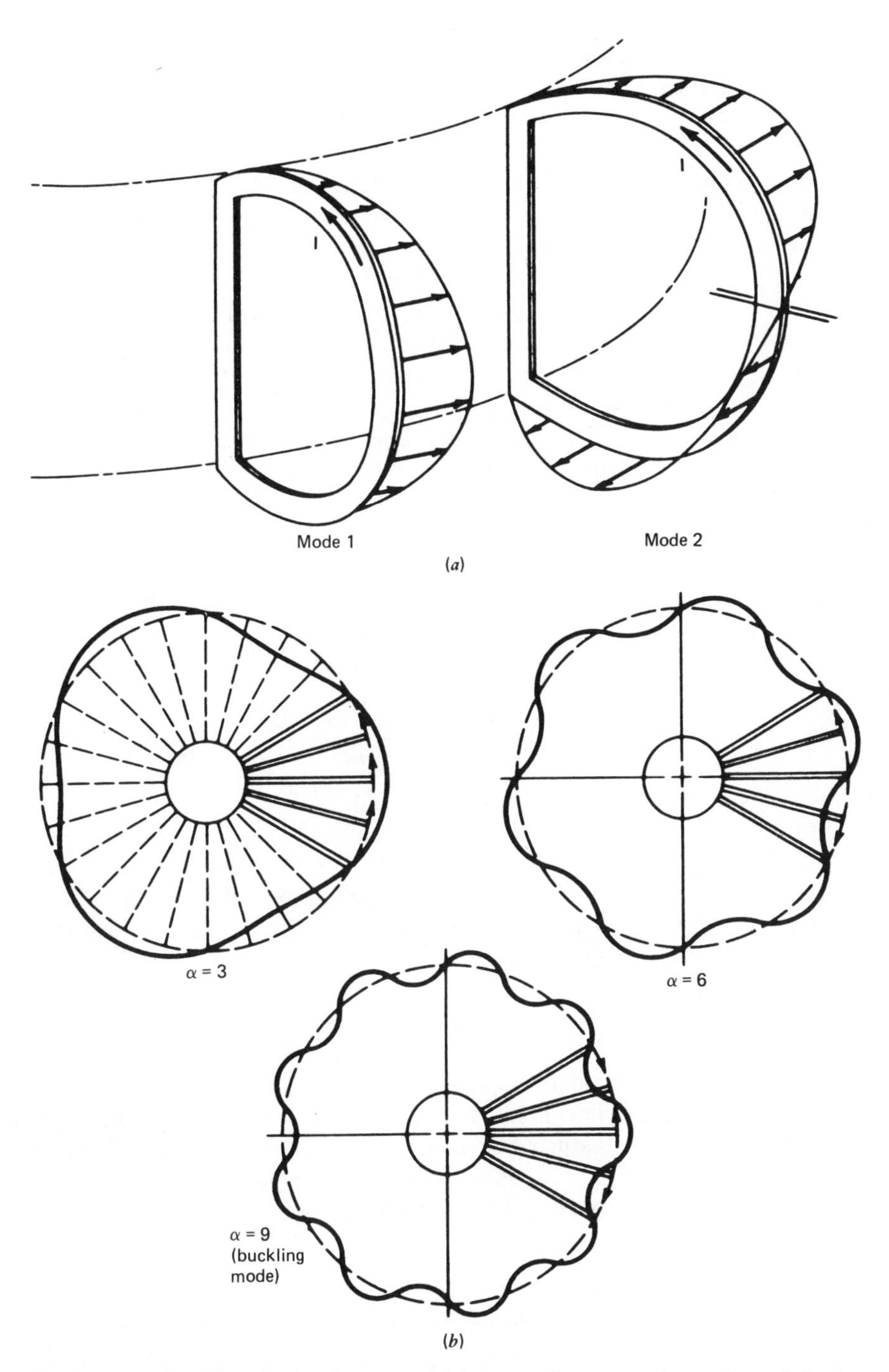

Figure 6-35 Buckling modes for a toroidal array of magnets in a tokamak fusion reactor: (*a*) lateral bending modes and (*b*) toroidal vibration modes.

buckles in an *alternating pattern* with the shortest possible wavelength between cells of equal phase. This buckling pattern is shown in Figure 6-35 for a toroidal array of TF coils. It should be noted that the other modal patterns also have critical currents at which $\omega_\alpha \to 0$, but these critical currents are higher than the value given in Eq. (6-7.62).

To calculate the magnetic stiffness constants λ_m, we note that the magnetic force represents the total force between the nth coil and its $n+m$ and $n-m$ neighbors; that is,

$$I^2\lambda_m 2u_n = I^2\frac{\partial}{\partial u_n}(L_{n(n+m)} + L_{n(n-m)}) \tag{6-7.63}$$

where L_{kl} is the mutual inductance between the kth and lth coils. In practice, this calculation is done numerically.

For nearest neighbor interactions, the frequency-current relation (6-7.61b) can be written in the form

$$\omega_\alpha^2 = \omega_{0\alpha}^2\left(1 - \frac{I^2}{I_{\alpha c}^2}\right) \tag{6-7.64}$$

where

$$\omega_{0\alpha}^2 = \frac{k_0 + 4k_1 \sin^2(\pi\alpha/N)}{m_0}$$

and

$$I_{\alpha c}^2 = \frac{k_0 + 4k_1 \sin^2(\pi\alpha/N)}{4\lambda_1 \sin^2(\pi\alpha/N)} \qquad (\lambda_1 > 0)$$

When $\lambda_1 < 0$, the frequency equation becomes

$$\omega_\alpha^2 = \omega_{0\alpha}^2(1 + \gamma_\alpha I^2)$$

It is clear that in these coupled magnetoelastic systems, the dynamic response of the set of magnets will depend on the current or the stored magnetic energy in the system.

Waves

Another phenomenon that appears in periodic magnet systems is the occurrence of waves. This is illustrated in Figure 6-36 for a toroidal set of coils with damping. A disturbance in one coil will propagate around the torus in a set of waves. The cell phase velocity is given by $v_\alpha = \omega_\alpha/q_\alpha$. This phenomenon has been observed experimentally in a superconducting 16-coil torus (Moon, 1980b).

Buckling Current Estimates for Full-Scale Reactors

The structural complexities of full-scale fusion tokamak reactor designs make it very difficult to determine the lateral stiffness of these magnets without using numerical methods, such as the finite-element method (see

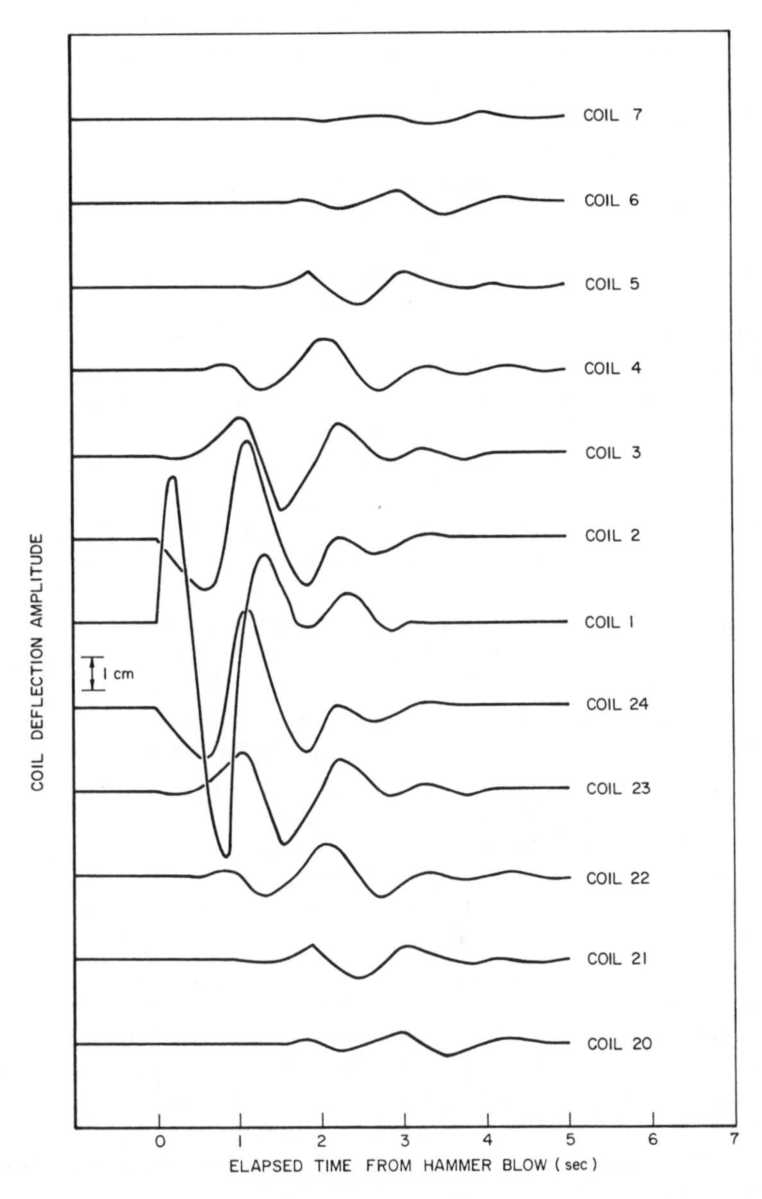

Figure 6-36 Wave propagation in a array of TF coils. (From Moon and Swanson, 1976, with permission. American Institute of Physics, copyright 1976.)

Miya et al., 1980; Miya and Uesaka, 1982). However, it is instructive to make estimates of buckling currents to see if the phenomenon will be important in full-scale fusion structures. In the example examined here we consider an early design for an experimental power reactor (EPR) from Argonne National Laboratory.

This design uses 16 toroidal field, "*D*"-shaped magnets. The coils are clamped along the inner straight legs and the number of pin supports along the outer circumference was varied from zero to two (see Table 6-3). The proposed magnet structure was a set of superconducting pancake magnets, in series, enclosed by a stainless-steel case. However, it has been found experimentally that a pancake array in an enclosed case does not act as a solid homogeneous beam in bending. For example, the bending stiffness of a stack of N separate plates of thickness d is $1/N^2$ that of a solid plate of thickness Nd. Thus, in our estimates of the lateral stiffness, we have used only the stiffness of the stainless-steel case surrounding the magnets.

The magnetic stiffness, or the second variation of the magnetic energy, was calculated numerically using a finite-element program for the inductance between one bent coil and two neighboring unbent coils for zero, one, and two pin constraints.

Estimates of the critical buckling current in ampere turns per magnet are shown in Table 6-3 for a clamped inner edge of the magnets. Estimates for a pinned inner edge are also given. The design current for the EPR design was 7×10^6 At per magnet. In all designs proposed thus far at least two pin constraints have been used. The buckling current for the two-pin constraint for the clamped inner edge is 16.4 MA, while that for the pinned inner edge is 10.8 MA. While these values are above the design current, it should be noted that small *misalignments* or *out-of-plane imperfections* will be amplified as one approaches the buckling current. Estimates of 1.5- to 8.4-mm lateral deflections at operating current for the two-pin mode were made for a 0.5% angular misalignment of the coils. Thus the combined stability and lateral imperfection analysis would suggest additional constraints between the coils.

Finally, it should be noted that magnetoelastic buckling of superconducting magnets is a *large-scale phenomena*. That is, it will not show up in laboratory magnets where low structural mass is not a premium. As magnets approach the 10-m-diam scale, however, the combination of higher magnetic forces (which increase as the square of the current) and the need to minimize structural mass will increase the possibility that magnetoelastic buckling will be a problem.

Experiments—Buckling of Toroidal Field Magnets

Experiments on magnetoelastic buckling of superconducting magnets have been performed by the author and co-workers and by Miya and co-workers in Japan. Almost all the predictions of the theory described above have been observed in small superconducting models of both an 8-coil and a 16-coil torus.

Table 6-3 Current Limitations on TF Coils in a 16-Coil Torus Due to Buckling and Misalignments

Bending Mode	Modal Bending Stiffness k_0 Clamped Base	Difference Modal Inductance λ_1 Clamped Base	Buckling Current Clamped at Base	Buckling Current Pinned at Base	Maximum Coil Deflection Due to Misalignments (0.5%)	
	(10^6 N/m)	(10^{-7} H/m^2)	(10^6 At per coil)	(10^6 At per coil)	Clamped (mm)	Pinned (mm)
1	33.3	4	4.56	n.a.[a]	—	—
2	147.7	5.70	8.57	4.94	—	—
3	780.3	7.12	16.4	10.8	1.5	8.4

[a]Not applicable.

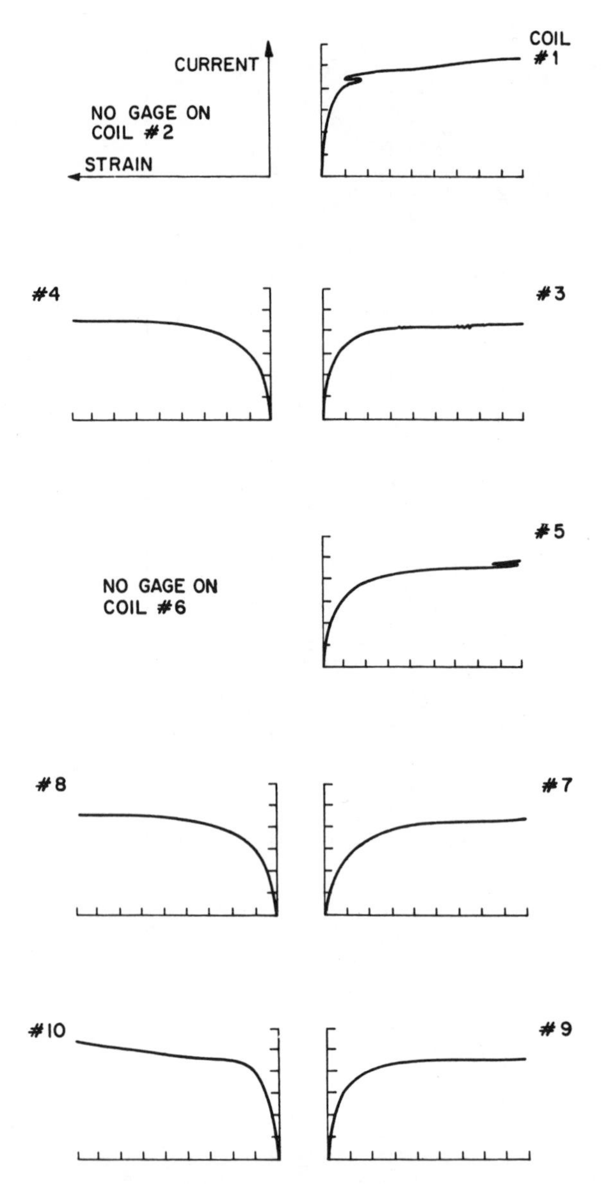

Figure 6-37 Experimental buckling in eight instrumented coils in a 16-coil torus: current vs. bending strain (from Moon, 1979b).

Table 6-4 Comparison of Theoretical and Experimental Buckling Currents for a 16-Coil Torus

Number of Coils	Mode	Coil Number	Experimental Values (At) Static	Experimental Values (At) Dynamic	Theoretical Values (At) Tension Correction Without	Theoretical Values (At) Tension Correction With
8	1	2	104	109	87.6	—
		3	114		88.5	—
16	1	3	41.8	44.8	46.5	—
		4	42.4	44.5	47.2	—
	2	3	—	117	96.9	107
		4	102–149	117	99.8	111
		7	161–213	—	—	—
		8	—	153	98.0	110
		9	125–161	—	—	—
	3	7	214	—	—	—
		8	214	—	186	239
		9	291	328 (9 and 10)	198	266
		10	248		189	246

Source: Moon (1979b).

First, simultaneous buckling of the complete set of coils was observed as the current approached the lowest critical buckling value as illustrated in Figure 6-37 (Swanson and Moon, 1977; Moon, 1979b). Second, the natural frequencies of almost all of the toroidal vibration modes were observed to decrease as shown in Figure 6-38. One mode, in which all magnets were in phase, was observed to suffer an increase in natural frequency. This effect is attributed to the tension in the coils produced by the magnetic pressure in the torus. As additional lateral constraints were added to coils, in the form of pin supports, this tension effect became larger. A summary of the comparison of experimental and theoretical values of the buckling current is given in Table 6-4 for a 16-coil torus. In this comparison, the bending stiffness of the coils was measured using the zero current natural frequency of the coils.

Miya et al. (1980) and Miya and Uesaka (1982) used a finite-element model to calculate the stiffness of coils in a superconducting experimental model and obtain reasonable agreement (within 6–22%) with experimental measurements of buckling current.

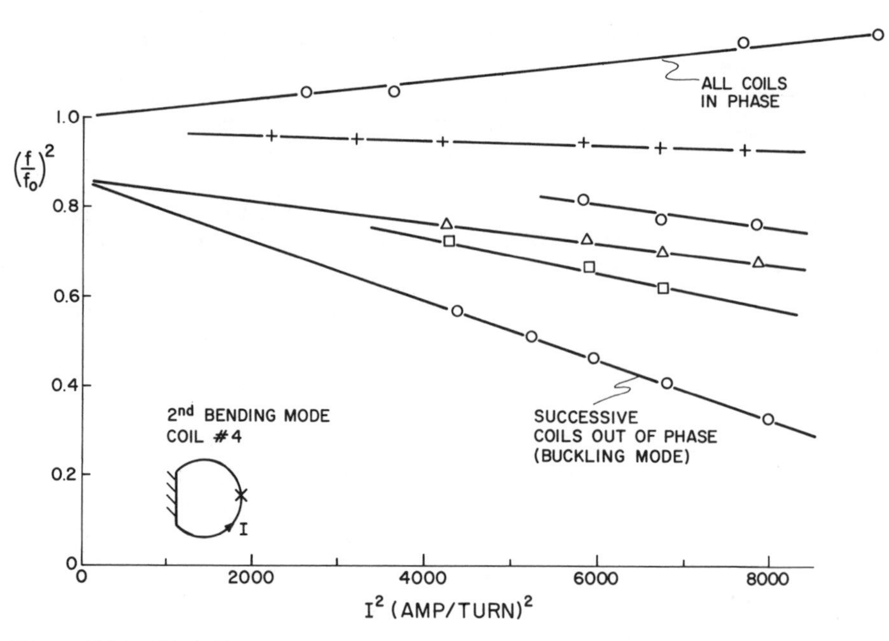

Figure 6-38 The effect of current on the vibration frequencies of toroidal field coils in a 16-coil torus (from Moon, 1980b).

Misalignment Sensitivity—Solenoid Coils for a Tandem Mirror Machine

One of the confinement schemes for magnetic fusion energy devices is a set of circular coils arranged in a discrete coil solenoid with end coils called a yin–yang pair (see Fig. 6-39). To design the support system for this set of coils, it is necessary to understand which modes are unstable under magnetic forces. These forces will act to move the coils out of alignment, even when the operating current is below the lowest buckling current. This problem is

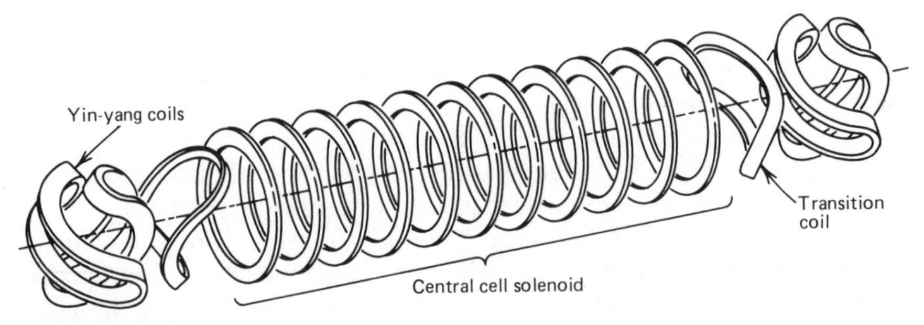

Figure 6-39 Sketch of solenoid coils for a tandem mirror fusion reactor (MFTF-B). (Illustration courtesy of Lawrence Livermore Laboratory, Livermore, California.)

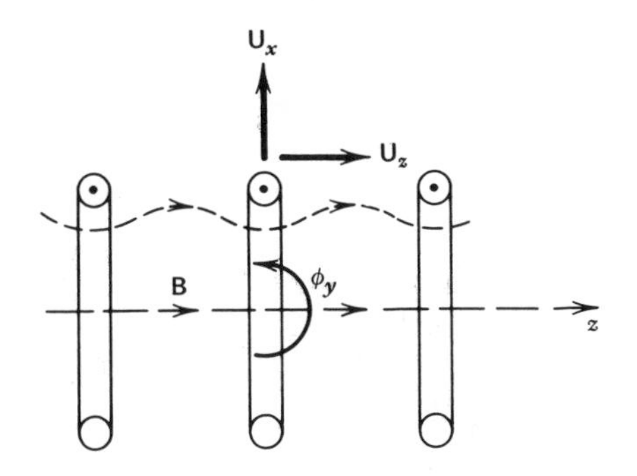

Figure 6-40 Cross section of three coils in a discrete coil solenoid.

illustrated for a set of coils designed for a tandem mirror fusion reactor to be built at Livermore National Laboratory.

Consider a circular coil symmetrically placed between two neighbors in a solenoid configuration (see Fig. 6-40). In general, the coil can have six rigid-body degrees of freedom. The resulting magnetic stiffness matrix relating the magnetic forces F and torques C with motion of the coil is diagonal as shown below:

$$\begin{bmatrix} F_x \\ F_y \\ F_z \\ C_x \\ C_y \\ C_z \end{bmatrix} = -\begin{bmatrix} \kappa_1 & & & & & \cdots 0 \\ & \kappa_1 & & & & \vdots \\ & & \kappa_3 & & & \\ \vdots & & & \kappa_4 & & \\ \vdots & & & & \kappa_4 & \\ 0\cdots & & & & & 0 \end{bmatrix} \begin{bmatrix} U_x \\ U_y \\ U_z \\ \phi_x \\ \phi_y \\ \phi_z \end{bmatrix} \tag{6-7.65}$$

where κ_1 represents the magnetic stiffness for lateral displacements; κ_3 represents the magnetic stiffness for axial displacements; and κ_4 represents the magnetic stiffness for rotation about either the x or y axes.

There is no change in inductance for rotations about the solenoidal axis and this mode is a neutral mode.

To determine the effect of misalignment, we assume that the coil is elastically constrained and that the current is below the buckling value. Also, we consider a single deformation mode described by a generalized coordinate U measured from the symmetric position of the coil. Let U_0 represent an initial misalignment. This means that the elastic restoring force will be proportional to $-k_e(U-U_0)$ while the magnetic force is given by $-\kappa U$ (κ is the magnetic stiffness; see, e.g., Section 4.6). For equilibrium, these forces must sum to zero, or

$$\kappa U + k_e(U - U_0) = 0$$

In the case of a magnetic instability, $\kappa < 0$, and we can write $\kappa = -I^2\lambda$, or

$$U = \frac{k_e U_0}{k_e - I^2 \lambda} \tag{6-7.66}$$

For this specific example, the differential inductance, which is proportional to the magnetic stiffness, was calculated by taking the difference between the inductance of the coil in the displaced position and the initial symmetric position using 200 straight elements to simulate each of the circular coils (see Section 4.6). Assumed coil displacements were on the order of 10^{-2} compared with the coil separation distance. The differential inductance was calculated according to the formula

$$\lambda = \frac{L - L_0}{\delta^2}$$

where δ is the coil displacement. For symmetric coil arrangements, the linear term in $L(\delta)$ is identically zero.

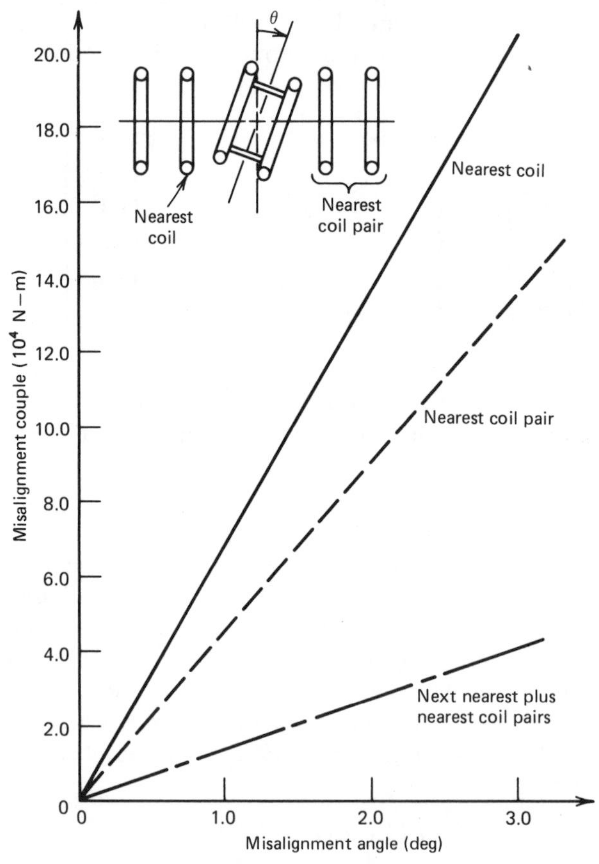

Figure 6-41 Couple on a pair of coils in the MFTF-B solenoid (see Fig. 6-39) due to angular misalignment.

The results of the calculation of magnetic stiffness and misalignment forces for a specific design are shown in Table 6-5, and Figures 6-41 to 6-43. The data are for the Mirror Fusion Test Facility (MFTF-B) solenoid coils (5-m diameter; 1.6×10^6 At). The calculations were performed for axial and lateral translation and rotation of a single coil or a coil pair with respect to neighboring coils.

Table 6-5 Magnetic Stiffness Matrix Elements for Rigid-Body Displacement of Solenoid Coils for a Tandem Mirror Fusion Reactor (MFTF-B)[a]

Mode	Single Coil: Nearest Coils	Coil Pair: Nearest Coils	Coil Pair: Nearest Coil Pair	Coil Pair: Next Nearest Coil Pair
λ_1	−0.843	−1.004	−1.210	
Lateral translation	10^{-6} N/m-A^2			
λ_3	1.687	2.008	2.420	
Axial translation	10^{-6} N/m-A^2			
λ_4	3.384	1.520	1.016	0.308
Rotation about diameter	10^{-6} N-m/rad-A^2			

[a]Parameters are: coil diameter, 5.0 m; coil spacing, 2.0 m; stiffness coefficients $\kappa = -I^2\lambda$; and $I = 1.6 \times 10^6$ At.

In brief, the results show that the solenoid coils are *unstable* under magnetic forces alone for axial translation and rotation about an axis parallel to a diameter. They are stable for lateral misalignments. This means that under mechanical constraints, misalignments or play in the supports will be amplified by the axial misalignment force or misalignment angle couple. In the lateral direction, the coils will self-align. These results suggest that redundant coil–vessel and coil–coil support schemes should be adopted to avoid catastrophic attraction of neighboring coils should a support fail.

In the MFTF-B design, pairs of coils are structurally tied together. When such a coil pair rotates about a diametral axis through the center of the pair, each coil translates and rotates simultaneously. A brief discussion of this problem was given in Section 4.6. The results in Figure 4-22 show the magnetic stiffness or differential inductance as a function of coil separation. Positive values indicate an unstable couple or torque, while negative values indicate self-alignment with respect to rotation. The MFTF-B coils are designed for a 2-m coil separation, which is fortuitously close to the crossover point as shown in Figure 4-22.

The magnitude of the misalignment couple on the magnet pair for a given

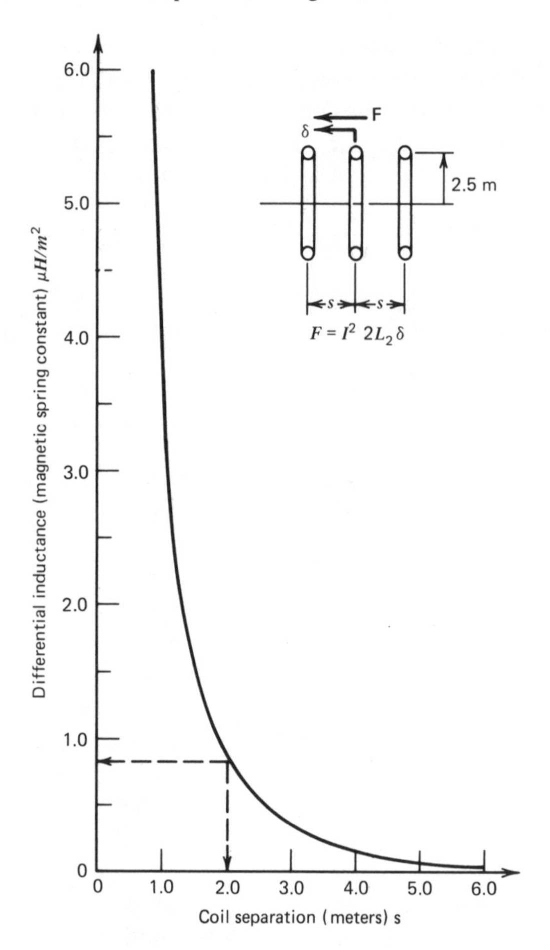

Figure 6-42 Magnetic spring constant for axial displacement of a coil in a discrete coil solenoid (MFTF-B, Fig. 6-39).

misalignment angle is shown in Figure 6-41 for three cases (1) nearest neighbor single-coil interaction, (2) nearest coil-pair interaction, and (3) next nearest and nearest coil-pair interaction. The figure shows that as more magnets are energized, the destabilizing couple decreases. The worst case, however, shows that 1.5° magnetic misalignment would produce about a 10^5 N-m (73.7×10^3 lbf-ft) moment on the coil. If this moment is balanced by forces separated by the 5-m diameter of the coil, the restraint forces must equal 20×10^3 N or 4.5×10^3 lbf. If the neighboring pairs also rotate, then this moment or couple will double and a 9000-lbf constraint force would be required.

The effect of axial translation of a coil between two neighbors and the separation distance between coils is shown in Figure 6-42. There is no

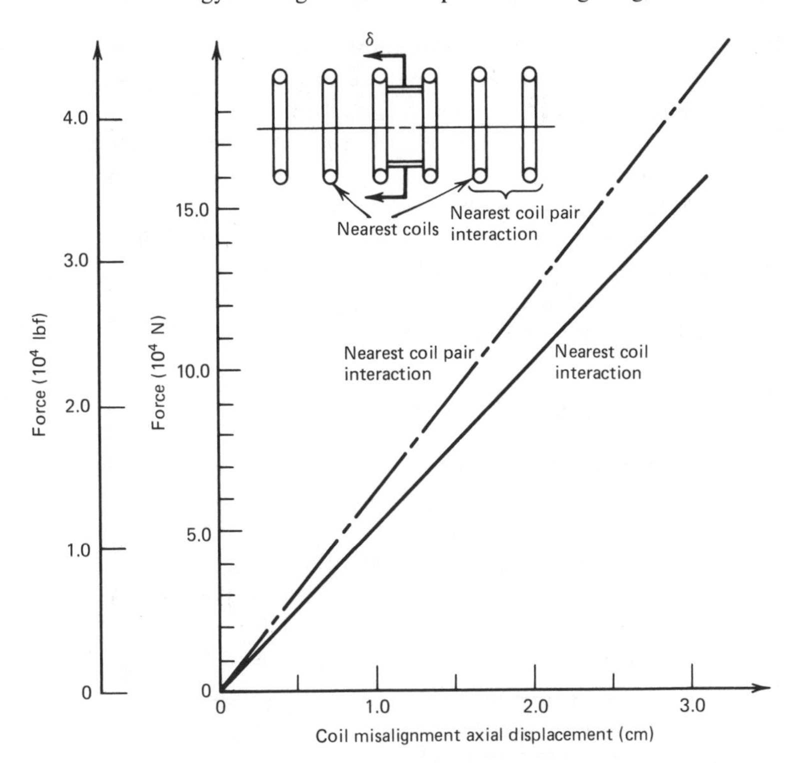

Figure 6-43 Axial misalignment force on a pair of coils in a discrete coil solenoid (MFTF-B, Fig. 6-39).

crossover distance and all the neighboring coils contribute to a destabilizing axial misalignment force. Values of the misalignment force for a MFTF-B solenoid coil pair are shown in Figure 6-43. A 1-cm axial misalignment of the magnetic center of the coil can produce a 14,000-lbf axial force on the coil pair. This load would act to produce lateral bending of the coils, which would further increase the force. If neighboring pairs also move, the force would double to about 28,000 lbf.

As noted above, the lateral misalignment force is *stabilizing*, which means that the magnetic forces act as a positive spring. The magnetic stiffness constants for a circular solenoid coil and coil pair are tabulated in Table 6-5 for the translation and rotation modes.

6.8 MASS-ENERGY SCALING LAWS FOR SUPERCONDUCTING MAGNETS

Mass in superconducting magnet systems is required to carry current, to dissipate heat, and to resist the magnetic body forces. For smaller magnets, the first two requirements are paramount, whereas for larger systems the structural loads determine a greater part of the mass requirements. One idea that has intrigued magnet designers is the concept of a *force-free* magnet in which the currents are arranged so that $\mathbf{J} \times \mathbf{B} = 0$. Chandrasekhar (1956), as well as Kuznetsov (1961b), found solutions which satisfy this condition. These force-free solutions involve magnetic systems of infinite extent. For finite current distributions, however, the virial theorem derived in Section 3.9 places a lower limit on the structural mass whenever the system stores magnetic energy. Recall that the theorem relates the trace of the stress tensor to the stored energy, that is [Eq. (3-9.9)],

$$\int \mathrm{Tr}(\mathbf{t})\,dv = \int_{\text{space}} \frac{B^2}{2\mu_0}\,dv \equiv \mathscr{E} \tag{6-8.1}$$

This theorem has been used by a number of authors to derive limits on the structural mass required for a magnetic energy storage system (Parker, 1958; Levy, 1962; Stekly, 1963; Moses, 1976; Eyssa, 1980).

The virial theorem is not limited to magnetic energy. It was first stated for a truss structure under gravitational loading by Maxwell (1869–1872). A

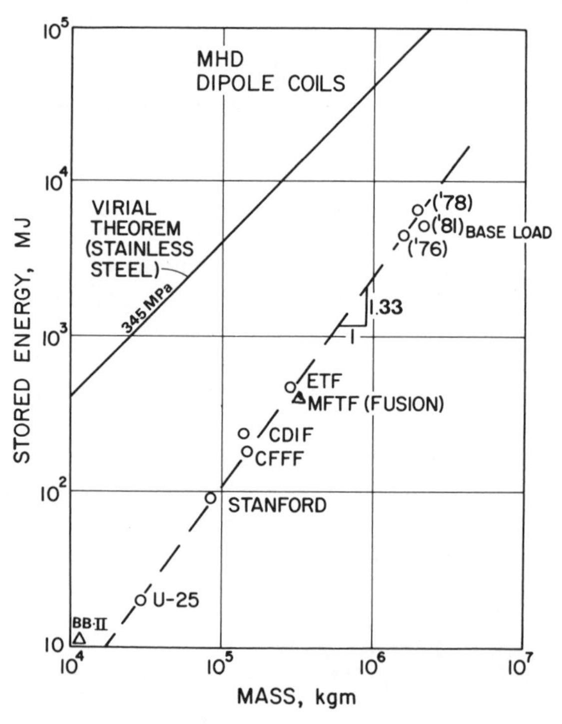

Figure 6-44 Stored energy vs. mass for MHD magnets (From Moon, 1982, with permission. American Institute of Physics, copyright 1982.)

general discussion of the virial theorem and its generalization can be found in Truesdell and Toupin (1960). The extension of the theorem to magneto-hydrodynamic systems was claimed by Chandrasekhar and Fermi (1953).

Application of the virial theorem to superconducting systems has usually assumed that the structure had separate tensile and compressive trusslike elements. If β represents the fraction of volume of material under compression and S_0 the design stress for both tension and compression, the theorem puts a lower bound on the mass $\mathcal{M}$, that is,

$$\mathcal{M} = \frac{\rho \mathscr{E}}{S_0(1 - 2\beta)} \tag{6-8.2}$$

This form has been used by Moses (1976) and Eyssa (1980) to analyze minimum mass structures for energy storage. For a long thin solenoid, however, one must consider at least a two-dimensional state of stress with

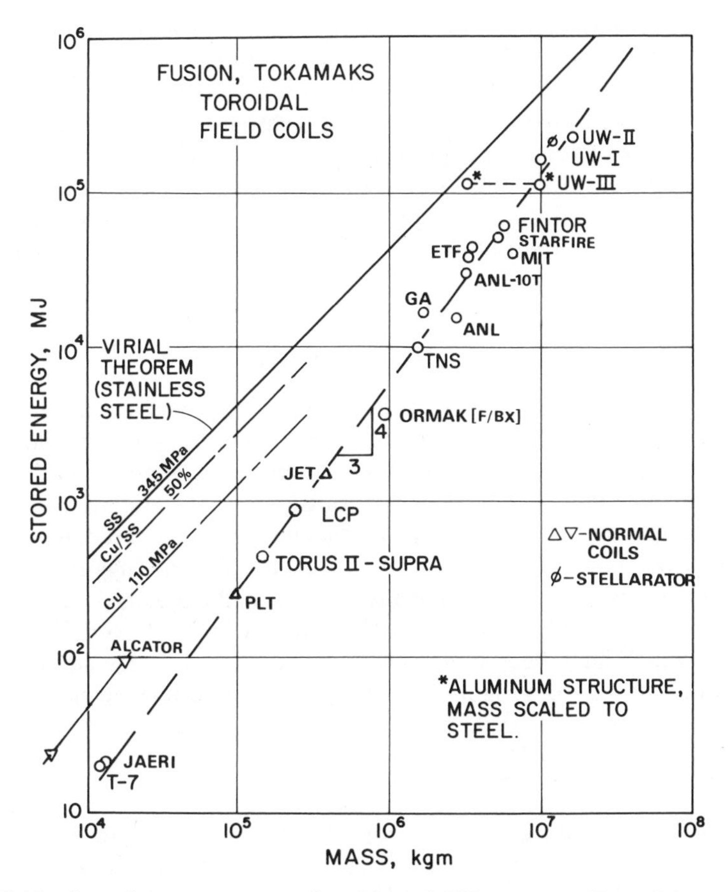

Figure 6-45 Stored energy vs. mass for tokamak TF magnets. (From Moon, 1982, with permission. American Institute of Physics, Copyright 1982.)

$t_{\theta\theta} = -2t_{zz}$ (see, e.g., Chapters 1 and 4). The trace of the stress tensor becomes

$$\mathrm{Tr}(\mathbf{t}) = \tfrac{1}{2}t_{\theta\theta} \tag{6-8.3}$$

If the structural material fails by yielding, then the maximum shear stress τ is given by $\frac{1}{2}S_Y$, where S_Y is the yield stress in tension. This leads to the condition

$$\mathrm{Tr}(\mathbf{t}) < \tfrac{1}{3}S_Y \tag{6-8.4}$$

and the virial theorem assumes the form

$$\mathcal{M} = \frac{3\rho\mathcal{E}}{S_Y} \tag{6-8.5}$$

This formula was derived by Sviatoslavsky and Young (1980) using the concept of separate axial and radial load structures.

Given this fundamental requirement for a minimum amount of structural mass to stored magnetic energy, it is instructive to examine how contemporary designs of superconducting structures compare with the virial limit. This question has been addressed by the author (Moon, 1982). The results are summarized in Tables 6-6 and 6-7 and in Figures 6-44 and 6-45. Two classes of magnets are examined—MHD and toroidal field (TF) magnets for fusion.

MHD Magnets

The stored energy mass comparisons for MHD magnets are shown in Table 6-6 and Figure 6-44. These magnets are essentially a pair of dipole magnets wound in a nonplanar configuration. Out-of-plane magnets generally involve considerable bending, which does not contribute to the minimum virial mass; these magnets are therefore less mass-efficient when compared, to say, planar TF coils. We can also include in this class out-of-plane magnets for mirror fusion machines, such as the yin–yang and baseball magnets developed at the Lawrence Livermore Laboratory. The designs in Table 6-6 include both magnets which have been built as well as those from design studies.

As one can see from Figure 6-44 most of these designs lie along a scaling law of the form

$$\mathcal{E} = C\mathcal{M}^n \tag{6-8.6}$$

where $n \simeq \frac{4}{3}$. This is remarkable considering the variety of materials and design stresses employed. We note that although all the magnets satisfy the minimum virial mass requirement [Eq. (6-8.2), ($\beta = 0$)], the mass does not scale linearly with the energy.

Table 6-6 Mass and Stored Magnetic Energy for MHD Magnets

Code Name	Design Group	Stored Energy (MJ)	Mass: Conductor and Structure (10^3 kg)	$\mathcal{M}/\mathcal{E}$ (kg/MJ)	References
U-25	Argonne National Laboratory	20	28	1400	*IEEE Trans. Mag.* **13**, No. 1, 632 (1977).
Stanford	MIT, General Dynamics	93	84[a]	903	*IEEE Trans Mag.* **17**, No. 1, 344 (1981).
CFFF	Argonne National Laboratory	168	145	863	*IEEE Trans. Mag.* **17**, No. 1, 529 (1981).
CDIF	MIT	240	144	600	A. Dawson, F. Bitter National Magnet Laboratory, MIT (personal communication).
ETF	MCA Corporation	483	290	600	*IEEE Trans. Mag.* **13**, No. 1, 636 (1977).
Base load ('76)	MCA Corporation	4480	1560	348	*IEEE Trans. Mag.* **13**, No. 1, 6 (1977).
Base load ('81)	MIT	5300	2150	406	A. Dawson, F. Bitter National Magnet Laboratory, MIT (personal communication).
Base load ('78)	MCA Corporation	6700	1880	269	*IEEE Trans. Mag.* **15**, No. 1, 306 (1979).
Baseball II (fusion)	Lawrence Livermore Laboratory	11.1	11.8	1060	"Fourth Symposium on Engineering Problems of Fusion Research," *IEEE Trans.*, **NS-18**, No. 4, 290–295 (1971).
MFTF (fusion)	Lawrence Livermore Laboratory	409	300	734	*IEEE Trans. Mag.* **15**, No. 1, 534 (1979).

Source: Moon (1982).
[a]May include Dewar mass.

Fusion Magnets

The designs for TF coils for fusion reactors are tabulated in Table 6-7 and shown in Figure 6-45. Again, all satisfy the absolute virial limit [Eq. (6-8.2), ($\beta = 0$)]. However, they too seem to scale according to $\mathscr{E} = C\mathscr{M}^{4/3}$.

This scaling is even more surprising considering the number and diversity of the designs, with designs from the United States, Japan, Europe, and the Soviet Union. The scaling law seems to be operative over four decades of stored energy. In almost all the designs, the conductor was Nb-Ti, with stainless steel as a structural material. Many of the designs employed the constant tension "*D*" shape to avoid bending in the plane of the magnet. Also shown are a few normal magnets with copper, such as the Joint European Torus (JET) and the Princeton University TFTR tokamak design. These also seem to fit on the scaling curve.

The data in Figures 6-44 and 6-45 for MHD and TF magnets suggest that some scaling law other than that based on stress is implicitly being used by different magnet groups around the world. One likely candidate is related to the *thermal stability*. To avoid quenching or becoming normal, the heat transfer system in a superconducting magnet must have sufficient capacity to take out energy input due to Joule heating. This condition may be expressed in the form (see, e.g., Stekly, 1973)

$$\frac{I^2 \rho_e}{A_0} \leq h_0 A_1 \Delta T \tag{6-8.7}$$

where h_0 is a heat transfer coefficient; A_1 is a surface area per unit length; ΔT is a temperature difference; A_0 is the cross-sectional area through which I flows; and ρ_e is the electrical resistivity of the superconducting composite when it carries normal current.

Suppose R is a global dimension of the magnet and r is representative of the cross section through which I flows. Then for a certain class of designs, we assume that R/r is constant. Thus $\mathscr{M}$ will scale as R^3. The stored energy is related to the inductance through $\mathscr{E} = \frac{1}{2}LI^2$. The inductance may be written in the form

$$L = \mu_0 R f\left(\frac{R}{r}\right)$$

Thus $\mathscr{E}$ scales as RI^2. We note that A_1 scales as r, but A_0 scales as r^2. If h_0, ρ_e, ΔT, and R/r all remain constant, the expression (6-8.7) leads us to conclude that I^2 scales as R^3. It follows that $\mathscr{E}$ has the form

$$\mathscr{E} \leq \eta \mathscr{M}^{4/3} \tag{6-8.8}$$

where η depends on the constant physical constants in the magnet. Thus thermal stability seems to be a likely candidate for scaling superconducting mass, especially in the lower-energy regime. Another proposal would scale the conductor mass on thermal stability and structural mass using the virial theorem.

Table 6-7 Mass and Stored Magnetic Energy for Tokamak TF Coil Magnets

Code Name	Design Group	Energy (MJ)	Mass[a] (10^3 kg)	kg/MJ	Number of Coils	References
T-7	Kurchatov Atomic Energy Institute	20	12	600	—	*IEEE Trans. Mag.* **15**, No. 1, 550 (1979).
JAERI—Cluster	Japan Atomic Energy Research Institute	20.7	13.2	638	2	*IEEE Trans. Mag.* **17**, No. 1, 494 (1981).
Torus II—Supra	EURATOM	440	144	327	24	*IEEE Trans. Mag.* **15**, No. 1, 542 (1979).
LCP	GD, GE, WESTH, JAERI, EURATOM, BB[b]	894	240	269	6	W. H. Gray, Oak Ridge National Laboratory (personal communication).
ORMAK-F/BX	Oak Ridge National Laboratory	3.66×10^3	960	262	24	Oak Ridge National Laboratory Report No. ORNL-TM-4634, June 1974.
TNS	General Atomic Corporation	10×10^3	1500	150	12	General Atomic Report No. GA-A15100, Vol. V, UC-2nd, October 1978, 5.3-144, 5.3-179.
ANL-EPR	Argonne National Laboratory	15.6×10^3	2800	180	16	Argonne National Laboratory Report No. ANL/CTR-75-2, June 1975.
GA-EPR	General Atomic Corporation	16.7×10^3	1700	102	16	General Atomic Report No. EPRI ER-289, Vol. II, December 1976, 5-70, Table 5.3-1.
ANL-10T	Argonne National Laboratory	30×10^3	3328	111	16	*IEEE Trans. Mag.* **13**, No. 1, 605 (1977).
ETF	MIT/FBNML-GE	38×10^3	3260	85.9	10	ETF Design Description, Oak Ridge National Laboratory, July 1980.

Table 6-7 (*Continued*)

Code Name	Design Group	Energy (MJ)	Mass[a] (10^3 kg)	kg/MJ	Number of Coils	References
ETF	MIT/FBNML-GE	45×10^3	3580	79.6	10	ETF Design Description, Oak Ridge National Laboratory, July 1980.
STARFIRE	Argonne National Laboratory	50×10^3	5300	106	12	Argonne National Laboratory Report No. ANL/FPP-80-1, Vol. 1, 1980.
HFCTR/MIT	FBNML/MIT	40×10^3	6800	170	16	*Seventh Symposium Engineering Problems of Fusion Research*, IEEE Publication No. 77CH1267-4-NPS, p. 629.
FINTOR	LNF del CNEN, Italy	60×10^3	5808	96.8	24	*IEEE Trans. Mag.* **13**, No. 1, 617 (1977).
UW-III	University of Wisconsin	108×10^3	3268	30.3	18	University of Wisconsin Report No. UWFDM-150, July 1976.
UW-I	University of Wisconsin	158×10^3	9960	63.0	12	University of Wisconsin Report No. UWFDM-150, July 1976.
UW-II	University of Wisconsin	223×10^3	16.1×10^3	72.2	24	University of Wisconsin Report No. UWFDM-150, July 1976.

STELLARATOR	University of Wisconsin	200×10^3	12×10^3	60.0	15	University of Wisconsin Report No. UWFDM-409, March 1981.
ALCATOR A (normal)	MIT/FBNML	24	5.75	240	—	*Seventh Symposium Engineering Problems of Fusion Research*, IEEE Publication No. 77CH1267-4-NPS, pp. 54–58 (1977).
ALCATOR C (normal)	MIT/FBNML	95	17.7	186	—	*Seventh Symposium Engineering Problems Fusion Research*, IEEE Publication No. 77CH1267-4-NPS, pp. 54–58 (1977).
PLT (normal)	Princeton Plasma Physics Laboratory	251	98	390	—	J. H. Schulz, MIT/FBNML (personal communication).
JET (normal)	Joint European Torus	1.45×10^3	384	265	32	*Seventh Symposium on Engineering Problems of Fusion Research*, IEEE Publication No. 77CH1267-4-NPS, Vol. 1, p. 28 (1977).

Source: Moon (1982).

[a]Mass includes conductor and structure.

[b]General Dynamics (GD), General Electric (GE), Westinghouse (WESTH), Hitachi (managed by JAERI), Siemens (for EURATOM), and Brown–Boveri (BB).

Whatever the implicit scaling principle, it seems clear that doubling the stored energy does not necessarily require twice the mass for actual designs as the virial theorem would lead one to conclude. One might also conclude from this exercise that almost all designs are so far from the virial limit that more efficient structural designs may offer significant mass savings.

Further discussion on this topic and comparison of virial mass with masses of solenoid magnets may be found in Moon (1982).

7

Mechanics of Ferromagnetic Solids and Structures

7.1 INTRODUCTION

To the layman, forces on magnetic materials are the most familiar effects of magnetomechanics, but to the mechanician they are the most difficult macroscopic force fields in classical physics. A brief review of the electromagnetic theory and equations of mechanics of ferromagnetic solids was given in Chapters 2 and 3. In this chapter we will explore the application of the basic theory to specific problems involving ferromagnetic solids and structures. These include forces and dynamics of ferromagnetic particles in magnetic fields; forces in electromagnet and permanent magnet circuits; and the bending of ferroelastic plates and shells.

By the term "ferromagnetic" we include all solids which can be strongly magnetized in a magnetic field, such as alloys of iron, nickel, and cobalt. The principal mechanism for magnetization in these materials is the alignment of electron spins below a critical temperature (Curie temperature—see Table B-3). In some materials, this spontaneous alignment of spins results in a net macroscopic magnetization **M** in the absence of an external magnetic field ("hard" magnetic material), while in other materials the alignment exists only in small domains where the net magnetization can only be maintained by application of an external field ("soft" magnetic material). For either hard or soft materials the net macroscopic body force on a solid is given by the expression

$$\mathbf{F} = \int (\mathbf{M} \cdot \nabla \mathbf{B}^0 + \mathbf{J} \times \mathbf{B}^0)\, dv \tag{7-1.1}$$

and the net body couple due to magnetization is given by

$$\mathbf{C} = \int \mathbf{M} \times \mathbf{B}^0 dv \tag{7-1.2}$$

An early technical application of ferromagnetic forces was the lifting magnet created by Joseph Henry in 1832. Conventional devices include solenoid actuators, switches and relays, magnetic brakes and clutches, magnetic separation systems, and field producing and shaping magnets for motors, generators, and physics research. In spite of the wide use of ferromagnetic forces, there are few references for the engineer on the mechanics of these devices except for rotating machinery. Early texts include Underhill (1924) and Roters (1941); the latter is almost a bible for solenoid-actuator design. McCaig (1977) offers a modern reference on permanent magnets while Cullity (1972) is a good text on the macroscopic physics of magnetic materials.

As reviewed in the first four chapters, interest in the mechanics of ferromagnetic *deformable* solids grew in the early 1960s. It was recognized that the definition of the local magnetic body force was not unique—that is, one could represent the forces as due to magnetic dipoles, poles, or magnetization currents, or other combinations of these. The difference between the stresses resulting from these different models was incorporated in the constitutive equations—that is, the equation relating the stresses to the strain, magnetic field, and temperature. An important reference to this theoretical treatment of deformable ferromagnetic solids is the work of Brown (1962, 1966). Many other authors followed, however, the list is too numerous to mention here. An early review is given by Paria (1967) and two other reviews by Pao (1978) and Moon (1978). A recent review of eastern European and Soviet research on magnetoelasticity may be found in English by Ambartsumyan (1982). While of great importance to the theoretical development of magnetoelasticity, the vast majority of these papers did not treat problems that were technically important to designers of ferromagnetic devices.

Several technically important classes of problems of ferromagnetic mechanics which are not treated here are those related to rotating machinery, aeromagnetoflutter, and electromagnetic braking devices. Hague (1929) is an early reference to electromagnetic forces in rotating machinery. Aeromagnetomechanics problems are important to magnetohydrodynamic (MHD) structures. A review of the combined effects on structures due to aerodynamics and magnetic forces is given by Librescu (1977).

In Section 7.2 we look at the problem of a ferromagnetic particle—an ellipsoid to be exact—in an inhomogeneous magnetic field. This problem has technical importance to the area of high-field gradient magnetic separation. Next, the classic single-degree-of-freedom, rigid-body electromagnetic solenoid-actuator is discussed in Section 7.3 and the mechanics of a permanent magnet switch or relay is treated in Section 7.4. An approximate theory for the bending of a ferroelastic plate in an inhomogeneous, static magnetic field is examined in Section 7.5 and experimental verification of this approximation is presented in Section 7.6. The bending of a ferroelastic shell in a uniform magnetic field is presented in Section 7.7. Magnetostriction effects will not be discussed in this chapter (see Section 2.5).

7.2 FERROMAGNETIC PARTICLES IN MAGNETIC FIELDS

In the absence of electric current, a ferromagnetic particle will only experience a force in an *inhomogeneous* magnetic field. If the magnetization of the particle is fixed and the sources of the external magnetic field are fixed, then Earnshaw's theorem (see Section 5.2) states that the magnetized particle cannot be in stable equilibrium under a static magnetic field. The result is still valid when the particle can be magnetized. (Earnshaw's theorem may not hold when a ferromagnetic particle is near diamagnetic material, such as a superconductor or in a ferrofluid.) For most problems, however, ferromagnetic particles will accelerate until they hit or come in contact with a boundary or surface. This behavior can be either useful or undesirable. In the former case, an inhomogeneous magnetic field can be used to remove ferromagnetic particles or ferrous-ore-bearing material from a stream of granular material. In the case of large high-field magnetic devices, such as MHD, fusion, or magnetic levitation superconducting magnets, the tendency of particles to be accelerated toward the magnet can be dangerous or, at the least, may damage the magnet.

The force on the particle is given by

$$\mathbf{F} = \int \mathbf{M} \cdot \nabla \mathbf{B}^0 \, dv$$

For fields with weak gradients, we can neglect the effect of the field gradient on the magnetization itself. One problem for which a solution is known is the soft magnetic ellipsoid (see Fig. 7-1). The solution may be found in Stratton (1941), p. 258:

$$\mathbf{M} = \chi \mathbf{H}^- \tag{7-2.1}$$

where $\mathbf{H}^-$ is the total field inside the ellipsoid. When the external field is uniform, it can be shown that the induced field inside the magnetized ellipsoid is uniform. If a, b, and c are the lengths of the principal semiaxis in the x, y, and z directions, respectively, then the induced magnetization may be related to the external field averaged over the particle volume

$$\mathbf{M} = \frac{\chi}{\mu_0} \left(\frac{B_x^0}{1 + \chi n_1}, \frac{B_y^0}{1 + \chi n_2}, \frac{B_z^0}{1 + \chi n_3} \right) \tag{7-2.2}$$

The numbers n_1, n_2, and n_3 are called demagnetization factors and are given by elliptic integrals of the form

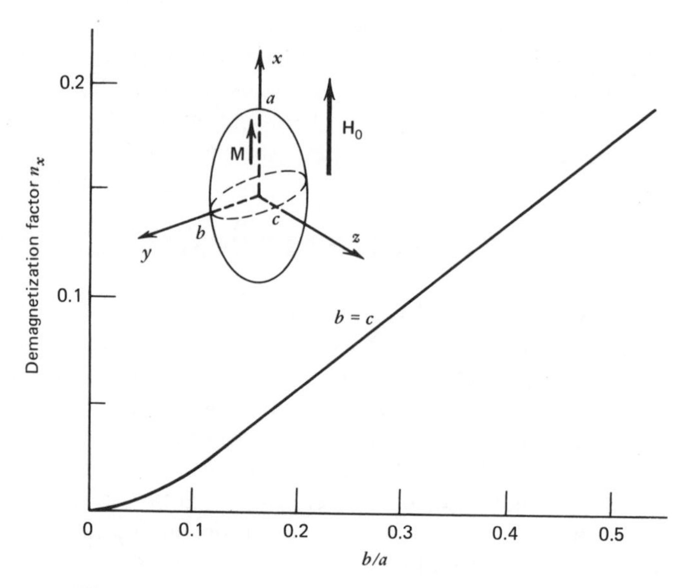

Figure 7-1 Demagnetization factor for an ellipsoid.

$$n_1 = \frac{abc}{2} \int_0^\infty \frac{ds}{(s+a^2)R} \tag{7-2.3}$$

where $R = [(s+a^2)(s+b^2)(s+c^2)]^{1/2}$.

In the special case of $b = c$, $a > b$,

$$n_1 = \frac{ab^2}{2} \int_0^\infty \frac{ds}{(s+a^2)^{3/2}(s+b^2)}$$

and

$$n_1 = \frac{1-e^2}{2e^3}\left(\ln\frac{1+e}{1-e} - 2e\right) \tag{7-2.4}$$

$$e^2 = 1 - \frac{b^2}{a^2}$$

This function is shown plotted in Figure 7-1.

For a sphere ($a = b = c$), $n_1 = n_2 = n_3 = \frac{1}{3}$ and for a cylinder, $n_1 = 0$ and $n_2 = n_3 = \frac{1}{2}$. For a long needlelike particle, the couple acting on the particle will orient the particle such that $\mathbf{M} \times \mathbf{B}^0 = 0$, or the long axis will lie along the field lines.

Acceleration of Particles near High-Field Magnets

For high-field magnet systems, such as MHD or fusion magnets, loose ferromagnetic objects, such as tools or nuts and bolts, may be accelerated

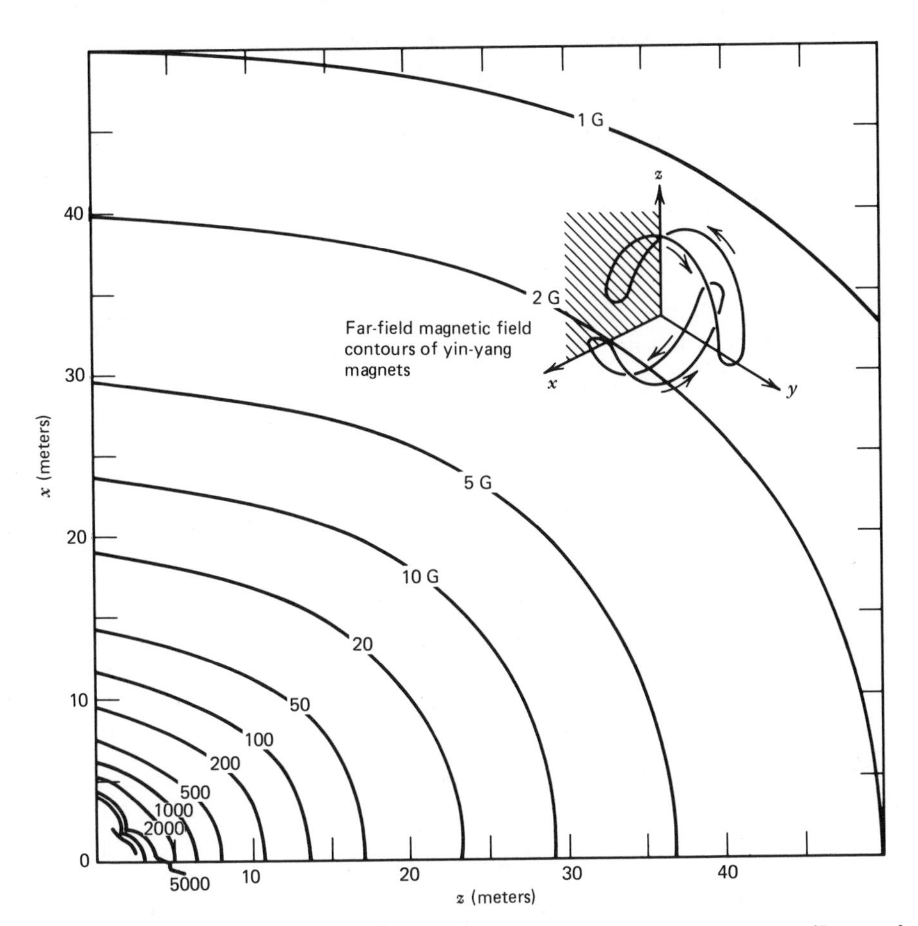

Figure 7-2 Magnetic field intensity near a pair of yin–yang magnets. (Personal communication from J. Horvath, Lawrence Livermore Laboratory, Livermore, California.)

toward the magnet when the field is turned on. For example, a fusion mirror reactor magnet, called a yin–yang at Livermore, California, produces a magnetic field environment as shown in Figure 7-2. Fields up to 100 times that of the earth may be found a distance of 15 m from the magnet center.

In low magnetic fields, the magnetization per unit volume along the field lines is given by

$$M_x = \frac{B_x^0}{\mu_0} \frac{\chi}{1 + \chi n_1} \tag{7-2.5}$$

and for high fields, the magnetization will saturate around 1 T, that is, $\mu_0 M = 1$.

There are two questions relating to the danger of loose objects accelerated by the magnet:

1. At what distance can the magnetic force overcome gravity and/or friction in order to accelerate the object?
2. If accelerated, what velocity will the object achieve as it moves toward the magnet?

Lifting Force. To lift the object, the magnetic force (which is proportional to the product of the field and field gradient) must be equal or greater than that due to gravity; that is, in terms of scalars,

$$VM\frac{dB^0}{dx} \geq \rho V g \tag{7-2.6}$$

where V is the volume and the magnetization M is given by Eq. (7-2.5),

$$M = \frac{\chi B^0/\mu_0}{1+\chi n}$$

When $\chi n \gg 1$, the criterion (7-2.6) can be written in the form

$$B^0\frac{dB^0}{dx} \geq \mu_0 \rho g n$$

or as a condition on the magnetic pressure,

$$P_m = \frac{(B^0)^2}{2\mu_0} \geq \frac{\rho n g B^0}{2(dB^0/dx)} \tag{7-2.7}$$

Values of B_x and dB_x/dx for a yin–yang magnet are shown in Figure 7-3. For a steel wrench or pipelike object with $b/a = 0.2$, $n_x = 0.056$ (see Figure 7-1), and $\rho = 7.8$ g/cm^3. The criterion on the field becomes

$$B^0\frac{dB^0}{dx} \geq 5.4 \times 10^{-3}\ \mathrm{T}^2/\mathrm{m} \tag{7-2.8}$$

From Figure 7-3, this criterion is easily met at distances of 6 m or less from the magnet center along the x axis. For a given magnet system, one would have to explore different directions along which the object could move to determine complete stability.

Impact Velocity. Having determined whether a ferromagnetic object can be lifted by a magnet, the next question is to what velocity can the body be accelerated. The kinetic energy of the object is equal to the work done by the field less that done by gravity. For estimation purposes, we neglect gravity or assume the object moves horizontally in the field of the magnet. The velocity is then given by (for one-dimensional motion)

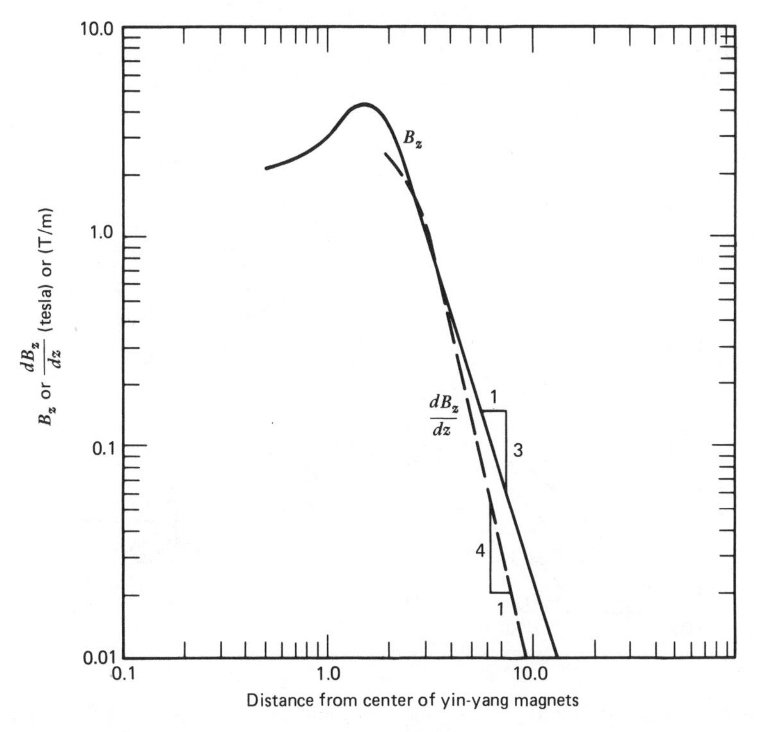

Figure 7-3 Magnetic field gradient near a pair of yin–yang magnets. (Personal communication from J. Horvath, Lawrence Livermore Laboratory, Livermore, California.)

$$\frac{1}{2}\rho v^2 = \int M \frac{dB^0}{dx}\, dx \qquad (7\text{-}2.9)$$

For low fields, we have $M = \chi H^-$ and we use Eq. (7-2.5). For $\chi n \gg 1$, Eq. (7-2.9) becomes

$$\rho v^2 = \frac{1}{\mu_0 n_x}[B^0(2)^2 - B^0(1)^2] \qquad (7\text{-}2.10)$$

When the starting field is small, that is, $B^0(1)^2 \ll B^0(2)^2$, the terminal velocity is proportional to the square root of the magnetic pressure. This relation is plotted in Figure 7-4, for $B^0(1) \cong 0$, and for different values of n_x or ratio a/b.

For high fields, the magnetization will saturate, and Eq. (7-2.9) becomes

$$\rho v^2 = 2M_s[B^0(2) - B^0(1)] \qquad (7\text{-}2.11)$$

If we choose $\mu_0 M_s = 1$ and $B^0(1) \cong 0$, then this relation becomes

$$\rho v^2 = \frac{2B^0}{\mu_0}$$

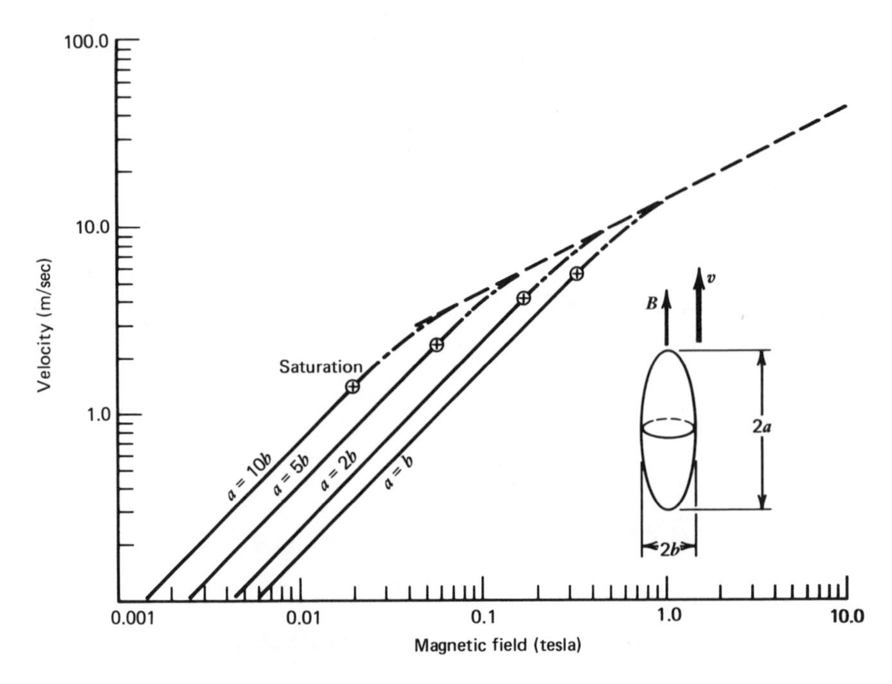

Figure 7-4 Terminal velocity vs. magnetic field for an ellipsoid in a field gradient [Eqs. (7-2.10) and (7-2.11)].

This relation is also plotted in Figure 7-4.

One can see that elongated objects reach higher velocities than spherelike objects for low fields. An ellipsoidal object with geometric ratio $a = 2b = 2c$ can reach a velocity of 10 m/sec in a 0.5-T field. Such a velocity would obviously be dangerous to people and could conceivably do damage to equipment.

Magnetic Separation Dynamics

The magnetic separator goes back to the time of Joseph Henry, when electromagnets were used to remove nails from horses's feed. Modern magnetic separators are used in solid-waste recycling systems to recover iron and steel waste from garbage. Recent developments in this technology have employed high gradient magnetic separation (HGMS) (see, e.g., Liu, 1979). This technique has been used to remove or filter minerals from a stream fed with small amounts of iron oxides or other weak ferromagnetic chemicals. For example, power-plant fly ash contains 15–20% by weight of iron oxides, part of which can be removed from the stack gases by dry magnetic separation. A discussion of industrial applications of magnetic separation may be found in a conference proceeding edited by Liu (1979). Kolm et al. (1975) have written a good descriptive article on HGMS.

While there are many techniques of magnetic separation, we will discuss the analysis of forces in the high field gradient method. Two classes of separators may be compared—those with single magnetic surface attractors and those employing matrix or multisurface attractors. Magnetic fields generally exhibit large gradients near sharp edges and corners. Also, the introduction of a large number of ferromagnetic spheroids, wires, or mesh into a uniform magnetic field will create regions of high field gradients near the wires (see Fig. 7-5). Thus a foreign-object ferroparticle moving past a magnetized collector wire will experience two fields—one from the background magnet $\mathbf{B}^0$ and the other from the collector wire $\mathbf{B}$, that is, the total external field in Eq. (7-1.1) is given by $\mathbf{B}^e = \mathbf{B}^0 + \mathbf{B}$.

Often the particles to be removed are contained in a fluid which moves past the high gradient attractor as shown in Figure 7-5. The competing forces acting on the foreign particle include the following:

$\mathbf{F}_m$ magnetic attractive force of the wire
$\mathbf{F}_d$ fluid viscous drag force
$\mathbf{F}_g$ gravity
$\mathbf{F}_b$ fluid buoyancy force

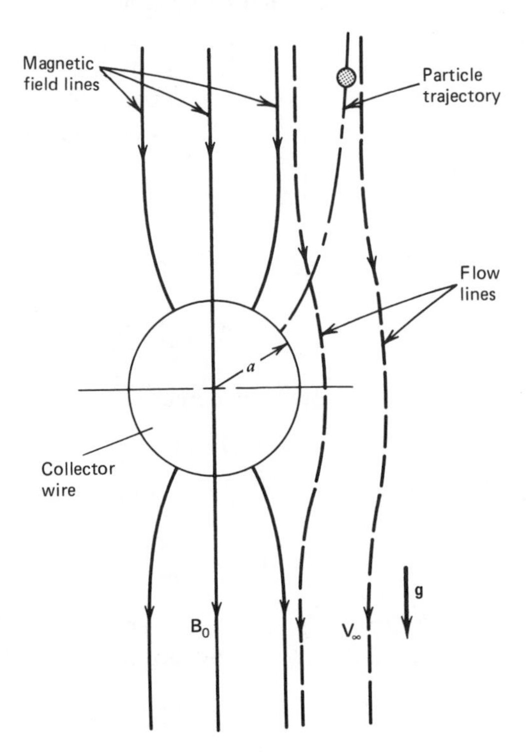

Figure 7-5 High gradient magnetic separation geometry.

The equation of motion of a particle (Newton's law) is given by

$$\rho V \frac{d^2\mathbf{r}}{dt^2} = \mathbf{F}_m + \mathbf{F}_d + \mathbf{F}_g + \mathbf{F}_b \tag{7-2.12}$$

When the particle has an elongated shape, the rotational dynamics may become important. If $\mathcal{A}$ represents the angular momentum of the object about its center of mass, the rotational dynamics must satisfy Euler's equations

$$\frac{d\mathcal{A}}{dt} = V\mathbf{M} \times \mathbf{B}^e + \mathbf{C}_d \tag{7-2.13}$$

where $\mathbf{C}_d$ represents a couple due to viscous drag. (The gravitational and buoyancy forces act through the center of mass and produce no moment.)

The dynamics of a spherical particle governed by Eq. (7-2.12) have been analyzed by a number of authors including Lawson et al. (1977). They studied the trajectories of particles in a uniform flow past a magnetized cylindrical wire (see Fig. 7-6) in order to determine under what conditions of field and flow the particles would hit the wire (i.e., the particles would be captured or removed from the flow). In these devices many particles collect on the magnetized wires and must be removed periodically. Photographs of particle build-up and analysis of the stability or ability of the attracted particle to remain attached to the wire under flow has been analyzed by Cowen et al. (1976a, 1976b).

For a uniform background field in the direction of flow, Lawson et al. (1976) assume that the wire is uniformly magnetized with a magnetic moment per unit length $\boldsymbol{\lambda}$ and the external field of the collector wire given by

$$\mathbf{H}^c = \nabla \times \left(\frac{\boldsymbol{\lambda} \times \mathbf{r}}{2\pi r^2}\right) \tag{7-2.14}$$

For a sphere, the magnetization per unit volume in the particle is given by

$$\mathbf{M} = \frac{\chi}{1 + \frac{1}{3}\chi}(\mathbf{H}^0 + \mathbf{H}^c) \tag{7-2.15}$$

and the magnetic force on the sphere becomes

$$\mathbf{F}_m = \mu_0 V \mathbf{M} \cdot \nabla \mathbf{H}^c \tag{7-2.16}$$

Lawson et al. assume a drag law for a sphere given by Stokes,

$$\mathbf{F}_d = 6\pi\eta b\left(\mathbf{v}^f - \frac{d\mathbf{r}}{dt}\right) \tag{7-2.17}$$

where η is the fluid viscosity, b the radius of the sphere, and $\mathbf{v}^f$ the velocity of the fluid. Finally, the gravitational and buoyant forces are given by

$$\mathbf{F}_g + \mathbf{F}_b = V(\rho_p - \rho_f)\mathbf{g} \tag{7-2.18}$$

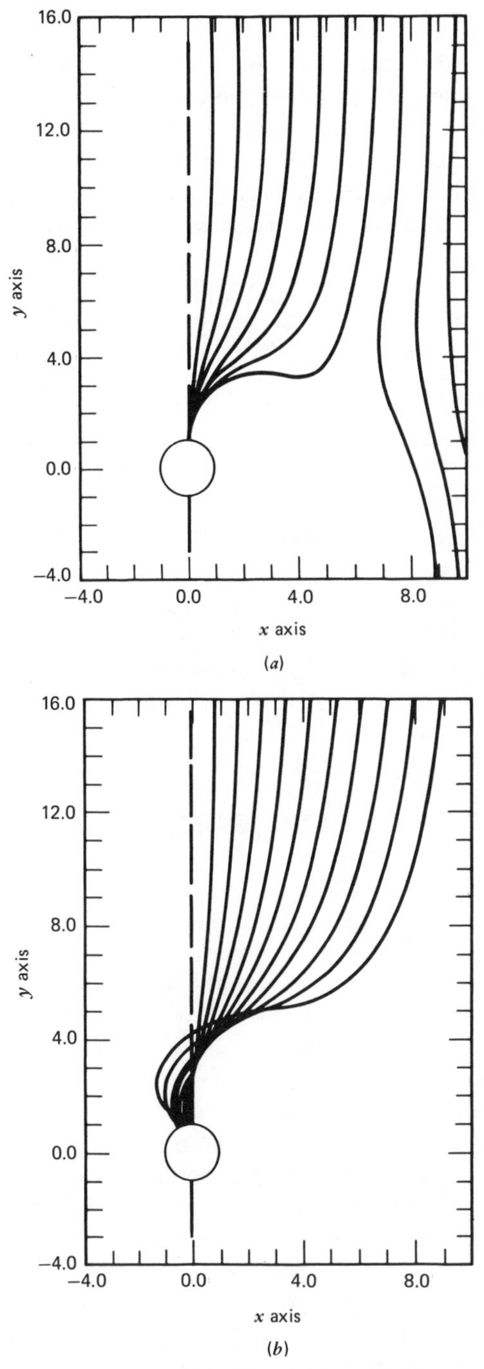

Figure 7-6 Typical capture trajectories for high gradient magnetic separation. (From Lawson et al., 1977, reprinted with permission. American Institute of Physics, copyright 1977.)

where ρ_p and ρ_f are the mass densities of particle and fluid, respectively, and $\mathbf{g}$ is a vector in the direction of gravity and magnitude 9.8 m/sec^2.

Lawson et al. assume further that the particle motion does not affect the basic flow around the wire. In their paper, they integrate Eq. (7-2.12) and plot trajectories of particle motion as reproduced in Figure 7-6.

By nondimensionalizing Eqs. (7-2.12) and (7-2.16)–(7-2.18), the number of independent parameters can be reduced to four nondimensional groups as defined below:

$$G_1 = \frac{\chi}{1+\frac{1}{3}\chi}\frac{B_0^2}{\mu_0}\frac{1}{\rho_p v_\infty^2}, \qquad G_2 = \frac{2b^2\rho_p v_\infty}{9a\eta} \tag{7-2.19}$$

and

$$G_3 = \frac{ag}{v_\infty^2}\left(1-\frac{\rho_f}{\rho_p}\right), \qquad G_4 = \frac{\lambda\mu_0}{2\pi a^2 B_0} \tag{7-2.20}$$

where the flow velocity far from the wire is v_∞ and a is the radius of the wire. G_2 is called the Stokes number and is a measure of the viscous force. For low-enough G_2, or low Stokes number, the viscous forces dominate the inertia effects and the trajectories are monotonic as regards capture of the particles (see Fig. 7-6*a*). For high Stokes number, the trajectories exhibit oscillations as shown in the capture trajectories in Figure 7-6*b*.

7.3 DYNAMICS OF ELECTROMAGNETIC ACTUATORS

Electromagnetic actuators form one of the largest classes of magneto-mechanical applications after rotating machinery. They are generally used to provide forces or motions to do work in machines or to position parts or close switches. Two classic books on the design of these devices are those by Underhill (1924) and Roters (1941). Two different kinds of actuators are shown in Figure 7-7. The left diagram shows a solenoid with axial plunger, while the right diagram illustrates the variable air-gap electromagnet with ferromagnetic flux paths. The force-distance relation for each electromagnet shows differences between the two basic actuators. The plunger type can sustain a large force over a considerable percentage of its distance, while the air-gap electromagnet produces large forces only when the gap is very small.

The principal forces in these devices are mainly ferromagnetic, although in some dynamic problems induced eddy currents may create additional forces. In Section 3.5 we discussed an energy method to derive the forces in electromagnetic devices. This method was used in Section 5.6 to study the mechanical stability of an air-gap electromagnet with an elastic keeper. Another method for examining magnetic forces on ferromagnetic devices is the Maxwell stress tensor discussed briefly in Chapter 1, and in more detail in

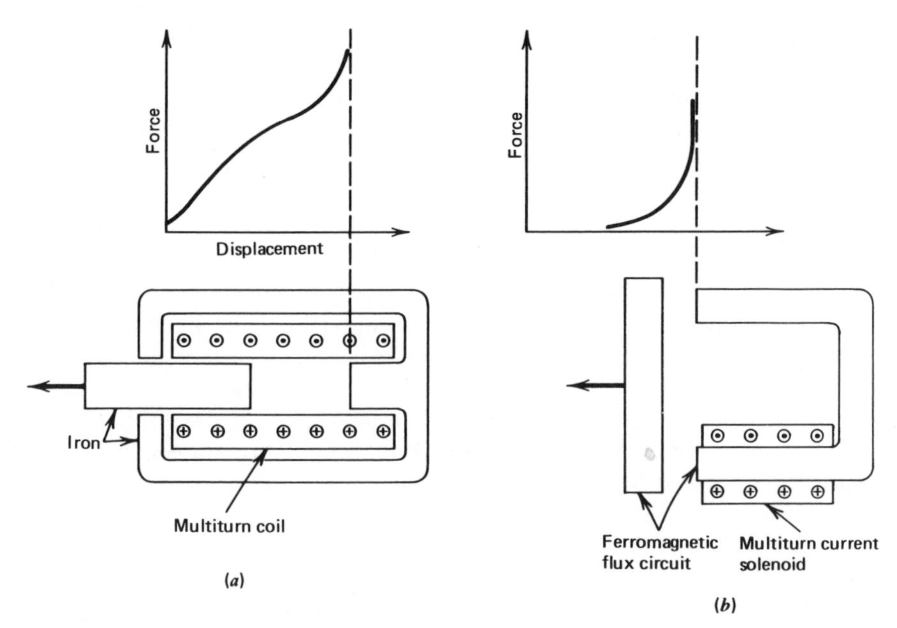

Figure 7-7 Force-displacement laws for actuators: (*a*) plunger electromagnet and (*b*) keeper-type electromagnet (after Underhill, 1924).

Sections 2.6 and 3.4.† To obtain the total force on a ferromagnetic object in this method, one integrates a stress vector around a closed surface surrounding the object; that is,

$$\mathbf{F} = \int \boldsymbol{\tau} dS \tag{7-3.1}$$

where $\boldsymbol{\tau}$ is given by the inner product of the Maxwell stress tensor and the normal $\mathbf{n}$ to the surface,

$$\boldsymbol{\tau} = \mathbf{n} \cdot [\mathbf{BB} - \boldsymbol{\delta}\tfrac{1}{2}B^2]/\mu_0 \tag{7-3.2}$$

We have assumed that the medium immediately surrounding the object is not ferromagnetic, that is, $\mu_r = 1$ outside the object. If we decompose $\mathbf{B}$ into components normal and tangential to the surface S, $\mathbf{B} = (B_n, B_t)$, and $\boldsymbol{\tau}$ becomes

$$\boldsymbol{\tau} = \frac{1}{\mu_0}\left(\frac{1}{2}[B_n^2 - B_t^2],\, B_n B_t\right) \equiv (\tau_n, \tau_t) \tag{7-3.3}$$

Thus the normal stress vector τ_n can be either tensile or compressive

†A discussion of magnetic forces on a plunger-type electromagnet may be found in Woodson and Melcher (1968), Part II, Section 8.5.

depending on the relative magnitude of B_n^2 or B_t^2. Also, a magnetic shear stress vector can act on a surface.

For many ferromagnetic problems, the fact that $\mu_r \gg 1$ implies that the tangential component of the field is small compared to the normal component so that the total force on the body is given by

$$\mathbf{F}_m = \frac{1}{2\mu_0} \int B_n^2 \mathbf{n} \, dA \tag{7-3.4}$$

where the integral is carried out over a surface just outside the body.

It is sometimes important to be able to understand the dynamics of an electromagnetic actuator. As a specific example, consider the air-gap electromagnet shown in Figure 7-8. The equations of motion for the mass and the circuit can be derived from either the energy method, that is, by using Lagrange's equations (see Section 3.5), or by a direct use of Newton's law for the mass and Maxwell's laws for the circuit. In this example we will use the direct method. As shown in Figure 7-8, a "U"-shaped soft ferromagnetic material with high permeability provides a flux path for the magnetic flux generated by the N turns of current I. We also assume that the keeper has high permeability and is elastically constrained such that a gap d_0 exists when the circuit is not activated. For a high permeability material, we can assume that the flux lines entering the movable keeper do so normal to the horizontal surface and we can therefore use the integral of the Maxwell stress vector $B_n^2/2\mu_0$ to calculate the magnetic force on the keeper. Thus

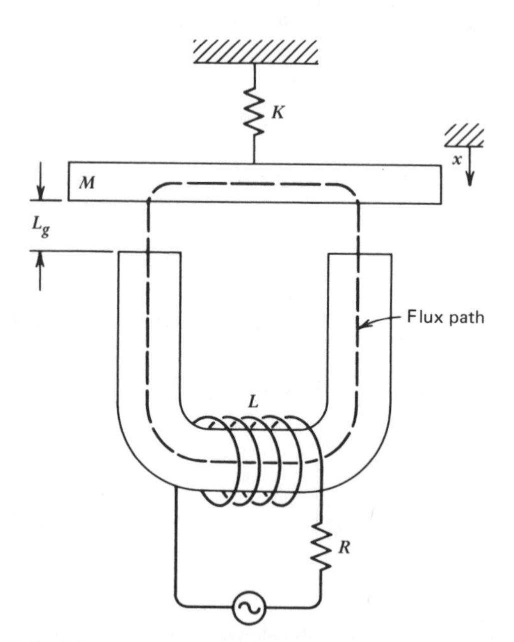

Figure 7-8 Electromagnet with elastically restrained keeper.

$$F_m = \frac{B_g^2}{2\mu_0} 2A \tag{7-3.5}$$

where B_g is the vertical field in the gap and A is the area of the electromagnet pole face. Implied in Eq. (7-3.5) is the assumption of a small gap, a uniform flux density across the pole face, and neglect of fringing fields. Eddy-current forces are also neglected in this example.

The equations for the dynamics of the mass are then

$$\frac{dx}{dt} = v \tag{7-3.6}$$

and

$$m\frac{dv}{dt} = -cv - kx + \frac{B_g^2}{\mu_0} A$$

For a circuit with applied voltage $V(t)$ and resistance R, the circuit equivalent of Faraday's law (2-2.5) is

$$N\frac{d\Phi}{dt} + RI = V(t) \tag{7-3.7}$$

Two other equations are required. First, we must relate the flux Φ to the field in the gap. Neglecting fringe or leakage flux,

$$\Phi = B_g A \tag{7-3.8}$$

Second, we must find an expression between the current I in the electric circuit and the flux Φ in the magnetic circuit. This can be done by applying Ampere's law in integral form around the center of the flux path, that is,

$$NI = \oint \mathbf{H} \cdot d\mathbf{s} \tag{7-3.9}$$

For a linear, soft ferromagnetic material, this can be expressed in the form

$$NI = \mathcal{R}\Phi \tag{7-3.10}$$

where $\mathcal{R}$ is called the reluctance and is given by [see Eq. (2-8.11)]

$$\mathcal{R} = \frac{2(d_0 - x)}{\mu_0 A} + \int \frac{ds}{\mu_0 \mu_r A(s)} \tag{7-3.11}$$

where the integral is carried out over the electromagnet path and the keeper.

It is useful to put the resulting equations into nondimensional form using d_0 for the characteristic length, $T_1 = (m/k)^{1/2}$ for the characteristic time, and $B_0 = V_0 T_1 / NA$ for the characteristic magnetic field value, where V_0 is the maximum value of the applied voltage.

The resulting set of coupled, nonlinear, first-order differential equations becomes

$$\frac{dx}{dt} = v$$

$$\frac{dv}{dt} = -2\eta v - x - \gamma B^2 \tag{7-3.12}$$

and

$$\frac{dB}{dt} = -\frac{T_1}{T_2}\left(\frac{1 - x + \alpha}{1 + \alpha}\right) B + V(t)$$

where

$$\gamma \equiv \frac{m}{k^2 d_0 A} \frac{V_0^2}{N^2 \mu_0}, \qquad \eta = \frac{c}{2(km)^{1/2}}$$

and

$$\alpha = \frac{A}{2d_0} \int_{\text{iron}} \frac{ds}{A\mu_r}, \qquad T_2^{-1} \equiv \frac{2Rd_0}{\mu_0 N^2 A}(1 + \alpha)$$

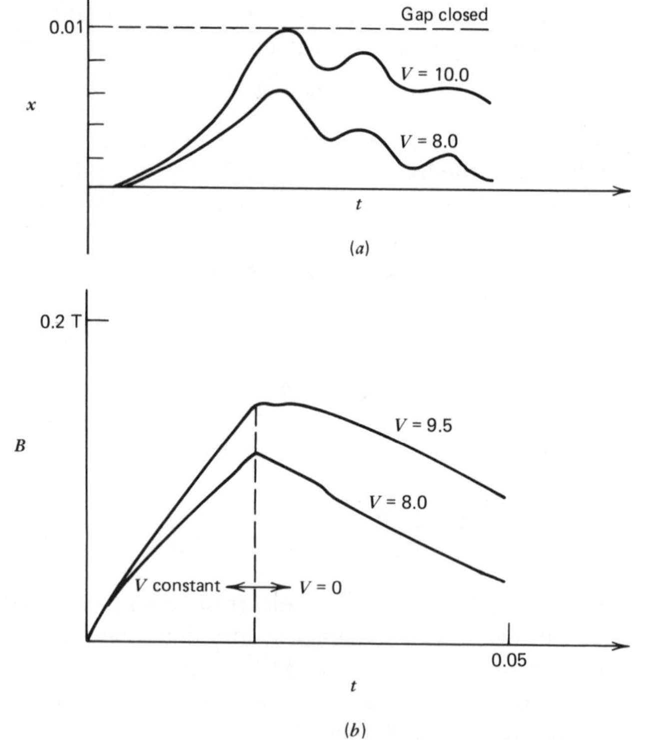

Figure 7-9 Dynamic response of electromagnet actuator: (a) keeper motion vs. time and (b) magnetic field vs. time.

Here T_2 is the decay time of the flux when $x = 0$. These equations can easily be integrated numerically. An example is shown in Figure 7-9, which shows time histories of the displacement x and the field B for a pulsed applied voltage. One observes that although the applied voltage may be discontinuous, the magnetic field is continuous in time and thus persists after $V = 0$. This means that the magnetic force continues to operate after the applied voltage is zero.

7.4 FEEDBACK-CONTROLLED ELECTROMAGNETIC LEVITATION

In Section 1.1, we discussed briefly two types of magnetic levitation. A magnetic levitation device provides a stable magnetic force equal and opposite to the gravitational force on a body. Such devices are used to suspend models in wind tunnels; to provide noncontacting bearings for gyroscopes and rotating machinery; and to suspend passenger-carrying vehicles (see, e.g., Geary, 1964; Rhodes and Mulhall, 1981). The electromagnetic levitation method (EML) uses magnetization forces that are induced by electric currents around a ferromagnetic circuit. For example in Figure 1-5, electromagnets in the vehicle are attracted to a ferromagnetic rail. EML can provide suspension at zero forward speeds in contrast to electrodynamic or repulsive levitation involving superconducting magnets (see, e.g., Chapters 1 and 8). However, in Section 5.6, we saw that the forces created by electromagnets are inherently unstable if the currents are constant. Therefore feedback forces are required to stabilize EML devices.

Several reviews of electromagnetic levitation are available, including Geary (1964), and an excellent U.S. Department of Transportation report by Meisenholder and Wang (1972). There are several EML methods that have met with success. These may be classified as direct, position feedback-control techniques and ac modulated or indirect feedback methods. In the latter, the magnet inductance is part of a tuned circuit whose natural frequency depends on the gap between the suspended mass and magnet. This method has been used to suspend gyroscopic devices for inertial sensors. It suffers from high eddy-current losses and a small range of stable air gaps. An analysis of ac tuned circuit methods may be found in Frazier et al. (1974) and Kaplan (1974).

In this section we present an analysis of a simple position feedback system for magnetic suspensions. This problem is an extension of the previous example where we have replaced the spring force with gravity and have added a position feedback-control system as shown in Figure 7-10. Following the approach used in Section 7.3, we write dynamical equations for both the suspended mass (using Newton's law) and the magnet circuit (using Maxwell's equation). The resulting equations have the same form as Eqs. (7-3.6) and (7-3.7), except that spring and damping terms are absent:

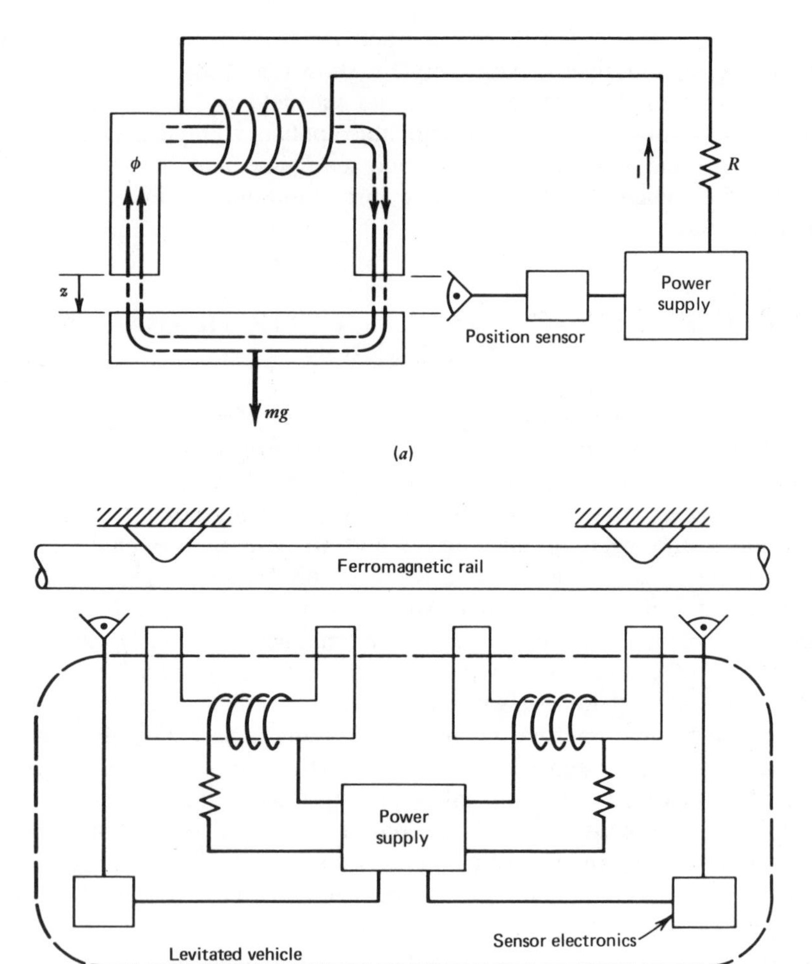

Figure 7-10 (*a*) Feedback-controlled EML system and (*b*) sketch of EML vehicle and support rail.

$$m\ddot{z} = mg + F$$

and

$$N\dot{\Phi} = -RI + V_0 + V_c \tag{7-4.1}$$

From Eq. (7-3.10) the flux Φ is related to the sum of the magnetic reluctances

$$\Phi = \frac{NI}{\mathcal{R}} \tag{7-4.2}$$

where

$$\mathcal{R} = \mathcal{R}_{\text{fe}} + \frac{2z}{\mu_0 A_g}$$

The magnetic force can be derived from either the Maxwell stress tensor, or the magnetic energy and follows from Eqs. (7-3.5) and (7-3.8),

$$F = \frac{-\Phi^2}{\mu_0 A_g} \tag{7-4.3}$$

In (7-4.1b) the constant voltage V_0 provides the primary current I_0.

The control voltage V_c generates the control current which will stabilize the suspended mass. In linear control theory, V_c is given by a linear operator on z. Following Meisenholder and Wang (1972), we choose the following control law:

$$V_c = G_1(z - z_0) + G_2\dot{z} + G_3\ddot{z} \tag{7-4.4}$$

The goal of a magnetic levitation design is to choose the gains G_i such that the system is dynamically stable for small perturbations from equilibrium. Toward this end we define the following perturbation variables:

$$\begin{aligned} z &= z_0 + s(t) \\ I &= I_0 + c(t) \\ \mathcal{R} &= \mathcal{R}_0 + \mathcal{R}_1 s \end{aligned} \tag{7-4.5}$$

where $c(t)$ represents the control current. The following equilibrium relations hold:

$$\begin{aligned} V_0 &= RI_0 \\ mg &= \frac{(NI_0)^2}{\mu_0 A_g \mathcal{R}_0^2} \end{aligned} \tag{7-4.6}$$

Next, we expand the nonlinear magnetic flux and force expressions (7-4.2) and (7-4.3) in a Taylor series in the perturbation variables, that is,

$$\Phi = \frac{NI_0}{\mathcal{R}_0}\left(1 + \frac{c}{I_0} - \frac{\mathcal{R}_1}{\mathcal{R}_0} s\right)$$

and

$$F = -mg\left(1 + \frac{2c}{I_0} - \frac{2\mathcal{R}_1}{\mathcal{R}_0} s\right) \tag{7-4.7}$$

The coupled linearized equations then take the form

$$\begin{aligned} \ddot{s} &= \frac{2g\mathcal{R}_1}{\mathcal{R}_0} s - 2g\frac{c}{I_0} \\ \frac{N^2}{\mathcal{R}_0}\dot{c} &= -Rc + \frac{N^2 I_0}{\mathcal{R}_0^2}\mathcal{R}_1\dot{s} + G_1 s + G_2\dot{s} + G_3\ddot{s} \end{aligned} \tag{7-4.8}$$

These equations can be rewritten in simplified notation:

$$\begin{aligned} \ddot{s} - \alpha^2 s &= -\beta c \\ \dot{c} + \gamma c &= \Gamma_1 s + (\Gamma_2 + \delta)\dot{s} + \Gamma_3 \ddot{s} \end{aligned} \tag{7-4.9}$$

where Γ_1, Γ_2, and Γ_3 are position, velocity, and acceleration feedback gains, respectively. This system of linear differential equations can be solved by either Laplace transforms or more simply in terms of the functions

$$\begin{bmatrix} s(t) \\ c(t) \end{bmatrix} = e^{\lambda t} \begin{bmatrix} \bar{s} \\ \bar{c} \end{bmatrix} \tag{7-4.10}$$

Substituting Eq. (7-4.10) into Eq. (7-4.9), we obtain an expression for the Laplace transform of the control current of $\bar{c}$,

$$\bar{c} = \frac{[\Gamma_1 + (\Gamma_2 + \delta)\lambda + \Gamma_3 \lambda^2]\bar{s}}{\lambda + \gamma} \tag{7-4.11}$$

The gains Γ_1, Γ_2, and Γ_3 (or G_1, G_2, and G_3) are chosen so that the Laplace transform of the control force $-\beta\bar{c}$ appears as a restoring spring and damper force,

$$-\beta\bar{c} = -a\bar{s} - b\lambda\bar{s} \tag{7-4.12}$$

where a is similar to a spring constant and b is a damping constant. Using Eq. (7-4.12) in Eq. (7-4.11), we obtain equations for the gains in terms of a and b; that is,

$$\begin{aligned} \Gamma_1 &= \frac{a\gamma}{\beta} \\ \Gamma_2 &= -\delta + \frac{a + \gamma b}{\beta} \\ \Gamma_3 &= \frac{b}{\beta} \end{aligned} \tag{7-4.13}$$

A necessary condition for stability is that $a > \alpha^2$; that is, the control stiffness should exceed the negative magnetic stiffness. In terms of our original physical variables,

$$G_1 > \frac{V_0 \mathscr{R}_1}{\mathscr{R}_0}$$

If only position and velocity feedback are used, that is, $G_3 = \Gamma_3 = 0$, then the linear dynamics cannot be made to correspond to a simple damped spring mass system and further analysis of the linear stability must be made. Such analysis would involve a root locus of all three roots λ as functions of the gains G_1 and G_2. Further discussion on the dynamics of electromagnetic levitation as applied to passenger-carrying vehicles may be found in Meisenholder and Wang (1972). They also report nonlinear simulation of transient vibrations using the nonlinear equations (7-4.1)–(7-4.4).

7.5 PERMANENT MAGNET CIRCUITS WITH ELASTIC COMPONENTS

Magnetization is a property of a material body that produces a magnetic field in addition to that produced by electric currents $\mu_0\mathbf{H}$ (see Section 2.5),

$$\mathbf{B} = \mu_0(\mathbf{H} + \mathbf{M}) \tag{7-5.1}$$

Magnetization can be induced by the **H** field (soft magnetic material) or may exist without external fields by spontaneous alignment of spins in the body.

Permanent magnetism does not mean that **B** is constant, but that $|\mathbf{M}|$ is fixed. The value of the field **B** depends on the reluctance of the flux paths around the magnet. In this section we examine the effect of an elastic magnetic circuit element on the magnetic forces between the permanent magnet and the elastic ferromagnetic body.

Imagine a magnetic circuit with a permanent magnet as shown in Figure 7-11. We assume that the flux lines are guided by soft ferromagnetic material of very high permeability and that the flux lines cross a gap of unit relative permeability and enter an elastic body with very high permeability. By assumption, we hold that the field lines enter the elastic body normal to the surface and we use the pole model for magnetic stresses to calculate the total force on the body. For simplicity, we replace the elastic behavior of the body with a spring. However, the elastic body could be a beam or plate as shown in the figure. Such problems are sometimes found in electric switches and relays.

If Φ_m is the flux created by the magnet and Φ_g is the flux that crosses the gap and enters the elastic body, the difference will be denoted by the leakage flux Φ_L where conservation of flux demands

$$\Phi_m = \Phi_g + \Phi_L \tag{7-5.2}$$

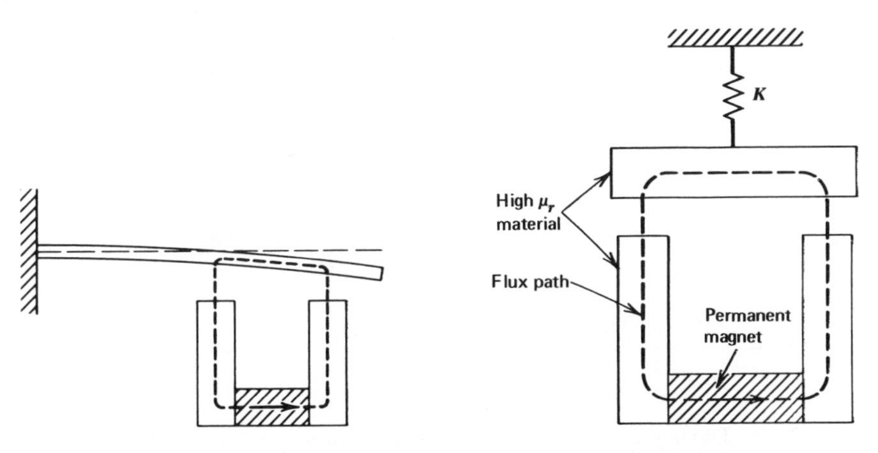

Figure 7-11 Permanent magnet circuits with elastic keepers.

In order to solve magnetic field problems involving permanent magnets, one must have a constitutive relation between **B** and **H** (see Section 2.5). In particular, for one-dimensional components of **B** and **H** a relation in the plane is given and one must use the second quadrant of the $B-H$ curve called the *demagnetization* curve. A demagnetization curve for a barium ferrite magnet is shown in Figure 7-12. The term demagnetization is used since $\mathbf{H}\cdot\mathbf{B}<0$ within the magnet when no electric currents are present. This is a consequence of Ampere's law in integral form,

$$\oint \mathbf{H}\cdot d\mathbf{l}=0$$

It will also be recalled that the $B-H$ curve for increasing demagnetization sometimes differs from the curve of decreasing demagnetization. The later curve is called the *recoil curve* (see Fig. 2-5). For some materials, such as ferrites and some rare-earth magnets, the recoil curve is identical to the demagnetization curve. For simplicity, we assume a linear relation between B and H in the second quadrant given by (see, e.g., McCaig, 1977)

$$B=B_r+\mu_0\mu_r H \tag{7-5.3}$$

In the magnetic flux circuit in Figure 7-11, B and H are not independent and depend on the deflection of the elastic keeper or the air-gap distance. This dependence can be derived from Ampere's law (2-2.4) applied to a closed path going through the magnet, gap, and keeper:

$$H_m L_m + 2\frac{B_g}{\mu_0}L_g=0 \tag{7-5.4}$$

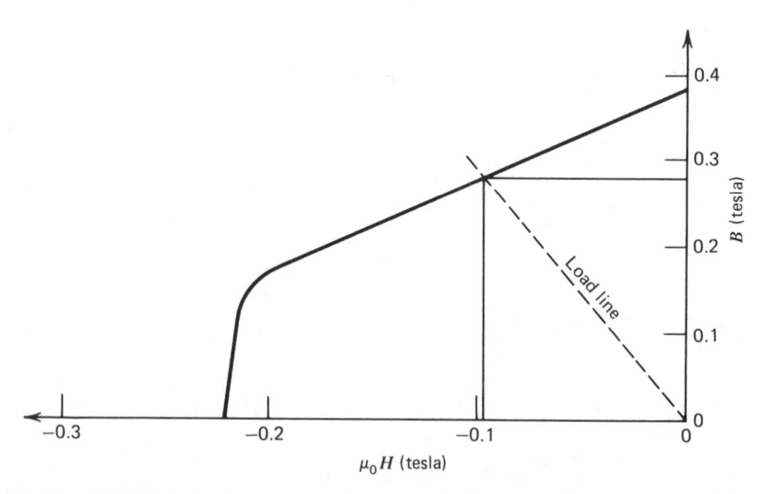

Figure 7-12 Demagnetization curves for permanent magnet material "Indox V", a barium ferrite ceramic permanent magnetic material. (Indiana General Corporation.)

where we have neglected $\int \mathbf{H} \cdot d\mathbf{l}$ in the high permeability circuit components. If A_g is the area of the air-gap pole face, this expression can be put into the form of a flux-reluctance relation similar to Eq. (7-3.10),

$$-H_m L_m = \mathcal{R}_g \Phi_g \tag{7-5.5}$$

where $\mathcal{R}_g = 2L_g/\mu_0 A_g$ is the air-gap reluctance. We can also write a similar relation for the leakage flux

$$\Phi_L = \frac{-H_m L_m}{\mathcal{R}_L} \tag{7-5.6}$$

$\mathcal{R}_L$ is the reluctance of the leakage flux path (see, e.g., Roters, 1941). Equations (7-5.5) and (7-5.6) can be combined with Eq. (7-5.2) to obtain a relation between the field in the permanent magnet

$$B_m = \frac{\Phi_m}{A_m}$$

and the demagnetization field H_m; that is,

$$B_m = -\frac{L_m}{A_m} \frac{1}{\mathcal{R}} H_m \tag{7-5.7}$$

where

$$\frac{1}{\mathcal{R}} = \frac{1}{\mathcal{R}_g} + \frac{1}{\mathcal{R}_L}$$

This relation is sometimes called a *load line* (see Fig. 7-12). Its intersection with the constitutive law (7-5.3) determines the operating fields in the magnet system; that is,

$$B_m = B_r + \mu_0 \mu_r \left(\frac{-A_m}{L_m} \mathcal{R} \right) B_m$$

or

$$B_m = \frac{B_r}{1 + \eta} \tag{7-5.8}$$

where

$$\eta = \frac{\mu_0 \mu_r A_m \mathcal{R}}{L_m} = \frac{\mathcal{R}}{\mathcal{R}_m}$$

Here we have defined a magnet reluctance

$$\mathcal{R}_m \equiv \frac{L_m}{\mu_0 \mu_r A_m} \tag{7-5.9}$$

For an ideal flux circuit with no leakage,

$$\eta = \frac{2\mu_r A_m}{A_g} \frac{L_g}{L_m}$$

We must remember, however, that the air-gap distance L_g is still unknown. To determine L_g, we require that B_m or B_g be compatible with the magnetic forces in Newton's law of equilibrium. Using the magnetic tension model for the total magnetic force on the keeper, we have

$$2A_g \frac{B_g^2}{2\mu_0} = kx \tag{7-5.10}$$

where k is the spring constant. The gap distance and the keeper displacement x are related by

$$L_g = x_0 - x$$

For an ideal circuit, $\Phi_m = \Phi_g$ and

$$B_g = \frac{A_m}{A_g} B_m$$

The resulting magnetic force is given by

$$F_m = \frac{B_r^2 A_m^2}{\mu_0 A_g} \left(1 + 2\frac{A_m}{A_g} \mu_r \frac{L_g}{L_m}\right)^{-2} \tag{7-5.11}$$

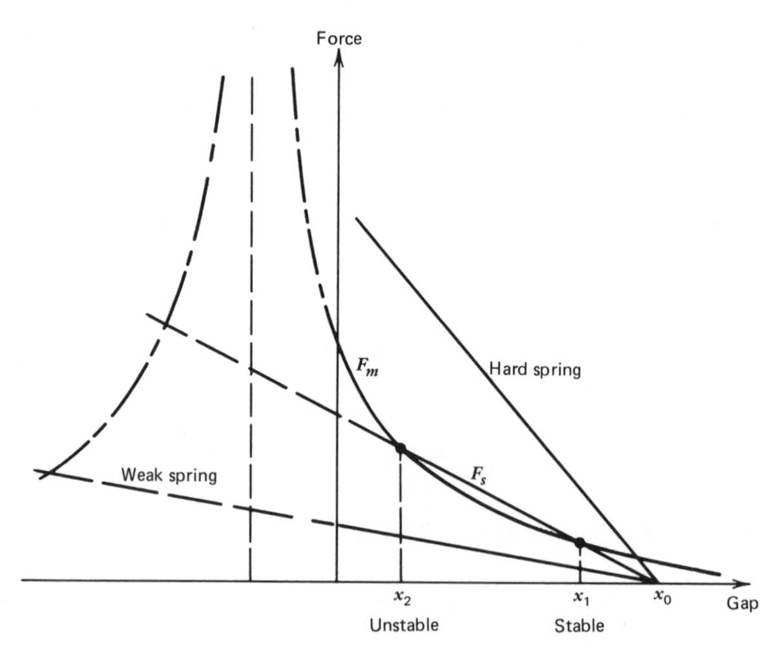

Figure 7-13 Force vs. gap for a permanent magnet circuit with elastic keeper.

This force has a strong dependence on the air-gap distance as shown in Figure 7-13. The resulting equilibrium condition for x then becomes

$$kx = \frac{B_r^2}{\mu_0} \frac{A_m^2}{A_g}\left(1 + 2\frac{A_m}{A_g}\mu_r\frac{x_0 - x}{L_m}\right)^{-2} \tag{7-5.12}$$

A graphical solution to this cubic equation in x is shown in Figure 7-13. This problem is typical of a cusp catastrophe (see Chapter 5). Mathematically, there can exist one, two, or three equilibrium solutions. In practice, there can exist only one, two, or no solution since the air-gap distance cannot be negative. Where two solutions exist, one is stable and the other is unstable. In Figure 7-13, x_1 is stable since for small $\Delta x > 0$ the spring force is greater than the magnetic force ($F_s > F_m$), while x_2 is unstable since for $\Delta x > 0$, $F_m > F_x$.

If the spring constant is not strong enough, there will be no practical solution except $L_g = 0$. In this case a contact force acts on the keeper whose value is the difference between the zero gap magnet force and the spring force. The maximum magnet force F_0 is given by

$$F_0 = \frac{B_r^2}{\mu_0} \frac{A_m^2}{A_g} \tag{7-5.13}$$

7.6 BENDING OF FERROMAGNETIC PLATES

The body forces on magnetized material can be represented by electromagnetic stresses acting on the surface of the body (see Section 2.6). These magnetic stresses are of the order of $\mu_0 M^2$, which for the best ferromagnetic material is around 10^2 N/cm^2 or less. Compared to the yield stress for typical ferromagnetic metals (greater than 10^4 N/cm^2), this stress is insignificant unless it acts on a structure where small loads applied at one point produce large stresses elsewhere in the solid. Such structural shapes include beams, plates, and shells, or any configuration for which the ratio of two characteristic dimensions is very large.

In Sections 3.4, 3.10, and 5.10, we discussed the fact that there are different formulations for representing magnetic body forces on magnetized materials, but for many structural problems the choice of a particular model over another was at most an error of $\mu_0 M^2$. One example was developed in Section 3.10 using the static theory of Pao and Yeh (1973) for a ferromagnetic plate in a static magnetic field. The resulting equation for bending of a classical thin plate was found to be

$$D\left(\frac{\partial^4 w}{\partial x^4} + 2\frac{\partial^4 w}{\partial x^2 \partial y^2} + \frac{\partial^4 w}{\partial y^4}\right) + \frac{[(\mathbf{B} \cdot \mathbf{n})^2]}{2\mu_0} = 0 \tag{7-6.1}$$

where w is the lateral plate deflection and the brackets $[\ ]$ are defined by

$[f] = f(\text{top surface}) - f(\text{bottom surface})$. D is the bending stiffness given in terms of the plate thickness Δ and the elastic constants Y and ν; $D = Y\Delta^3/12(1-\nu^2)$. In this model the effect of the body forces are replaced by magnetic tensions on the top and bottom surfaces of the plate.

To correctly use Eq. (7-6.1), one must evaluate the magnetic stress term very carefully. For example, the change in the plate normal vector due to rotation of the plate midsurface must be accounted for in the magnetic boundary conditions (see Sections 3.10, 5.10). Also, in inhomogeneous magnetic fields, the magnetic stresses should be evaluated at the *deformed* position of the plate. Such effects introduce *magnetic stiffness* terms into the equilibrium equations for the plate (see Section 4.6).

In this section we will present a simple application of the approximate theory presented in Section 3.10 to the problem of bending of a cantilevered ferromagnetic plate by a pair of current filaments.

There are several ways to proceed in these problems short of finding an exact solution. The first is to solve for the magnetic field in and around the *rigid* structure and calculate the initial magnetic loading. This loading is then used to calculate the deformation. This is straightforward and by using a standard finite-element method, along with a numerical method for static magnetic fields, one can use the method to solve many problems.

However, if the deformation is large enough to appreciably change the magnetic stresses, then the magnetic stiffness effects must be included. If the magnetic forces are localized, one may use the structural matrix method as given in the example to follow. Here one uses only the deformed structure in the neighborhood of the magnetic force to find the perturbed field.

When the magnetic forces are distributed over the structure, the determination of magnetic stiffness is more difficult since the perturbed magnetic forces depend on the deformation in the entire structure. In general, this problem leads to an integodifferential equation. As yet no general methods for these problems have been given in the literature.

Example Using the Structural Matrix Method

Consider a long rectangular plate with two current filaments placed under the plate and parallel to the plate surface as shown in Figure 7-14. We assume that the plate has properties of soft ferromagnetic material and that the relative permeability is very large, that is, $\mu_r \gg 1$.

The boundary conditions on the magnetic field at the plate surface require continuity of the normal component of **B** and the tangential component of **H**. The assumption $\mu_r \gg 1$ leads to the result that the ratio of the tangential to the normal component of **B** outside the plate is very small, that is, $B_t^+/B_n^+ \sim 0$. One can also show that the field on the top of the plate is almost zero.

With these assumptions one can use the *image method* to calculate the normal field outside the plate as was done in Section 5.5 (see also Section 4.2). However, we wish to determine the effect of the deflection and change

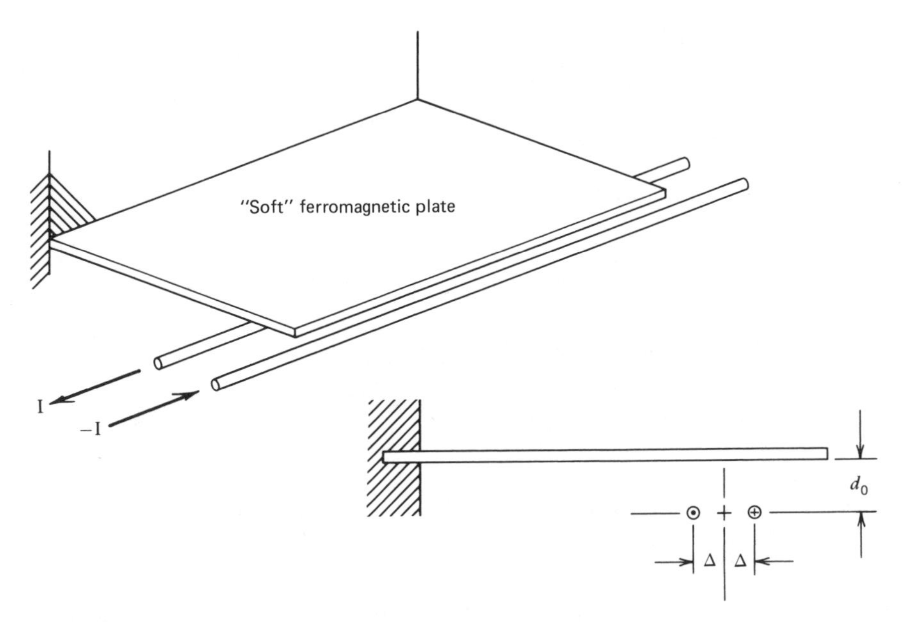

Figure 7-14 Sketch of current filaments near an elastic ferromagnetic plate.

of slope on the over-all stiffness of the plate so the magnetic forces are calculated using the deformed geometry as shown in Figure 7-15. The normal component of the field at the deformed surface can be shown to be

$$B_n = \frac{\mu_0 I}{\pi}\left(\frac{x - \Delta \cos \beta}{(x - \Delta \cos \beta)^2 + d_2^2} - \frac{x + \Delta \cos \beta}{(x + \Delta \cos \beta)^2 + d_1^2}\right) \tag{7-6.2}$$

where x is a local coordinate tangential to the deformed plate, Δ is the distance between the filaments, and d_1 and d_2 are the filament-plate gaps given by

$$d_1 = \frac{(d_0 - u)}{\cos \beta} + \Delta \sin \beta \tag{7-6.3}$$

and

$$d_2 = \frac{(d_0 - u)}{\cos \beta} - \Delta \sin \beta$$

Here d_0 is the initial gap, u the displacement of the beam, and β is the slope of the bent beam near the filaments. We assume that Δ/L is small so that u and $\beta = \partial u/\partial x$ represent deformation at the midpoint between the filaments.

For many problems this magnetic force distribution is sufficiently concentrated with respect to the length of the beam that we can replace the distributed force problem by a discrete force and couple acting on the beam.

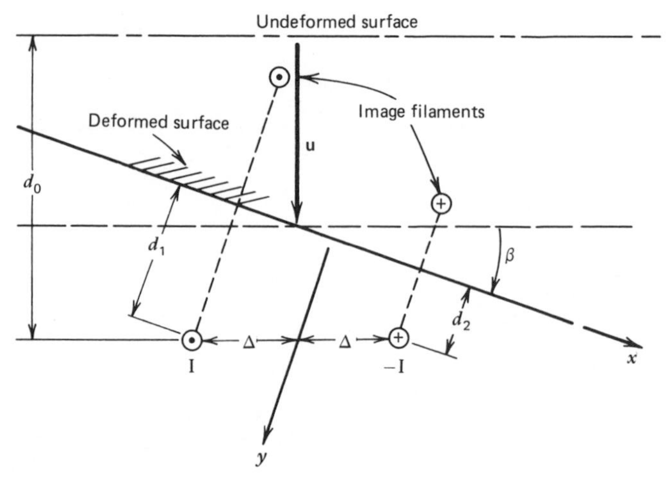

Figure 7-15 Geometry of two current filaments near the rotated ferromagnetic surface.

The concentrated magnetic force and couple per unit filament length are given by

$$F = \int_{-\infty}^{\infty} \frac{B_n^2}{2\mu_0} \, dx, \qquad C = \int_{-\infty}^{\infty} x \frac{B_n^2}{2\mu_0} \, dx \tag{7-6.4}$$

This approximation allows one to use standard structural matrix methods to find the solution $u(x)$ and $\beta(x)$. For example, let $[k_e]$ represent a 2×2 elastic stiffness matrix for the beam with a force F and couple C on it:

$$\begin{Bmatrix} F \\ C \end{Bmatrix} = [k_e] \begin{Bmatrix} u \\ \beta \end{Bmatrix} \tag{7-6.5}$$

This expression assumes that u and β are small. The elements of the matrix will depend on the boundary conditions on the plate edges.

We next expand the magnetic force and couple in terms of the deformation variables which for this problem assumes the form

$$F = F_0 + \kappa u, \qquad C = \gamma \beta \tag{7-6.6}$$

Equating the elastic and magnetic force and couple we obtain

$$[k_e] \begin{Bmatrix} u \\ \beta \end{Bmatrix} = F_0 \begin{Bmatrix} 1 \\ 0 \end{Bmatrix} - [k_m] \begin{Bmatrix} u \\ \beta \end{Bmatrix} \tag{7-6.7}$$

where

$$[k_m] = \begin{bmatrix} -\kappa & 0 \\ 0 & -\gamma \end{bmatrix}$$

is the magnetic stiffness matrix. The deflection and slope are then given by

$$\begin{Bmatrix} u \\ \beta \end{Bmatrix} = F_0[k_e + k_m]^{-1}\begin{Bmatrix} 1 \\ 0 \end{Bmatrix} \tag{7-6.8}$$

The two wire example can be approximated by a two-dimensional dipole field if we take the limit $\Delta \to 0$ and $I \to \infty$, such that $\Delta I \to \Gamma$. The resulting magnetic field is found to be

$$B_n = \frac{2\mu_0 \Gamma}{\pi(x^2 + d^2)^2}(x^2 \cos\beta + 2xd \sin\beta - d^2 \cos\beta) \tag{7-6.9}$$

where $d = (d_0 - u)/\cos\beta$. The distribution of magnetic tension along the bottom of the plate is shown in Figure 7-16. From Eq. (7-6.4) we can obtain expressions for the total magnetic force and couple on the plate. For small slopes, $\cos\beta \cong 1$ and $\sin\beta \cong \beta$, these expressions are approximately given by

$$F = \int_{-\infty}^{\infty} \frac{B_n^2}{2\mu_0}\, dx_1 = \frac{2\mu_0 \Gamma^2}{\pi^2 d^3} \int_{-\infty}^{\infty} \frac{(\eta^2 - 1)^2}{(\eta^2 + 1)^4}\, d\eta = \frac{\mu_0 \Gamma^2}{2\pi d^3} \tag{7-6.10}$$

and

$$C = \int_{-\infty}^{\infty} \frac{x_1 B_n^2}{2\mu_0}\, dx_1 = \frac{8\mu_0 \Gamma^2 \beta}{\pi^2 d^2} \int_{-\infty}^{\infty} \frac{\eta^2(\eta^2 - 1)}{(\eta^2 + 1)^4}\, d\eta = 0 \tag{7-6.11}$$

The expression for F can be checked by calculating the force between the

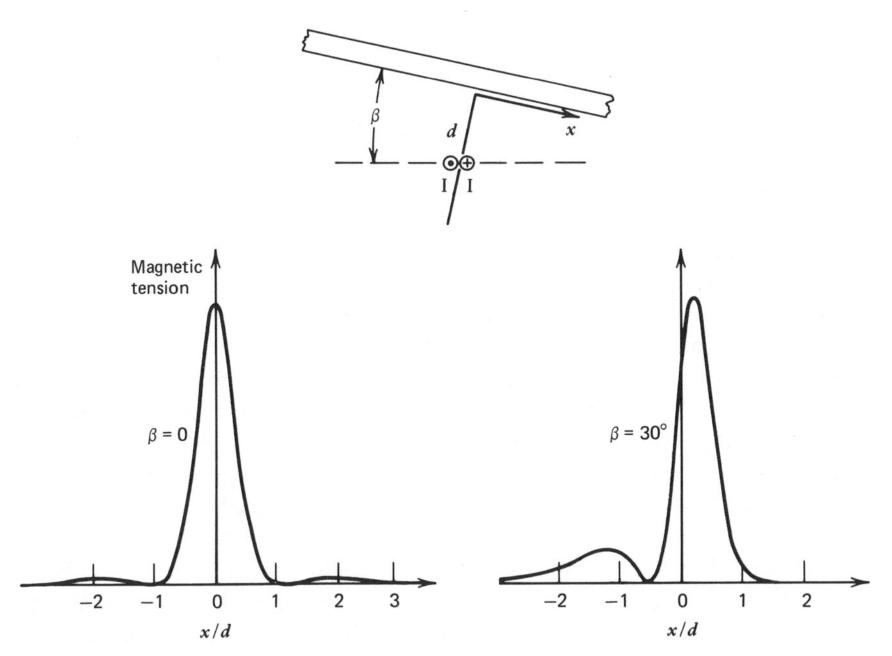

Figure 7-16 Magnetic tension on ferromagnetic surface due to two current filaments: (*a*) zero plate rotation and (*b*) rotated plate surface.

dipole pair and its image in the plate as was done in Section 5.5. For small deflections and slopes F takes the form

$$F = \frac{\mu_0 I^2 \Delta^2}{2\pi d_0^3}\left(1 + \frac{3u}{d_0}\right)$$

In the simple example of a two-filament dipole centered under a simply supported plate of length L and bending stiffness D per unit width, the matrix equation (7-6.5) becomes a scalar equation for the midspan displacement u, with $k_e = 48D/L^3$. Using the results in Eq. (7-6.10), we find

$$u = \frac{\mu_0 I^2 \Delta^2}{2\pi d_0^3} \frac{1}{(48D/L^3 - 3\mu_0 I^2 \Delta^2 / 2\pi d_0^4)}$$

We note the *nonlinear* relation between the deflection u and the force parameter $\mu_0 I^2$. This is another example of the importance of the magnetic stiffness effect (Section 4.6). Also the deflection becomes unbounded as the current reaches a critical value. A related example which also exhibits a limit point instability has been observed experimentally and is discussed in the next section.

7.7 EXPERIMENTS ON BENDING OF FERROELASTIC PLATES

Most of the experiments reported in the literature on ferromagnetic plates in magnetic fields concern bifurcation-type stability or buckling problems. The object of most of these studies was to measure the critical buckling field. Few studies exist in the modern literature on the problem of the prediction or measurement of stresses in flexible ferromagnetic structures, such as beams and plates, in magnetic fields.

An approximate theory for predicting the bending deformation in plates was presented in Chapter 3 using the Maxwell stress tensor—specfically using the magnetic tension $B_n^2/2\mu_0$ on the lateral surfaces of the plate to calculate the transverse load. An example was discussed in the previous section. In this section we present some simple experimental results which support the approximate theory and illustrate an experimental technique for measuring distributed magnetic loads on structures.

The structure chosen for this example is a simply supported beam-plate which is bent by magnetic forces from a strong permanent magnet (see Fig. 7-17). The magnet was a 2.5-cm-diam rare-earth magnet with a surface field of around 0.21 T with one face of the magnet on a steel keeper. The beam shown in Figure 7-17 has a thickness of 0.64 mm and a width of 27 mm. The distance between supports was 23 cm and the gap between the plate and the magnet face varied from 7 to 20 mm.

As the magnet is brought near the plate, the beam becomes magnetized

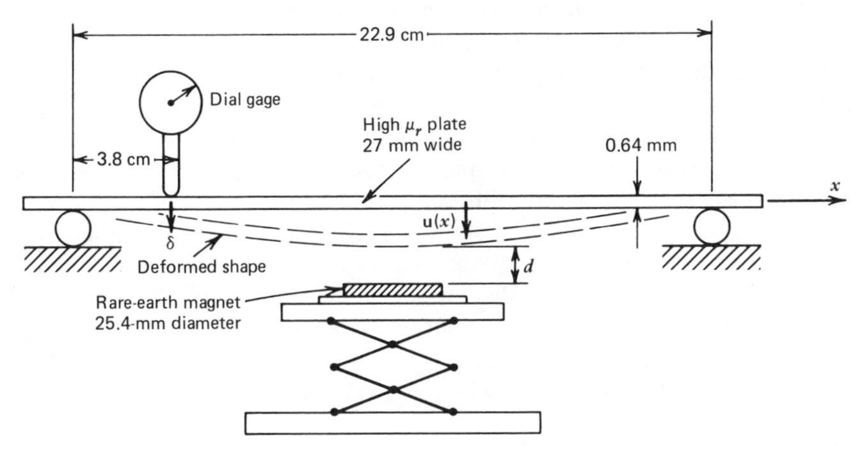

Figure 7-17 Sketch of experimental apparatus for bending of a ferromagnetic beam near a permanent magnet.

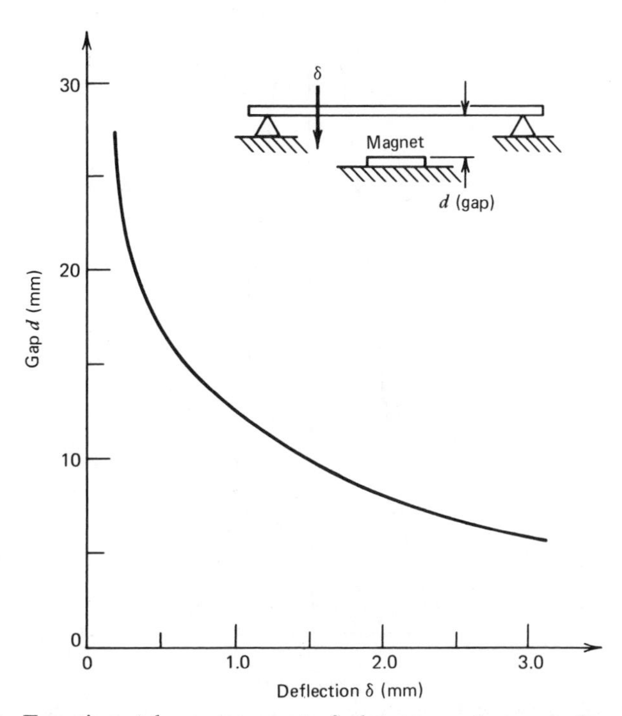

Figure 7-18 Experimental measurement of plate–magnet gap vs. beam deflection (see Fig. 7-17).

(the material was a high μ_r material with a small hysteresis loop). The resulting magnetic forces between the magnetic dipoles in the permanent magnet and the induced dipoles in the plate produce bending. Since the bending decreases the plate-magnet gap, which in turn increases the force, the system is potentially unstable. A limit point or critical gap exists at which the beam snaps down onto the magnet face. A plot of deflection versus plate-magnet gap is shown in Figure 7-18.

Our goal in this example is to show that by measuring the magnetic field on the lateral surfaces of the plate one can use the calculated magnetic tensions to correctly predict the measured deflections of the beam. The equation of equilibrium for the deflection of beam is

$$D\frac{d^4u}{dx^4} = f(x) \tag{7-7.1}$$

where the bending stiffness is given by

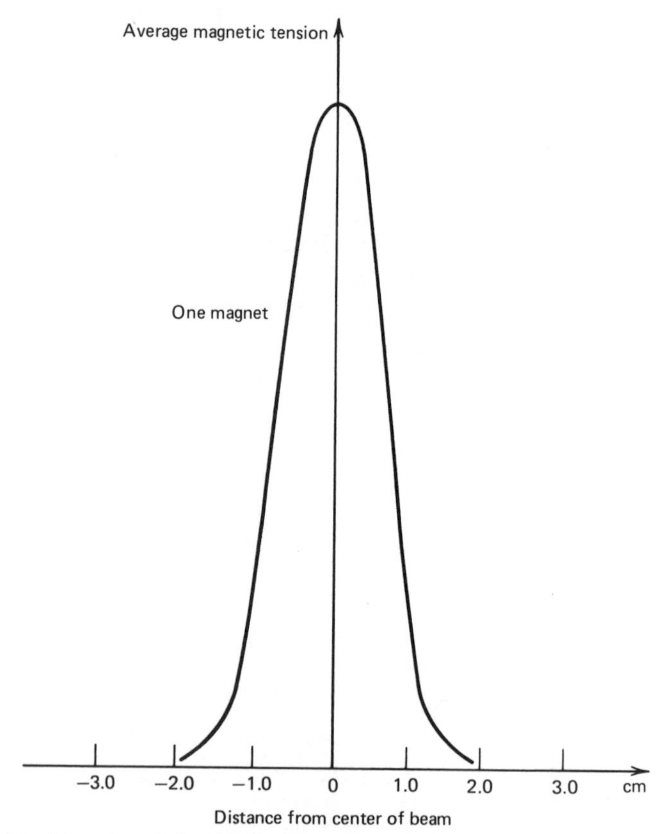

Figure 7-19 Experimental distribution of magnetic tension along ferromagnetic beam-plate due to a permanent magnet (see Fig. 7-17).

$$D = \frac{Yb\Delta^3}{(1-\nu^2)12}$$

Here Y is Young's modulus, ν is Poisson's ratio, b is the width of the plate, and Δ is the thickness.

The lateral magnetic load per unit length is the integral of the magnetic tensions across the width of the plate,

$$f(x) = \frac{1}{2\mu_0}\int_0^b (B_+^2 - B_-^2)\,dy \qquad (7\text{-}7.2)$$

where the normal field component is implied; that is, $B = \mathbf{B} \cdot \mathbf{n}$, "+" indicates the side nearest the magnet, and "−" indicates the side away from the magnet.

The measurement of $\mathbf{B} \cdot \mathbf{n}$ can be easily performed with a small hand-held Hall-effect gaussmeter (see Chapter 9). The measured field distribution across the beam face and the distribution of $f(x)$ along the beam is shown plotted in Figure 7-19. Integration of $f(x)$ along the beam then gives the total magnetic force on the beam. In these experiments, the magnet produced between 0.5 to 3.0 N of force. By integrating the magnetic tensions over the plate for four different values of gap, a magnetic force-gap relation was obtained which showed that the force was inversely proportional to the gap distance squared as shown in Figure 7-20:

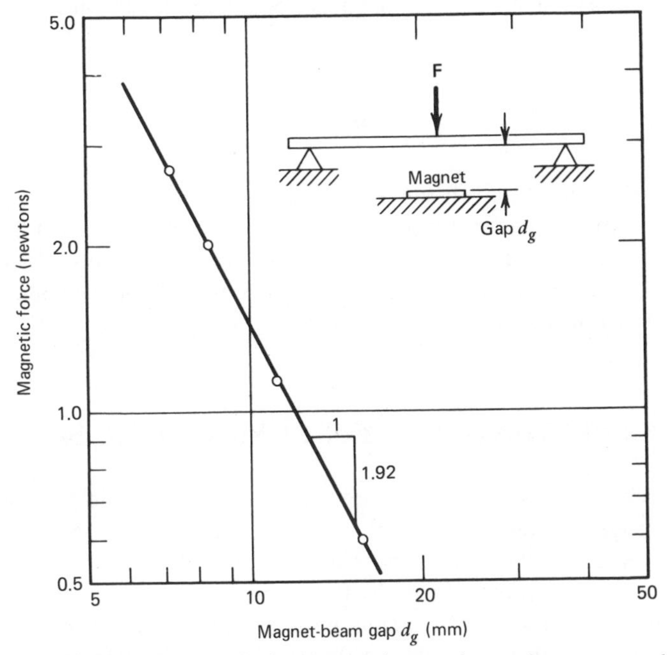

Figure 7-20 Calculated magnetic force on a beam-plate using measured magnetic fields (see Fig. 7-17).

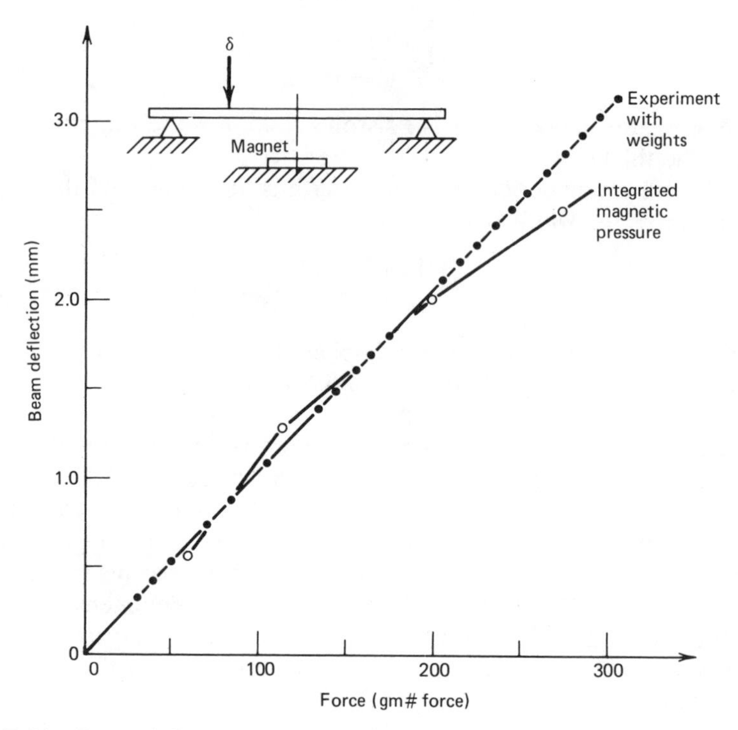

Figure 7-21 Beam deflection vs. magnetic force and gravitational force—experimental data.

$$F = \frac{c}{d^2} \tag{7-7.3}$$

where d is the actual gap distance after deflection.

A plot of the magnetic force versus deflection is shown in Figure 7-21. Also shown for comparison is an experiment with mass-gravity loading. The two curves are remarkably close for large gaps, which confirms the use of the magnetic tension method.

7.8 BENDING OF A FERROMAGNETIC SHELL

In the previous two sections we showed how the magnetic tension method could be used to predict the deflection of a straight beam assuming the magnetic force produced a narrow distribution of magnetic tension equivalent to a concentrated force. In this section we present the solution of the bending of a thin elastic shell with a wide distribution of magnetic tension. In particular, we consider a thin, circular, soft ferromagnetic shell in a uniform magnetic field (see Fig. 7-22). (A ferromagnetic shell or enclosure

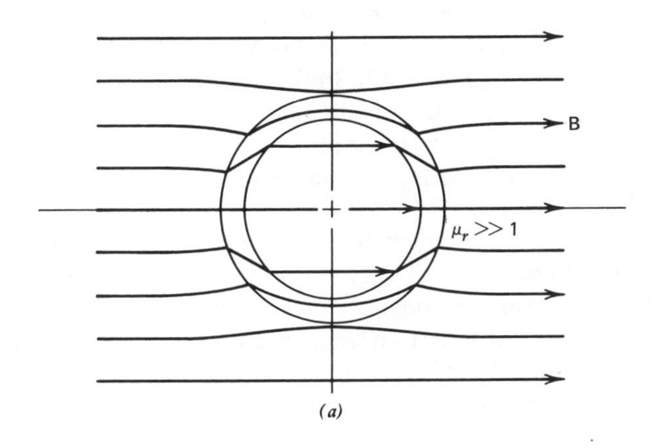

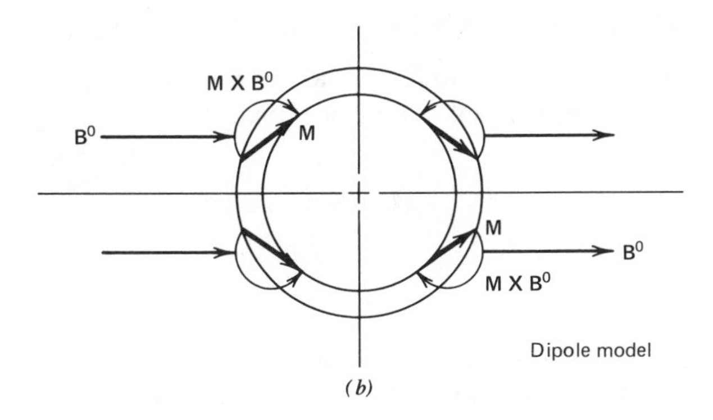

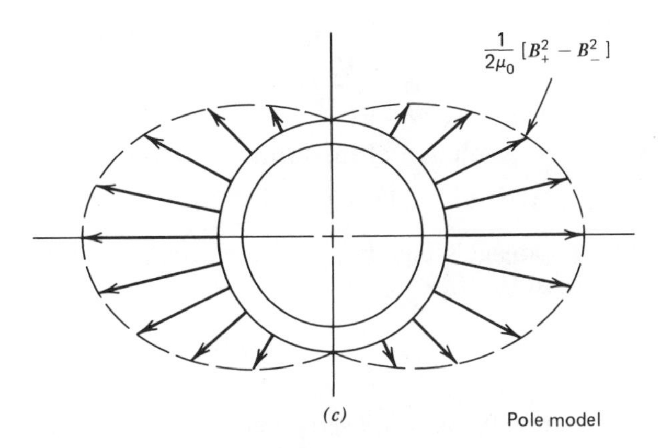

Figure 7-22 Ferromagnetic cylindrical shell in a uniform magnetic field: (*a*) field pattern, (*b*) couple distribution—dipole model, and (*c*) magnetic tension distribution—pole model.

is often used to shield a volume containing sensitive electronic equipment from magnetic fields.) The shell acts to displace the flux lines of the uniform field from its interior (see Fig. 7-22). In the magnetic dipole model, the body force is zero, but a distribution of body couples acts to bend the shell. A pole model interpretation produces a distribution of magnetic tension around the shell which also leads to bending. A comparison of these two models is shown in Figure 7-22.

In this section we will neglect the magnetic stiffness effect. That is, we calculate the magnetic forces as they act on the rigid cylindrical shell. In the first part of this section, the magnetic field distribution on the rigid shell is determined. Next, the magnetic tension is calculated and the equations of equilibrium for the shell are solved for the radial displacement. Finally, we describe an experiment which demonstrates qualitatively some of the results of the analysis.

Magnetic Field in a Cylindrical Shell in a Uniform Field

When the fields are steady in time and no electric currents are present, the field $\mathbf{H}$ can be derived from a potential ϕ, which satisfies Laplace's equation, that is,

$$\mathbf{H} = \nabla \phi \tag{7-8.1}$$

and

$$\nabla^2 \phi = 0 \tag{7-8.2}$$

Consider a cylindrical shell of outer radius b and inner radius a. In polar coordinates the applied uniform field can be written as

$$\mathbf{B}^0 = B_0(\cos\theta \mathbf{e}_r - \sin\theta \mathbf{e}_\theta) \tag{7-8.3}$$

which can be derived from a potential

$$\phi^0 = \frac{B_0}{\mu_0} r \cos\theta \tag{7-8.4}$$

Outside we assume a solution of the form

$$\phi^{\mathrm{III}} = \phi^0 + \phi^1 \tag{7-8.5}$$

where ϕ^1 satisfies Laplace's equation and the condition $\phi^1 \to 0$ as $r \to \infty$. These conditions lead to

$$\phi^1 = \frac{A}{r} \cos\theta \tag{7-8.6}$$

In the other two regions one can show that

$$\phi^{\mathrm{II}} = \left(B_1 r + \frac{C_1}{r}\right) \cos\theta \tag{7-8.7}$$

and

$$\phi^{\mathrm{I}} = D_1 r \cos\theta$$

The second solution implies a uniform field inside the cylinder as shown in Figure 7-22. The magnetic boundary conditions at $r = a$ are

$$H_\theta^{\mathrm{I}} = H_\theta^{\mathrm{II}}, \qquad B_r^{\mathrm{I}} = B_r^{\mathrm{II}} \text{ at } r = a$$

or

$$\phi^{\mathrm{I}} = \phi^{\mathrm{II}}, \qquad \frac{\partial \phi^{\mathrm{I}}}{\partial r} = \mu_r \frac{\partial \phi^{\mathrm{II}}}{\partial r} \text{ at } r = a \tag{7-8.8}$$

Similar conditions must hold at $r = b$. These conditions lead to four algebraic equations for A, B_1, C_1, and D_1. The solution can be shown to be

$$\begin{aligned} B_1 &= \frac{\mu_r + 1}{2\mu_r} D_1 \\ C_1 &= \frac{2a^2(\mu_r - 1)B^0}{\mu_0 \Delta} \\ D_1 &= \frac{4\mu_r B^0}{\mu_0 \Delta} \end{aligned} \tag{7-8.9}$$

and

$$\Delta = \left(1 - \frac{a^2}{b^2}\right)(\mu_r^2 + 1) + 2\mu_r\left(1 + \frac{a^2}{b^2}\right)$$

Of particular interest is the term $\mathbf{M} \times \mathbf{B}^0$ which gives the magnetic body couple on the solid. Integrating across the thickness we get the magnetic body couple distribution around the shell:

$$\mathbf{C} = \int_a^b \mathbf{M} \times \mathbf{B}^0 dr = \mathbf{e}_r \times \mathbf{e}_\theta \frac{b-a}{ab} (\mu_r - 1) B_0 C_1 \sin 2\theta \tag{7-8.10}$$

For large μ_r, one can show that $\mathbf{C}$ is independent of the permeability μ_r. Again, since there is no gradient in the external field $\mathbf{B}^0$ there is zero net force on the cylinder.

Magnetic Tension

To obtain the distribution of magnetic tension, the radial field inside and outside the shell must be calculated since

$$T_{rr} = \frac{1}{2\mu_0}[B_r^2(r = b) - B_r^2(r = a)] \tag{7-8.11}$$

In terms of the constants given in Eq. (7-8.9), the radial field inside the shell thickness is found to be

$$B_r = \mu_0 \frac{\partial \phi^{II}}{\partial r} = \mu_0 \mu_r \left(B_1 - \frac{C_1}{r^2} \right) \cos \theta$$

Using Eq. (7-8.9), one obtains the following expression for T_{rr},

$$T_{rr} = \frac{2B_0^2 \mu_r^2 \cos^2 \theta}{\mu_0 \Delta^2} \left\{ \left[\mu_r \left(1 - \frac{a^2}{b^2}\right) + \left(1 + \frac{a^2}{b^2}\right) \right]^2 - 4 \right\} \qquad (7\text{-}8.12)$$

where Δ is given in Eq. (7-8.9).

For a thin shell we assume $b = a + \delta$ and $\delta/a \ll 1$. Also for a high permeability material, we assume that $\mu_r \delta/a \gg 2$. Under these assumptions, the magnetic tension becomes

$$T_{rr} = 4P_m \cos^2 \theta = 2P_m(1 + \cos 2\theta) \qquad (7\text{-}8.13)$$

where P_m is the magnetic pressure in the external field, that is, $P_m = B_0^2/2\mu_0$.

Equilibrium Equations for a Shell

The magnetic forces are assumed to be independent of the axial direction and only planar deformations of the shell are examined. Under these assumptions, the equilibrium conditions for a differential shell element are identical to those of a curved beam or ring as given in Section 6.7. Let N and T represent the radial shear and circumferential tension forces per unit axial length, and G represent the bending moment about an axis parallel to the axis of the shell. Then the equations of equilibrium in the radial and circumferential direction and in the moment equilibrium are given by

$$\begin{aligned} \frac{dN}{dS} + \frac{T}{R} &= T_{rr} \\ \frac{dT}{dS} - \frac{N}{R} &= 0 \\ \frac{dG}{dS} + N &= 0 \end{aligned} \qquad (7\text{-}8.14)$$

where $dS = R d\theta$, and R is the undeformed average radius $\frac{1}{2}(b + a)$. By choosing the undeformed radius, we are implicitly neglecting the change in stiffness due to tension.

In addition to the equilibrium equation, one needs a relation between the bending moment and the deformation. In general, there will be two independent components of displacement of the midsurface of the shell, u_r and u_θ. For many practical problems, however, the bending effects dominate the change in circumferential length. Thus we adopt the assumption of an *inextensible circumference*. It can be shown that this leads to the relation (see, e.g., Love 1922)

$$\frac{du_\theta}{d\theta} = u_r \tag{7-8.15}$$

Under this assumption, the bending moment is related to the change in curvature, κ, as given below:

$$G = \frac{Y\Delta^3}{12(1-\nu^2)}\left(\kappa - \frac{1}{R}\right) \tag{7-8.16a}$$

where

$$\kappa = \frac{1}{R} + \frac{d^2u}{dS^2} + \frac{u}{R^2} \qquad (u \equiv -u_r) \tag{7-8.16b}$$

Now to find a solution, given the magnetic tension loading, we eliminate T from the first two equations of (7-8.14) and obtain an equation for the shear force N; that is,

$$\frac{d^2N}{dS^2} + \frac{N}{R^2} = \frac{d}{dS}T_{rr} \tag{7-8.17}$$

Using Eq. (7-8.13), it is straightforward to show that

$$N = \tfrac{4}{3}RP_m \sin 2\theta \tag{7-8.18}$$

Integrating the last equation of (7-8.14), we find a relation for the bending moment G; that is,

$$G = \tfrac{1}{3}R^2P_m \cos 2\theta \tag{7-8.19}$$

Finally, using the bending moment-displacement relation, one can show that the radial displacement is given by the expression

$$u = -U_0 \cos 2\theta$$

and

$$\frac{U_0}{\Delta} = \frac{8}{3}\left(\frac{R}{\Delta}\right)^4 \frac{P_m(1-\nu^2)}{Y} \tag{7-8.20}$$

As an example, for a steel shell with $R/\Delta = 100$, in a uniform magnetic field of $B_0 = 0.1\,\text{T}$, $U_0 = 4.7\Delta$. This example illustrates that a rather low magnetic tension $P_m = 0.4\,\text{N/cm}^2$ can result in a nontrivial deformation of the ferromagnetic structure.

Neglecting stresses in the thickness or radial direction of the shell, it can be shown that the bending stress is linear across the thickness and is given by

$$t_{\theta\theta} = \frac{Y}{1-\nu^2}(r-R)(\kappa-\kappa_0) \tag{7-8.21}$$

Using Eqs. (7-8.16b) and (7-8.20), one finds that the maximum bending stress Σ_0 is found at $r - R = \pm\frac{1}{2}\Delta$ and $\theta = 0$, $\pm\frac{1}{2}\pi$, and $\pm\pi$ and is given by

$$\Sigma_0 = 4\left(\frac{R}{\Delta}\right)^2 \frac{B^2}{2\mu_0} \tag{7-8.22}$$

Under the conditions of $R/\Delta = 10^2$ and $B = 0.1$ T, the stress in a steel shell would be 15.9 kN/cm^2 (23,000 psi). Thus a small magnetic pressure of 0.40 N/cm^2 can be magnified by the geometric factor $(R/\Delta)^2$ to produce a nontrivial stress or deformation in the structure.

Experiments

To see if a low magnetic field could produce bending in a ferromagnetic shell, a very thin shell, $R = 12.7$ mm and $R/\Delta = 50$, was cut from an ordinary steel rod of length equal to 7.6 cm. A semiconductor strain gage was cemented to the outside of the shell sensitive to circumferential strain at the midsection along the length. The shell was supported at its ends by circular nylon plates which were themselves attached to a brass rod through the middle of the shell as shown in Figure 7-23. Two different experiments were performed. In one experiment the strain gage was fixed at a specific angular location θ relative to the applied field and the field was increased and decreased. In the second experiment the value of the field was fixed and the shell was rotated so that the gage measured strain at different angles to the applied field. The magnet used was a 17.8-cm-diam electromagnet with a 5-cm pole gap. The applied field ranged from $B = 0$ to 0.06 T.

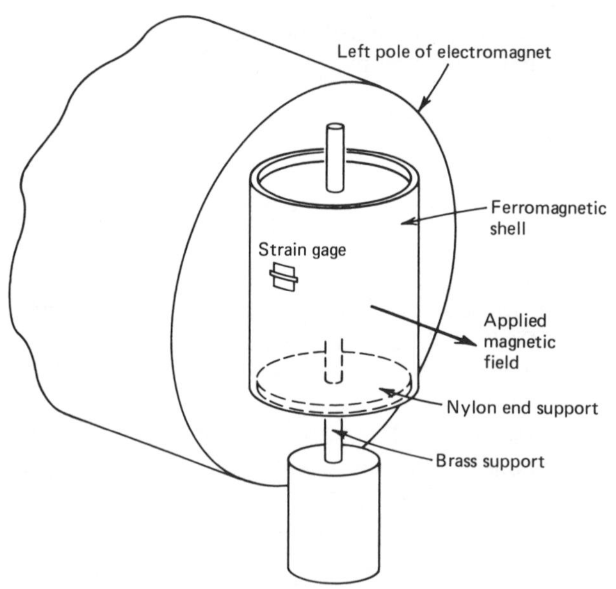

Figure 7-23 Sketch of experimental apparatus for measuring deformation of a ferromagnetic shell.

The results for the strain at different angles at a fixed magnetic field are shown in Figure 7-24. The results showed the predicted cos 2θ behavior with the exception that the $\theta = 0$ and $\theta = \frac{1}{2}\pi$ values were not equal. There was very little hysteresis, however, in rotating the shell in the field. The measured strain versus magnetic field is shown plotted in Figure 7-25 for different angles. Strains at very low field do not show a B^2 dependence. However, for large fields, the B^2 behavior becomes evident. After the field was removed, the shell had a residual magnetic field of 2–3 10^{-4} T.

Direct comparison of induced stresses was not possible because of a number of differences between the experimental conditions and the idealized model. First, the cylinder was finite in length which meant that flux exclusion was not as effective compared to the infinite cylinder. In fact, the measured ratio of field inside to that outside was only 0.8. For an infinite-length cylinder of high permeability, this ratio should be of the order of $2R/\mu_r\Delta$. Another difference between theory and experiment was the end constraints. Although they were soft plastic, they no doubt constrained the bending of the shell at the ends and hence limited the bending at the middle. In spite of

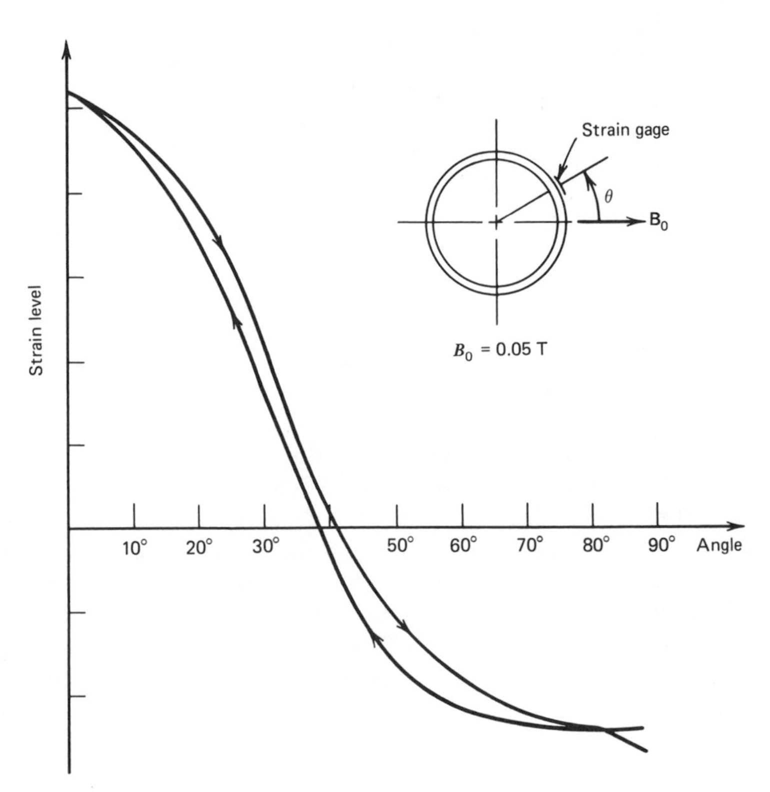

Figure 7-24 Circumferential bending strain vs. angle for a cylindrical shell in a magnetic field (see Fig. 7-23).

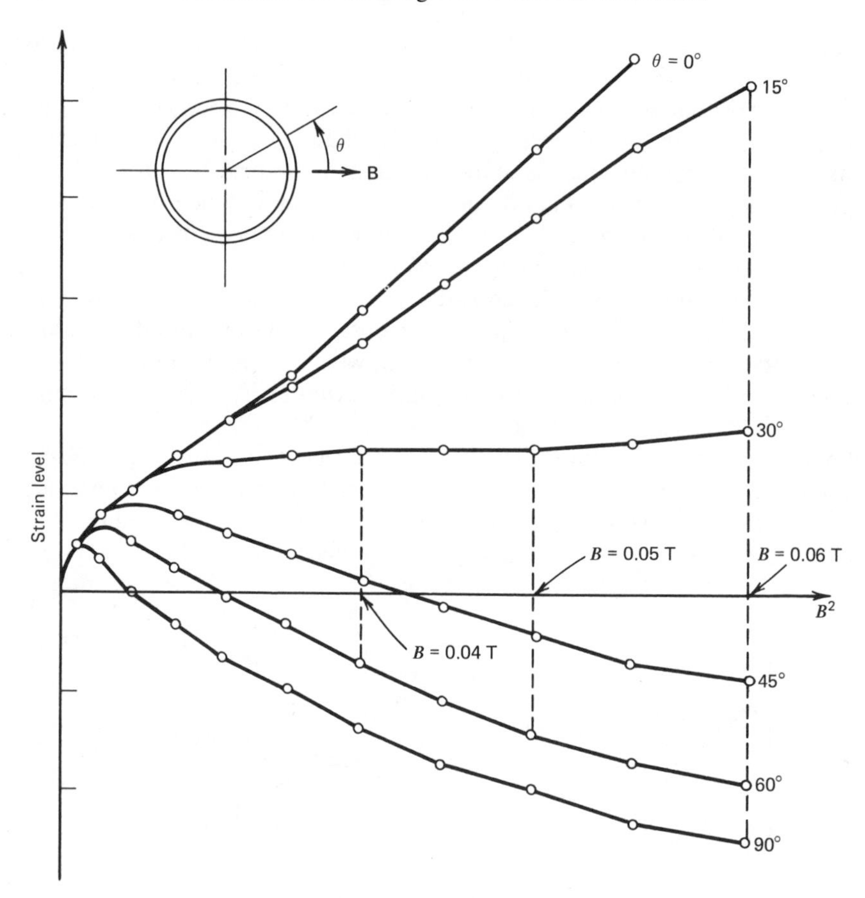

Figure 7-25 Circumferential bending strain vs. magnetic field for a cylindrical shell (see Fig. 7-23).

these differences, the predicted maximum stress at 0.05 T was 990 N/cm^2 (1440 psi) and the measured stress was 520 N/cm^2 (750 psi).

It is clear that more-refined experiments as well as more sophisticated shell analysis, taking into account the finite-length and end constraints, are needed to assess the accuracy of the analytical method. However, to the author's knowledge no similar study can be found in the literature. In spite of the difference in stress levels, however, all the qualitative results of the experiments support the theoretical model using magnetic tensions based on the pole model.

8

Magnetically Induced Currents and Forces

8.1 INTRODUCTION

In this chapter we discuss methods for calculating the induced currents, forces, and temperature fields resulting from the interaction of pulsed or time-varying magnetic fields and electrically conducting solids. These problems may be classified into two groups (1) those problems for which the deformation or motion of the solid does not appreciably affect the induced currents and (2) those problems in which the interaction between the induced currents and deformation is strong enough to require the simultaneous solution of both current and deformation fields. In the first set of problems, one treats the conducting solid as rigid when seeking the solution for the induced currents. These calculated currents and magnetic field are then used to find the magnetic forces and the resulting deformation or motion of the solid. Such problems are said to be hierarchically coupled. An example of this type of problem is the magnetic induction of stress waves in an elastic half-space by a pulsed magnetic field.

An example of the second type of problem is the acceleration of a coil in an induction-type mass driver. Here the continuously changing velocity and position of the accelerating coil affects the induced currents, which, in turn, affect the acceleration force on the coil. These problems involve concepts of magnetic stiffness (see Section 4.6) and magnetic damping.

The study of forces on electrically conducting bodies due to induced currents has a long history going back to the time of Maxwell. The work of Jeans (1925) follows the methods of Maxwell. Smythe (1968) presents solutions to a number of classic problems, including eddy currents in cylindrical and spherical bodies, and plane sheet conductors. Specialized monographs on the subject of eddy currents include Lammeraner and Stafl

(1966) and Stoll (1974). Almost all of these authors use analytical methods based on the magnetic field vector potential. Most of these problems involve spacially uniform magnetic fields so that only torques can be generated on the bodies. Also, the majority of papers in the literature deal with harmonically varying fields.

However, many important applications require inhomogeneous excitation fields, such as magnetic suspension of vehicles, linear induction motors, tokamak fusion reactors, and magnetic forming. There is substantial literature on induction machines. Most books such as Slemon and Straughen (1980), deal with rotating machinery, but a few deal with linear induction devices, such as in Yamamura (1979), or Nasar and Boldea (1976). These applications involve induction fields with sinusoidal variations in both space and time.

A recent renewal of interest in eddy-current research has been sparked by the development of numerical methods, such as the finite-element and boundary integral methods. At the time of this writing, a number of international conferences have been held on the subject and the interested reader is encouraged to seek out the proceedings of these meetings for further information. (See, e.g., the Proceedings of COMPUMAG, 1979, Grenoble, France, and 1981, Chicago, Illinois. Papers from the latter conference have been published in *IEEE Trans. Mag.* **18**, No. 2, 1982.)

8.2 BASIC EQUATIONS FOR EDDY CURRENTS

The Quasistatic approximation

The equations used to determine the induction of eddy currents in conductors are a limiting form of Maxwell's equations. The approximate theory neglects the electric displacement currents $\partial \mathbf{D}/\partial t$. This assumption is based on two arguments. First, let us examine the conservation-of-charge equation (2-2.1):

$$\boldsymbol{\nabla} \cdot \mathbf{J} + \frac{\partial q}{\partial t}$$

For a linear material, one can write $\mathbf{J}$ in terms of $\mathbf{D}$ using Eqs. (2-4.5) and (2-4.7) and then use the relation between charge density and $\mathbf{D}$, $q = \boldsymbol{\nabla} \cdot \mathbf{D}$, to obtain an equation for the charge density

$$\frac{\partial q}{\partial t} + \frac{\sigma}{\varepsilon} q = 0 \tag{8-2.1}$$

This equation has a solution proportional to $\exp[-(\sigma/\varepsilon)t]$, so that any net charge will decay exponentially in a characteristic time, $\tau = \varepsilon/\sigma$ (for aluminum $\tau < 10^{-18}$). Thus the almost-instantaneous decay of charge in good conductors leads one to write

$$\nabla \cdot \mathbf{J} = 0 \tag{8-2.2}$$

This suggests that the current density has a vector potential

$$\mathbf{J} = \nabla \times \mathbf{H}$$

which differs from Maxwell's equation (2-2.4) by the absence of $\partial \mathbf{D}/\partial t$.

In the second argument, one writes Maxwell's equations for time harmonic fields of frequency ω. The equation for the magnetic field $\mathbf{B}$ in a linear isotropic conductor becomes

$$\nabla^2 \mathbf{B} = i\mu\sigma\omega\left(1 + \frac{i\omega\varepsilon}{\sigma}\right)\mathbf{B} \tag{8-2.3}$$

It is clear that ω can be as high as 10^8 and the second term on the right-hand side of Eq. (8-2.3) will still be very small. Under the assumption that $\omega\varepsilon/\sigma \ll 1$, the nonharmonic form of Eq. (8-2.3) will become

$$\nabla^2 \mathbf{B} = \mu\sigma \frac{\partial \mathbf{B}}{\partial t} \tag{8-2.4}$$

This can also be derived directly from Maxwell's equations by dropping the displacement current $\partial \mathbf{D}/\partial t$.

The assumption of zero charge changes the form of Maxwell's equations from a wave equation to a *diffusion equation*. This has important mathematical consequences with regards to finding analytical or numerical solutions. The neglect of free volume charge density does not mean that net charge cannot reside in a good conductor; quite the contrary since a good conducting body in an electric field will experience regions of positive and negative charge density. However, the charge will reside on the surface of the conductor. Thus while the presence of charge in the interior is neglected in the quasistatic form of Maxwell's equations, it will affect the boundary conditions on the surface. This can be important in arcing problems between conductors.

The forces on a nonferromagnetic conductor depend on the current distribution $\mathbf{J}$. The current density depends on the electric and magnetic fields in the conductor through a generalized Ohm's law $\mathbf{J}(\mathbf{E}, \mathbf{B}, \mathbf{v})$, where $\mathbf{v}$ is the velocity of the conductor. There are three classes of eddy-current problems. In the first set of problems one starts with an initial current density $\mathbf{J}^0$ and induces secondary currents $\mathbf{J}^1$ and field $\mathbf{B}^1$. In the second class a primary time-varying magnetic field $\mathbf{B}^0(t)$ is given and one induces the current $\mathbf{J}$ and a secondary field $\mathbf{B}^1$. In the third set one can have a stationary field $\mathbf{B}^0$ and an initial velocity $\mathbf{v}$. This again leads to induced currents and a secondary field $\mathbf{B}^1$.

The basic set of equations consists of the quasisteady form of Maxwell's equations, which describe the evolution of the electric and magnetic fields, and a constitutive relation (Ohm's law) for the current density $\mathbf{J}(\mathbf{E}, \mathbf{B}, \mathbf{v})$. In ferromagnetic conductors one needs an additional constitutive relation $\mathbf{B}(\mathbf{H})$.

Constitutive Relation for J

The simplest constitutive law for **J** for a linear, isotropic material neglects the effect of **B**, that is, for a stationary conductor,

$$\mathbf{J} = \sigma \mathbf{E} \tag{8-2.5}$$

where σ is the electric conductivity (see also Section 2.4). A more general relation is required in anisotropic media, such as in a superconducting fiber composite material operating in the normal regime; that is,

$$\mathbf{J} = \boldsymbol{\sigma} \cdot \mathbf{E} \quad \text{or} \quad \mathbf{E} = \boldsymbol{\rho} \cdot \mathbf{J} \tag{8-2.6}$$

where $\boldsymbol{\sigma}$ and $\boldsymbol{\rho}$ are second-order symmetric tensors and $\boldsymbol{\rho} = \boldsymbol{\sigma}^{-1}$. According to the theory of second-order tensors, a set of orthogonal directions can be found in the material such that Eq. (8-2.6) can be written in the form

$$\mathbf{J} = \sigma_1 E_1 \mathbf{e}_1 + \sigma_2 E_2 \mathbf{e}_2 + \sigma_3 E_3 \mathbf{e}_3$$

where $\mathbf{e}_1$, $\mathbf{e}_2$, and $\mathbf{e}_3$ are orthogonal unit vectors and σ_i are the principal conductivity coefficients in these directions. In a moving conductor, one must replace **E** with $\mathbf{E} + \mathbf{v} \times \mathbf{B}$ in Eq. (8-2.5) or (8-2.6).

At high magnetic fields and low temperatures (or very high conductivity) the Hall effect must be added to Ohm's law (see also Sections 3.7 and 8.7). For isotropic materials this takes the form

$$\mathbf{E} = \frac{1}{\sigma} \mathbf{J} + R_H \mathbf{B} \times \mathbf{J} \tag{8-2.7}$$

R_H is called the Hall constant.

Temperatures can enter the constitutive relation for **J** in two ways. First, the electric conductivity is a strong function of temperature, where for most materials $\partial\sigma/\partial T < 0$. Second, temperature gradients can induce current flow. This phenomenon is known as the "thermoelectric effect" [Eq. (3-7.6] (see, e.g., Ziman, 1964) and includes the reciprocal effect of the current flow on the heat flux vector.

Direct Formulation

The quasistatic equations of Faraday and Ampere are first-order linear partial differential equations in **E**, **B**, **J** and **H**. They are

$$\nabla \times \mathbf{E} + \frac{\partial \mathbf{B}}{\partial t} = 0 \tag{8-2.8a}$$

and

$$\nabla \times \mathbf{H} - \mathbf{J} = 0 \tag{8-2.8b}$$

These equations are often reduced to a second-order partial differential equation by using constitutive equations for **B**(**H**) and **J**(**E**, **B**, **v**). The classic treatment of the subject assumes a linear ferromagnetic material and Ohm's law for linear isotropic material,

$$\mathbf{B} = \mu\mathbf{H} \tag{8-2.9a}$$

$$\mathbf{J} = \sigma(\mathbf{E} + \mathbf{v} \times \mathbf{B}) \tag{8-2.9b}$$

Using Ohm's law (8-2.9b) and Ampere's law (8-2.8b) to eliminate **J** and **E** one can easily show that

$$\frac{1}{\mu\sigma}\nabla^2\mathbf{B} + \nabla \times (\mathbf{v} \times \mathbf{B}) = \frac{\partial \mathbf{B}}{\partial t} \tag{8-2.10}$$

For a constrained rigid conductor, $\mathbf{v} = 0$ and

$$\nabla^2\mathbf{B} = \mu\sigma\frac{\partial \mathbf{B}}{\partial t} \tag{8-2.11}$$

For a rigid conductor one can derive similar equations for the fields **E** and **J**; for example,

$$\nabla^2\mathbf{J} = \mu\sigma\frac{\partial \mathbf{J}}{\partial t} \tag{8-2.12}$$

In Cartesian coordinates, each of the components of **B**, **E**, or **J** satisfies a diffusion equation. For time harmonic problems with frequency ω, there is a common characteristic length or skin depth for all the fields $\delta = (2/\mu\sigma\omega)^{1/2}$. For bodies whose smallest dimension $L \gg \delta$, all the time-varying fields and currents will be confined to a layer in the solid of the order of the skin depth (see also Chapters 1 and 2)

For two- and three-dimensional problems, the direct formulation in terms of **B** and **J**, that is Eqs. (8-2.11) and (8-2.12), involves several partial differential equations coupled through boundary conditions. An alternative approach is to use potential methods which in certain classes of two- and three-dimensional problems involves a single partial differential equation for a scalar potential function.

Anisotropic Conductors

The basic induction equation for a rigid, linear anisotropic conductor is given by

$$\nabla \times [\boldsymbol{\rho} \cdot \nabla \times \mathbf{B}] = -\mu\frac{\partial \mathbf{B}}{\partial t} \tag{8-2.13}$$

To illustrate the consequences of an anisotropic Ohm's law consider a material which is isotropic in a plane (x_1, x_2) but has different resistivity normal to the plane. This problem is similar to an untwisted, filamentary superconductor in the normal state (see Chapter 6).

There are three classes of two-dimensional problems: (1) axial field along the filament or x_3 direction, or $B_1 = B_2 = 0$; (2) axial current flow, $J_1 = J_2 = 0$, and (3) planar current flow with $J_2 = 0$ and $B_1 = B_3 = 0$.

1. *Axial field case.* In this problem we assume that $B_1 = B_2 = 0$ and $\partial/\partial x_3 = 0$. This leads to $J_3 = 0$ and the equation

$$\nabla_1^2 B_3 = \mu\sigma_1 \frac{\partial B_3}{\partial t}$$

where $\sigma_1 = 1/\rho_1$ and $\nabla_1^2 = \partial^2/\partial x_1^2 + \partial^2/\partial x_2^2$. Thus this case does not differ mathematically from the isotropic case.

2. *Axial current.* Here we assume $J_1 = J_2 = 0$, $B_3 = 0$, and $\partial/\partial x_3 = 0$. This leads to the following equations:

$$\nabla_1^2 B_i = \mu\sigma_3 \frac{\partial B_i}{\partial t}$$

where $i = 1, 2$. Again this case is identical to the isotropic problem.

3. *Planar flow across the filaments.* Here we assume $J_2 = 0$, $B_1 = B_3 = 0$, and $\partial/\partial x_2 = 0$. These assumptions lead to the following equation for the nonzero field component $B_2(x_1, x_3, t)$:

$$\frac{\sigma_1}{\sigma_3}\frac{\partial^2 B_2}{\partial x_1^2} + \frac{\partial^2 B_2}{\partial x_3^2} = \mu\sigma_1 \frac{\partial B_2}{\partial t} \tag{8-2.14}$$

This equation differs from the isotropic case when $\sigma_1/\sigma_3 \neq 1$.

To obtain a diffusionlike equation one can define a new distance variable $\bar{x}_1 = (\sigma_3/\sigma_1)^{1/2} x_1$. Using this stretched coordinate, the induction equation (8-2.14) assumes the form of the classic isotropic case

$$\frac{\partial^2 B_2}{\partial \bar{x}_1^2} + \frac{\partial^2 B_2}{\partial x_3^2} = \mu\sigma_1 \frac{\partial B_2}{\partial t} \tag{8-2.15}$$

8.3 POTENTIAL METHODS

There are two vector potential methods for solving eddy-current problems: One is based on the *conservation of flux*

$$\nabla \cdot \mathbf{B} = 0, \qquad \mathbf{B} = \nabla \times \mathbf{A} \tag{8-3.1}$$

and the other on the *conservation of charge*

$$\nabla \cdot \mathbf{J} = 0, \qquad \mathbf{J} = \nabla \times \boldsymbol{\Psi} \tag{8-3.2}$$

Magnetic Field Potential

When Eq. (8-3.1) is substituted into Faraday's law (2-2.5), one finds the relation

$$\nabla \times \left(\mathbf{E} + \frac{\partial \mathbf{A}}{\partial t} \right) = 0$$

This implies that

$$\mathbf{E} = -\frac{\partial \mathbf{A}}{\partial t} - \nabla \phi \tag{8-3.3}$$

where ϕ is the scalar potential.

The three components of **A** in addition to ϕ do not lead to a unique solution. One must provide an additional condition on **A**. One possibility is to require that **A** be divergence-free, that is,

$$\nabla \cdot \mathbf{A} = 0 \tag{8-3.4}$$

This relation is called a "gauge condition." A discussion of the choice of different gauge conditions in eddy-current problems is given by Carpenter (1977). When Eq. (8-3.1) is used in Ampere's law (2-2.4), along with Eq. (8-3.4), the following relation results:

$$\nabla^2 \mathbf{A} = -\mu \mathbf{J} \tag{8-3.5}$$

where a linear ferromagnetic material is assumed; that is, $\mathbf{B} = \mu \mathbf{H}$. When **J** is known, a particular solution of this equation can be found (see, e.g., Smythe, 1968):

$$\mathbf{A} = \frac{\mu}{4\pi} \int \frac{\mathbf{J}}{R} \, dv \tag{8-3.6}$$

The volume integral is carried out over all regions of space where $\mathbf{J} \neq 0$ and R is the distance from the source point of **J** to the field point at which **A** is measured.

To complete the set of equations, a constitutive relation between **J** and the fields **E** and **B** must be given. The vector potential **A** is associated with the current density **J**, while the scalar potential ϕ is related to the charge on the surface of the conductor.

For linear homogeneous material we have

$$\nabla \cdot \mathbf{J} = \sigma \nabla \cdot \mathbf{E} = 0$$

The second relation coupled with Eqs. (8-3.3) and (8-3.4) leads to an equation for ϕ,

$$\nabla^2 \phi = 0 \tag{8-3.7}$$

Using the potentials **A** and ϕ one can derive other forms of the induction law. For example, for an isotropic conductor with no Hall effect, Eqs. (8-3.3) and (8-3.6) lead to

$$\mathbf{J} + \frac{\mu\sigma}{4\pi} \int \frac{1}{R} \frac{\partial \mathbf{J}}{\partial t} \, dv = -\sigma \nabla \phi \tag{8-3.8}$$

This integral formulation is sometimes useful for approximate or numerical methods, such as the finite-element method. In these problems the conductor is represented by a set of elements or nodes where J_i represents the current flow in a certain direction at a given node. The set of unknowns $\{J_i\}$ is represented as a column vector and the resulting coupled set of algebraic equations for time harmonic problems ($e^{i\omega t}$ time dependence) assumes the form

$$\{J_i\} + i\omega[C_{ij}]\{J_j\} = \{J_i^s\} \tag{8-3.9}$$

where the vector $\{J_i^s\}$ represents source current or source field terms (Silvester et al., 1971). The matrix $[C_{ij}]$ depends on the geometry and electromagnetic properties of the conductor (see also Chari and Silvester 1980).

Current Potential or Stream Function

An alternate potential method is based on $\nabla \cdot \mathbf{J} = 0$. This leads to a current vector potential

$$\mathbf{J} = \nabla \times \boldsymbol{\Psi} \tag{8-3.10}$$

In general, $\boldsymbol{\Psi}$ is not the magnetic field $\mathbf{H}$. This can be seen by substituting Eq. (8-3.10) into Ampere's law,

$$\nabla \times (\mathbf{H} - \boldsymbol{\Psi}) = 0$$

or

$$\mathbf{H} = \boldsymbol{\Psi} - \nabla \eta \tag{8-3.11}$$

For a linear homogeneous ferromagnetic conductor, substitution of Eqs. (8-3.10) and (8-3.11) into Faraday's law (2-2.5) leads to an equation for $\boldsymbol{\Psi}$; that is,

$$\nabla \times \nabla \times \boldsymbol{\Psi} = -\mu\sigma \frac{\partial \boldsymbol{\Psi}}{\partial t} + \sigma\mu \nabla \frac{\partial \eta}{\partial t} \tag{8-3.12}$$

If one assumes a gauge condition $\nabla \cdot \boldsymbol{\Psi} = 0$, then the following equations result:

$$\begin{aligned} \nabla^2 \boldsymbol{\Psi} &= \mu\sigma \frac{\partial \boldsymbol{\Psi}}{\partial t} - \sigma\mu \nabla \frac{\partial \eta}{\partial t} \\ \nabla^2 \eta &= 0 \end{aligned} \tag{8-3.13}$$

The latter equation follows from $\nabla \cdot \mathbf{B} = 0$ for linear homogeneous materials.

8.4 FOUR ELEMENTARY EDDY-CURRENT PROBLEMS

Induced Current and Torque on a Rigid Circuit

A simple example of the magnetic induction of eddy currents is the problem of a rigid planar circuit in a time-varying magnetic field (see Fig. 8-1).

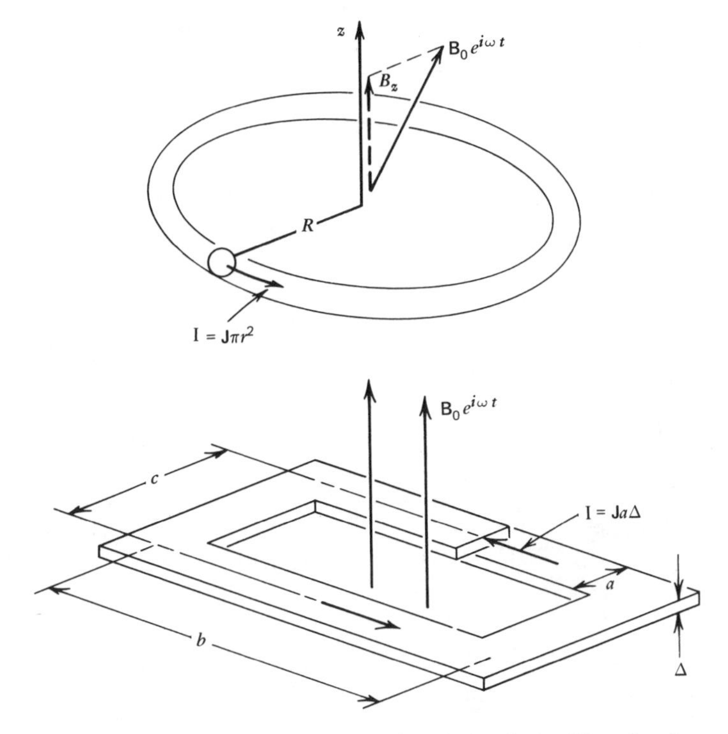

Figure 8-1 Induction of currents in wire and platelike circuits.

Consider a planar circuit with resistance R, self-inductance L, and planar area A.

The circuit form of Faraday's law (2-8.4) relates the change in flux through the circuit to the voltage or line integral of **E** around the circuit. The total flux is the sum of the applied flux B_zA and the induced flux LI, and the voltage drop is given by the product of resistance and current; that is,

$$\frac{d}{dt}(LI + AB_z) = -RI \tag{8-4.1}$$

where B_z is the component of the approximately uniform magnetic field vector normal to the plane of the circuit. When $B_z(t)$ is sinusoidal, that is, $B_z = B_0\gamma e^{i\omega t}$ (γ is a direction cosine), the induced current in the circuit is given by

$$I = -\frac{B_0\gamma A}{L}\frac{\Omega}{1+\Omega^2}(\Omega + i)\,e^{i\omega t} \tag{8-4.2}$$

where $\Omega = \omega L/R$ is a nondimensional frequency. The real and imaginary parts of I are plotted in Figure 8-2 and show that $\mathrm{Re}(I)$ approaches an asymptote for $\Omega \to \infty$, while $\mathrm{Im}(I)$ has a maximum at $\Omega = 1$.

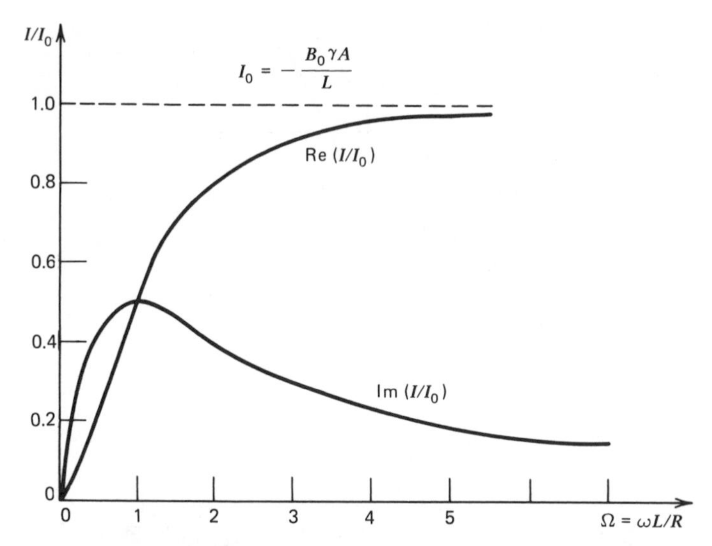

Figure 8-2 Real and imaginary parts of induced current in a circular current loop [Eq. (8-4.2)].

The net force on a circuit in a uniform magnetic field is zero (see Chapter 3). However, there will be a torque that will vary in time and depends on the magnetic-moment vector for the circuit $\mathbf{m}$ [Eq. (3-3.4)] and the relative direction of the oscillating field; that is,

$$\mathbf{C} = \mathbf{m} \times \mathbf{B}^0$$
$$\mathbf{m} = \frac{1}{2} I \oint \mathbf{r} \times d\mathbf{r} \tag{8-4.3}$$

where $\mathbf{r}$ is a position vector in the plane of the coil. If we orient the x–y axes so that $\mathbf{B}^0 = (0, B_y^0, B_z^0)$ and $\mathbf{m} = (0, 0, I\mathcal{M})$, then $\mathbf{C} = (C_x, 0, 0)$ and

$$C_x = -\mathcal{M} I B_y^0$$

Two special cases of interest are (1) a fixed field direction with oscillating magnitude and (2) a constant magnitude field but a rotating vector about an axis in the plane of the circuit. These cases are described below.

Fixed Field Direction. For this case the magnetic field components are related by the following equations:

$$B_z^0 = b(t) \cos \theta, \qquad B_y^0 = b(t) \sin \theta$$

where $b(t) = B_0 \operatorname{Re}(e^{i\omega t})$. [Note $\gamma = \cos \theta$ in Eq. (8-4.2).] In order to calculate the torque, we must use *real variables* and not complex quantities. Also, since the field is sinusoidal, it is of interest to calculate the *time-averaged torque*

$$\langle C_x \rangle = \frac{\omega}{2\pi} \int_0^{2\pi/\omega} C_x dt \tag{8-4.4}$$

This leads to the expression

$$\langle C_x \rangle = \frac{\mathcal{M} B_0^2 A \sin 2\theta}{4L} \frac{\Omega^2}{1+\Omega^2} \tag{8-4.5}$$

From Eq. (8-4.5) one can see that an unsupported coil in a uniform oscillating field is unstable about the $\theta = 0$ direction. A small perturbation leads to a torque about the x axis which tends to increase θ.

Rotating Magnetic Field. In this case the magnetic field components are related by

$$B_z^0 = B^0 \cos \omega t, \qquad B_y^0 = B^0 \sin \omega t$$

The time-average torque is given by

$$\langle C_x \rangle = -\mathcal{M} B^0 \langle \mathrm{Re}(I) \, \mathrm{Im}(e^{i\omega t}) \rangle$$

or

$$\langle C_x \rangle = -\mathcal{M} B^0 \tfrac{1}{2} \, \mathrm{Im}(I e^{-i\omega t}) \tag{8-4.6}$$

This expression becomes ($\gamma = 1$),

$$\langle C_x \rangle = -\frac{\mathcal{M} B_0^2 A}{2L} \frac{\Omega}{1+\Omega^2} \tag{8-4.7}$$

Thus the maximum torque occurs when the frequency of rotation equals the characteristic frequency $\omega = R/L$. These problems are related to the magnetomechanics of rotating machinery of which there is a large literature (see, e.g., Slemon and Straughen, 1980).

Magnetic Diffusion and Skin Depth

While wave propagation is a distinctive feature of time-dependent electromagnetic fields in dielectric media or free space, in good electrical conductors, time-dependent magnetic fields exhibit diffusive behavior. The general equation for a moving conductor is given by Eq. (8-2.10). When the velocity of the conductor is a constant ($\nabla \mathbf{v} = 0$), the equations for $\mathbf{B}$ take the form

$$\frac{1}{\mu\sigma} \nabla^2 \mathbf{B} = \left(\frac{\partial}{\partial t} + \mathbf{v} \cdot \nabla \right) \mathbf{B}, \qquad \nabla \cdot \mathbf{B} = 0 \tag{8-4.8}$$

Stationary Conductors

Consider the case when the conductor is not moving relative to the observer and the field is one-dimensional, that is, $\mathbf{B} = (B(y, t), 0, 0)$. Then Eq. (8-4.8) becomes

$$\frac{1}{\mu\sigma}\frac{\partial^2 B}{\partial y^2} = \frac{\partial B}{\partial t} \tag{8-4.9}$$

This is the classic diffusion equation, identical to that found in heat transfer describing the flow of heat energy into a solid.

Consider the example of a conducting half-space $y > 0$, where $B\ (y = 0)$ is a prescribed harmonically varying field $B_0 e^{i\omega t}$. The solution to Eq. (8-4.9) for the field inside the conductor is found to be

$$B = B_0 e^{i(\omega t - y/\delta)} e^{-y/\delta}$$

where $\delta^2 = 2/\mu\sigma\omega$. Thus the magnetic field decays exponentially to a value of B_0/e in a characteristic distance δ called the skin depth. The conductor effectively shields the interior from the applied field by the induction of eddy currents $\mathbf{J} = (0, 0, J)$ whose own magnetic field opposes the applied field in the interior of the conductor. Values of the skin depth for various conductors and frequencies are shown in Figure 8-3.

For transient applied fields a formal solution may be found in the literature on the heat or diffusion equation. [see, e.g., R. V. Churchill, *Operational Mathematics* (2nd ed.), McGraw-Hill, New York, 1958, pp. 132–134]. If $B_0(t)$ is a known field at $y = 0$, then

$$B(y, t) = \frac{2}{\sqrt{\pi}}\int_{\eta}^{\infty} B_0\left(t - \frac{\mu\sigma y^2}{\lambda^2}\right) e^{-\lambda^2}\, d\lambda \tag{8-4.10}$$

where $\eta = \frac{1}{2} y(\mu\sigma/t)^{1/2}$. If $B_0(t)$ is a step increase from zero to B_0 at $t = 0$, then

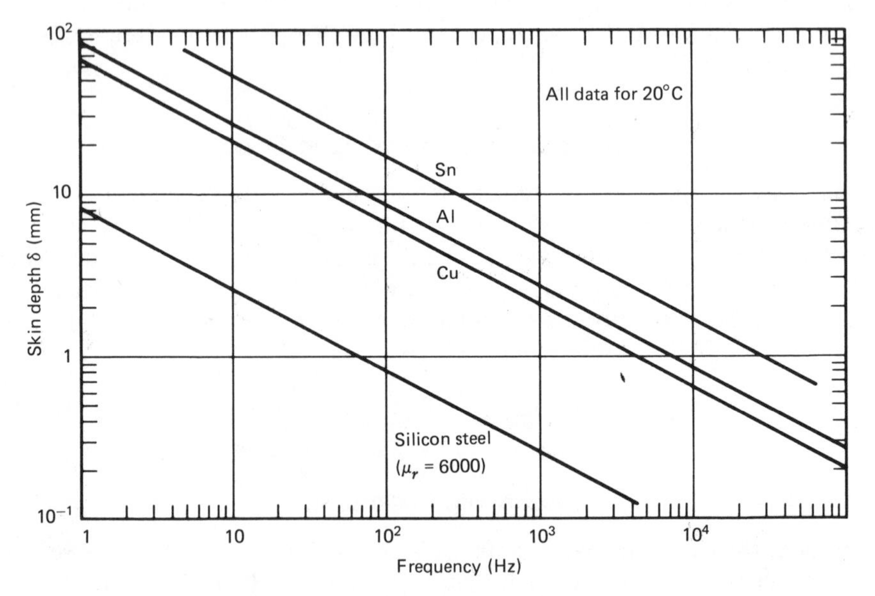

Figure 8-3 Skin depth vs. frequency.

$$B = B_0 \operatorname{erfc}(\eta) \tag{8-4.11}$$

where $\operatorname{erfc}(\eta)$ is the complementary error function. This solution is shown in Figure 8-4. The induced current is given by the expression

$$\mu_0 J = \frac{\partial B}{\partial y} = -B_0 \left(\frac{\mu\sigma}{\pi t}\right)^{1/2} e^{-\mu\sigma y^2/4t} \tag{8-4.12}$$

In this transient example we can define a time-dependent skin depth

$$\delta_t = \left(\frac{4t}{\mu\sigma}\right)^{1/2} \tag{8-4.13}$$

which grows in time.

If one has a conducting plate of finite thickness Δ, one can use this formula to estimate the time for a magnetic field to diffuse through the plate after a sudden change in the field, that is, set $\delta_t = \Delta$,

$$t_0 = \tfrac{1}{4}\mu\sigma\Delta^2$$

In both of these examples it is easy to show that the magnetic body force on the solid is *compressive*. Further one can prove that the total force per unit area depends only on the magnetic pressure on the surface; that is, for a nonferromagnetic conductor,

$$F_y = \int_0^\infty J_z B_x dy = -\int_0^\infty \frac{B}{\mu_0}\frac{\partial B}{\partial y} dy$$

and

$$F_y = \frac{1}{2\mu_0} B^2(y = 0)$$

For a finite plate one has

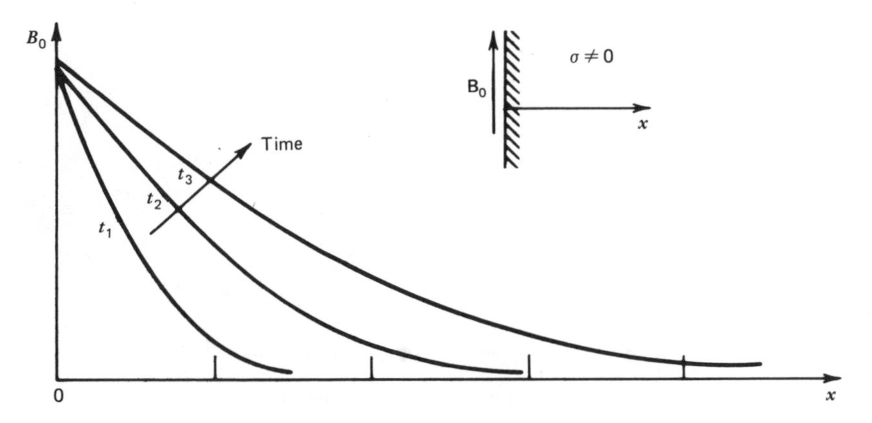

Figure 8-4 Magnetic field diffusion in a conductor.

$$F_y = \frac{1}{2\mu_0}[B_x^2(y=0) - B_x^2(y=\Delta)] \tag{8-4.14}$$

Thus there is a net force on the plate away from the source field B_0 as long as the magnetic field has not diffused through the plate. If the source field is used to accelerate the plate with no restraining forces present, we can estimate the impulse given to the plate for a step change in magnetic field. If ρ is the density of the plate and v its final velocity then

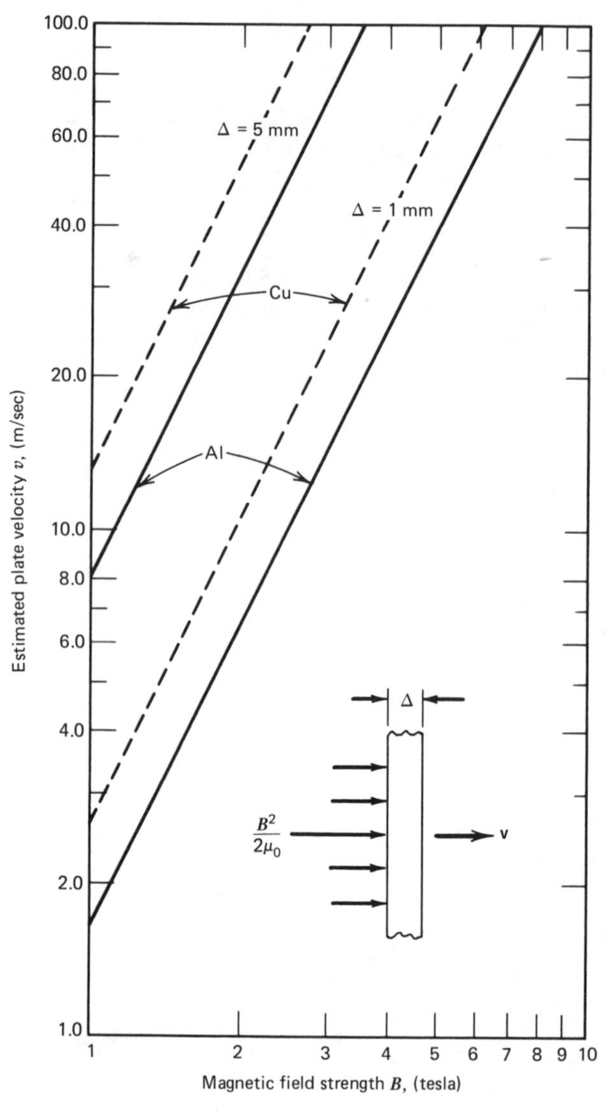

Figure 8-5 Magnetically induced velocity in a conducting plate [Eq. (8-4.15)].

$$v\rho\Delta \sim t_0 F_y = \tfrac{1}{8}\sigma\Delta^2 B_0^2$$

or

$$v \sim \frac{\sigma \Delta B_0^2}{8\rho} \tag{8-4.15}$$

Thus we arrive at the interesting result that thicker plates end up with a higher terminal velocity than thinner plates. Figure 8-5 shows this estimated velocity for various conductors and magnetic fields.

Moving Conductor; Traveling Wave Diffusion

An application of great technical importance is the case of a moving conductor in a magnetic field, such as rotary and linear electric induction motors (see, e.g., Nasar and Boldea, 1976). Consider, for example, the case of the *linear electric motor*. As in the previous example, we consider a conducting half-space, $y > 0$, and assume it is moving in the x direction with velocity v. Parallel to this half-space we imagine an array of linear conductors, each with electric current (see Fig. 8-6). Each conductor carries a current proportional to $e^{i\omega t}$, but is out of phase with its neighboring current. We further assume that we can treat this array of surface currents as a continuous distribution $K(x, t)$ in amperes per meter. Since any distribution $K(x)$ can be represented as a sum of harmonic functions $e^{\pm ikx}$, we choose to look at one particular wavelength $\lambda = 2\pi/k$. The applied surface current then has the form

$$K(x, t) = K_0 e^{i(\omega t - kx)} \tag{8-4.16}$$

This function varies in space and time. However, if one moves in a reference

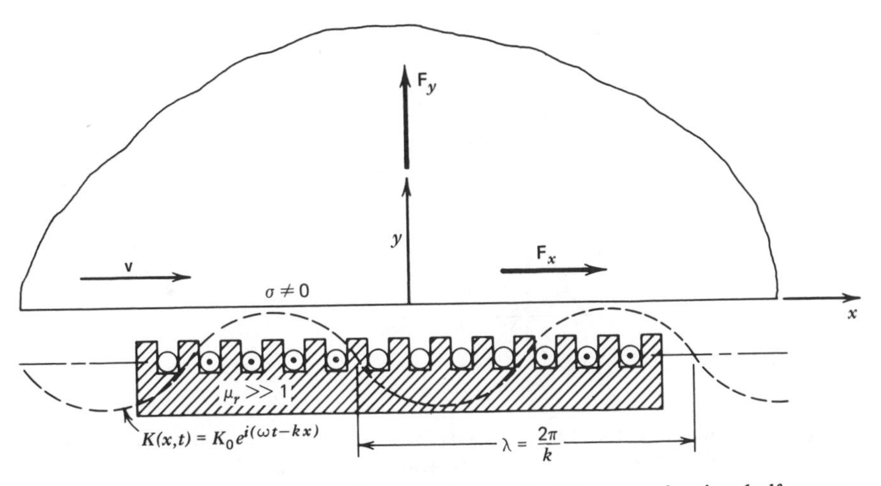

Figure 8-6 Sketch of linear induction motor effect in a conducting half-space.

frame that travels to the right with velocity $c = \omega/k$, the shape of this function will be stationary. The velocity of the current wave, c, is assumed to be different than the velocity of the moving solid, v, that is, $c - v \neq 0$.

For this case one must have both components of $\mathbf{B} = (B_x, B_y, 0)$ as required by the conservation of flux $\nabla \cdot \mathbf{B} = 0$. The solution can be shown to have the form (see Woodson and Melcher, 1968, p. 368)

$$\mathbf{B} = \mathbf{B}_1 e^{-\alpha y} e^{i(\omega t - kx)} \tag{8-4.17}$$

For $\mathbf{B}$ to satisfy the general diffusion equation, the constant $\mathbf{B}_1$ and α must have the values

$$\mathbf{B}_1 = \mu_0 K_0(1, -ik/\alpha) \tag{8-4.18}$$

$$\alpha = k(1 + iS)^{1/2}$$

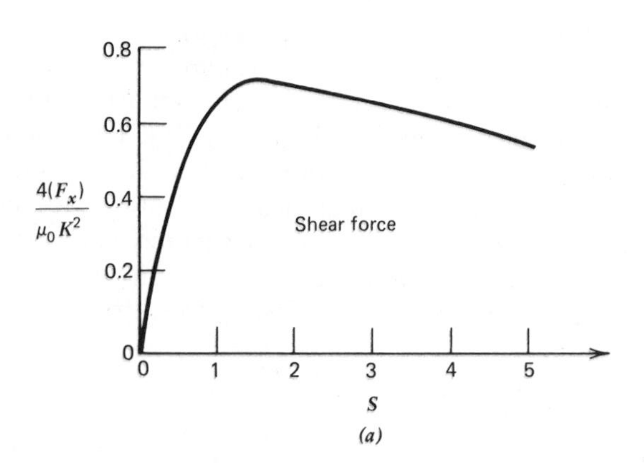

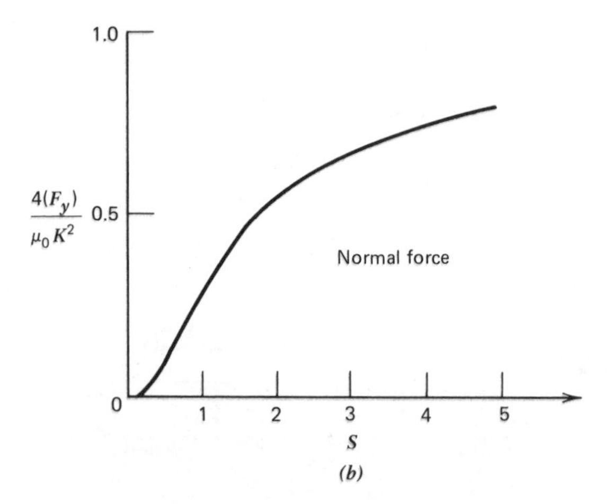

Figure 8-7 Magnetically induced forces for a linear motor: (*a*) magnetic shear force vs. slip velocity and (*b*) normal magnetic force vs. slip velocity. (From Woodson and Melcher, 1968, with permission. John Wiley & Sons, copyright 1968.)

and

$$S = \frac{\mu_0 \sigma(\omega - kv)}{k^2}$$

The body force density is given by

$$\mathbf{J} \times \mathbf{B} = J_z(-B_y, B_x, 0)$$

Thus one can see that a body force *tangential* to the surface as well as one *normal* to the surface can be generated. Integrating the body forces in the y direction and averaging these forces over a cycle we obtain average values of the lift and thrust forces; that is,

$$\langle F_y \rangle = \frac{\mu_0 K_0^2}{4} \frac{(1+S^2)^{1/2} - 1}{(1+S^2)^{1/2}} \tag{8-4.19}$$

and

$$\langle F_x \rangle = \frac{\mu_0 K_0^2}{4} \frac{S}{(1+S^2)^{1/2} \operatorname{Re}(1+iS)^{1/2}}$$

These forces are shown in Figure 8-7. It is clear that when the relative velocity of the solid and field is zero, the forces are zero. Also, the horizontal force can reverse sign depending on whether the phase velocity c is greater or less than the solid velocity v. The normal or lift force is independent of the sign of S, or $c - v$. This solution forms part of the theoretical basis for magnetic levitation and propulsion.

Magnetic Field Due to Current in a Thin Plate—Stream Function

While many of the basic principles of magnetics and mechanics have been well-understood for over a century, few analytical solutions for coupled problems can be found in the literature. One of the reasons for this is that geometries which lead to simple solutions in mechanics do not often lead to simple electromagnetic problems and vice versa. One example is the thin plate with current flowing in it, as shown in Figure 4-18. While simple force distributions lead to simple stress patterns in the plate, a uniform current distribution can be shown to lead to a highly nonuniform magnetic field density in and around the vicinty of the plate. Steady currents and fields in a thin plate were discussed in Section 4.5. In this section we consider induced or time-dependent problems.

We assume that the current is uniform across the thickness Δ, and define a current density per unit length $\mathbf{I}$,

$$\mathbf{I} = \Delta \mathbf{J} = \sigma \Delta \mathbf{E}$$

The continuity condition (8-2.2) and the absence of free charge lead to

$$\nabla \cdot \mathbf{I} = 0 \tag{8-4.20}$$

Following the analog in hydrodynamics, we define a "stream function" [Eq. (8-3.10)] ψ such that

$$\mathbf{I} = \nabla \times (\psi \mathbf{n}) = -\mathbf{n} \times \nabla \psi \tag{8-4.21}$$

where $\mathbf{n}$ is normal to the plate. From Eq. (8-4.21) one can interpret lines of constant ψ, at any instant of time, as the direction of the current flow. The stream function for a rigid stationary plate must satisfy Faraday's law (8-2.8a), which leads to

$$\nabla^2 \psi = \sigma \Delta \frac{\partial}{\partial t} (\mathbf{B}^0 + \mathbf{B}^1) \cdot \mathbf{n} \tag{8-4.22}$$

where $\mathbf{B}^0 \cdot \mathbf{n}$ is the normal component to the plate of the applied field and $\mathbf{B}^1 \cdot \mathbf{n}$ is the normal component of the induced field.

The induced magnetic field $\mathbf{B}^1$ is created by the current in the plate. Inside the plate, $\mathbf{B}^1$ and $\mathbf{I}$ are related by

$$\nabla \times \mathbf{B}^1 = \frac{\mu_0}{\Delta} \mathbf{I}$$

Outside the plate,

$$\nabla \times \mathbf{B}^1 = 0$$

The induced field is given by the Biot–Savart integral (4.2.1)

$$\mathbf{B}^1(r) = \frac{\mu_0}{4\pi\Delta} \int_{\text{Plate}} \frac{\mathbf{I}(\mathbf{r}') \times (\mathbf{r} - \mathbf{r}')}{|\mathbf{r} - \mathbf{r}'|^3} \, dv' \tag{8-4.23}$$

The self-body force distribution per unit area of the plate is given by

$$\mathbf{I} \times \mathbf{B}^1 = -B_z^1 \nabla \psi \tag{8-4.24}$$

where the tangential field is zero at the plate midsurface. Note that $\nabla \psi$ gives the direction of the in-plane force. Let $R = r - r'$, then B_z^1 can be written in the form

$$B_z^1 = \frac{\mu_0}{\pi\Delta} \int \left[\frac{\partial \psi}{\partial x'} \frac{\partial}{\partial x'} \left(\frac{1}{|\mathbf{R}|} \right) + \frac{\partial \psi}{\partial y'} \frac{\partial}{\partial y'} \left(\frac{1}{|\mathbf{R}|} \right) \right] dv'$$

It has been shown (see Moon, 1978) that the thickness integration can be effectively carried out by using the divergence theorem and $\partial\psi/\partial z = 0$ in the plate, and setting $\psi = 0$ on the edges of the plate.

For a two-dimensional plate we have

$$B_z^1(x, y, 0) = \frac{\mu_0 \psi}{\Delta} - \frac{\mu_0}{4\pi} \int \frac{\psi(x', y') \, da'}{[(x - x')^2 + (y - y')^2 + \frac{1}{4}\Delta^2]^{3/2}} \tag{8-4.25}$$

When I is independent of the variable x, the integral (8-4.25) becomes

$$B_z^1(y, 0) = \frac{\mu_0 \psi}{\Delta} - \frac{\mu_0}{2\pi} \int \frac{\psi(y') \, dy'}{(y' - y)^2 + \frac{1}{4}\Delta^2} \tag{8-4.26}$$

A special case is the induction of one-dimensional flow in a long rectangular strip shown in Figure 8-8. For steady-state harmonic applied magnetic fields an integrodifferential equation for the stream function is found from Eqs. (8-4.22) and (8-4.26). In nondimensional form this equation becomes

$$\frac{d^2\phi}{dx^2} - i\mathcal{R}\phi + \frac{\mathcal{R}}{2\pi}\int_{-a}^{a} \frac{\phi d\xi}{(\xi - x)^2 + \frac{1}{4}} = i\mathcal{R}B(x) \qquad (8\text{-}4.27)$$

where $\psi = \phi e^{i\omega t}$, $B_z^0 = (\mu_0/\Delta)Be^{i\omega t}$, and $\mathcal{R} = \mu_0\sigma\omega\Delta^2$ (a magnetic Reynolds number), and $2a$ is the width of the plate. Analytical solutions to this equation are not available in the literature. However, numerical solutions using the finite-element method have been obtained (Yuan et al., 1981). A numerical solution showing the induced current distribution in a long rectangular plate due to two current filaments is shown in Figure 8-8. The local solution in Fig. 8-8 neglects the self-field effects in Eq. (8-4.27) by setting $\mathcal{R} = 0$ on the left side of Eq. (8-4.27).

For two-dimensional time-dependent magnetic fields, the equation for the stream function in a thin conducting plate becomes

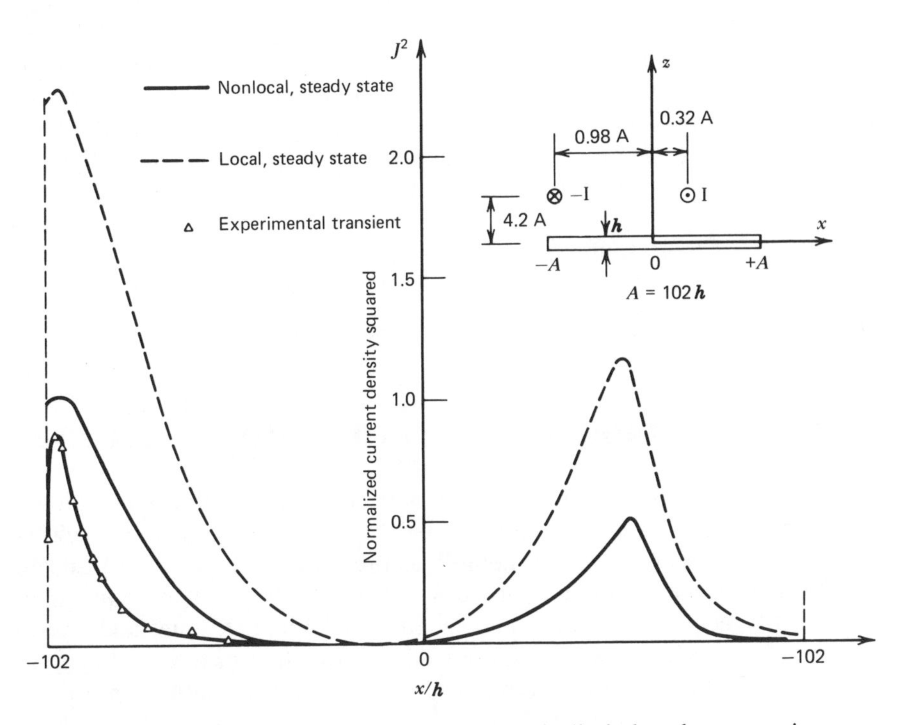

Figure 8-8 Finite-element calculation of magnetically induced currents in a one-dimensional conducting plate. Comparison of self-field effect (nonlocal) and experimental measurement using infrared scanning (from Yuan et al., 1982).

$$\nabla^2\psi - \mu_0\sigma\frac{\partial\psi}{\partial t} + \frac{\mu_0\sigma\Delta}{4\pi}\int\frac{(\partial\psi/\partial t)\,dx'\,dy'}{[(x'-x)^2+(y'-y)^2+\frac{1}{4}\Delta^2]^{3/2}} = \sigma\Delta\frac{\partial B_z^0}{\partial t} \tag{8-4.28}$$

Solutions to this equation are not available in the literature. However, numerical solutions using the finite-element method have been obtained by Yuan et al. (1981).

If the plate motion is taken into account, we must use Ohm's law for moving media,

$$\mathbf{I} = \nabla \times \psi\mathbf{n} = \sigma\Delta[\mathbf{E} + \mathbf{v} \times \mathbf{B}]$$

When this expression is used in Faraday's law (8-2.8a), the equation for ψ is modified as

$$\nabla \times \nabla \times \psi\mathbf{n} = -\sigma\Delta\frac{\partial}{\partial t}(\mathbf{B}^0 + \mathbf{B}^1) + \sigma\Delta\,\nabla \times (\mathbf{v} \times \mathbf{B}^0) \tag{8-4.29}$$

where $\mathbf{v}$ is the velocity of the plate surface and only linear terms are included.

For *low frequencies*, the equation for ψ takes the simpler form

$$\nabla^2\psi = \sigma\Delta\frac{\partial}{\partial t}B_n^0 \tag{8-4.30}$$

The self-field term has been neglected in Eq. (8-4.30). For time harmonic fields, this equation becomes a Poisson equation and can readily be solved with standard analytic and numerical methods. Morjaria et al. (1981) have obtained solutions for eddy currents in a conducting plate with a crack using the boundary integral method. Contours of constant ψ, which give the current flow direction, are shown in Figure 8-9. This numerical solution shows that concentrations of current exist at the crack tip which can lead to increased temperature, melting, and even fracture at the crack tip (see, e.g., Yagawa and Horie, 1982).

8.5 EDDY-CURRENT MAGNETIC-MOMENT METHOD

For structural conductors of complex geometry, it is impossible to completely characterize the local eddy-current behavior everywhere and a global approach must be taken. One method, as presented in Laudau and Lifshitz (1960), is to treat the circulating eddy currents as an equivalent magnetic dipole $\mathbf{m}$. Once this dipole is known, the total force, torque, and energy dissipated in the body can be determined as a function of the exciting field.

The magnetic dipole for eddy currents is related to the current-density vector by the equation,†

†The definition of $\mathbf{m}$ is independent of the origin of $\mathbf{r}$ provided $\int \mathbf{J}\,dV = 0$.

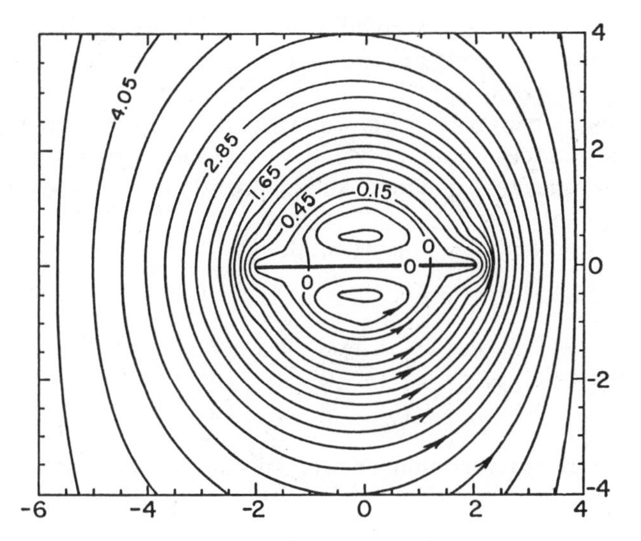

Figure 8-9 Induced eddy-current flow around a crack in a thin plate (from Morjaria et al., 1981).

$$\mathbf{m} = \frac{1}{2} \int \mathbf{r} \times \mathbf{J}\, dv \tag{8-5.1}$$

where it is assumed that the net current through the body is zero [See also Eq. (4-2.9)]. In a time-varying magnetic field, the eddy currents in a nonferromagnetic body are linearly related to the magnetic field. To distinguish steady fields from the time-varying field, we use the notation $\mathbf{B}^0$ for the steady and $\mathbf{B}^1$ for the time-varying field, or

$$\mathbf{B} = \mathbf{B}^0(r) + \mathbf{B}^1(r, t) \tag{8-5.2}$$

Further, we can synthesize $\mathbf{B}^1(t)$ by a Fourier integral

$$\mathbf{B}^1(r, t) = \frac{1}{2\pi} \int_{-\infty}^{\infty} \hat{\mathbf{B}}^1(r, \omega) e^{i\omega t} d\omega \tag{8-5.3}$$

For simplicity, we shall assume that the gradients in $\mathbf{B}^1$ are small so that each part of the conducting test specimen sees the same field $\mathbf{B}^1$. Since there is a linear relationship between $\mathbf{J}$ and $\mathbf{B}^1$ we can write (following Landau and Lifshitz, 1960)

$$\hat{\mathbf{m}} = \frac{V \boldsymbol{\alpha} \cdot \hat{\mathbf{B}}^1}{\mu_0} \tag{8-5.4}$$

where $\boldsymbol{\alpha}$ is a complex valued symmetric tensor ($\alpha_{ij} = \alpha_{ji}$) called the *magnetic polarizability tensor* or matrix, V is the volume, and the symbol ($\hat{}$) indicates a Fourier-transformed variable.

Consider the two examples given below (see Figure 8-10).

For a flat plate (see Fig. 8-10a) the magnetic moment only has one component, $\hat{m}_z = \alpha \hat{B}_z^1 V/\mu_0$.

Consider now a bent plate as shown in Figure 8-10(b) with the bend along the y axis and surfaces in the $x-y$ and $y-z$ planes. Neglecting circulating currents in the thickness direction, the magnetic-moment vector $\mathbf{m}$ has the form

$$\mathbf{m} = (m_x, 0, m_z) \tag{8-5.5}$$

and the matrix takes the form

$$\begin{bmatrix} m_x \\ m_y \\ m_z \end{bmatrix} = \begin{bmatrix} \alpha_{11} & 0 & \alpha_{13} \\ 0 & 0 & 0 \\ \alpha_{13} & 0 & \alpha_{33} \end{bmatrix} \begin{bmatrix} B_x^1 \\ B_y^1 \\ B_z^1 \end{bmatrix} \frac{V}{\mu_0} \tag{8-5.6}$$

since only the x and z components of $\mathbf{B}^1$ can generate significant eddy-current flow.

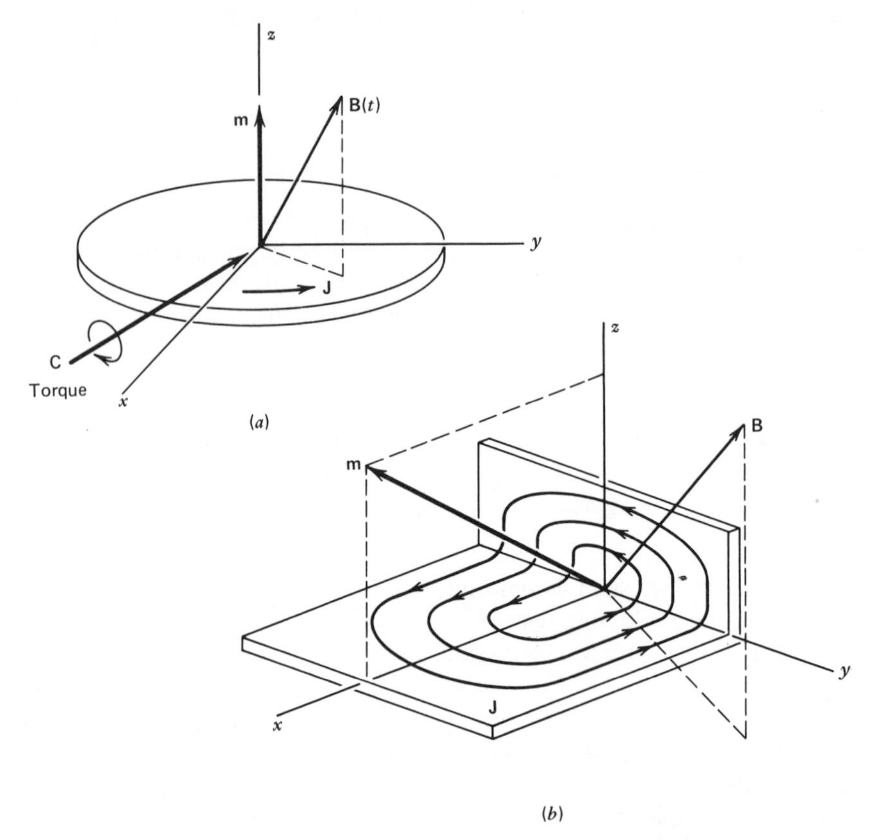

Figure 8-10 Equivalent magnetic moment for thin plates: (a) circular plate and (b) bent rectangular plate.

In the example of the flat plate, only one function of freqency $\alpha(\omega)$ must be determined, while in Eq. (8-5.6) three functions α_{11}, α_{13}, and α_{33} must be determined.

Once $\mathbf{m}$ or $\boldsymbol{\alpha}$ is known for a conducting body, the torque on the body is given by

$$\mathbf{C} = \mathbf{m} \times (\mathbf{B}^0 + \mathbf{B}^1) \tag{8-5.7}$$

For induction fields with a weak gradient (i.e., $\mathbf{B}$ relatively constant over the object), the total force is given by

$$\mathbf{F} = \mathbf{m} \cdot \nabla (\mathbf{B}^0 + \mathbf{B}^1) \tag{8-5.8}$$

In both Eqs. (8-5.7) and (8-5.8), the field $\mathbf{B} = \mathbf{B}^0 + \mathbf{B}^1$ does not include the field of $\mathbf{m}$ on the principle that the self-field of a set of currents in a body cannot create a net force or torque on the body.

To calculate the energy dissipated, we write $\mathbf{M}$ in terms of its Fourier components (to simplify notation we drop the superscript on $\mathbf{B}^1$),

$$\mathbf{m} = \frac{1}{2\pi} \int_{-\infty}^{\infty} \hat{\mathbf{m}} e^{i\omega t} \, d\omega \tag{8-5.9}$$

where

$$\hat{\mathbf{m}} = \frac{V}{\mu_0} \boldsymbol{\alpha} \cdot \hat{\mathbf{B}}$$

The instantaneous power dissipated for a magnetic dipole in a time-varying field is (see, e.g., Landau and Lifshitz, 1960)

$$P = -\mathbf{m} \cdot \frac{\partial \mathbf{B}}{\partial t}$$

For a harmonic field, the average power dissipated is given by

$$\langle P \rangle = \frac{1}{2} \frac{V}{\mu_0} \omega \alpha''_{ij} \hat{\mathbf{B}}_i \hat{\mathbf{B}}_j^* \tag{8-5.10}$$

where the asterisk indicates the complex conjugate and $\alpha = \alpha' + i\alpha''$.

The average torque in a harmonic field is given by

$$\langle \mathbf{C} \rangle = \tfrac{1}{2} \operatorname{Re}(\hat{\mathbf{m}} \times \hat{\mathbf{B}}^*) \tag{8-5.11}$$

For transient or pulsed fields we define a function $\mathsf{A}(t)$ related to the Fourier transform of the polarizability tensor, that is,

$$\mathsf{A}(t) = \frac{1}{2\pi} \int_{-\infty}^{\infty} \frac{\boldsymbol{\alpha}}{i\omega} e^{i\omega t} \, d\omega \tag{8-5.12}$$

or

$$T[\mathbf{A}] = \frac{\boldsymbol{\alpha}}{i\omega}$$

where $T[\ \ ]$ stands for Fourier transform. Then it can be shown that

$$\mathbf{m} = \frac{V}{\mu_0} \int_{-\infty}^{t} \mathbf{A}(\tau) \cdot \dot{\mathbf{B}}(t-\tau)\, d\tau \tag{8-5.13}$$

where $\mathbf{B}$ is zero for $t<0$. This relation is similar to that in the field of viscoelasticity and $\mathbf{A}(t)$ might be called the *magnetic relaxation function*.

For example, consider a *ramp* function $\mathbf{B} = \mathbf{B}^0\ t/\tau_0$ for $t>0$. Then $\dot{\mathbf{B}} = \mathbf{B}^0/\tau_0$ is a constant and we can write

$$\mathbf{m} = \frac{V}{\mu_0 \tau_0} \mathbf{G} \cdot \mathbf{B}_0 \tag{8-5.14}$$

where

$$\mathbf{A}(\tau) = \frac{d\mathbf{G}}{dt}$$

Consider the one-component problem

$$\hat{m}_z = \frac{V\alpha}{\mu_0} \hat{B}_z \tag{8-5.15}$$

and suppose we measure $m_z(t)$ by some torque device. Then the polarizability function $\alpha(\omega)$ can be found from

$$\alpha(\omega) = -\frac{\omega^2 \tau_0 \mu_0}{V B_0} T[m_z(t)] \tag{8-5.16}$$

Thus it is not necessary to perform harmonic tests to measure $\alpha(\omega)$ or $\hat{\mathbf{m}}(\omega)$ provided one has sufficiently noise-free measurements of $\mathbf{m}(t)$ (or the torque) with which to take a Fourier transform.

In the literature there are only a few examples of explicit calculations of $\mathbf{m}$ or $\boldsymbol{\alpha}(\omega)$—one is the spherical conductor and the other is the thin-wall conducting cylinder in a transverse field. Both examples (discussed below) serve to illustrate some important scaling problems.

1. *Sphere.* The first example is that of a solid conducting sphere in a harmonic magnetic field. In this case the matrix $\boldsymbol{\alpha}$ is reduced to a scalar α times the identity matrix. A natural nondimensional group is the ratio of radius a to skin depth

$$\delta = \left(\frac{2}{\mu_0 \sigma \omega}\right)^{1/2}$$

$$\alpha = \alpha' + i\alpha''$$

$$\alpha' = -\frac{3}{8\pi}\left(1 + \frac{3}{x}\frac{\sin x - \sinh x}{\cosh x - \cos x}\right) \tag{8-5.17}$$

and

$$\alpha'' = -\frac{9}{4\pi}\frac{1}{x^2}\left(1 - \frac{x}{2}\frac{\sinh x + \sin x}{\cosh x - \cos x}\right)$$

here $x = 2a/\delta$.

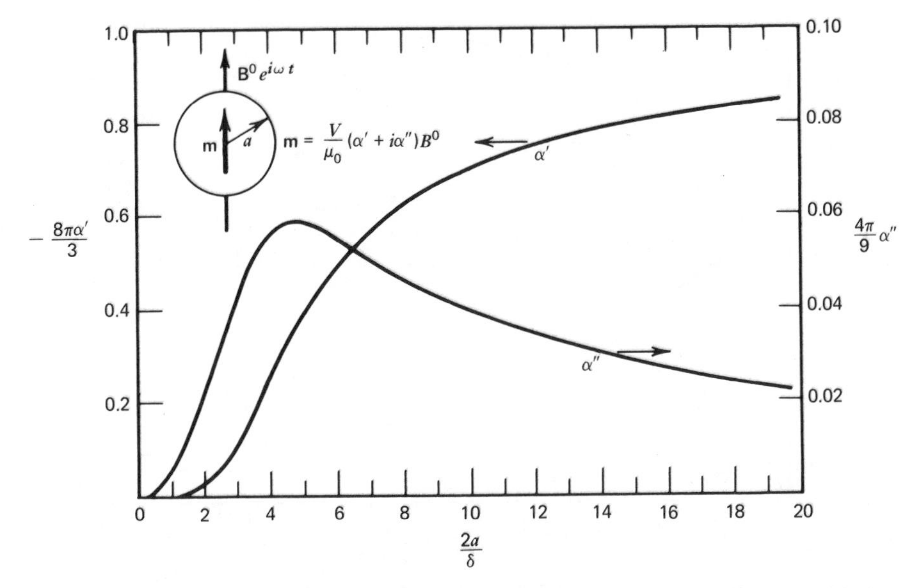

Figure 8-11 Real and imaginary parts of induced magnetic moment in a conducting sphere vs. the square root of nondimensional frequency [Eq. (8-5.17)].

The solution to this problem may be found in Landau and Lifshitz (1960) and Smythe (1968). The functions α' and α'' are shown plotted in Figure 8-11. This figure shows that α'' attains a maximum at a characteristic value of a/δ. This is similar to the behavior of the induced currents in a circuit (see Figs. 8-1 and 8-2).

2. *Thin-walled cylinder.* The second example is that of a thin-walled conducting cylinder in a transverse harmonic magnetic field (see Figure 8-12). The solution is given in Moullin (1955).

In this case there are two length scales, the radius of the cylinder a and the wall thickness Δ. What is unexpected is that the natural nondimensional group that emerges involves the product of Δ and a.

We define

$$\mathcal{R}_m = \frac{a\Delta}{\delta^2}$$

Then

$$\hat{\mathbf{m}} = \frac{Via}{\mu_0 \Delta} \frac{2\mathcal{R}_m}{(1 + i\mathcal{R}_m)} \hat{\mathbf{B}}^0 \tag{8-5.18}$$

where $\hat{\mathbf{m}}$ is along the direction of $\hat{\mathbf{B}}^0$ and $V = 2\pi a\Delta$. The units of $\mathbf{m}$ are magnetic moment per unit length of the cylinder. The real and imaginary parts of $\hat{\mathbf{m}}$ or α are shown plotted in Figure 8-12. Again, the part associated with the dissipation, α'', reaches a maximum at a characteristic value of $a\Delta/\delta^2$. This value indicates the value of frequency at which the self-field

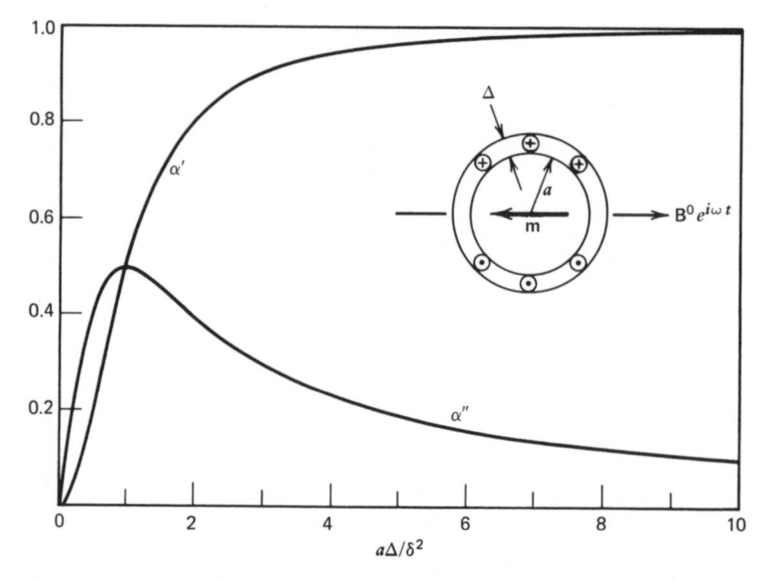

Figure 8-12 Real and imaginary parts of induced magnetic moment in a thin cylindrical shell vs. nondimensional frequency (magnetic Reynolds number) [Eq. (8-5.8)].

effects begin to get large. The nondimensional group $\mathcal{R}_m = \frac{1}{2}a\Delta\mu_0\sigma\omega$ is sometimes called a *magnetic Reynolds number* in analogy with fluid mechanics (see Section 8.6).

Experimental work on eddy currents in thin-walled cylinders in rotating magnetic fields had been reported (see Boyle et al., 1964). A rotating field can be shown to be equivalent to two sinusoidal fields orthogonal to each other and out of phase by $\frac{1}{2}\pi$ radians. The measured torque is then proportional to α''. The theoretical solution may be found in Smythe (1968).

8.6 SCALING PRINCIPLES IN EDDY-CURRENT PROBLEMS

The number of variables in a magnetomechanical problem can be reduced by using nondimensional groups. A nondimensional approach allows one to perform experiments on a small model specimen and to scale the results up to a full-scale device.

As an example, consider an eddy-current problem with the following variables:

$B^0(t)$ applied magnetic field
τ time constant associated with $B^0(t)$

$\mathbf{J}$	current density
σ	electrical conductivity
μ	permeability
D_1, D_2, and D_3	geometric parameters

Suppose, for instance, a particular problem has three geometric parameters D_1, D_2, and D_3, for example, a hollow cylinder of finite length in a uniform magnetic field. Then one can show that the induced current density must have the form

$$\frac{J\mu D_1}{B_0} = f\left(\frac{\mu\sigma D_1^2}{\tau}, \frac{D_2}{D_1}, \frac{D_3}{D_1}\right)$$

where B_0 is the maximum value of $B^0(t)$ and J is the current density at a given location in the body. Thus a problem with eight variables has been reduced to four variables.

In the following discussion, we will assume that an applied field $B^0(t)$ is given and its rate of change is described by a time constant τ or frequency ω. If $B^0(t)$ has a significant gradient, then an additional length variable must be added to the list.

Scaling Rules—General Principles

We assume that the conductors obey a linear magnetization law. When displacement currents are neglected, **B** and **E** can be shown to obey the vector, partial differential equations (8-2.11) and (8-2.12). If one nondimensionalizes the distance by a length scale λ and the time by τ, then these equations take the form

$$\bar{\nabla}^2\mathbf{B} = \mathcal{R}_m \frac{\partial \mathbf{B}}{\partial \bar{t}} \tag{8-6.1}$$

where bars indicate derivatives with respect to the nondimensional quantities $\bar{r} = r/\lambda$ and $\bar{t} = t/\tau$. Here $\mathcal{R}_m$ is a nondimensional group sometimes called the magnetic Reynolds number that is,

$$\mathcal{R}_m = \frac{\mu\sigma\lambda^2}{\tau} \tag{8-6.2}$$

$\mathcal{R}_m$ can be viewed either as the ratio of the square of two lengths λ^2/δ^2, where $\delta = (\mu\sigma/\tau)^{1/2}$ is called the skin depth, or the ratio of two times τ_d/τ, where $\tau_d = \mu\sigma\lambda^2$ is a characteristic decay time.

In a typical problem of induced currents, with no flow into or out of the body, the initial electrical field is zero and the currents are induced by an applied time-varying field $\mathbf{B}^0(t)$ (relative to the conductor time frame). The induced currents $\mathbf{J}$ create their own magnetic field $\mathbf{B}^1(t)$ given by Eq. (2-6.3) so that the total field is

$$\mathbf{B} = \mathbf{B}^0 + \mathbf{B}^1(t)$$

If B_0 (scalar) represents the maximum value of the vector field $\mathbf{B}^0(t)$ and τ is the characteristic rise or decay time of $\mathbf{B}^0(t)$, then typically $\mathbf{J}$ can be written in the form

$$\mathbf{J} = \frac{B_0}{\tau}\mathbf{G}(\mathbf{r}, t, \mu\sigma) \tag{8-6.3}$$

However, $\mathbf{G}$ has dimensions. To nondimensionalize this expression we examine Faraday's law for the case when $\mathbf{B}^1 \to 0$; that is, (Eq. 2-2.10)

$$\oint \mathbf{J} \cdot d\mathbf{l} = -\sigma \frac{\partial}{\partial t} \int \mathbf{B}^0 \cdot d\mathbf{a}$$

This suggests writing $\mathbf{J}$ in the following form:

$$\mathbf{J} = \mathcal{R}_m \frac{B_0}{\mu\lambda}\mathbf{g}(\bar{\mathbf{r}}, \bar{t}, \mathcal{R}_m)$$

The group $J_0 \equiv B_0/\mu\lambda$ has dimensions of current density so that a complete nondimensional representation for $\mathbf{J}$ is

$$\frac{J}{J_0} = \mathcal{R}_m \mathbf{g}(\bar{\mathbf{r}}, \bar{t}, \mathcal{R}_m) \tag{8-6.4}$$

SCALING RULE NO. 1

If two similar eddy-current experiments (i.e., same nondimensional shapes and time history $\mathbf{B}^0/B_0$) have the same magnetic Reynolds number, then the current densities scale as $B_0/\mu\lambda$.

Scaling Rules—Long Prismatic Bars

A long bar or tube in a uniform magnetic field at an angle to the bar axis can be separated into an axial field problem and a transverse field problem. The axial field $\mathbf{B}^0 = (0, 0, B_z^0)$ generates a two-dimensional current-density field $\mathbf{J} = (J_x, J_y, 0)$. The transverse field problem $\mathbf{B}^0 = (B_x^0, B_y^0, 0)$, in turn, generates a one-dimensional current density $\mathbf{J} = (0, 0, J_z)$.

Consider the case of the long bar in an axial field. Since $\mathbf{J}$ satisfies a continuity equation, that is, $\nabla \cdot \mathbf{J} = 0$, we can write $\mathbf{J}$ in terms of a stream function [Eq. (8-3.10)], where ψ satisfies the equation

$$\nabla^2 \psi = \sigma\mu \frac{\partial \psi}{\partial t} - \sigma \frac{\partial B_z^0}{\partial t}$$

In nondimensional form this becomes

$$\bar{\nabla}^2 \bar{\psi} = \mathcal{R}_m \frac{\partial \bar{\psi}}{\partial \bar{t}} - \frac{\partial \bar{B}^0}{\partial \bar{t}} \tag{8-6.5}$$

where $\bar{\psi} = \psi/\psi_0$, $\bar{B}^0 = B_z^0/B_0$, and $\psi_0 = B_0\sigma\lambda^2/\tau$. The current density is thus proportional to ψ_0/λ.

The transient decay behavior is determined from the solutions to the homogeneous equation for ψ. For example, for a rectangular-cross-sectional bar of dimensions $a \times b$, the eigenvalues of the problem τ_{nm} for a solution of the form (n and m are related to the number of half wavelengths in the a and b directions respectively)

$$\psi = e^{-t/\tau_{nm}}\Psi(x, y)$$

with $\psi = 0$ on the boundary are given by

$$\tau_{nm} = \frac{\mu\sigma a^2}{\pi^2}\left[n^2 + m^2\left(\frac{a}{b}\right)^2\right]^{-1}$$

If we use the time constant of the applied field τ to nondimensionalize τ_{nm}, we come up with the result

$$\bar{\tau}_{nm} = \frac{\mathcal{R}_m}{\pi^2}\left[n^2 + m^2\left(\frac{a}{b}\right)^2\right]^{-1} \tag{8-6.6}$$

where $\lambda = a$. For bars of arbitrary cross section with no holes, we generalize this result in the following scaling rule:

SCALING RULE NO. 2

If the magnetic Reynolds number $\mathcal{R}_m$ is the same for both model and full-scale experiment, then the nondimensional time constants of both model and full-scale experiment will be equal when both have similar shapes (e.g., equal aspect ratios a/b for the rectangular bar).

Energy-Loss Scaling

For a long prismatic bar (of length L), in a parallel magnetic field, the energy loss is given by

$$\mathscr{E} = \frac{L}{\sigma}\int_0^\infty \int J^2\, da\, dt \tag{8-6.7}$$

where the applied transient field $\mathbf{B}^0(t)$ is assumed to be steady or zero for $t < 0$. Using $\mathbf{J}$ in the form

$$\mathbf{J} = \frac{\mathcal{R}_m B_0}{\mu\lambda}\mathbf{g}(\bar{x}, \bar{y}, \bar{t}, \mathcal{R}_m)$$

one can show that $\mathscr{E}$ has the form

$$\mathscr{E} = \lambda^2 L\left(\frac{B_0^2}{2\mu}\right)\mathcal{R}_m H(\mathcal{R}_m) \tag{8-6.8}$$

where H is some function of the magnetic Reynolds number and $B_0^2/2\mu$ represents the magnetic pressure or energy density in the applied field. Thus we are led to the following scaling law:

SCALING RULE NO. 3

If the model (M) and full-scale (FS) experiments for two prismatic bars in axial fields have the same magnetic Reynolds number, then the energy loss per unit length of conductor is given by the ratio

$$\frac{(\mathscr{E}/L)_{\mathrm{M}}}{(\mathscr{E}/L)_{\mathrm{FS}}} = \frac{(\lambda^2 B_0^2/2\mu)_{\mathrm{M}}}{(\lambda^2 B_0^2/2\mu)_{\mathrm{FS}}}$$

Torque Scaling for Prismatic Bars

The current flow around the cross section of a prismatic bar can be thought of as a magnetic moment $\mathcal{M}$ distributed along the bar [see Eq. (8-5.1)]. If the bar is simultaneously in the axial field $\mathbf{B}^A(t)$ and a steady uniform transverse field $\mathbf{B}^T$, then there will be a torque on a length L of the bar given by

$$\mathbf{C} = L\mathcal{M} \times \mathbf{B}^T \qquad (8\text{-}6.9)$$

One can then show that the torque in the model and full-scale experiments obey the following rule:

SCALING RULE NO. 4

If the model and full-scale experiments for two prismatic bars have the same magnetic Reynolds number, then the torque or couple scales according to the rule

$$\frac{(C)_{\mathrm{M}}}{(C)_{\mathrm{FS}}} = \frac{(L\lambda^2 B_A B_T)_{\mathrm{M}}}{(L\lambda^2 B_A B_T)_{\mathrm{FS}}}$$

where B_A is the maximum value of the axial field and B_T is the maximum value of the transverse field.

Thus for two bars with the same diameter to length ratios, the torque scales as the *cube* of the length or diameter.

Scaling Rules—Thin-Walled Cylinders

Many structural and heat transfer elements in fusion reactors will be made of conductors in the form of thin-walled plates, curved sheets, and cylindrical tubing.

Circular Cylinder. This solution has been discussed in Section 8.5, Eq.

(8-5.18). Referring to Figure 8-12, the transverse magnetic field $B_0 e^{i\omega t}$ generates axial current flow. The solution for the induced current is given by

$$J = J_0 \sin\theta$$

and (8-6.10)

$$|J_0| = \frac{2B_0}{\mu\Delta} \frac{\mathcal{R}_m}{(1+\mathcal{R}_m^2)^{1/2}}$$

where $\mathcal{R}_m = \frac{1}{2}\mu\sigma\omega\Delta a$ or $\mathcal{R}_m = a\Delta/\delta^2$, where δ is the skin depth.

This solution is interesting in that the magnetic Reynolds number depends on the product of a and Δ, and also since the distribution with angle is independent of $\mathcal{R}_m$. A similar solution is obtained if one considers a thin-walled cylinder in a harmonic uniform axial field where **J** now flows in the circumferential direction.

SCALING RULE NO. 5

The eddy-current densities of two thin cylindrical tubes scale as the ratio of $B/\mu\Delta$ when the magnetic Reynolds numbers $\mathcal{R}_m = \frac{1}{2}\mu\sigma\omega a\Delta$ are matched, that is,

$$\frac{(J)_{\mathrm{M}}}{(J)_{\mathrm{FS}}} = \frac{(B_0/\mu\Delta)_{\mathrm{M}}}{(B_0/\mu\Delta)_{\mathrm{FS}}}$$

It is not difficult to show that this principle can be extended to other thin-walled closed cylinders of noncircular shapes, such as ellipses.

Scaling Rules—Thin Plates

Low-Frequency Limit. Consider a thin flat conducting plate and assume a stream function ψ for **J** [Eq. (8-3.10)]. Assume that the induced magnetic field is small compared with the applied field. Also, the applied field is assumed to be uniform in space and harmonic in time, that is, $\mathbf{B}^0 = B_0 \mathbf{e}_z e^{i\omega t}$. Then **J** is given by

$$J_x = -i\sigma\omega B_0 \frac{\partial\psi}{\partial y}, \qquad J_y = i\sigma\omega B_0 \frac{\partial\psi}{\partial x} \tag{8-6.11}$$

Using Faraday's law (all quantities are assumed to depend on $e^{i\omega t}$), one can show that [see Eq. (8-4.28)],

$$\nabla^2\psi = -1 \tag{8-6.12}$$

The average power dissipation per unit area for a plate of thickness Δ is given by

$$\langle P\rangle = \frac{\sigma\omega^2 B_0^2 \Delta}{2A} \int (\nabla\psi)^2 \, da \tag{8-6.13}$$

For a circular plate of radius a, $\partial\psi/\partial r = -\frac{1}{2}r$, $A = \pi a^2$, and

$$\langle P\rangle = \tfrac{1}{16}\sigma\omega^2 B_0^2 \Delta a^2$$

Dresner (1979) has solved a similar problem for a linear time function for $B^0(t)$ and has shown that for a fixed A, the circular plate has the largest P. The energy dissipated per unit cycle per unit area is then of the form

$$\mathscr{E} = C_2 \mathscr{R}_m \frac{B^2}{2\mu}\Delta \qquad (8\text{-}6.14)$$

where $\mathscr{R}_m = \frac{1}{2}\mu\sigma\omega a^2$ is the ratio of radius to skin depth squared.

Rectangular Plates. When one geometric length is involved such as a solid cylinder or sphere, there are two nondimensional groups and the scaling laws take the form

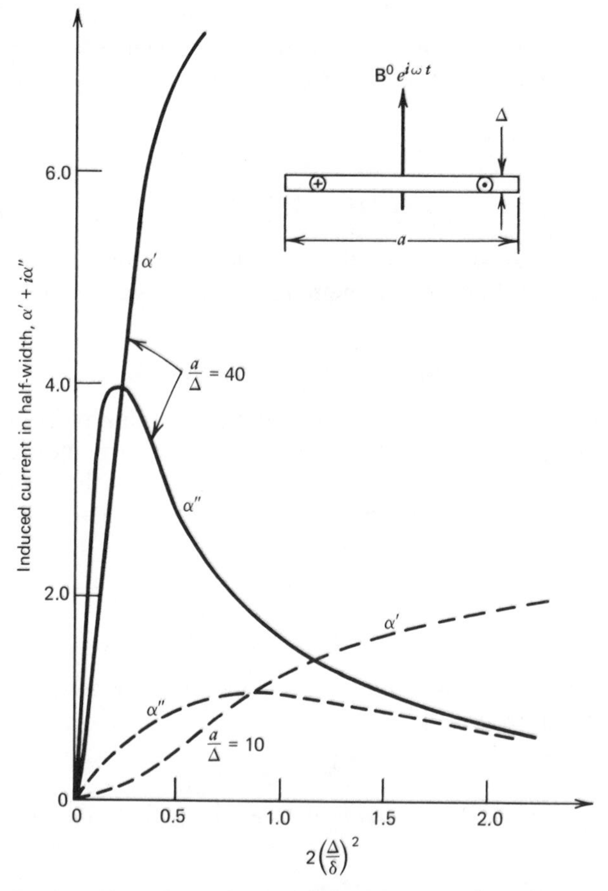

Figure 8-13 Real and imaginary parts of induced current in a thin conducting plate vs. nondimensional frequency for two values of width-to-thickness ratio (from K. Hara, Cornell University, unpublished report).

$$\frac{J\mu\lambda}{B_0} = F_1(\mu\sigma\omega\lambda^2)$$

where $\lambda = a$ is the radius of the cylinder or sphere.

For a rectangular plate of plane dimensions $a \times b$ and thickness Δ we have an additional nondimensional group a/b and we should expect the nondimensional form of J to depend on the variables in the following way:

$$\frac{J\mu\Delta}{B_0} = F_2\left(\mu\sigma\omega\Delta^2, \frac{a}{\Delta}, \frac{a}{b}\right) \tag{8-6.15}$$

However, for long rectangular plates, we have discovered by finite-element calculations and analogy to the thin cylinder [Eq. (8-6.10)] that only one of the three gorups is significant; that is, for $b/a \to \infty$,

$$\frac{J\mu\Delta}{B_0} = \mathcal{R}_m F_2(\mathcal{R}_m) \tag{8-6.16}$$

where $\mathcal{R}_m = \mu\sigma\omega a\Delta$.

This scaling rule is illustrated in Figures 8-13 and 8-14. Figure 8-13 shows the induced current density in a long rectangular plate in a transverse uniform magnetic field $B_0 e^{i\omega t}$, for two different width-to-thickness ratios a/Δ and different frequencies in the form $\mathcal{R}_\Delta = \mu\sigma\omega\Delta^2$. However, if we change the frequency scale using $\mathcal{R}_m = \mu\sigma\omega a\Delta$ and plot $J\mu\Delta/B_0$, both curves fall on the same curve as shown in Figure 8-14.

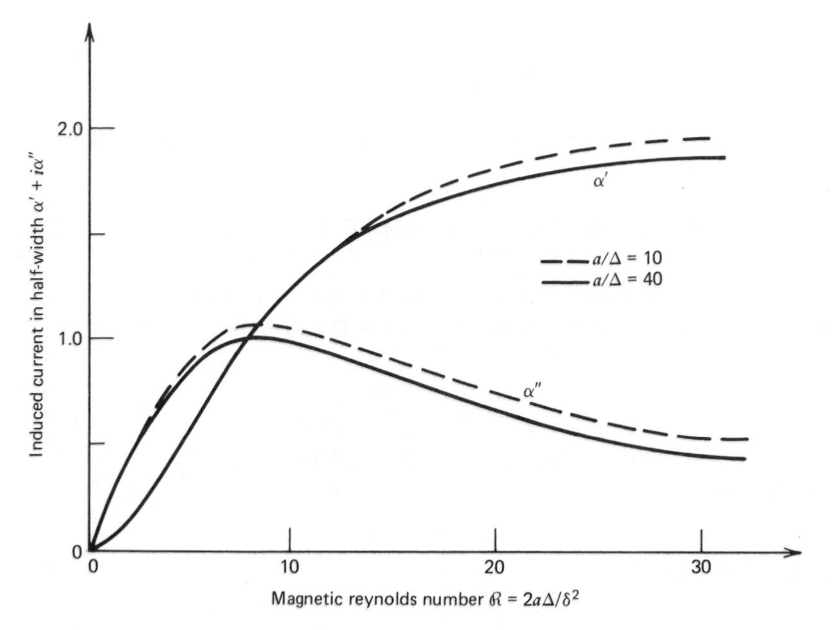

Figure 8-14 Real and imaginary parts of induced current in a thin conducting plate; data of Figure 8-13 rescaled.

8.7 HALL POTENTIAL EFFECTS ON EDDY CURRENTS

In classical electrodynamics of macroscopic metallic conductors the current density $\mathbf{J}$ is usually specified as a function of the electric field $\mathbf{J}(\mathbf{E})$ and for linear isotropic materials is given by Ohm's law (8-2.5). However, at the microscopic level, the motion of free electrons can be described in part by a classical momentum balance law that involves the magnetic field vector $\mathbf{B}$ as well as $\mathbf{E}$; that is,

$$mN\frac{d\mathbf{v}}{dt} = -eN(\mathbf{E} + \mathbf{v} \times \mathbf{B}) + \mathbf{f}(\mathbf{v}) \tag{8-7.1}$$

where $\mathbf{v}$ is the velocity of the electron, m the effective electron mass, e the electronic charge, N the number density of electrons, and f is the average force due to scattering of the electrons by the ionic lattice and other electrons in the free-electron gas. Typically, $\mathbf{f}$ is assumed to be a linear function of the velocity $\mathbf{v}$. For steady motions, the term on the left-hand-side in Eq. (8-7.1) is dropped. If one defines the current density as

$$\mathbf{J} \equiv -eN\mathbf{v}$$

then Eq. (8-7.1) becomes

$$\mathbf{J} - R_H\sigma\mathbf{J} \times \mathbf{B} = \sigma\mathbf{E} \tag{8-7.2}$$

This generalization of Ohm's law accounts for the effect of the magnetic field. R_H is called the *Hall constant*. Using Ampere's law (8-2.8b) one can write

$$\mathbf{J} = \sigma\mathbf{E} - \frac{\sigma R_H}{\mu_0}\mathbf{B} \times (\nabla \times \mathbf{B})$$

Thus the magnetic field diffusion equation (8-2.10) becomes (neglecting convection, $\mathbf{v} = 0$)

$$\nabla^2\mathbf{B} - \mu_0\sigma\frac{\partial\mathbf{B}}{\partial t} = -\sigma R_H \nabla \times (\mathbf{B} \cdot \nabla\mathbf{B}) \tag{8-7.3}$$

The nonlinear nature of this equation is clear in contrast to the classical equation (8-2.10). In MKS units, R_H has a theoretical value given by

$$R_H = \frac{-1}{Ne} \tag{8-7.4}$$

For real metals, the constant R_H differs from the value given by Eq. (8-7.4) and depends on temperature. For example, for copper at 20°C,

$$R_H Ne = -0.8$$

but for aluminum,

$$R_H Ne = -0.4$$

(see, e.g., Kittel, 1968, p. 243).

To illustrate one of the consequences of this law, we assume that $\mathbf{J}$ is forced to flow in a transverse magnetic field, that is, $\mathbf{J}\cdot\mathbf{B}=0$. Then Eq. (8-7.2) shows that there is a parallel and transverse electric field induced; that is

$$E_{\parallel}=\rho J_{\parallel}$$
$$E_{\perp}=R_H J_{\parallel} B \tag{8-7.5}$$

or

$$E_{\perp}=\frac{R_H B}{\rho}E_{\parallel}$$

where $\mathbf{E}_{\parallel}\times\mathbf{J}_{\parallel}=0$, and $\mathbf{E}_{\perp}\cdot\mathbf{J}_{\parallel}=0$. The transverse field $\mathbf{E}_{\perp}$ is sometimes referred to as the *Hall potential* and is used in a number of devices to measure magnetic fields.

The importance of the Hall potential in eddy-current problems may be estimated by examining the magnitude of the nondimensional group

$$\frac{R_H B_T}{\rho}\equiv\sigma R_H B_T$$

where B_T is transverse to the current-density vector. At room temperatures for $B_T<1$ T this number is very small and the term $\mathbf{J}\times\mathbf{B}$ may be dropped in Eq. (8-7.2). However, at cryogenic temperatures, the resistivity may decrease significantly (see Fig. 6-2) to where $R_H B_T/\rho$ is of order unity.

For example, for $\frac{1}{8}$–$\frac{1}{2}$ hard OFHC (oxygen-free high-conductivity) copper at 4.2°K,

$$\rho=2.5\times10^{-10}\ \Omega\ \text{m} \tag{8-7.6}$$

(At room temperature, pure copper has a resistivity $\rho=1.69\times10^{-8}\ \Omega$ m or $\rho_{300}/\rho_{4.2}=67.6$). The value of R_H does not change significantly from room temperature to 4.2°K.

Thus we will use the value of R_H for copper at room temperature,

$$R_H=-0.55\times10^{-10}\ \text{m}^3/\text{C} \tag{8-7.7}$$

With the data from Eqs. (8-7.6) and (8-7.7), we estimate for $\frac{1}{8}$–$\frac{1}{2}$ hard copper at 4.2°K,

$$\left|\frac{R_H B_T}{\rho}\right|\geq\left(\frac{0.55\times10^{-10}}{2.5\times10^{-10}}\right)B_T=0.22B_T \qquad \text{at } 4.2°\text{K}$$

At room temperature, we have

$$\left|\frac{R_H B_T}{\rho}\right|=(3.25\times10^{-3})B_T \qquad \text{at } 300°\text{K}$$

Thus one can see that the Hall effect or magnetoresistance can become important at low temperatures and high magnetic fields greater than 1 T.

These conditions may be satisfied in superconducting magnet systems, such as in magnetic fusion reactors.

To derive the basic equations for eddy currents with Hall effects we divide the magnetic field into three parts: a steady part $\mathbf{B}_T$ (e.g., the toroidal magnetic field in fusion reactors), the induction field $\mathbf{B}_P$ (usually the poloidal field in fusion reactors), and the self-induced field $\mathbf{B}^1$, or

$$\mathbf{B} = \mathbf{B}_T + \mathbf{B}_P(t) + \mathbf{B}^1(t) \tag{8-7.8}$$

As an approximation, we neglect $\mathbf{B}_P$ and $\mathbf{B}^1$ in the constitutive law Eq. 8-7.2 and write

$$\mathbf{E} = \rho \mathbf{J} - R_H \mathbf{J} \times \mathbf{B}_T \tag{8-7.9}$$

This, in effect, linearizes the magnetic field diffusion problem.

To simplify the problem we consider a long cylinder of arbitrary cross section with the axis aligned with the z axis such that all quantities satisfy $\partial/\partial z = 0$ (see Fig. 8-15). We divide the fields $\mathbf{E}$, $\mathbf{J}$, and $\mathbf{B}$ into transverse and axial or parallel components, that is,

$$\begin{aligned} \mathbf{E} &= \mathbf{E}_\perp + \mathbf{E}_\parallel \\ \mathbf{J} &= \mathbf{J}_\perp + \mathbf{J}_\parallel \end{aligned} \tag{8-7.10}$$

and

$$\mathbf{B} = \mathbf{B}_\perp + \mathbf{B}_\parallel$$

We assume $\mathbf{B}_P(t)$ is axial and $\mathbf{B}_T$ is transverse, that is, $\mathbf{B}_P \cdot \mathbf{B}_T = 0$. The constitutive equation can be written as two equations:

$$\begin{aligned} \mathbf{E}_\parallel &= \rho \mathbf{J}_\parallel - R_H \mathbf{J}_\perp \times \mathbf{B}_T \\ &\text{and} \\ \mathbf{E}_\perp &= \rho \mathbf{J}_\perp - R_H \mathbf{J}_\parallel \times \mathbf{B}_T \end{aligned} \tag{8-7.11}$$

To solve the problem we use vector potentials for $\mathbf{B}^1$ and $\mathbf{J}$. From (16), we have

$$\nabla \cdot \mathbf{J}_\perp = 0, \qquad \nabla \cdot \mathbf{B}_\perp = 0$$

Thus the transverse parts of $\mathbf{J}$ and $\mathbf{B}$ are related to potentials ψ and A by the equations

$$\begin{aligned} \mathbf{J}_\perp &= \nabla \times \psi \mathbf{e}_z = -\mathbf{e}_z \times \nabla \psi \\ \mathbf{B}_\perp^1 &= \nabla \times A \mathbf{e}_z = -\mathbf{e}_z \times \nabla A \end{aligned} \tag{8-7.12}$$

Using Faraday's and Ampere's laws (8-2.8a) and (8-2.8b), one can derive partial differential equations which the potentials A and ψ must satisfy. These equations can be shown to be

$$\nabla^2 A - \mu_0 \sigma \frac{\partial A}{\partial t} = \mu_0 \sigma R_H [(\mathbf{e}_z \times \nabla \psi) \times \mathbf{B}_T] \cdot \mathbf{e}_z$$

$$\nabla^2 \psi - \mu_0 \sigma \frac{\partial \psi}{\partial t} = \frac{\sigma R_H}{\mu_0} \mathbf{e}_z \cdot \nabla \times (\nabla^2 A \mathbf{e}_z \times \mathbf{B}_T) + \sigma \frac{\partial B_P}{\partial t} \tag{8-7.13}$$

Exact solutions to these coupled equations will not be sought. Instead a low-frequency analysis will be presented.

In the low-frequency limit, we neglect the effects of the self-field $\mathbf{B}^1$. Under this assumption we obtain

$$\nabla \times \mathbf{E}_\| = 0$$

or

$$E_\| = 0$$

This leads to an explicit relation between the in-plane and axial currents $\mathbf{J}_\perp$ and $\mathbf{J}_\|$, that is,

$$J_\| - \sigma R_H \mathbf{e}_z \cdot (\mathbf{J}_\perp \times \mathbf{B}_T) = 0 \tag{8-7.14}$$

Neglect of $\mathbf{B}^1$ also leads to the following equation for the stream function ψ,

$$\nabla^2 \psi = -\sigma R_H \mathbf{e}_z \cdot \nabla \times (\mathbf{J}_\| \times \mathbf{B}_T) + \sigma \frac{\partial B_P}{\partial t} \tag{8-7.15}$$

Using Eqs. (8-7.14) and (8-7.15) one obtains an equation for ψ,

$$\nabla^2 \psi + (\sigma R_H B_T)^2 (\mathbf{e}_T \cdot \nabla)^2 \psi = \sigma \frac{\partial B_P}{\partial t} \tag{8-7.16}$$

where $\mathbf{e}_T$ is a unit vector in the direction of the steady transverse field $\mathbf{B}_T$ and $\mathbf{B}_T$ is assumed to be uniform. For example, if $\mathbf{e}_T \equiv \mathbf{e}_x$ (Fig. 8–15) then Eq. (8-7.16) takes the form

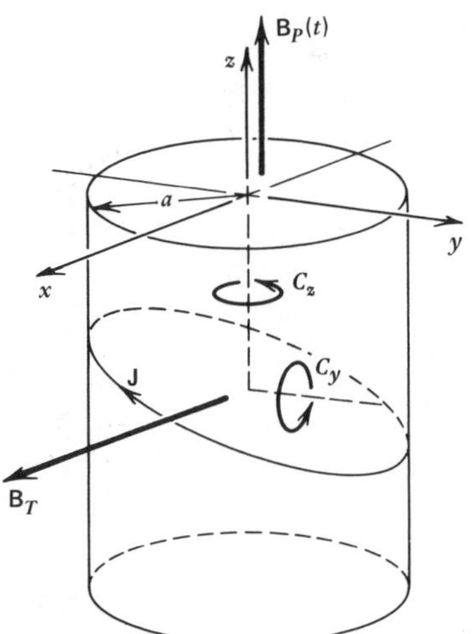

Figure 8-15 Hall potential effects on induced current in a conducting cylinder.

$$[1+(\sigma R_H B_T)^2]\frac{\partial^2 \psi}{\partial x^2}+\frac{\partial^2 \psi}{\partial y^2}=\sigma\frac{\partial B_P}{\partial t} \tag{8-7.17}$$

and the three components of the current-density vector are given by

$$(J_x, J_y, J_z)=\left(\frac{\partial \psi}{\partial y}, -\frac{\partial \psi}{\partial x}, \sigma R_H B_T \frac{\partial \psi}{\partial x}\right) \tag{8-7.18}$$

The boundary conditions for zero current crossing the surface of the cylinder lead to

$$\psi = \text{const (on the boundary)}$$

By changing coordinates such that

$$x_1=\frac{x}{(1+\beta^2)^{1/2}}, \qquad (\beta=\sigma R_H B_T)$$

Eq. (8-7.17) takes the form of a Poisson equation. For a harmonic induction field

$$B_P = B_0 e^{i\omega t}$$

we assume that

$$\psi = -\tfrac{1}{2}\Psi e^{i\omega t} i\sigma\omega B_0 \tag{8-7.19}$$

and Eq. (8-7.17) becomes

$$\frac{\partial^2 \Psi}{\partial x_1^2}+\frac{\partial^2 \Psi}{\partial y^2}=-2 \tag{8-7.20}$$

The form of this equation is identical to that for the stress function for the torsion of a shaft of noncircular cross section. The solution to Eq. (8-7.20) may be found in a number of elasticity texts, such as Love (1922).

Eddy Currents in an Elliptic Cylinder

Suppose the boundary equation for the cylinder is in the shape of an ellipse (see Fig. 8-15), that is,

$$\frac{x^2}{a^2}+\frac{y^2}{b^2}=1$$

then in the (x_1, y) coordinates we have

$$\frac{x_1^2}{a_1^2}+\frac{y^2}{b^2}=1 \tag{8-7.21}$$

where

$$a_1=\frac{a}{(1+\beta^2)^{1/2}} \tag{8-7.22}$$

Then the solution which satisfies Eqs. (8-7.20) and (8-7.21) can be shown to be

$$\Psi = \frac{1}{2}\left[\frac{(\gamma-1)x^2}{1+\beta^2} - (\gamma+1)y^2\right] \tag{8-7.23}$$

where

$$\gamma = \frac{a^2 - b^2(1+\beta^2)}{a^2 + b^2(1+\beta^2)}$$

The current-density components are then found to be ($e^{i\omega t}$ implicit):

$$\begin{aligned} J_x &= \frac{i\sigma\omega B_0}{2}(\gamma+1)y \\ J_y &= \frac{i\sigma\omega B_0}{2}\frac{\gamma-1}{1+\beta^2}x \\ J_z &= -\beta J_y \end{aligned} \tag{8-7.24}$$

When $a = b$, $\gamma = -\beta^2/(2+\beta^2)$ and

$$\begin{aligned} J_x &= i\sigma\omega B_0 \frac{y}{2+\beta^2} \\ J_y &= -i\sigma\omega B_0 \frac{x}{2+\beta^2} \\ J_z &= -\beta J_y \qquad (\beta = \sigma R_H B_T) \end{aligned} \tag{8-7.25}$$

The current flow is sketched in Figure 8-15. The flow is circulating, but is tilted about the B_T axis. In polar coordinates, we have

$$J_r = 0, \qquad J_\theta = -\frac{i\sigma\omega B_0 r}{2+\beta^2}, \qquad J_z = i\sigma\omega B_0 \frac{\beta r\cos\theta}{2+\beta^2}$$

The in-plane component J_θ is the same as without the Hall effect, $\beta = 0$, but with an effective resistivity $\rho^* = \rho(1+\frac{1}{2}\beta^2)$.

Since there is no assumed gradient in either the poloidal or toroidal fields, the *net magnetic force* per unit length *is zero*. However, there is a *nonzero torque* given by

$$\begin{aligned} \mathbf{C} &= \int \mathbf{r}\times(\mathbf{J}\times\mathbf{B})\,da \\ \mathbf{C} &= \int \mathbf{r}\times[\mathbf{J}_\perp\times(\mathbf{B}_P+\mathbf{B}_T)]\,da + \int \mathbf{r}\times(\mathbf{J}_\parallel\times\mathbf{B}_T)\,da \end{aligned} \tag{8-7.26}$$

This leads to two components of the torque

$$\mathbf{C} = \tfrac{1}{4}J_0 B_T \pi a^4 \mathbf{e}_y - \tfrac{1}{4}J_0 B_T^2 \sigma R_H \pi a^4 \mathbf{e}_z \tag{8-7.27}$$

where $J_0 = -i\sigma\omega B_0/(2+\beta^2)$. The first term is the classical torque that would act on an Ohmic conductor. The second term, however, is new and is due to the Hall potential. One can see that it is proportional to the square of the steady field $\mathbf{B}_T$. (See Fig. 8-15.)

8.8 LEVITATION FORCES ON MOVING CONDUCTORS

Eddy currents may be induced when a conductor *moves* in an otherwise steady source of magnetic field. This phenomenon has been exploited to create magnetic forces of sufficient magnitude to levitate or suspend moving vehicles over electrically conducting guideways. A photograph of a Japanese magnetically levitated vehicle is shown in Figure 8-16 and a laboratory model levitated on a rotating wheel guideway is shown in Figure 8-17. Vehicles have been designed to travel at speeds of more than 130 m/sec (see, e.g., Laithwaite, 1977); Rhodes and Mulhall, 1981). In these applications, the conducting guideway can be a periodic set of passive coils or a continuous

Figure 8-16 Magnetically levitated vehicle using superconducting magnets—repulsive method. (Photo courtesy of Japan National Railways.)

sheet "track", as shown in Figure 8-18. The techniques for calculating the magnetic forces are numerous and employ circuit models (Iwamoto et al., 1974), Fourier analysis (Reitz and Davis, 1972), image methods (Reitz, 1970), and numerical methods. One important model that is also used in the study of linear induction motors is the continuous thin track approximation. We will describe this analysis technique in the next section for long current filaments moving over a thin sheet conductor. We will extend the analysis to a vibrating filament moving over a conducting sheet. This problem illustrates a little-known phenomenon in magnetomechanical systems, that is, negative damping.

The dynamics of magnetically levitated vehicles involve some interesting problems, including static and dynamic instabilities (see, e.g., Davis and Wilkie, 1971); Wilkie, 1972; Moon, 1977; Chu and Moon, 1983). However, we do not have the space in this book to properly treat these problems.

Figure 8-17 Magnetically levitated model on a rotating wheel using permanent magnets—repulsive method, at 40 m/sec. (Cornell University.)

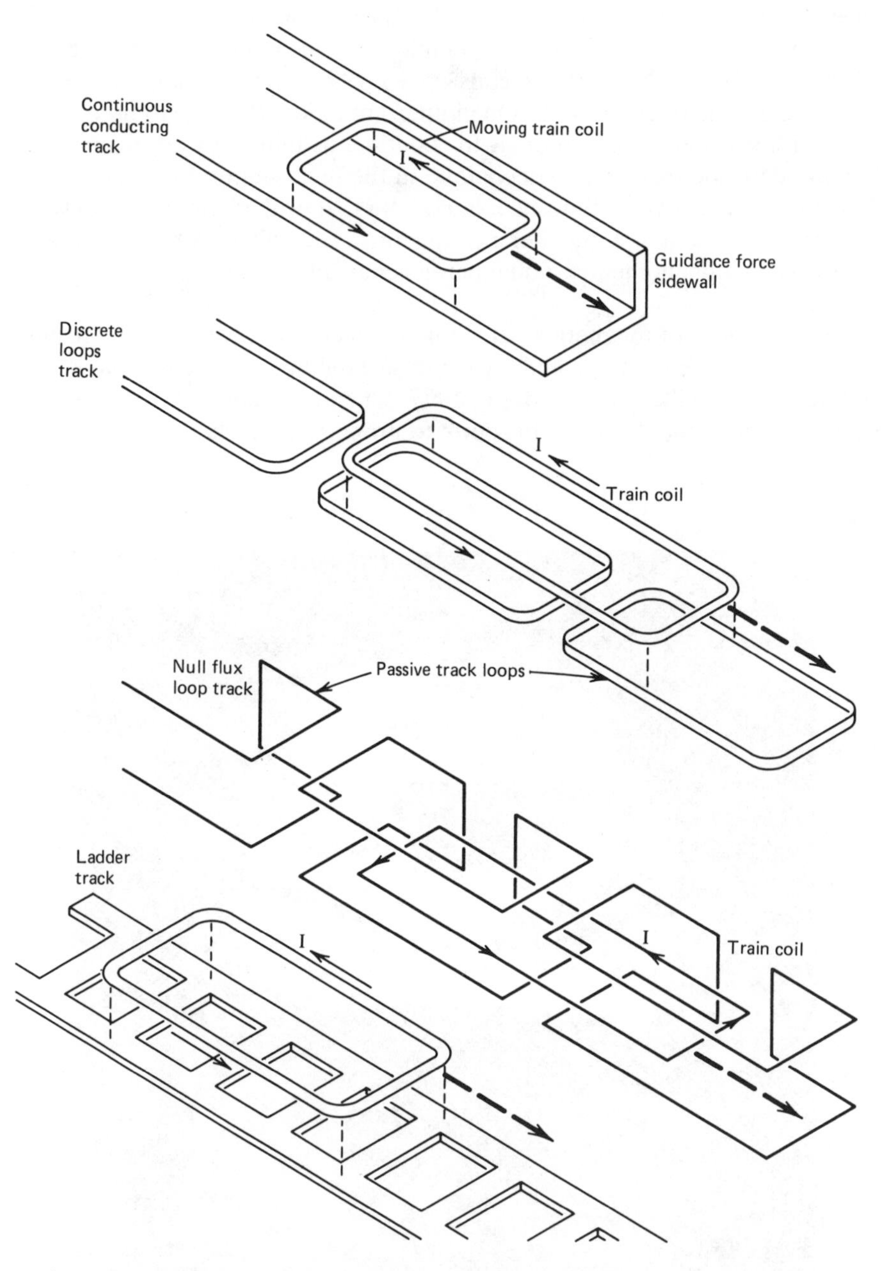

Figure 8-18 Track configurations for repulsive magnetic levitation of vehicles.

Moving Filament over a Conducting Sheet Track—Steady Theory

We imagine a rectangular coil moving over a flat conducting sheet track with constant speed v at a constant height h. To simplify the analysis, we focus only on the filaments moving transverse to the motion. As a further approximation, we consider only one-dimensional current flow in this conducting plate, that is, $\mathbf{J} = J\mathbf{e}_z$, as shown in Figure 8-19.

Although the filament is moving to the right with speed v, the current distribution in the track and the magnetic field pattern are steady relative to the current filament. Thus, in the moving system, $\partial B/\partial t = 0$. One could use Eq. (8-2.10), but we derive a simpler equation for the induced currents in this section.

This resulting form of Faraday's law (8-2.8a) under the above assumptions becomes

$$\frac{\partial E_z}{\partial x} = 0, \qquad \frac{\partial E_z}{\partial y} = 0$$

We assume that as $x \to \pm\infty$, $E_z \to 0$, so that $E_z = 0$. Using Ohm's law (8-2.9b) for a moving conductor, we have

$$J_z = -\sigma v B_y \tag{8-8.1}$$

It is further assumed that J_z is uniform across the track thickness. We introduce a current density per unit length $K = J_z \Delta$, where Δ is the track thickness.

The total magnetic field B_y is the sum of the induction field of the filament B^0 and an induced field B^1 due to the distribution $K(x)$. Thus Eq. (8-8.1) assumes the form

$$K = -\sigma \Delta v (B^0 + B^1) \tag{8-8.2}$$

To complete the equation necessary to derive the current density $K(x)$, we need a relation between J and B^1 or Ampere's law (8-2.8b). The integrated form of Ampere's law is called the Biot–Savart integral [Eq. (8-4.23)]. This relation may be simplified by replacing the three-dimensional current distribution by a distribution of current filaments. The vertical component of the induced field at position x due to a current filament at η of strength $K d\eta$ is then given by

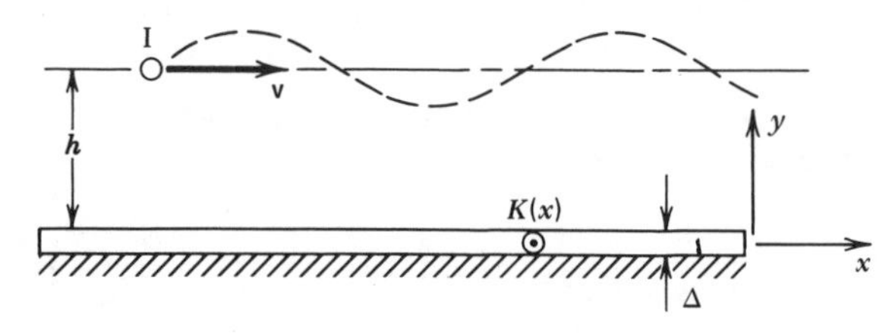

Figure 8-19 Oscillating current filament moving over a conducting plate guideway.

$$dB^1 = \frac{\mu_0 K d\eta}{2\pi(x-\eta)}$$

The total integral of B^1 therefore involves a singular integrand $1/(x-\eta)$. [Compare this with Eq. (8-4.26) where a different method was used.) It is assumed that $K(\eta)$ is everywhere continuous so that the Cauchy principal value is taken; that is,

$$B^1 = \frac{\mu_0}{2\pi} \mathrm{P}\int_{-\infty}^{\infty} \frac{K(\eta)d\eta}{x-\eta} \tag{8-8.3}$$

Substituting this expression into Eq. (8-8.2) we obtain an integral equation for $K(\eta)$

$$K(x) + \frac{v}{w\pi} \mathrm{P}\int_{-\infty}^{\infty} \frac{K(\eta)d\eta}{x-\eta} = -\frac{2v}{w}\frac{B^0(x)}{\mu_0} \tag{8-8.4}$$

where $w = 2/\mu_0\sigma\Delta$ is a characteristic velocity and the ratio v/w is sometimes called the magnetic Reynolds number (see Section 8.6).

This equation has a simple solution which may be obtained by using the technique of Fourier transforms. If we define the Fourier transform of K as $T[K]$, then the following relations are useful:

$$T[K] = \int_{-\infty}^{\infty} e^{ikx} K(x)\,dx$$

$$T\left[\int_{-\infty}^{\infty} K \frac{1}{x-\eta}\,d\eta\right] = T[K]T\left[\frac{1}{x}\right]$$

$$T\left[\frac{1}{x}\right] = i\pi, \qquad k>0 \tag{8-8.5}$$

$$T\left[\frac{1}{x}\right] = -i\pi, \qquad k<0$$

$$T\left[\frac{1}{x}\right]T^*\left[\frac{1}{x}\right] = \pi^2$$

where the asterisk indicates the complex conjugate of a function. Using these relations, the transform of Eq. (8-8.4) becomes

$$T[K] = -\frac{2v}{\mu_0 w}\frac{T[B_y^0]}{1+v^2/w^2}\left(1 - \frac{v}{\pi w}T\left[\frac{1}{x}\right]\right) \tag{8-8.6}$$

Using the convolution theorem for Fourier transforms, an expression for $K(x)$ is obtained in terms of B_y^0,

$$K(x) = -\frac{2}{\mu_0}\frac{v}{w}\frac{1}{1+v^2/w^2}\left(B_y^0(x) - \frac{v}{\pi w}\int_{-\infty}^{\infty}\frac{B_y^0}{n-x}\,d\eta\right) \tag{8-8.7}$$

This expression is not restricted to a single moving filament but is valid for all systems of moving field sources provided they are of limited extent, produce a one-dimensional current pattern, and travel at the same velocity.

If we use the fact that $\mathbf{B}^0$ can be derived from a potential function, it can be shown that the integral in Eq. (8-8.7) can be related to the x component of the field; that is,

$$K(x) = -\frac{2}{\mu_0}\frac{v}{w}\frac{B_{0y} + B_{0x}v/w}{1 + v^2/w^2} \tag{8-8.8}$$

For a single current filament of strength I_0 at height h above the track,

$$B_y^0 = \frac{\mu_0 I}{2\pi}\frac{x}{x^2+h^2}, \qquad B_x^0 = \frac{\mu_0 I}{2\pi}\frac{h}{x^2+h^2} \tag{8-8.9}$$

The lift and drag forces on the moving filament are equal and opposite to the resultant forces on the track. The vector force distribution on the track is

$$\mathbf{f} = K(-B_y^0, B_x^0)$$

The vertical component is sometimes called the magnetic pressure and is shown plotted in Figure 8-20. The integral of these forces over the track gives the total lift and drag on the filament; that is,

$$F_y = \frac{\mu_0 I_0^2}{4\pi h}\frac{v^2}{w^2+v^2}$$
$$F_x = \frac{w}{v}F_y \tag{8-8.10}$$

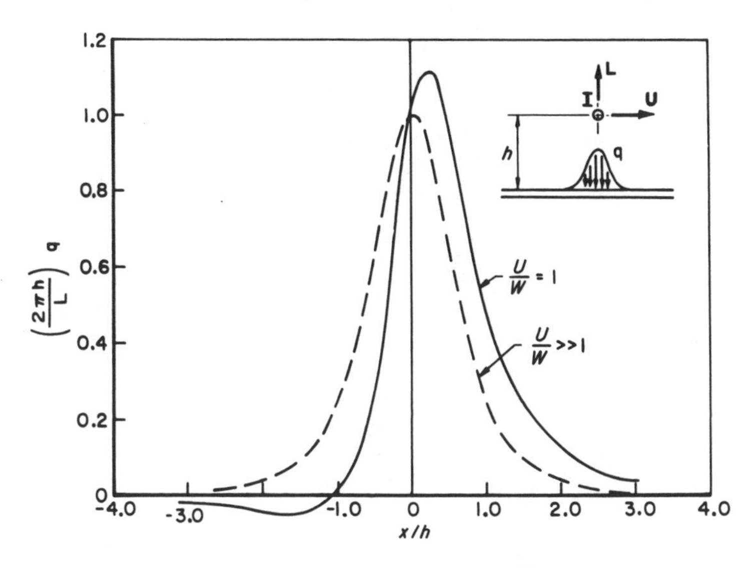

Figure 8-20 Magnetically induced pressure on a conducting plate due to a moving current filament.

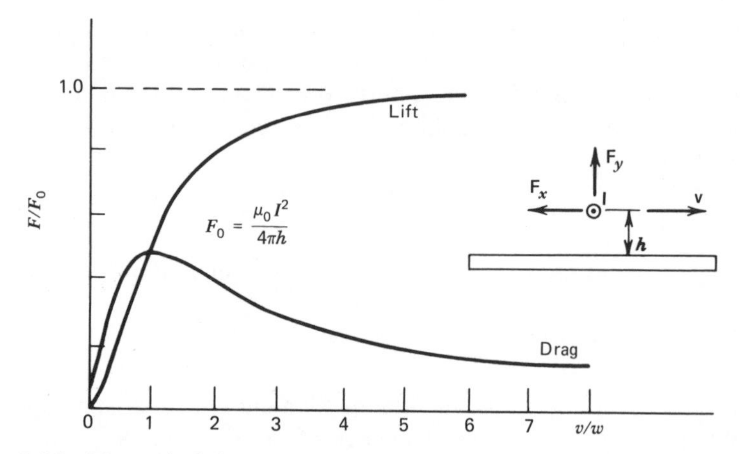

Figure 8-21 Magnetic left and drag forces vs. normalized velocity for a current filament moving over a conducting plate track.

The speed dependence of the lift and drag forces is shown in Figure 8-21 and shows a drag maximum at $v = w$. The high velocity lift value may be derived from an image method in which a filament of opposite current is placed directly under the track at a distance h. The resulting field pattern above the track at high speeds has zero vertical field component as shown in Figure 8-22.

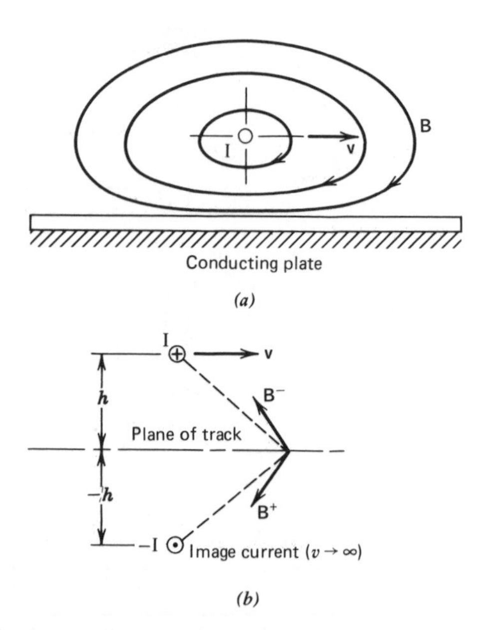

Figure 8-22 (*a*) High-speed magnetic field configuration near a conducting plate and (*b*) current filament and image current, high-speed limit.

Vibrating Current Filament—Negative Damping

To study the dynamics of levitated bodies, it is necessary to understand the dynamic magnetic forces resulting from oscillations of the vehicle magnets with respect to the conducting guideway. In the case of the previous example of a moving filament, we wish to study the induced currents and forces when the height varies with time; that is,

$$h = h_0 + u_0 e^{i\omega t}, \qquad u_0 \ll h_0$$

The perturbed magnetic field to terms linear in u_0 becomes

$$\mathbf{B}(x, t) = \mathbf{B}^0(x) + \mathbf{b}^0(x) e^{i\omega t} \tag{8-8.11}$$

where

$$b_y^0 = -\frac{\mu_0 I_0 x u_0 h_0}{\pi(h_0^2 + x^2)^2}$$

Note, higher-order terms in u_0 have been dropped.

This perturbation field will induce a time-varying current distribution

$$K(x, t) = K_0(x) + K_1(x) e^{i\omega t} \tag{8-8.12}$$

where K_0 was derived in the previous section.

This problem was studied by the author (Moon, 1977) and we will only summarize the results here. Other studies of negative damping due to a vibrating magnet have been done by Davis and Wilkie (1971), Iwamoto et al. (1974), and Baiko et al. (1980).

The perturbation current density K_1 can be shown to satisfy the following integrodifferential equation under similar assumptions used to find K_0; that is,

$$\frac{dK_1}{dx} + \frac{\mu_0 \sigma \Delta}{2\pi} \mathrm{P}\!\int_{-\infty}^{\infty} \left(v \frac{dK_1}{d\eta} - i\omega K_1(\eta) \right) \frac{d\eta}{x - \eta} = \sigma \Delta \left(i\omega b_{0y} - v \frac{db_{0y}}{dx} \right) \tag{8-8.13}$$

A Fourier-transform technique can again be used to solve this equation. The resulting expression for $T[K]$ is given by

$$T[K_1] = -\frac{2}{\mu_0 w} \frac{(v + \omega/k) T[B_{0y}][1 - (v + \omega/k) T[1/x]/\pi w]}{1 + (v + \omega/k)^2/w^2} \tag{8-8.14}$$

The object of this exercise is to demonstrate that negative damping is possible in convective magnetomechanical systems. However, with ω real we have a forced vibration problem. To obtain a free vibration problem, we consider ω as a Fourier-transform parameter resulting from a transform parameter on the time domain. K_1 should then be interpreted as a time Fourier transform of the desired current distribution, that is, $K_1 = K_1(x, \omega)$. From Eq. (8-8.14) we observe that the inversion of $K_1(\omega)$ is a complicated convolution integral since ω appears in both the denominator as well as the numerator. Thus the dissipation is not simply related to the velocity of the vertical oscillation $u = u_0 e^{i\omega t}$. To obtain an approximation, we expand $T[K_1]$ in powers of ω,

$$T[K_1] \cong T[K_1]\Big|_{\omega=0} + \omega \frac{dT[K_1]}{d\omega}\Big|_{\omega=0} \tag{8-8.15}$$

It can be shown that the damping force is proportional to the second term on the right-hand side of Eq. (8-8.15) where

$$\frac{d}{d\omega} T[K_1]\Big|_{\omega=0} = \frac{\sigma\Delta[1-(v/w)^2]}{[1+(v/w)^2]k} T[b_{0y}] \tag{8-8.16}$$

Thus the effective damping will change sign depending on whether the velocity v is greater or less than the characteristic velocity $w = 2/\mu_0\sigma\Delta$.

To calculate the perturbed force on the wire F'_y we proceed as in previous section, using the force on the track. The resulting expression shows clearly the possibility of *negative damping*:

$$F'_y = -\frac{\mu_0 I^2}{4\pi h_0}\left(\frac{v^2}{v^2+w^2}\frac{u}{h_0} + \frac{w^2-v^2}{(w^2+v^2)^2}\frac{w\,du}{dt}\right) \tag{8-8.17}$$

where $u = u_0 e^{i\omega t}$.

The damping force is shown plotted in Figure 8-23. Experiments by

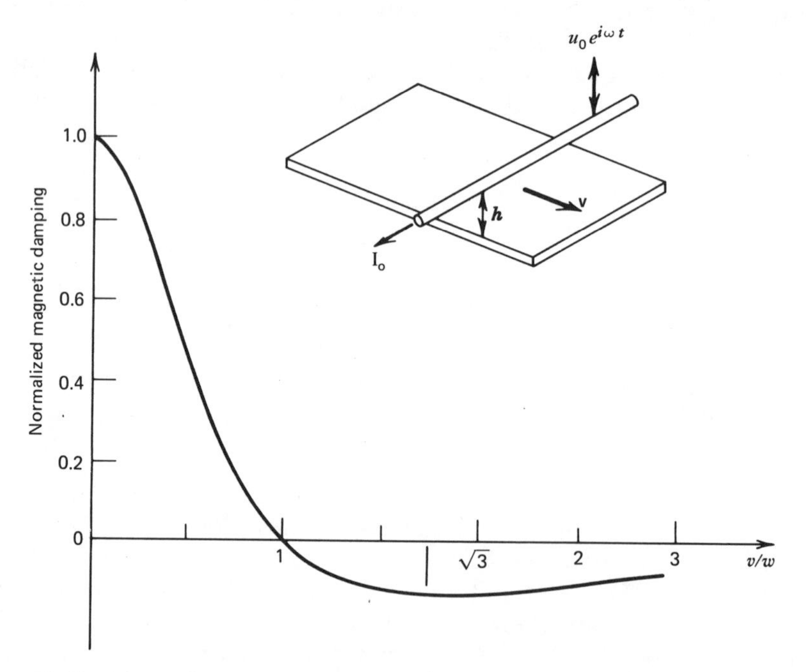

Figure 8-23 Magnetic damping constant vs. speed for a vibrating wire moving over a conducting plate track.

Iwamoto et al. (1974) and Moon (1977) have demonstrated this phenomenon. In real systems the total damping of a vehicle will involve aerodynamic as well as magnetic damping so that while the total damping may decrease with speed, it will not become negative because of the increase of aerodynamic damping forces.

8.9 ELECTROMAGNETIC LAUNCHERS

Much of the development of energy-intensive magnetomechanical devices in the late 19th and early 20th centuries was devoted to rotating machinery. In the 1960s, however, increased research and development efforts were directed toward linear machines related to the propulsion of high-speed vehicles, including linear induction and linear synchronous motors for propelling vehicles up to 500 km/h (see, e.g., Nasar and Boldea, 1976; Laithwaite, 1977; Rhodes and Mulhall, 1981). In the late 1970s a new phase of this research focused on electromagnetic launchers, magnetic "cannons," and magnetic *mass drivers*. The purpose of these devices is to accelerate massive objects.† Although several papers on this subject had appeared in the literature before this period, for example, Winterberg (1966), one of the early efforts involved space applications, such as launching material from the moon, or moving asteroids around in space. This effort was led by O'Neill from Princeton University and Kolm of MIT and co-workers (see, e.g., Chilton et al., 1977; Snow et al., 1982). They named these devices mass drivers. In one design a superconducting coil would interact with a linear series of pulsed thruster coils (see Fig. 8-24). Each thruster pulse is designed to be initiated when the force between the driving and moving coils is at a maximum. The designers thus claim a very high-energy conversion efficiency.

Another device developed in this period was the *rail gun* (see, e.g., Rashleigh and Marshall, 1978; Hawke and Scudder, 1979; Deadrich et al., 1982). Research on this launcher was motivated by potential military applications (see Fig. 8-25). In one design a 0.3-kg projectile in a 4-m-long rail gun is to be accelerated to 3 km/sec using a current pulse of 1.5-MA along the rails (see Deis and McNab, 1982). The accelerating magnetic force uses the flux trapped between two parallel conductors. There are several design variations. In one design the rails are shorted by an arc which moves behind the projective. The $\mathbf{J} \times \mathbf{B}$ force on the moving arc accelerates the projectile. In another design the rails are shorted by the projectile itself.

A third variation, based on the O'Neill/Kolm mass driver, is called a *θ gun* and is a linear induction device (see Fig. 8-26; also Burgess et al., 1982). As in the mass driver, the accelerator has a series of pulser coils. But the currents

†In the 1960s NASA had a program to develop electromagnetic propulsion devices (see Jahn, 1968). These were mainly plasma devices and differ from those discussed in this section.

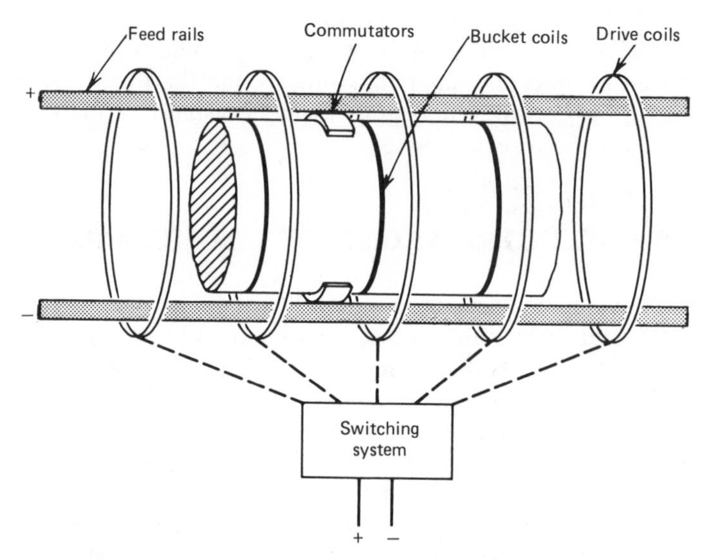

Figure 8-24 Coaxial magnetic accelerator or mass driver. (From Kolm and Mongeau, IEEE Spectrum, April 1982, reprinted with permission. IEEE, copyright 1982.)

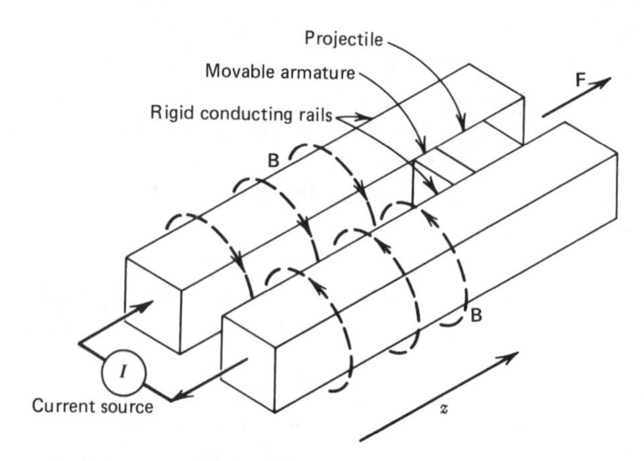

Figure 8-25 Sketch of rail gun geometry. (From Deadrich et al., 1982, reprinted with permission. IEEE, copyright 1982.)

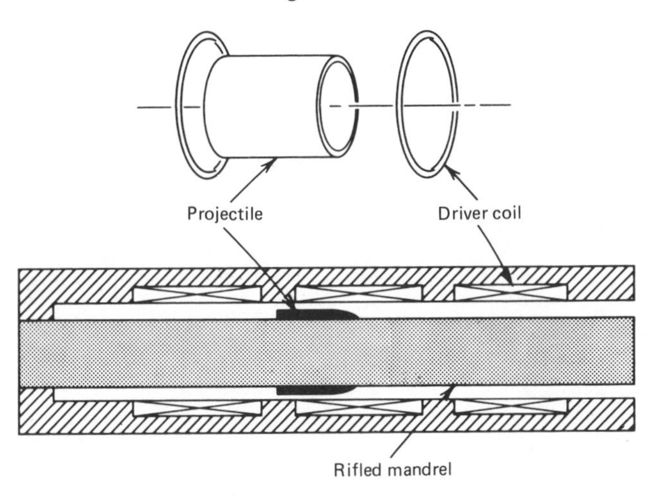

Figure 8-26 Sketch of θ-gun mass driver. (From Burgess et al., 1982, reprinted with permission. IEEE, copyright 1982.)

in the projectile are induced by the motion and the time-changing driver current.

Magnetic accelerators are also of interest for impact fusion (see, e.g., Kostoff et al., 1982).

Inductive Magnetic Accelerator Equations

To illustrate the analysis required to design an accelerator device, we consider the special case of an inductive circular coil driver shown in Figure 8-27*a*. In this example we assume a voltage is applied to the circuit of coil 1, while the current in the accelerating coil is induced by the rate of change of I_1 and the motion of the coil itself. A discrete circuit model is shown in Figure 8-27*b* in which the mutual inductance between the two coils is a function of coil separation z. To analyze the dynamics we can employ Newton's law coupled with Kirchhoff's equations for circuits, or we can use a Lagrangian approach.

The Lagrangian is the sum of the kinetic energy and magnetic energy functions (see Section 3.5),

$$\mathscr{L} = \mathscr{T} + \mathscr{W}_m$$

The dissipation function for the system will incorporate the resistive losses, but will neglect mechanical damping losses such as air drag. No energy is assumed to be stored in elastic energy. The equations become

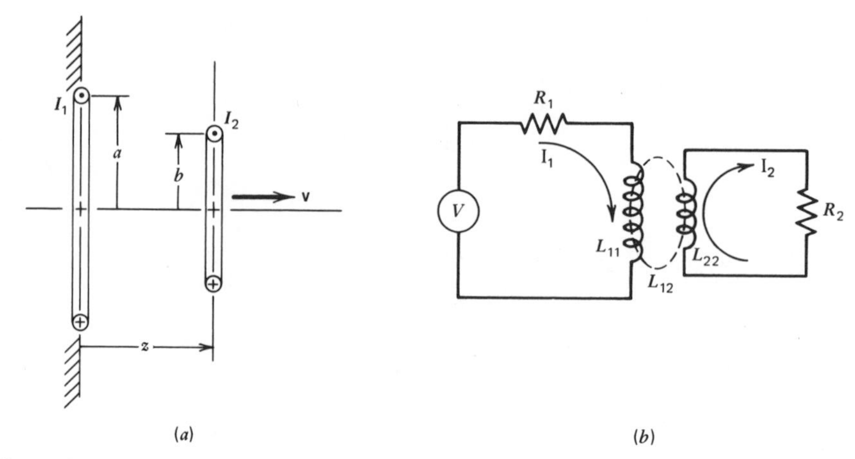

Figure 8-27 Two-coil magnetic accelerator: (*a*) geometry of physical system and (*b*) equivalent circuit of driver and accelerating coils.

$$\frac{d}{dt}\frac{\partial \mathscr{L}(z, \dot{z}, I_1, I_2)}{\partial \dot{z}} - \frac{\partial \mathscr{L}}{\partial z} = -\frac{\partial \mathscr{R}}{\partial \dot{z}}$$

$$\frac{d}{dt}\frac{\mathscr{L}(z, \dot{z}, I_1, I_2)}{\partial I_1} = -\frac{\partial \mathscr{R}}{\partial I_1} + V \tag{8-9.1}$$

and

$$\frac{d}{dt}\frac{\partial \mathscr{L}}{\partial I_2} = -\frac{\partial \mathscr{R}}{\partial I_2}$$

where

$$\mathscr{T} = \tfrac{1}{2}m\dot{z}^2, \qquad \mathscr{W}_m = \tfrac{1}{2}(L_{11}I_1^2 + 2L_{12}I_1I_2 + L_{22}I_2^2)$$

and

$$\mathscr{R} = \tfrac{1}{2}R_1I_1^2 + \tfrac{1}{2}R_2I_2^2$$

The equations then take the form of coupled nonlinear equations

$$m\ddot{z} = I_1I_2L'_{12}$$

$$L_{11}\dot{I}_1 + L_{12}\dot{I}_2 + R_1I_1 + I_2L'_{12}\dot{z} = V \tag{8-9.2}$$

and

$$L_{12}\dot{I}_1 + L_{22}\dot{I}_2 + R_2I_2 + I_1L'_{12}\dot{z} = 0$$

These equations show that the induced current is due to two effects—one proportional to the change of flux or current in the driver circuit $\dot{I}_1$ and the other due to the motion of the coil $\dot{z}$. (Note $L'_{12} \equiv dL_{12}/dz$.)

When the currents in the two coils are fixed, the magnetic force between the coils as a function of axial displacement is shown in Figure 8-28. The nonlinear nature of the mutual force and the existence of a force maximum at a particular coil separation are important characteristics of these systems.

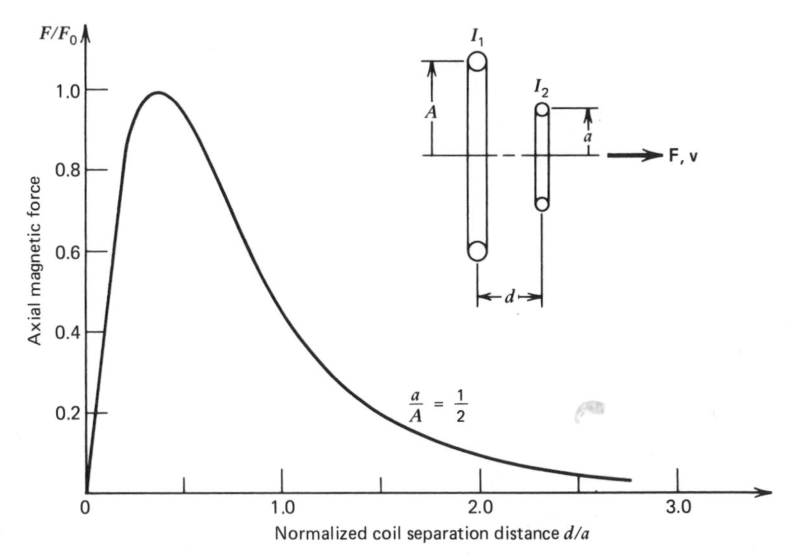

Figure 8-28 Normalized magnetic force between two coaxial circular coils vs. axial distance.

The special case when the current in the drive coil is specified is considered. The equations for a single accelerator stage are given by

$$I_1 = I_0 \sin\frac{\pi t}{\tau_0}, \qquad 0 \le t \le \tau_0, \qquad I_1 = 0, \qquad t > \tau_0 \tag{8-9.3a}$$

$$\frac{dz}{dt} = v \tag{8-9.3b}$$

$$m\frac{dv}{dt} = I_1 I_2 L_{12}' \tag{8-9.3c}$$

and

$$L_{22}\frac{dI_2}{dt} = -R_2 I_2 - L_{12}\frac{dI_1}{dt} - L_{12}' v I_1 \tag{8-9.3d}$$

These nonlinear equations may be integrated numerically. The induced current and velocity history for a particular example is shown in Figure 8-29. Of particular interest is the current reversal in the induced current which will produce a negative acceleration at the end of the drive coil pulse.

An approximate analytical solution may be obtained when the pulse time is short compared with the distance the coil moves. During the pulse we assume $\dot{z} = v_0$ is a constant, as well as the quantities L_{12} and L_{12}'. At the end of the pulse, the impulse imparted to the coil is calculated and the velocity is updated for the next accelerator stage using the relation

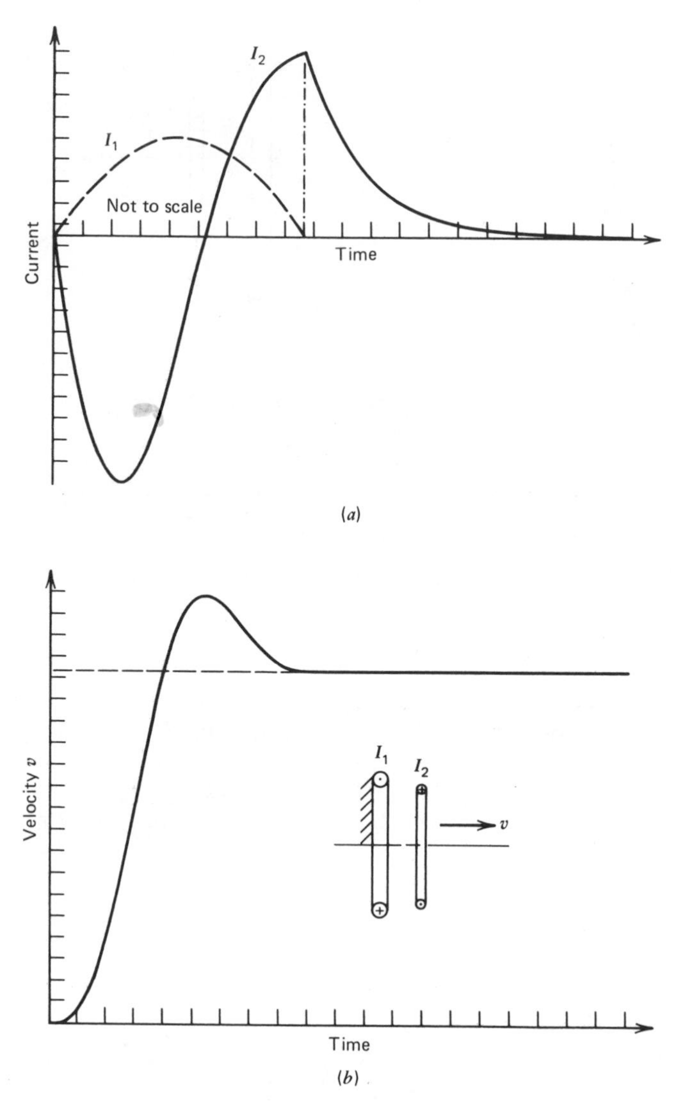

Figure 8-29 Dynamic response of an accelerated coil (see Fig. 8-27) due to a half-sine pulsed current in driver coil: (*a*) induced current in accelerated coil vs. time and (*b*) axial velocity of accelerated coil vs. time.

$$\Delta v = \frac{1}{m} \int_0^{\tau_0} I_1 I_2 L'_{12} \, dt \qquad (8\text{-}9.4)$$

Under these assumptions, the first-order differential equation for I_2 may be integrated to obtain

$$I_2 = -\frac{I_0\beta}{(1+\beta^2)^{1/2}}\left[\left(\frac{L_{12}}{L_{22}}\right)^2 + \hat{v}^2\right]^{1/2}\left[\cos\left(\frac{\pi t}{\tau_0} + \phi_0\right) - \cos\phi_0 e^{-t/\tau_2}\right] \tag{8-9.5}$$

where the following expressions are defined:

$$\tau_2 = \frac{L_{22}}{R_2}, \qquad \beta = \frac{\pi\tau_2}{\tau_0}$$

$$\tan\phi_0 = -(\beta - \hat{v})(1 + \beta\hat{v})^{-1}$$

and

$$\hat{v} = \frac{-v_0\tau_0 L'_{12}}{L_{12}\pi}$$

One can show that the velocity reduces the impulse. There is, in fact, a limiting velocity when $\hat{v} = 1$ for which the time-changing flux and the velocity-produced flux change cancel each other out in the right-hand side of Eq. (8-9.3d).

Simplified Rail Gun Analysis

Another magnetic mass accelerator device to which one can apply analytical techniques is the parallel rail gun shown in Figure 8-25. In this device, current is stored in a coil of inductance L and resistance R. The moving mass travels between two conducting rails whose inductance $L_r(x)$ and resistance $R_r(x)$ change with the displacement of the mass. The equations of motion for the circuit and mass are given by

$$L_0\frac{dI}{dt} + R_0 I + \frac{d}{dt}(L_r I) + R_r I = 0$$

and (8-9.6)

$$m\frac{dv}{dt} = \frac{I^2}{2}\frac{dL_r}{dx} - cv$$

For simplicity, we will neglect the resistive losses in the rails and the mass damping, that is, set $R_r = c = 0$. We also make the following approximations:

$$L_r = L_1 x$$

and

$$\frac{L_1}{L_0} \ll 1$$

The resulting simplified equations become

$$\frac{dI}{dt} = -\frac{R_0}{L_0} I - \frac{L_1}{L_0} Iv$$

and (8-9.7)

$$\frac{dv}{dt} = \frac{1}{2m} L_1 I^2$$

with initial conditions $I(0) = I_0$ and $v(0) = 0$.

This set of nonlinear differential equations may be solved in the $I^2 - v$ plane or phase plane. It is easy to see that the $I = 0$ axis is an infinite set of equilibrium points. From Eq. (8-9.7) one can show that trajectories in the (I^2, v) plane have slopes given by

$$\frac{dI^2}{dv} = -\frac{4m}{L_0 L_1}(R_0 + L_1 v)$$

This may be easily integrated to obtain

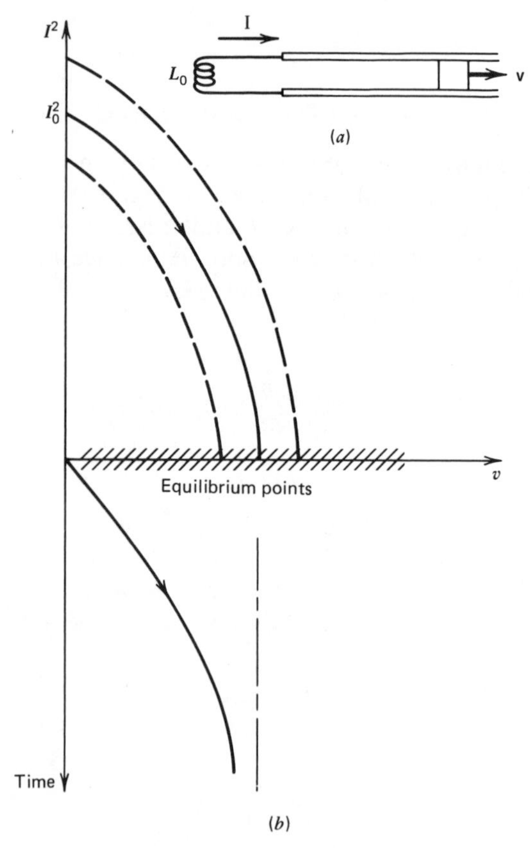

Figure 8-30 Current vs. velocity for an idealized rail gun driven by an inductive energy source [Eq. (8-9.8)].

$$I^2 = I_0^2 - \frac{4m}{L_0 L_1}\left(R_0 v + \frac{L_1 v^2}{2}\right) \tag{8-9.8}$$

The trajectories are parabolas in the phase plane as shown in Figure 8-30(*a*). The time history of the velocity of the projectile may be obtained by solving the equation

$$\frac{dv}{dt} = \frac{1}{2}\frac{L_1 I_0^2}{m} - \frac{2R_0 v}{L_0} - \frac{L_1}{L_0}v^2 \tag{8-9.9}$$

It can readily be shown that the solution is given by

$$v = \frac{L_0}{2L_1}\left(\Gamma \tanh\frac{\Gamma(t-t_0)}{2} - \frac{2R_0}{L_0}\right) \tag{8-9.10}$$

where

$$\Gamma = \left[4\left(\frac{R_0}{L_0}\right)^2 + \frac{2I_0^2 L_1^2}{mL_0}\right]^{1/2}$$

and

$$t_0 = \frac{2}{\Gamma}\tanh^{-1}\left(\frac{-2R_0}{L_0\Gamma}\right)$$

This function is plotted in Figure 8-30(*b*) and shows an asymptotic velocity as $t \to \infty$ given by

$$v_\infty = \frac{L_0}{2L_1}\left(\Gamma - \frac{2R_0}{L_0}\right) \tag{8-9.11}$$

Of course the analysis of an actual rail gun must account for the rail resistance, including skin effects and rail friction. In an arc-driven rail gun, the dynamics of arc behavior must also be determined (see, e.g., Powell and Batteh, 1982). In one version of rail gun designs, the rails are accelerated toward one another with explosives behind the mass in order to compress the flux and obtain higher magnetic pressure on the projectile (see Deadrich et al., 1982; also Fig. 9-3).

8.10 DYNAMICS OF DEFORMABLE SOLIDS IN MAGNETIC FIELDS

For transient or pulsed magnetic field problems, the use of rigid-body mechanics can only be applied if the characteristic time constant of the field pulse t (rise time or pulse length) satisfies the following identity,

$$t \gg \frac{L}{C_E}$$

where C_E is one of the wave speeds associated with the elastic solid and L is a characteristic length of the solid. For shorter-field pulse times, the deformable properties of the solid become important. The effects of magnetic forces take a finite time to propagate through the solid in the form of stress waves.

The study of dynamic magnetic forces in deformable solids is of importance to the following problems:

1. Stress wave propagation in steady magnetic fields.
2. Magnetic forming of conducting solids.
3. Impulse testing of solids and structures.
4. Induced stress waves using pulsed magnetic fields.

A guide to some of the literature in the first three areas will be presented. A more detailed discussion of the calculation of stress waves using pulsed fields will then be discussed. In this problem, the coupling between field and deformation is one way; that is, we calculate the induced eddy currents as if the conductor were rigid, and then use the calculated magnetic forces to determine the nature of the stress waves.

Mutual coupling between the magnetic field and the deformable solid is of some importance in the dynamic testing of flexible structures where both the velocity of the solid and the change of relative geometry of conductor and field source affect the induced currents. One problem of this nature has been solved by Yuan (1981) using numerical methods (see also Yuan et al., 1982).

Stress Waves and Magnetic Fields

The study of stress waves in ferromagnetic materials for solid-state electronic devices has received a great deal of attention. Theoretical models for this phenomenon incorporate the dynamics of the spin continuum and the elastic continuum. The subject has been treated extensively elsewhere and will not be reviewed here (see, e.g., Tiersten, 1964).

One dynamic magnetoelastic problem that has a relatively large literature is the propagation of stress waves in the presence of a large magnetic or electric field. A review of many of these works has been given by Paria (1967), which includes references to the Polish school under Kaliski (see, e.g., Kaliski and Nowacki 1965). One of the earliest studies was by Knopoff (1955), who attempted to estimate the effect of the earth's magnetic field on the propagation of seismic waves. His conclusion was that the effect was negligible. Experiments on stress wave propagation in nonferromagnetic conductors have demonstrated that only a small change in wave-speed results (Robey, 1953; Galkin and Koroliuk, 1958; Alers and Fleury, 1963). A more significant effect is the damping of elastic waves due to the induction of eddy currents (see, e.g., Dunkin and Eringen, 1963). Wave propagation in a soft ferromagnetic elastic conductor was studied by Pao and Hutter (1973) from the general nonlinear equation of continuum mechanics. An earlier study of waves with discontinuities (singular surfaces) in nonlinear magnetoelastic continua was done by McCarthy (1966a, b) and later by Chen and McCarthy (1975).

Another class of problems involves the generation of elastic waves and vibrations by a time-varying or pulsed magnetic field. Two works by the author and students involve an applied transient magnetic field at the boundary of an elastic conductor. The resulting phenomena are eddy-current induction in a boundary layer near the surface, magnetic body force generated waves, Joule heating, and thermoelastic waves. In Moon and Chattopadhyay (1974) the problem of a step change in magnetic field at the surface of an elastic half-space was analyzed, and in Chian and Moon (1981) the magnetic generation of elastic waves in a cylindrical space was analyzed. In both works experiments confirmed the prediction of $\mathbf{J} \times \mathbf{B}$ generated waves. However, the predicted thermoelastic waves were not observed. This subject will be discussed below. The magnetic field rise time and characteristic thermal time in the material are found to be important in determining the size of the thermoelastic wave. A description of magnetic field stress wave generation has been given by Snell et al. (1973).

Magnetically generated vibrations in a conducting cylinder were studied by Miya et al. (1980) using the finite-element method for the field and deformation. Experimental confirmation was also reported.

Magnetic Forming

Pulsed magnetic field devices to deform metal were developed in the 1960s. Transient currents in a multiturn coil were used to produce a time-varying field pattern near a metallic structure. The induced currents in the workpiece created a magnetic pressure which was used to deform the part or accelerate the piece against a forming die (see Fig. 8-31). The method worked well for nonferromagnetic metals; however, the conversion of magnetic energy to work of deformation was at best only 15% efficient. The devices usually stored electric energy in a large capacitance bank (20–100 kJ). When the bank was shorted by the forming coil, large currents would flow in the circuit. Other energy storage devices for magnetic forming include rotating motors or flywheels and superconducting energy storage magnets.

An early study of the physics of pulsed magnets and the effects of high transient magnetic fields on solid conductors was given by Furth et al., (1957a, b). A description of the magnetic forming process may be found in Birdsall et al. (1961). The development of the technique was funded by NASA and is reported in a number of obscure reports. Little work on the magnetomechanics of the process was done in this period, however. Lippman and Schreiner (1964) gave an analytical estimate of the velocity of magnetically imploded tubes. An experimental study of forming efficiency and stresses in flat plates and cylinders has been given by Baines et al. (1965–

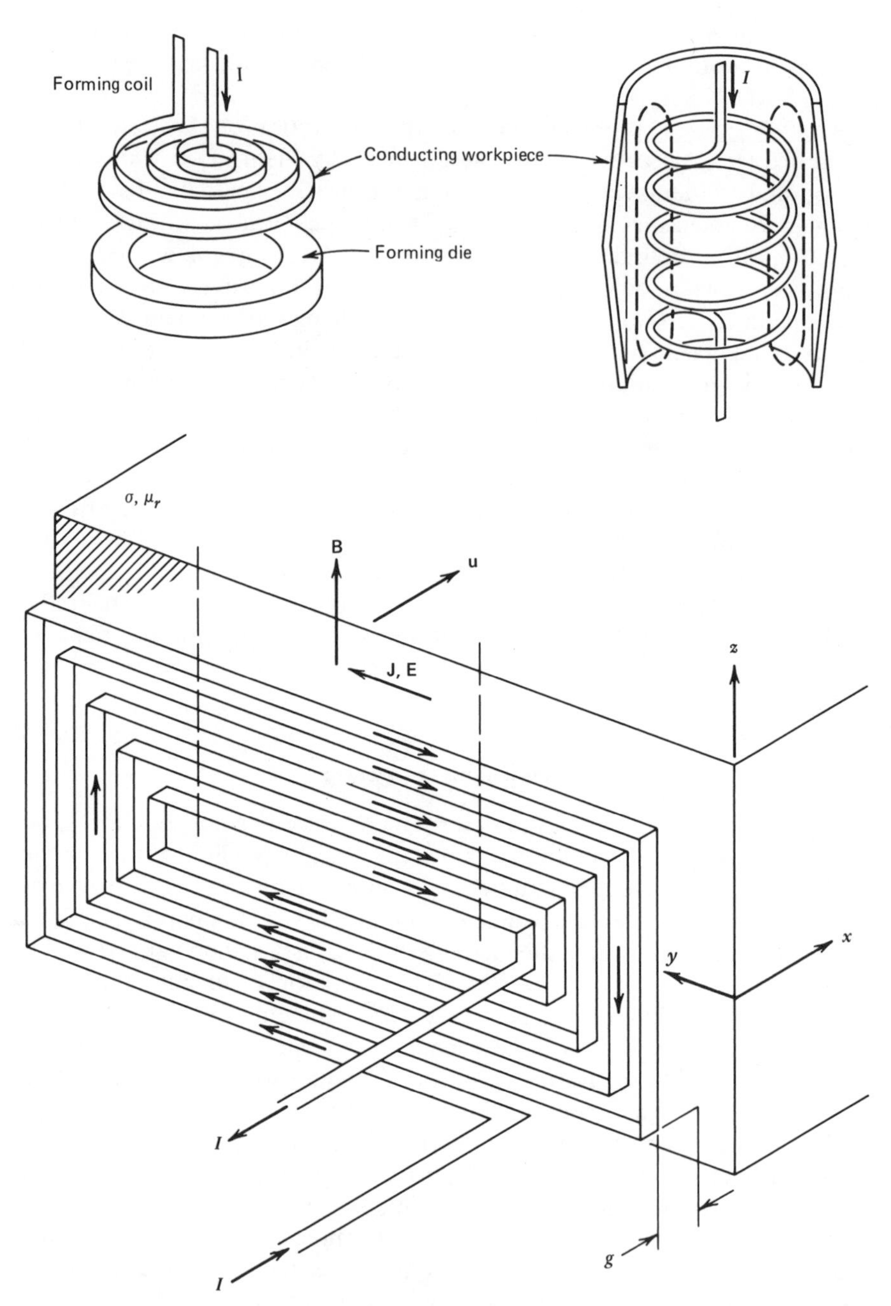

Figure 8-31 Magnetic forming coil-workpiece configurations.

1966). Their estimates of the ratio of plastic work to stored electrical energy were found to be between 6 to 14%.

Details of the magnetomechanics, including calculations of eddy currents, field diffusion, and force distributions, may be found in Hillier and Lal (1968). Another study of the magnetic forming process is by Al-Hassani et al. (1974), which includes a number of references. Other references are given in the 1978 review on magneto-solid mechanics by the author. Although a promising application, little work on the subject of magnetic forming from 1976–1984 has been reported.

Magnetic Impulse Testing of Solids

Another application of pulsed magnetic fields is the testing of the dynamics of solids and structures under distributed impulsive forces. For example, Al-Hassani and Johnson (1970) studied the deformation of beams and frame structures under pulsed forces generated by discharging a capacitance bank though a coil placed near the structure. Plastic buckling of thin-walled tubes was also studied using this technique (Al-Hassani, 1974) as well as large deformations of frame structures under pulsed forces (Hashmi and Al-Hassani 1975). Walling and Forrestal (1973) investigated the elastoplastic behavior of aluminum rings using pulsed magnetic forces of microsecond duration. A short-duration force on the side of a thin shell was generated in a similar manner by Forrestal and Overmier (1974). The strength of a single-turn conducting ring under a pulsed current within the ring itself was investigated by Herlach and McBroom (1973) using pulsed fields of the order of 100 T.

The effects of the deformation and velocity of the structure on the magnetic forces and motion of a cantilevered plate was recently published by Yuan et al. (1983). In this work a finite-element method was employed which calculated both the induced currents and the deformation. Experiments were also performed which verified the calculation.

The technique of using pulsed magnetic fields to test the response to short-duration stresses and forces has the advantage over other methods of being noncontacting and capable of rise times of the order of microseconds. For stress rise times of shorter duration, however, the laser has proved to be more efficacious.

Magnetically Induced Stress Waves

When a transient magnetic field is applied to the surface of an elastic conductor the following physical phenomena can occur:

1. Induction of eddy currents near the surface (skin effect).
2. Heat generation near the surface.

3. Propagation of stress waves from the skin effect zone due to $\mathbf{J} \times \mathbf{B}$ body forces.
4. Propagation of stress waves due to thermal gradients.
5. If melting occurs, ablation at the surface may result in additional stress waves.

Additional phenomena may also occur if atomic or molecular disassociation occurs at the surface. Such phenomena are possible under interaction with laser radiation and the surface continuum must be treated as a plasma. In this section we assume that no melting or plasma phenomena occur at the surface. The equations for a nonferromagnetic elastic conducting solid were derived in Chapter 3.

The equations for both temperature and magnetic field exhibit diffusion-type solutions. Both the induced temperature and eddy currents are confined to a boundary layer or skin depth near the surface for small times after the transient field is initiated. The thickness of the thermal and magnetic boundary layers are different [see 1-2.5; Table B-6].

The equation for the displacements and stresses in the solid, however, represent three coupled wave equations with forcing terms proportional to the thermal and field gradients. For an isotropic elastic solid, two types of body waves may occur—a compressional wave with wave speed determined by the elastic constants and density

$$C_L = \left(\frac{\lambda + 2\mu}{\rho}\right)^{1/2}$$

and a shear wave with speed given by

$$C_S = \left(\frac{\mu}{\rho}\right)^{1/2}$$

where λ and μ are called Lamé constants. μ is the shear modulus and both are related to Young's modulus by the formula

$$Y = \frac{\mu(3\lambda + 2\mu)}{\lambda + \mu}$$

The subscript on C_L indicates that the direction of wave propagation is longitudinal with respect to the displacement. The shear wave, however, is transverse to the displacement and is similar to electromagnetic waves. When the wave travels down a rod with stress-free surface, the longitudinal wave speed is modified and is given by

$$C_0 = \left(\frac{Y}{\rho}\right)^{1/2}$$

Another wave may travel on the surface of the solid and is called a

Rayleigh surface wave with a speed denoted by C_R. These wave speeds are ordered as follows for all isotropic materials:

$$C_L > C_0 > C_S > C_R$$

Wave speeds for different conductors are given in the Table B-5, in Appendix B.

In general a transient, surface magnetic field with nonzero gradient will produce both compressional, shear, and Rayleigh waves similar to an impact force applied to the surface. However, the solution of the complete set of Eqs. (3-7.9)–(3-7.11) for a three-dimensional problem is very difficult and the author has found none published in the literature. An axisymmetric two-dimensional solution has been given by Chain and Moon (1981).

To illustrate the phenomena and the technique for predicting the stress waves we examine the one-dimensional example of a conducting half-space with a uniform time-varying surface magnetic field.

One-Dimensional Stress Waves

Consider a nonferromagnetic half-space defined by $x \geq 0$ with elastic constants λ and μ, electrical conductivity σ, and thermal conductivity and heat capacity κ and c_v, respectively. At $x = 0$, we impose a time-varying magnetic field parallel to the surface with no gradients along the z direction (see Fig. 8-32). Under these assumptions, the vector fields assume the following form

$$\mathbf{B} = (0, 0, B(x, t)), \qquad \mathbf{J} = (0, J(x, t), 0), \qquad \mathbf{u} = (u, 0, 0)$$

The stress component of interest will be a tensile or compressive stress t_{xx}.

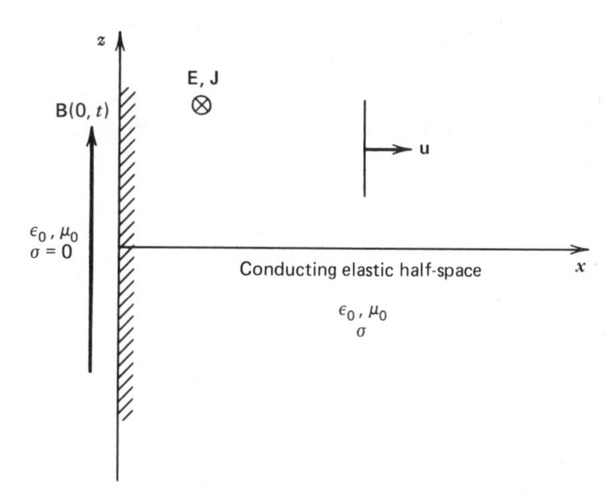

Figure 8-32 Pulsed magnetic field near a conducting half-space.

The governing one-dimensional form of the equations of momentum, energy, and electromagnetics are written below (see Chapter 3):

$$\rho\frac{\partial^2 u}{\partial t^2}-\frac{\partial t_{xx}}{\partial x}=JB \tag{8-10.1}$$

$$c_v\rho\frac{\partial\theta}{\partial t}+\alpha\theta_0(3\lambda+2\mu)\frac{\partial^2 u}{\partial x\partial t}-\kappa\frac{\partial^2\theta}{\partial x^2}=\frac{1}{\sigma}J^2 \tag{8-10.2}$$

and

$$\frac{\partial E}{\partial x}+\frac{\partial B}{\partial t}=0,\qquad -\frac{\partial B}{\partial x}=\mu_0 J \tag{8-10.3}$$

The constitutive equations for the stress and the current density are given by

$$t_{xx}=(\lambda+2\mu)\frac{\partial u}{\partial x}-\alpha(3\lambda+2\mu)\theta \tag{8–10.4}$$

and

$$J=\sigma\left(E-B\frac{\partial u}{\partial t}\right) \tag{8-10.5}$$

The right-hand side of Eq. (8-10.1) is the $\mathbf{J}\times\mathbf{B}$ body force term. The source term for the heat equation (8-10.2) is the Joule heating $\mathbf{J}'\cdot\mathbf{E}'$, where primes indicate fields measured relative to the moving material point. With neglect of free charge, $J=J'$ and $E'=E-B(\partial u/\partial t)$. Thus there are three nonlinear terms in this set of equations, that is, JB, J^2, and $B(\partial u/\partial t)$. It can be shown that for certain classes of problems the latter term may be dropped without much error. Also, the thermoelastic coupling term may also be neglected in a first estimate of the induced stress waves.

It is helpful to nondimensionalize the coupled set of equations. The characteristic length and time scales chosen are

$$x_0=(\sigma\mu_0 C_L)^{-1},\qquad t_0=(\sigma\mu_0 C_L^2)^{-1}$$

The magnetic field is nondimensionalized by the maximum value of the surface magnetic field B_0 and the stress with respect to the magnetic pressure $B_0^2/2\mu_0$. A temperature unit is

$$T_0=\frac{B_0^2}{\mu_0}\frac{1}{\mu_0\sigma\kappa}$$

The values of these characteristic dimensions for aluminum and copper are listed in Table 8-1. Two nondimensional groups appear in the equations

$$a=\frac{\sigma\mu_0\kappa}{\rho c_v},\qquad m=\frac{2\alpha(3\lambda+2\mu)}{\sigma\mu_0\kappa}$$

The nondimensionalized equations for a one-dimensional conducting elastic half-space are given by

Table 8-1 Characteristic Units for Elastic Waves in a Conducting Solid

	Al	Cu
x_0	3.8×10^{-3} mm	2.7×10^{-3} mm
t_0	0.63×10^{-3} μsec	0.54×10^{-3} μsec
T_0 [a]	76.0°K	27.5°K
θ_s [a]	0.63°K	0.41°K
a	4.13×10^{-3}	8.46×10^{-3}
m	800	560

Source: Moon and Chattopadhyay (1974).
[a] For $B_0 = 1$ T.

$$\frac{\partial B}{\partial t} - \frac{\partial^2 B}{\partial x^2} = 0 \tag{8-10.6}$$

$$\frac{1}{a}\frac{\partial \theta}{\partial t} - \frac{\partial^2 \theta}{\partial x^2} = f \tag{8-10.7}$$

and

$$\frac{\partial^2 t_{xx}}{\partial x^2} - \frac{\partial^2 t_{xx}}{\partial t^2} = \frac{\partial g}{\partial x} + \frac{\partial h}{\partial t} \tag{8-10.8}$$

where

$$f = \left(\frac{\partial B}{\partial x}\right)^2, \qquad g = \frac{\partial B^2}{\partial x}, \qquad h = m\frac{\partial \theta}{\partial t}$$

This is a set of nonlinear partial differential equations. However, with the neglect of the term $\partial[B(\partial u/\partial t)]\partial x$ in Eq. (8-10.3), Eqs. (8-10.6)–(8-10.8) are hierarchically coupled. That is, one first solves for B, then solves for θ, and finally uses these solutions to find the stress t_{xx}.

The boundary conditions on the stress are

$$t_{xx}(x = 0) = 0$$

If the boundary has convective losses, then in dimensional units

$$\kappa\frac{\partial \theta}{\partial x} = -\eta_0(\theta - T_s) \quad \text{at} \quad x = 0$$

where T_s is the temperature of the fluid outside the surface and η_0 is the convection coefficient. Finally, the magnetic field is a prescribed function at $x = 0$, that is,

$$B\,(x = 0, t) = B_0\phi(t)$$

where the maximum value of $\phi(t) = 1$.

In dimensional variables, the solution for $B(x, t)$ is given by the following integral of the surface field $\phi(t)$; that is,

$$B(x, t) = \frac{2B_0}{\sqrt{\pi}} \int_{(x/2)(\mu_0\sigma/t)^{1/2}}^{\infty} \phi\left(t - \frac{\mu_0 \sigma x^2}{\lambda^2}\right) e^{-\lambda^2}\, d\lambda \tag{8-10.9}$$

When $\phi(t)$ is a step function in time the magnetic field in the half-space is given by the complementary error function

$$B(x, t) = B_0 \operatorname{erfc}\left[\frac{x}{2}\left(\frac{\mu_0\sigma}{t}\right)^{1/2}\right] \tag{8-10.10}$$

When the Joule-heating source term is calculated from $B(x, t)$, the general solution for the nondimensional temperature is given by the following double integral for the case when the surface $x = 0$ is insulating; that is,

$$\theta = \left(\frac{a}{4\pi}\right)^{1/2} \int_0^{\infty} d\xi \int_0^t \frac{f(\xi, \tau)}{(t-\tau)^{1/2}} \left[\exp\left(-\frac{(x+\xi)^2}{4a(t-\tau)}\right) + \exp\left(-\frac{(x-\xi)^2}{4a(t-\tau)}\right)\right] d\tau \tag{8-10.11}$$

For a step magnetic field the source term in the heat equation (8-10.7) becomes

$$f = \left(\frac{\partial B}{\partial \xi}\right)^2 = \frac{1}{\pi\tau} \exp\left[\frac{-\xi^2}{2\tau}\right]$$

Substituting this expression into Eq. (8-10.11), one can show that the temperature on the surface jumps to a constant value for $t > 0$. This surface temperature is given by

$$\theta_s = \frac{a}{\pi(1-2a)^{1/2}} \ln \frac{1 + (1-2a)^{1/2}}{1 - (1-2a)^{1/2}} \tag{8-10.12}$$

Values of θ_s for aluminum and copper are given in Table 8-1.

The stress wave solution for the half-space may be solved by the method of characteristics. The solution consists of two parts. For the stresses due to $\mathbf{J} \times \mathbf{B}$ forces, one has the following expressions:

$$\begin{aligned} t_{xx} &= \frac{1}{2}\int_0^t g(t+x-\tau, \tau)\, d\tau - \frac{1}{2}\int_0^{t-x} g(t-x-\tau, \tau)\, d\tau \\ &\quad - \frac{1}{2}\int_{t-x}^{t} (g(\tau - t + x, \tau)\, d\tau \qquad (x < t) \end{aligned} \tag{8-10.13}$$

$$t_{xx} = \frac{1}{2}\int_0^t g(t+x-\tau, \tau)\, d\tau - \int_0^t g(x-t+\tau, \tau)\, d\tau \qquad (x > t)$$

The second stress wave is due to thermoelastic stresses resulting from Joule heating and is given by

$$t_{xx} = \frac{1}{2}\int_0^x h(\xi, t-x+\xi)\,d\xi + \frac{1}{2}\int_x^{t+x} h(\xi, t+x-\xi)\,d\xi$$

$$-\frac{1}{2}\int_0^{t-x} h(\xi, t-x-\xi)\,d\xi \qquad (x < t) \qquad (8\text{-}10.14)$$

$$t_{xx} = \frac{1}{2}\int_{x-t}^{x} h(\xi, \xi-x+t)\,d\xi + \frac{1}{2}\int_x^{x+t} h(\xi, x+t-\xi)\,d\xi \qquad (x > t)$$

Numerical results for a step increase in field at the surface show that the magnetic field and temperature are confined to a boundary layer (see Moon and Chattopadhyay, 1974) with the current density given by

$$\mu_0 J = \frac{\partial B}{\partial x} = -B_0\left(\frac{\mu_0\sigma}{\pi t}\right)^{1/2} e^{-\mu_0\sigma x^2/4t}$$

At a given time we can define a skin depth given by

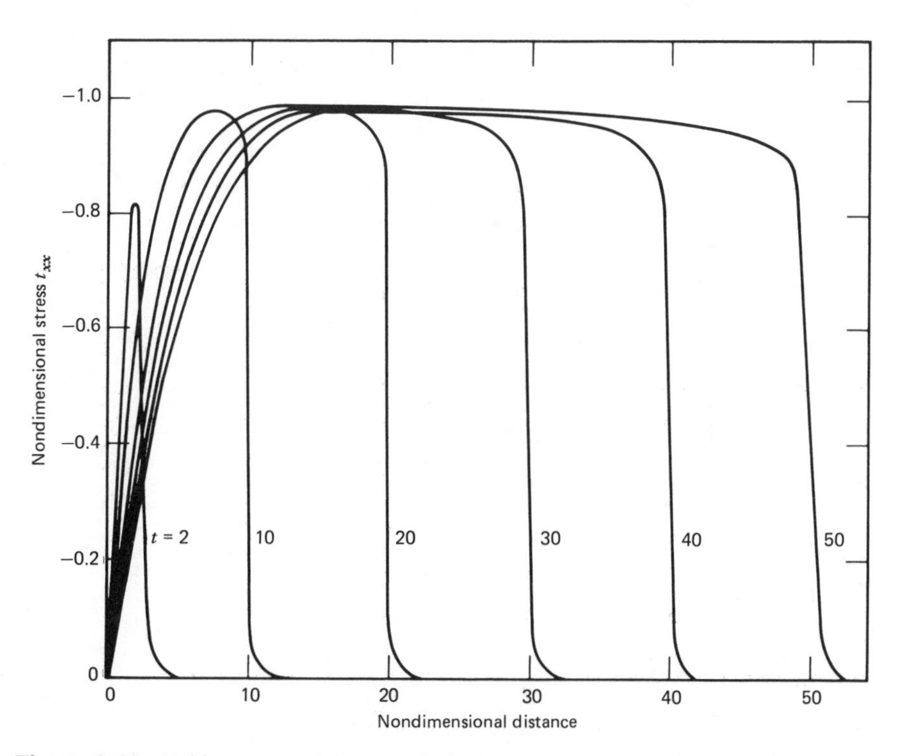

Figure 8-33 Eddy-current induced elastic stress waves vs. distance for various nondimensional times (from Moon and Chattopadhyay, 1974).

$$\delta = \left(\frac{4t}{\mu_0 \sigma}\right)^{1/2}$$

At this depth, $J = \bar{e}^1 J_0$ where J_0 is the surface current density. The $\mathbf{J} \times \mathbf{B}$ stress wave is shown plotted in Figure 8-33. After a short time it reaches a value given by the magnetic pressure. A smooth wave front propagates into the solid with speed C_L. As discussed in Section 3.8, one could replace this body force problem with an approximate boundary value problem in which the stress at $x = 0 + \varepsilon$ is given by the magnetic pressure $B^2/2\mu_0$.

The thermoelastic Joule-heating wave is shown in Figure 8-34. After a short distance, the wave carries zero momentum with both compression and tension parts. The discontinuity, which travels at the longitudinal curve speed C_L, is characteristic of thermoelastic problems in which a sudden rise in temperature appears on the boundary. Boley and Tolins (1962) have shown that when the temperature has a finite rise time of the order of 1 μsec, the thermoelastic wave amplitude is reduced considerably for aluminum or copper. However, in laser radiation problems sharp rise times are possible at the surface and this wave may become important.

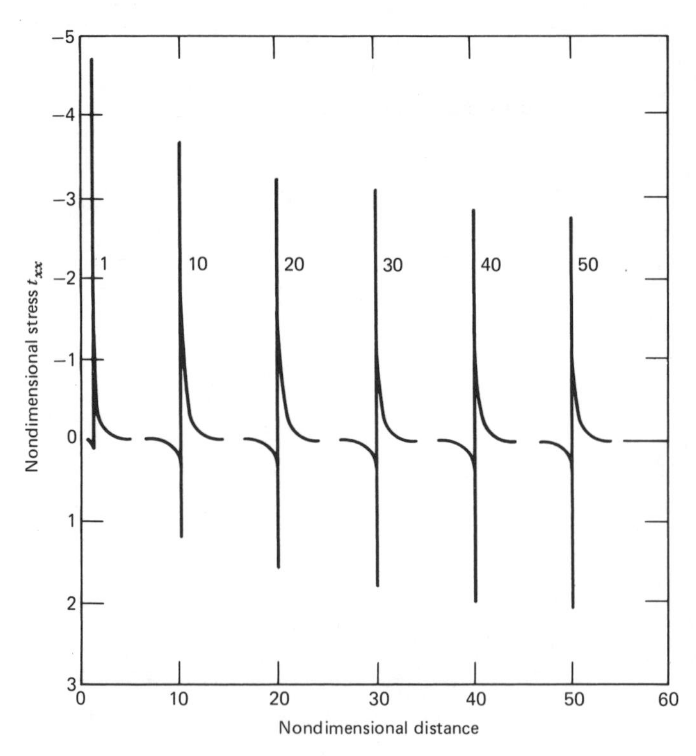

Figure 8-34 Joule-heating induced thermoelastic stress waves vs. distance for various times (from Moon and Chattopadhyay, 1974).

Experiments

Three experimental studies related to the above theoretical analysis of magnetically induced stress waves were published by the author and his students. In the first study (Moon and Chattopadhyay, 1974), a flat circular pancake coil was placed near the end of a circular rod. The 400-μsec-long half-sine magnetic pulse generated a compressional wave with amplitude equal to the magnetic pressure (see Fig. 8-35). No thermoelastic wave was observed, presumably because of the relatively long rise time of the field pulse. (See also Fig. 9-12.)

In the second study (Moon, 1978), the interaction of the pulsed magnetic field with a rod of ferromagnetic material showed that additional forces come to play besides the eddy-current $\mathbf{J} \times \mathbf{B}$ forces. The magnetization of the steel led to tensile stresses in the stress wave after the initial eddy-current-related compressional stresses (see Fig. 8-36).

The third study involved the generation of cylindrical stress waves in an aluminum plate by generating a 60-μsec half-period damped-sine-function

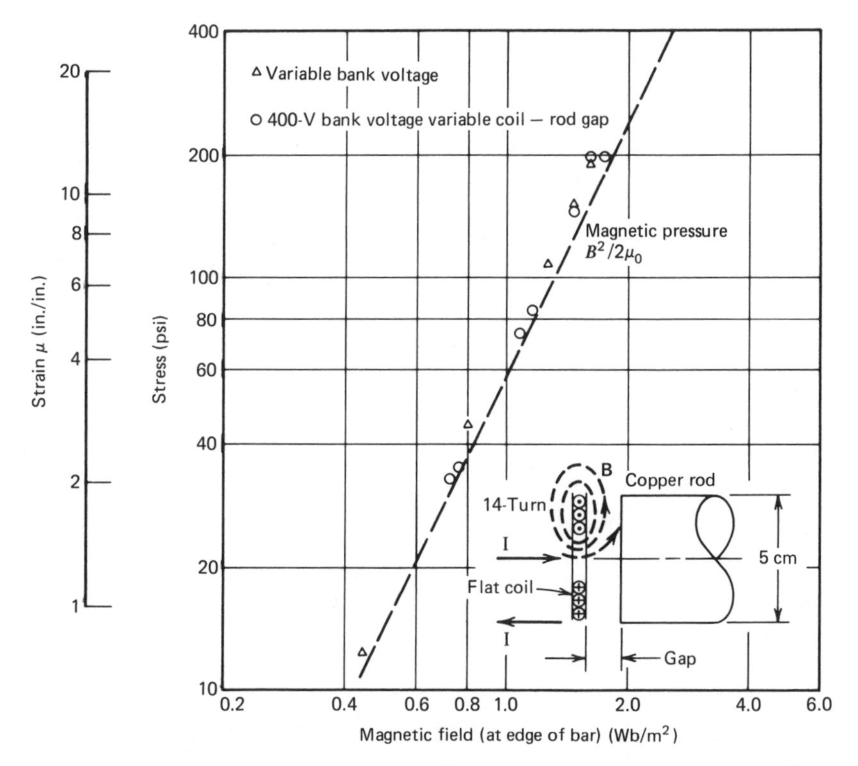

Figure 8-35 Maximum strain in a magnetically induced stress wave in a plate vs. magnetic field (from Moon and Chattopadhyay, 1974).

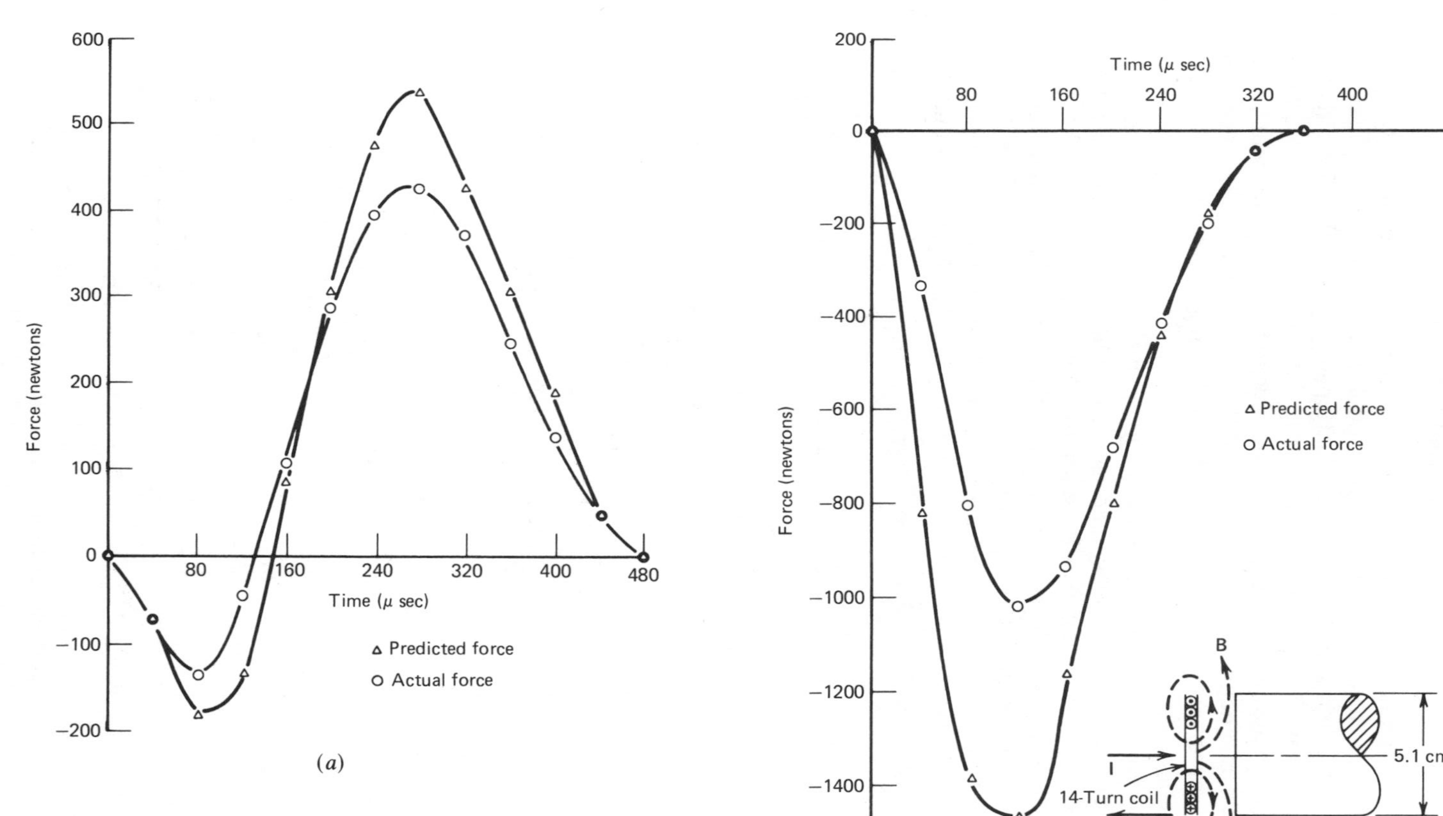

Figure 8-36 Magnetically induced force history in a bar by a pulsed coil: (*a*) steel bar—eddy-current and magnetization effects and (*b*) aluminum bar—eddy-current effects only. (From Moon, 1978, reprinted with permission. Pergamon Press Inc., copyright 1978.) (Predicted stress in both cases based on measured fields and the Maxwell stress tensor.)

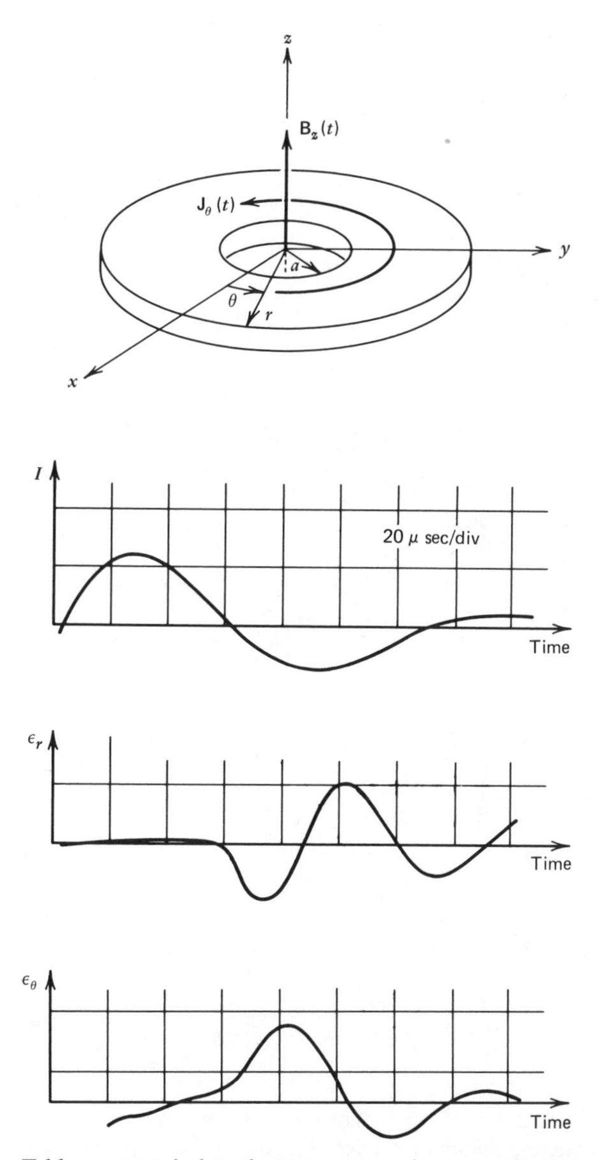

Figure 8-37 Eddy-current induced stress waves in a conducting annular plate. (Reprinted with permission from *Int. J. of Solids and Structures* **17**(1), C. T. Chian and F. C. Moon. Pergamon Press Ltd., copyright 1981.)

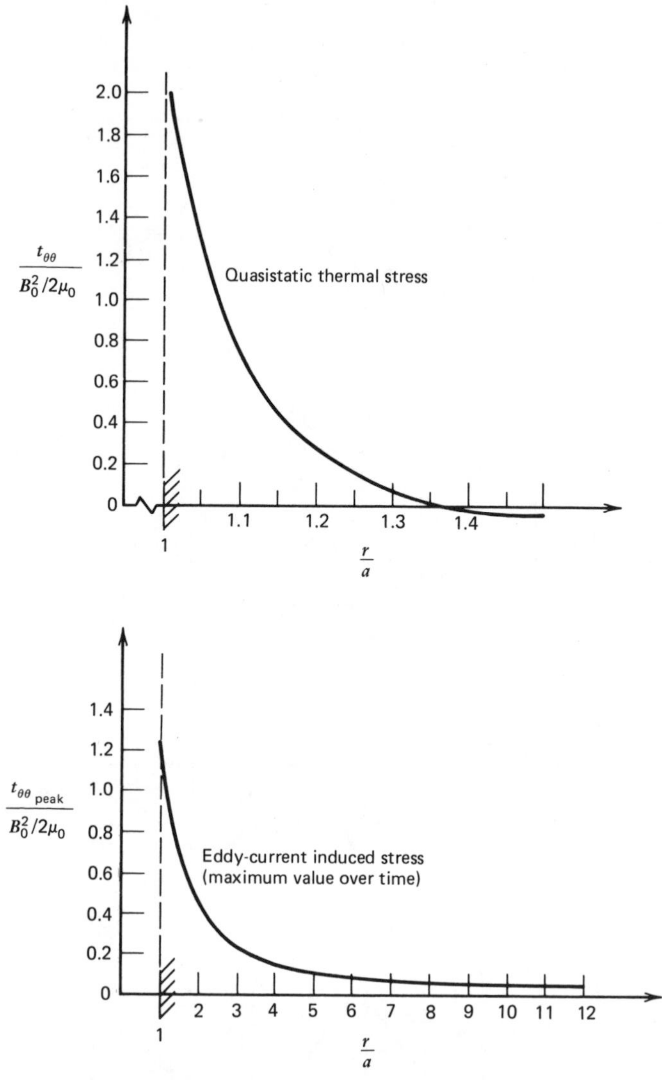

Figure 8-38 Comparison of maximum eddy-current induced stresses and thermal stresses in a conducting annular plate (see Fig. 8-37). (Reprinted with permission from *Int. J. of Solids and Structures* **17**(1), C. T. Chian and F. C. Moon. Pergamon Press Ltd., copyright 1981.)

axial magnetic field in a hole in the plate (Chian and Moon, 1981). Cylindrical stress waves, unlike plane waves in solids, change their shape as they propagate. Theoretical predictions of these waves agreed remarkably well with the experimental data shown in Figure 8-37. These calculations were based on an analysis similar to that in the previous section, using only $\mathbf{J} \times \mathbf{B}$ forces. No thermoelastic waves were detected, presumably because of the long rise time. However, a temperature rise at the surface of the hole in the plate of about 6.0°K was measured for a 5-T pulse. This temperature rise near the hole boundary results in quasisteady thermoelastic stresses which are a factor of 2 times the magnetic pressure in the hole as shown in Figure 8-38.

9

Experimental Methods in Magneto-Solid Mechanics

9.1 INTRODUCTION

While the major emphasis of this book has been theoretical modeling of magnetomechanical systems, such analysis is best done in parallel with experimental research. This chapter cannot pretend to offer a complete course on measurement techniques of magnetic phenomena. However, it is hoped that the novitiate will find that the methods discussed in this survey are not difficult to understand, though, as in most experimental work, specific skills and experience may be required to master these techniques.

There are a number of modern references on the generation of magnetic fields, including Montgomery (1980) and Brechna (1973), in which the interested reader may find more detailed information.

In this chapter we review a few methods for creating steady and time-varying magnetic fields (see Section 9.2), as well as describe techniques for measuring fields and currents (see Sections 9.3 and 9.4). Force and strain measurement techniques are discussed in Sections 9.5 and 9.6. Force measurements in magnetomechanical problems are complicated by the presence of magnetic fields which can often cause unwanted electrical noise in the data. Special techniques to avoid spurious force and strain measurements are presented. Two methods for detecting small vibrations using magnetic fields are discussed in Section 9.7, including an application of a SQUID magnetometer.

In Section 9.8 we review some of the unique problems associated with experimental work in superconductivity, including those connected with cryogenic temperatures. Finally, in Section 9.9, we discuss the effects of magnetic fields on humans and safe exposure limits for magnetic fields.

As in any experimental science, we should not overlook the importance of nondimensional groups in magneto-solid mechanics. Since these have been discussed in some of the theoretical chapters (see, e.g., Section 8.6), no special presentation will be made here. However, if one is to predict the behavior of a large-scale magnetomechanical system from measurements on small-scale models, the experiments must be designed to cover the range of the relevant nondimensional groups in the large-scale system.

9.2 GENERATION OF MAGNETIC FIELDS AND CURRENTS

The principal methods for generating magnetic fields are permanent ferromagnets, electromagnets, that is, soft ferromagnetic iron or steel encircled by many turns of electric current, and so-called air-core magnets, which are simply many turns of wire arranged in space to produce a specified magnetic field distribution. Permanent magnets and laboratory electromagnets are usually static devices. In rotary and linear motors the electromagnet produces oscillatory magnetic fields. Air-core magnets can produce either steady or pulsed magnetic fields. Cryogenically cooled, superconducting solenoids are used to generate a uniform steady high magnetic field (>1 T) in a cylindrical volume.

The use of permanent magnetic materials for producing a magnetic field environment is usually limited to a working space of a few cubic centimeters or less and to fields below 1 T.

Soft ferromagnetic electromagnets have two major applications (1) to create a specified field intensity environment with high uniformity and (2) to provide a specified field gradient environment. The latter usually involve high-energy physics experiments or high-field gradient separation applications. Commercially available uniform field electromagnets usually have a maximum working environment of less than 5000 cm^3 and field intensity less than 1.2 T.

Pulsed Fields and Currents

Pulsed magnetic fields and high currents are used in magnetic forming devices, plasma and fusion machines, and electrohydraulic forming and occur during power system transients and in lightning strikes. Perhaps the most widely used device for generating transient currents is the stored charge capacitance bank. Electric energy in the form of charges separated by a voltage across a dielectric is converted to magnetic energy by shorting the two plates of the capacitor with a resistive or inductive load. A schematic for one such system is shown in Figure 9-1. The capacitor bank is connected to

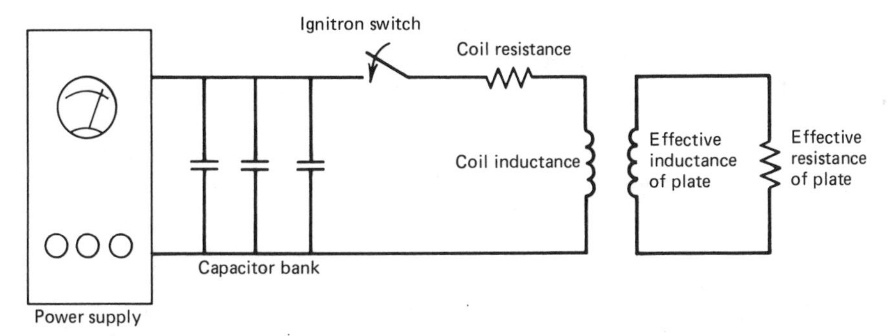

Figure 9-1 Sketch of capacitance bank apparatus for creating pulsed magnetic fields.

an inductor or coil by means of a mechanical rat trap switch or an electronic switch such as an ignitron.

The design of a capacitance bank depends on the desired peak current output and the duration of the pulse. For a simple LC circuit, the current, charge, and voltage are sinusoidal functions of time. The peak values of these variables, denoted by a subscript zero, are related as follows:

$$I_0 = \frac{\pi}{\tau} Q_0 = \frac{\pi}{\tau} CV_0 \tag{9-2.1}$$

where τ is the half-period. The half-period is given by $\tau = \pi\sqrt{LC}$. Using this expression in Eq. (9-2.1), the peak current–voltage relation is found

$$I_0 = \sqrt{\frac{C}{L}}\, V_0 \tag{9-2.2}$$

Often the inductance is determined by the application, such as the size of a magnetic forming tool. Thus the choice of discharge time will determine the capacitance, leaving only the voltage as the major parameter in achieving the desired peak current.

Pulse Shaping

The output from a simple L–R–C circuit capacitance bank is a damped sine wave as shown in Figure 9-2a. When other pulse shapes are desired, a transmission line design is used as shown in Figure 9-2b. An actual bank is comprised of a number of separate capacitors with their common terminals tied together in parallel. When the inductance between each capacitor is relatively low, the capacitors all release their charge almost simultaneously producing the damped sine wave in Figure 9-2a. However, when inductance is placed between the capacitors the discharge of each successive capacitor from the load is delayed thereby producing a flatter pulse as shown in Figure 9-2b. In effect, the bank acts as a lumped transmission line.

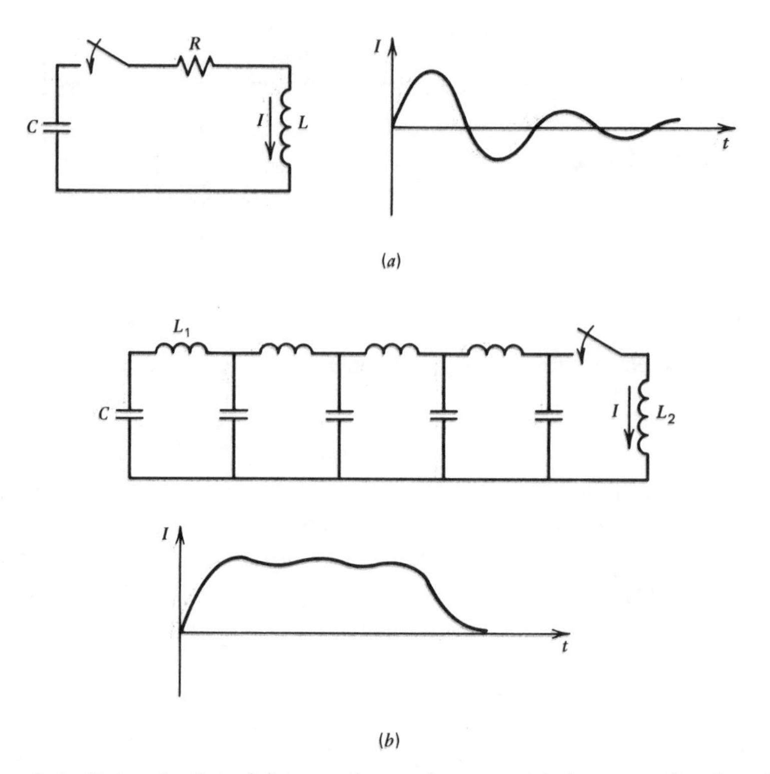

Figure 9-2 Pulse shaping: (*a*) lumped capacitance and inductance circuit and time response and (*b*) distributed inductance capacitance bank and circuit time response.

Of course, current pulses can be actively shaped and time-controlled by electronically switching in and out different capacitors to the load. These techniques will not be discussed here.

Two other methods for creating pulsed magnetic fields involve the rotary inertia method and flux compression.

A *rotating pulsed field device* consists of a high inertia rotor connected to a voltage or current generator. The rotor is accelerated to a high speed without any electromagnetic load by either an electric motor or gas turbine. When the load is switched on, the kinetic energy of the rotor is converted to either voltage or current, thus decelerating the rotor.

In one version the rotor contains moving conductors that cut the magnetic field lines created by external, stationary coils. In another application the rotor turns a constant field magnet. The moving field lines then generate voltages in stationary conductors.

In the *flux compression method* shown in Figure 9-3, a steady field magnet creates a relative low magnetic field in a large volume. The initially low flux density can be increased substantially by decreasing the cross-sectional area

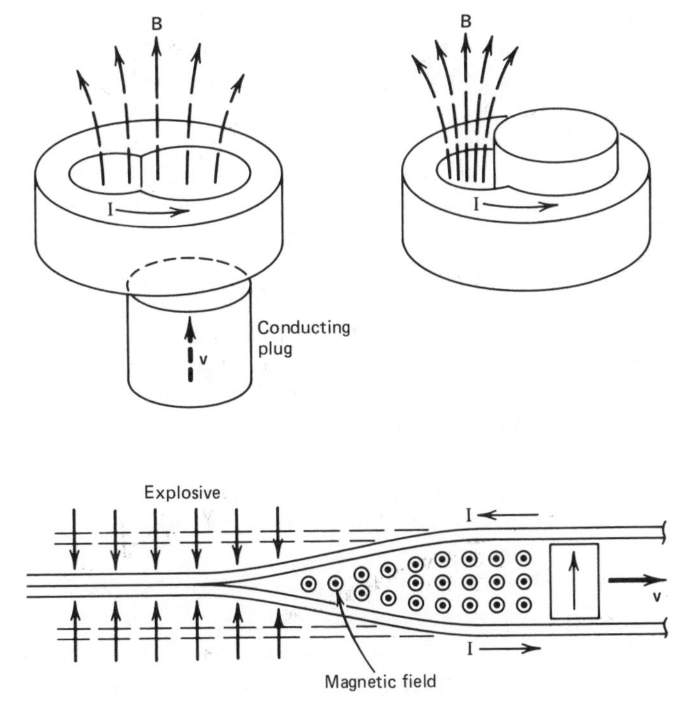

Figure 9-3 Flux compression methods.

through which the flux flows. This can be accomplished by placing a superconducting conductor into the volume (see Fig. 9-3) or by accelerating a good conductor. The induced eddy currents then prevent flux leakage and conservation of flux results in a higher flux density in the smaller volume.

In the lower part of Figure 9-3, this flux compression is accomplished by explosively moving the pair of current-carrying rails toward one another. This technique is used in the design of magnetic rail guns (see Chapter 8).

9.3 METHODS OF MEASURING MAGNETIC FIELDS AND CURRENTS

Three standard techniques for measuring magnetic flux and flux density are:

1. Rotating coil gaussmeter.
2. Search coil.
3. Hall effect.

Other methods involving the change of physical properties may also be used such as piezoresistance, or magnetoresistive effects. A very sensitive method called the SQUID—superconducting quantum interference device—can measure extremely small magnetic fields. All four methods are available commercially.

The rotating coil gaussmeter measures a steady magnetic field by rotating a small, multiturn coil about an axis through the coil diameter. The changing flux in the coil produces a voltage which can be accurately measured.

The Magnetic Field Search Coil

Like the rotating coil gaussmeter, the search coil is based on the law of induction. For time-varying magnetic fields, one need only encircle the changing flux with a conductor to generate an electric field in the conductor and a voltage at its ends. To avoid the inhomogeneities in the field, many turns of wire are wound in a small coil. The voltage generated at the ends of the leads to the coil is proportional to the cross-sectional area A, number of turns N, and the rate of change of magnetic field, that is,

$$V = NA\frac{\partial B_n}{\partial t} \tag{9-3.1}$$

where B_n is the component of the magnetic field along the axis of the coil. To determine $B_n(t)$, a passive or active electronic integrator is used with the coil

$$B_n(t) = \frac{1}{NA}\int V(t)\,dt \tag{9-3.2}$$

Hall-Effect Method

The Hall potential was discussed in Section 8.7. Many types of Hall-effect probes are available commercially. They are small and can measure high as well as low fields and can operate in a differential mode. These commercial units usually have limited transient response below several kilohertz.

The Hall gaussmeter is based on the fact that the resistance of certain metals changes in a magnetic field (magnetoresistance). In particular, if current flow along a wire, J_x, is placed in a transverse magnetic field B_z (see Fig. 9-4), a voltage appears across the metal E_y perpendicular to both $\mathbf{B}$ and $\mathbf{J}$. When $\mathbf{J} = (J_x, 0, 0)$, $E = (E_x, E_y, 0)$, and $\mathbf{B} = (0, 0, B_z)$, Eq. (8-7.5) gives

$$E_y = R_H J_x B_z$$

or

$$B_z = \frac{E_y}{R_H J_x} \tag{9-3.3}$$

Thus by accurately measuring J_x and E_y one can obtain B_z. The Hall constant R_H has been measured for different metals (see Chapter 8).

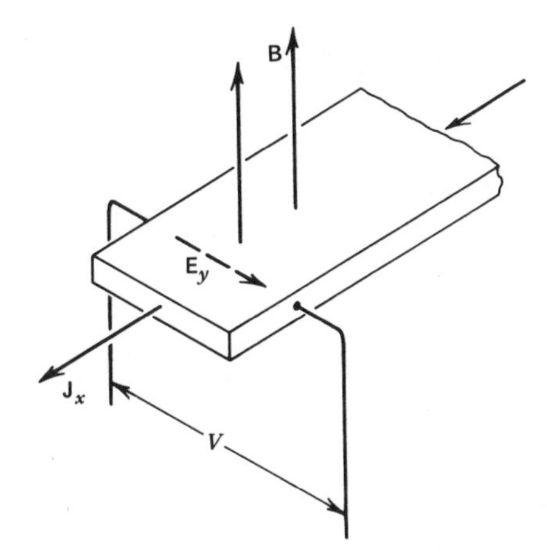

Figure 9-4 Hall-effect probe.

Current Measurements

For the measurement of the total induced current flowing through a conductor two techniques are recommended:

1. Rogowski coil.
2. Modified Hall-effect probe.

To measure high transient currents, a simple device known as a Rogowski coil is used. A toroidal coil encircles the conductor through which the input current will pass (see Fig. 9-5) and the output of this transducer is a voltage. The device is based upon two principles of electromagnetism (1) a current filament generates a circular magnetic field concentric with the electric current and (2) a changing magnetic flux piercing a loop of wire generates a voltage across the ends of the wire.

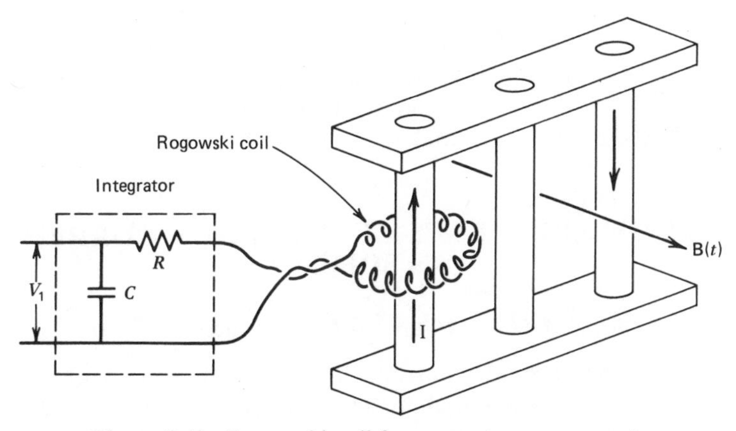

Figure 9-5 Rogowski coil for current measurement.

The magnetic field generated by the current is given by

$$B = \frac{\mu_0 I}{2\pi r}$$

If toroidal loops of wire encircle B at a distance r from the current filament, the flux penetrating the toroidal loops is approximately

$$\Phi = NB\pi a^2$$

where N is the number of toroidal loops in the Rogowski coil and a is the minor radius of the torus.

The induced voltage is thus given by

$$V = \frac{d\Phi}{dt} = \frac{Na^2\mu_0}{2r}\frac{dI}{dt} \tag{9-3.4}$$

A passive or active integrator is used to obtain I from the measured voltage V. For example, if I has a pulse length of τ seconds and the product of the resistance and capacitance of a passive integrator is such that $RC \gg \tau$ (see Fig. 9-5), then the current is given approximately by

$$I \cong \frac{2r}{Na^2}\frac{RCV_1}{\mu_0} \tag{9-3.5}$$

where V_1 is the output voltage from the integrator. Thus a 1-kA pulse with $RC = 10$ msec, $N = 100$, $r = 3$ cm, and $a = 0.5$ cm, will produce a peak voltage from the Rogowski-coil integrator of about 5 mV.

Of course the formula given in Eq. (9-3.5) is only approximate since the flux across the toroidal minor diameter varies; the resistance of the wire in the coil was neglected; and the integration is approximate. None the less, calibration of Rogowski coils against other current transducers give values within 10% of that given in Eq. (9-3.5).

Modified Hall probes for current measurement are available commercially. These devices encircle the conductor with an open ferrite ring. The ferrite concentrates the magnetic flux on a standard Hall probe. These devices are not as adaptable to irregular-shaped or large conductors as the Rogowski coil. However, the modified Hall probes can measure steady currents while the Rogowski coils require time-varying currents.

9.4 METHODS OF MEASURING EDDY CURRENTS IN SOLIDS

Differential Search Coil Eddy-Current Probe

One method to measure eddy currents in thin sheetlike conductors uses the relation between the tangential magnetic field $\mathbf{H}_t$ and the eddy-current density

$$\mathbf{H}_t^+ - \mathbf{H}_t^- = \mathbf{J} \times \mathbf{n}^+ \Delta \tag{9-4.1}$$

where $\mathbf{n}^+$ is the normal to the surface. We have assumed that $\mathbf{J}$ is uniform across the sheet thickness Δ.

The jump in $\mathbf{H}_t$ across the sheet can be measured by placing search coils on either side of the sheet and measuring the difference in voltage. This voltage is proportional to $\dot{\mathbf{H}}$ so that an integrator is required. A sketch of the device is shown in Figure 9-6.

If B^+ and B^- are the tangential components of the magnetic field on the top and bottom surfaces of the plate, then the current density in the plate is given by the following equations:

$$B^+ - B^- = \mu_0 \Delta J$$

$$V = A\left(\frac{\partial B^+}{\partial t} - \frac{\partial B^-}{\partial t}\right) \tag{9-4.2}$$

and

$$J = \frac{1}{\mu_0 \Delta A} \int_0^t V(\tau)\, d\tau$$

Poynting Vector Eddy-Current Probe

The dissipation of energy in a conductor due to eddy currents is related to the surface integral of the Poynting vector $\mathbf{S} = \mathbf{E} \times \mathbf{H}$. For harmonic fields $\mathbf{E} = \hat{\mathbf{E}} e^{i\omega t}$ and $\mathbf{H} = \hat{\mathbf{H}} e^{i\omega t}$, the average power dissipated is given by Eq. 2-2.7:

$$\langle P \rangle = \frac{1}{2} \mathrm{Re}\left(\int_A \hat{\mathbf{H}} \times \hat{\mathbf{E}}^* \cdot \mathbf{n}\, da\right) \tag{9-4.3}$$

where the asterisk denotes the complex conjugate and the area A encloses the body. A sensor designed on this principle has been designed and tested by Köhler (1980). The device consists of a voltage-measuring circuit for $\mathbf{E}$ and a search coil to measure $\mathbf{H}$. This device was designed to measure local eddy-current densities near the surface of metals.

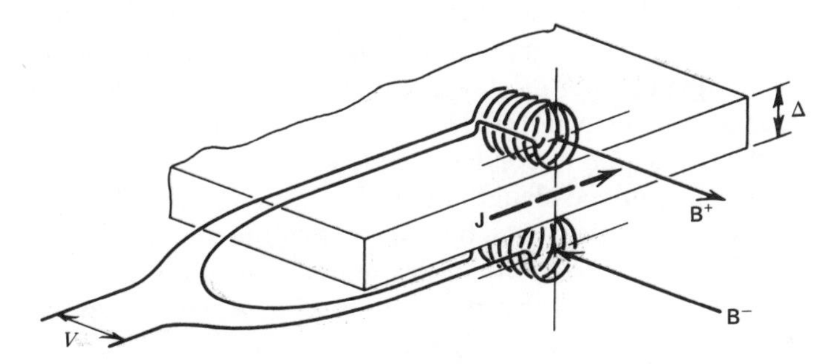

Figure 9-6 Differential search coils for eddy-current measurement.

Infrared Techniques for Eddy-Current Density Measurement

In normal conductors, eddy currents generate heat. When heat conduction can be neglected, the rise in temperature may be interpreted as a measure of the square of the eddy-current density J^2. The basic equations are given below:

$$-\kappa\nabla^2\theta + c\frac{\partial\theta}{\partial t} = \mathbf{J}\cdot\mathbf{E} = J^2/\sigma \tag{9-4.4}$$

When heat transfer can be neglected,

$$c\frac{\partial\theta}{\partial t} \cong \frac{J^2}{\sigma}$$

or

$$\theta(t) = \frac{1}{c\sigma}\int^t J^2(\tau)\,d\tau \tag{9-4.5}$$

When these conditions exist, infrared scanning of the surface of a conductor may provide a measure of J^2 at the surface. This technique has been successfully employed by Moon (1974). A sketch of the experimental setup used is shown in Figure 9-7 along with a black and white isotherm picture. Commercial systems come with a color quantizer to provide color-coded isotherms. Commercial manufacturers make two-dimensional scanners which can be focused on a plane at a given distance. A technical description of these devices is given below. Less-expensive line scanners are also available.

Whatever unit is used, an infrared transparent window between the scanner and the test specimen must be provided. The scanner may be placed several meters from the object. However, as one moves the scanner further away from the object, other radiating objects will come into the field of view.

This method measurement can be very useful if one wants an over-all view of the eddy-current patterns in complex shapes. Once critical areas are identified using the infrared technique, other local measurements of **J** or **B** can be made with more standard measuring devices.

The scanner consists of three main subsystems (1) the detector, (2) the continuous grey coding of temperatures on a black and white monitor, and (3) a color coding and color display monitor. In one design, the detector is a mercury cadmium telluride (HgCdTe) crystal maintained at 78°K by liquid nitrogen. HgCdTe is a photoconductive element sensitive to infrared radiation in the 2- to 12-μm-wavelength range. Radiation from different points in the field are focused on the crystal by two rotating mirrors for both horizontal and vertical scans. The sensitivity of this unit is as low as 0.2°C. The output from the crystal detector is used to produce a variable grey spot on a cathode-ray display tube.

In one commercial unit, the two-dimensional grey infrared picture has 525

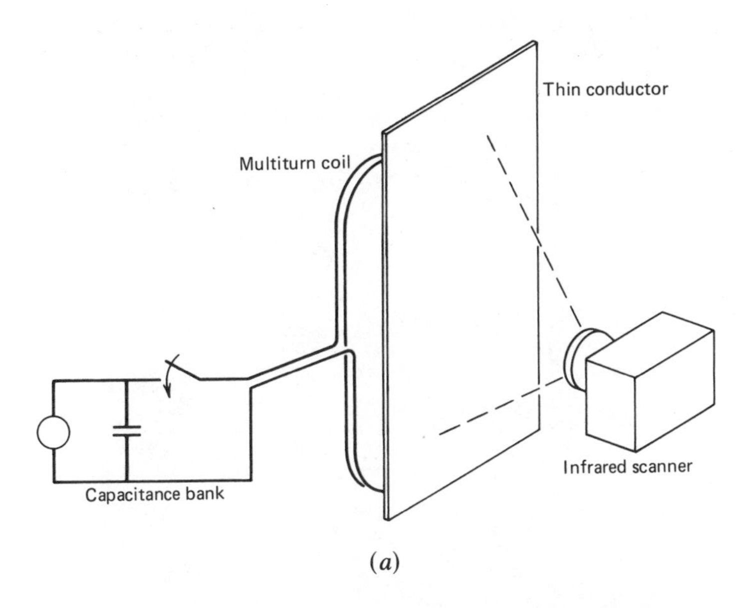

(*a*)

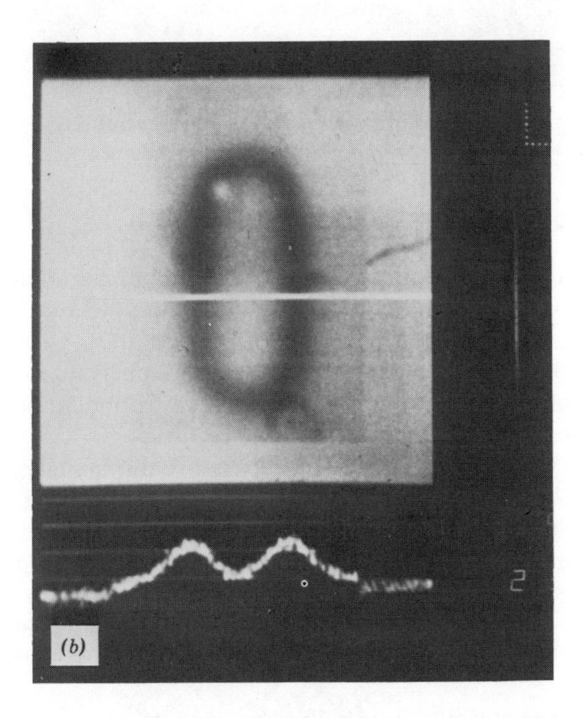

Figure 9-7 (*a*) Sketch of infrared eddy-current detection experiment and (*b*) grey scale, infrared thermogram of induced eddy currents in a thin aluminum plate (bottom trace is temperature across white line, 1°C maximum temperature).

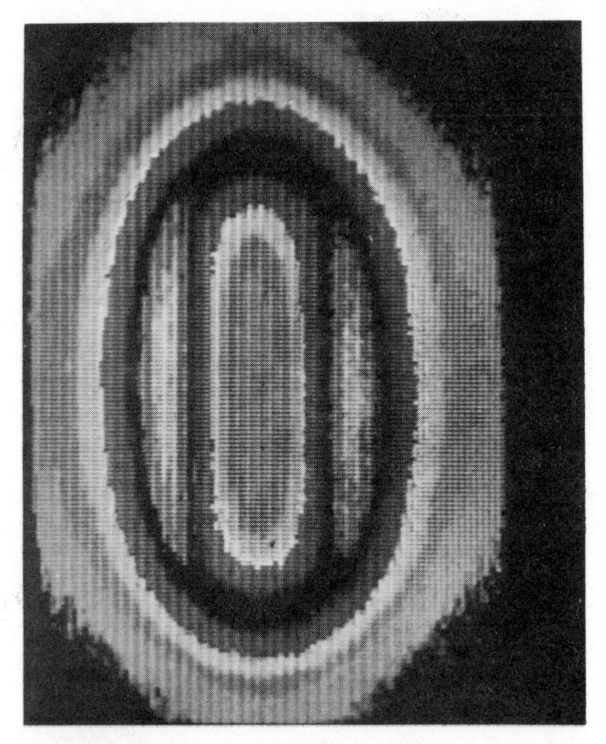

Figure 9-8 Infrared thermogram of induced eddy currents in a thin aluminum plate—black and white photograph of color quantized data.

raster lines and 600 elements per line on a 7.6-cm × 7.6-cm screen. The field can be focused from 10 cm to 6 m. The depth of field is 15 cm at a distance of 80 cm.

An isotherm option is available as well as a graphics temperature display. One horizontal scan of temperatures is graphed below the two-dimensional grey image of the temperature field.

In one unit the scan and graphic temperature display takes 1–2 sec. The data can be stored electronically so that pictures can be taken of the display tube. The data can be color quantized in a separate unit. Each color represents a temperature interval. The resulting display is a beautiful color set of isotherms (see Fig. 9-8).

In another commercially available unit the scan time is shorter (16 per second), but the image is coarser (210 horizontal lines, 140 elements per line).

The amount of radiation transmitted to the detector depends not only on the temperature of the transmitting object but on its emissivity. To avoid spurious data due to differences in emissivity, the conductors should be sprayed with a high-emissivity, black coating. These coatings also decrease spurious infrared reflections from other objects.

9.5 STRAIN GAGE TECHNIQUES IN MAGNETIC FIELDS

Strain measurement in magnetic fields with wire or foil strain gages poses four problems for the experimentalist:

1. Magnetostrictive effects.
2. Magnetoresistive effects.
3. Spurious voltage signals in time-varying magnetic fields.
4. Resistance changes in cryogenic environments.

Foil or wire strain gages are passive resistive elements in which one tries to correlate the change in resistance $\Delta R/R$ with the strain ε. However, in magnetic field and cryogenic environments, this relationship becomes more complex as represented by the following expression:

$$\frac{\Delta R}{R} = g\left(\varepsilon, B, \frac{\partial B}{\partial t}, \theta\right)$$

Because these resistance elements often have Ni or Fe as a constituent, they sometimes exhibit magnetostriction (see Chapter 2) which can produce a change in length of the gage. The other effect is related to the Hall potential (see Section 8.7) and is called *magnetoresistance*, which, as the name implies, means that the resistance of the gage depends on the magnetic field, usually in a nonlinear way.

Early work on the problem of magnetoresistance of strain gages was done by Gunn and Billinghurst (1957) and Takaki and Tsuji (1958). Review of the effect of magnetic fields on strain gages may be found in Stein (1962) and Freynik et al. (1978). Walstrom (1975) has also presented data on strain gage magnetoresistance at 4.2°K (liquid helium) in fields up to 8 T.

Experimental data usually plot the change in $\Delta R/R$ with field as if it were an apparent strain. Walstrom (1975), for example, has shown that at low temperatures "Karma"-type gages (a nickel-chromium alloy) exhibit a positive magnetoresistance effect at high fields, while "Constantan"-type gages (a copper-nickel alloy) have a negative magnetoresistance-induced apparent strain (see Fig. 9-9). At room temperature and liquid nitrogen (78°K), the magnetoresistive effect is small for "Karma"-type gages.

Freynik et al. (1978) studied a modified Karma (Ni/Cr) alloy gage at 4.2°K from 0 to 12 T. They observed a small negative magnetoresistive effect at fields below 2 T, and a strong positive effect at fields up to 12 T. These effects were found to be insensitive to the field direction relative to the plane

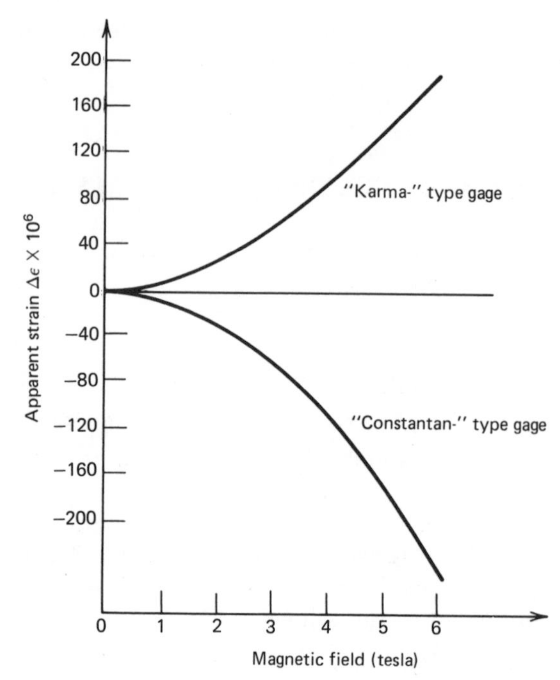

Figure 9-9 Induced apparent strain vs. magnetic field for "Karma"-(nickel-chromium) and "constantan"-(Cu-Ni) type strain gages. (From Walstrom, 1975, reprinted with permission. Butterworth Scientific, Ltd., copyright 1975.)

of the gage. They also found only a 5% increase change in the actual gage factor at 4.2°K compared with room temperature.

In superconducting device experiments, one must also deal with the problem of large temperature changes from 300 to 4.2°K. The effects of cryogenic environments on strain gage performance has received a great deal of study (see, e.g., Kaufman, 1963; Telinde, 1970). Part of the apparent strain effect involves thermal contraction as well as inherent changes in resistivity of the material and depends on the material to which the gage is bonded. Typical curves from Telinde (1970) are shown in Figure 9-10 for modified Karma (Ni/Cr) gages.

To avoid induced noise, three methods have been employed—shielding, time delay, and low inductance gages. A low inductance gage and circuit is shown in Figure 9-11. The gage is folded back on itself to minimize inductance. This has been successfully used by Miya et al. (1980). A dummy gage was also placed near the active one, and copper shielding and proper grounding were employed. Miya et al. obtained good strain data due to a 14-kA pulse with a 20-μsec rise time near a circular cylinder. A shield room for the amplifier and recording equipment and oscilloscopes is also recommended to minimize noise.

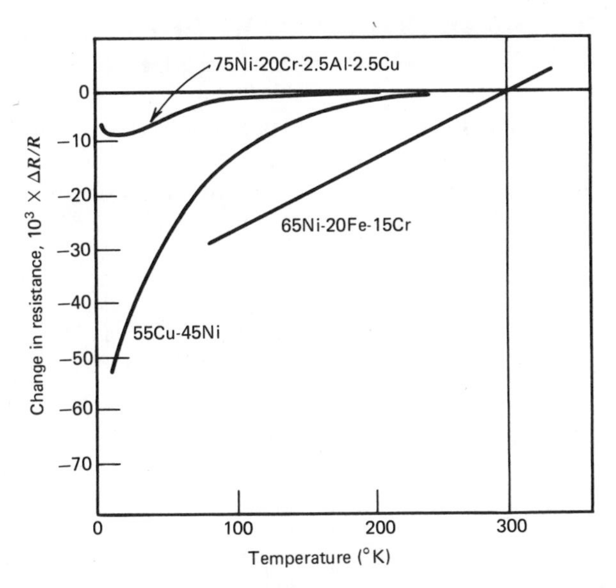

Figure 9-10 Apparent strain vs. temperature for Ni/Cr strain gages. (Reprinted from Telinde, 1970, *Experimental Mechanics*, Vol. 10, No. 9, p. 396, Sept. 1970.)

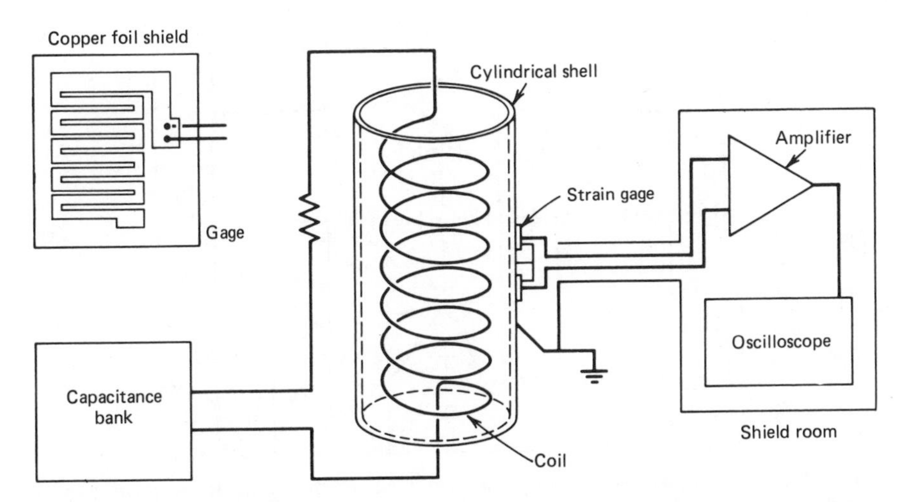

Figure 9-11 Low-inductance strain gage and application. (From Miya et al., 1980, reprinted with permission. North-Holland Publishing Co., copyright 1980.)

9.6 FORCE MEASUREMENTS IN PULSED MAGNETIC FIELDS—STRESS WAVE TECHNIQUE

The induction of current in one solid by pulsed currents in another, the flow of current from one solid to another, and the interaction of laser pulses with a solid surface all generate transient forces and deformations. Measurement of these forces, however, is not easy since large amounts of electrical noise

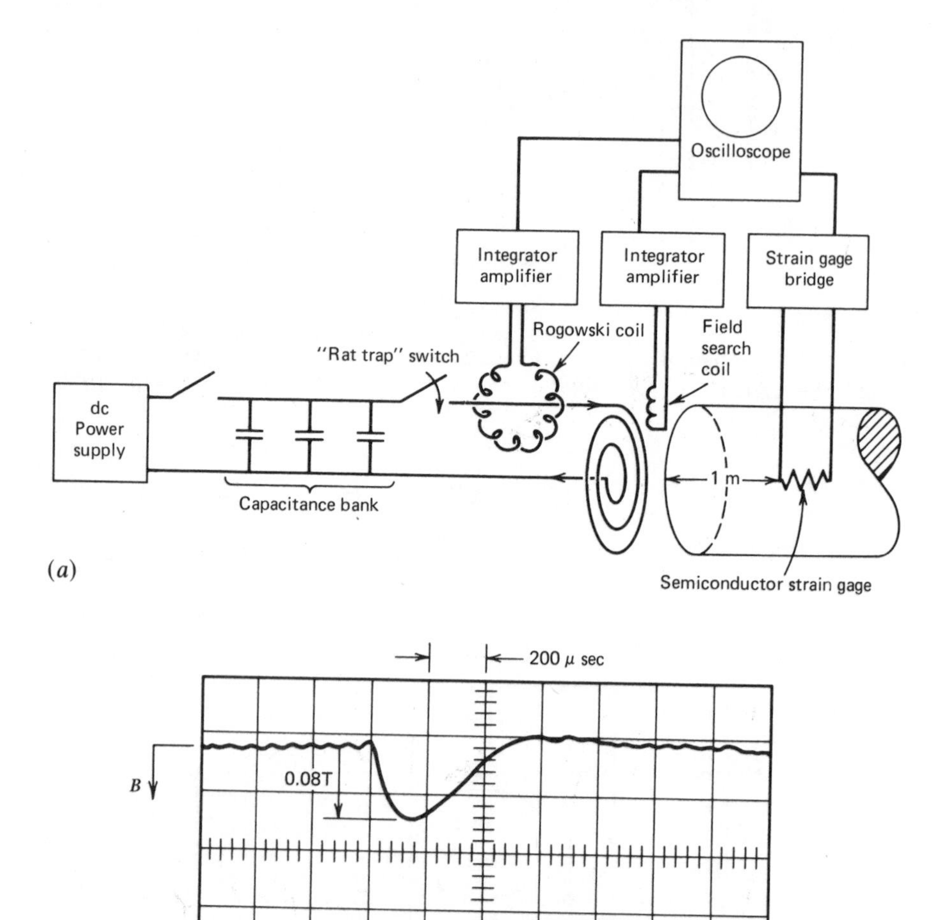

Figure 9-12 Stress wave—technique for measuring magnetic forces: (*a*) experimental apparatus and (*b*) current pulse and delayed force history. (From Moon and Chattopadhyay, 1974.)

usually accompany the currents and forces. Since modern force transducers usually involve electronic devices such as strain gages or electric transducers, one is faced with the problem of separating the electrical noise in these transducers from the force generated output voltages. One technique to separate the noise and force signals is to temporarily delay the force signal until the electrical noise is over, and then convert the force to an electrical voltage. This can be accomplished by storing the force-time history in the form of a stress wave. One such device is shown in Figure 9-12.

For example, if the magnetic force between an exciter coil and a magnetic forming workpiece is desired, the workpiece is attached to a long metal bar. When the current is pulsed in the forming coil, the force on the workpiece generates a stress wave in the bar. Strain gages at the center of the bar begin to record the average longitudinal stress in the bar Δt seconds after the current is turned on; that is,

$$\Delta t = \frac{L}{2C_L}$$

where C_L is the wave speed in the bar and L is the length of the bar (see Section 8.10). The time delay should be greater than the duration of the current pulse.

The stress pulse in the bar will, of course, reflect back and forth so the bar should be long enough to prevent the force signal from interacting with its reflection. Also, for very-short-duration pulses the wavelength in the bar should be greater than the diameter of the bar by about six times to avoid distortion of the force signal due to dispersion. (See e.g., Moon and Chattopadhay, 1974.)

This technique can also be used to measure pulsed magnetic torque generation by converting the torque into a shear wave in a bar.

9.6 VIBRATION DETECTION USING PASSIVE MAGNETIC FIELDS

Conventional methods for detecting structural motions include strain gages and accelerometers, which must be directly applied to the solid, and noncontacting devices such as capacitance and inductive and eddy-current impedance detectors which usually are placed in *near* contact with the solid (1–10 mm). However, vibrations in structures can be detected at large distances by using the vibrating structure as a transmitting antenna whose time-varying magnetic fields can be detected by a passive inductance coil. There are a number of mechanisms by which this can be accomplished:

1. *Vibrating magnetic dipoles.* If a material can retain some permanent magnetism (such as iron or steel) or has induced magnetism due to say the

earth's magnetic field, then as the structure vibrates, these moving dipoles will produce changing flux in the vicinity of the sensing coil. A diamagnetic material such as bismuth or a superconductor in a weak magnetic field can produce the same effect.

2. *Vibrating current-carrying or induced eddy-current conductors.* If a structure is carrying current, the magnetic field associated with these currents will change when the structure vibrates. If there is no initial current in the conductor, circulatory currents (eddy currents) can be induced when the structure vibrates in a magnetic field with nonzero gradient. The associated time-varying magnetic fields can be detected by a sensing coil.

3. *Magnetostriction effects.* The change of magnetization in a ferromagnetic solid such as iron or steel can be effected by a change in strain (Villari–Matteucci effect; see Chapter 2). Thus, even when the displacements of a structure are small, changes in stress due to vibration can change the local magnitude and direction of the magnetic dipoles in the strained

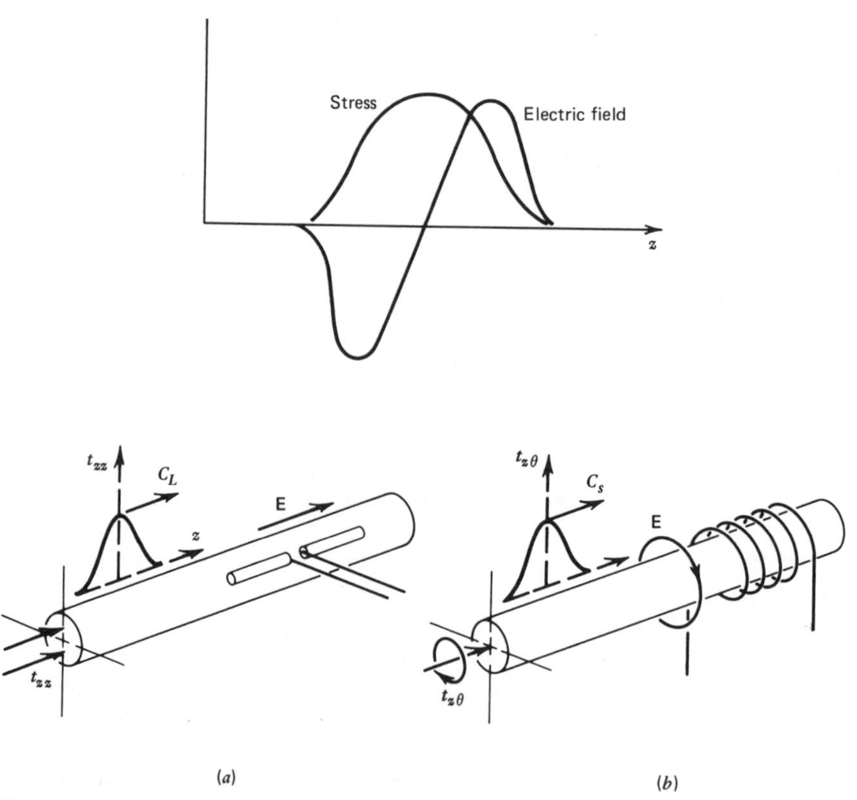

Figure 9-13 Stress wave detection using electron inertia effect (*a*) longitudinal or dilatational wave (dipole sensor) and (*b*) torsional wave multiturn coil sensor (from Moon, 1971).

solid. The resulting changes in magnetic field outside the body can be detected by a sensing coil.

4. *Barkhausen effect.* Ferromagnetic materials are known to have a fine structure of magnetic domains or spin-aligned regions. Changes in macroscopic magnetization are found to be a result of a realignment of these magnetic domains. The switching of spin alignment in domains has been found to be accompanied by a time-varying magnetic field outside the solid which can be converted to voltage or audio signals. (This was discovered by Barkhausen in 1919). Thus vibration-induced strains could initiate changes in domain alignment and produce a sequence of high-frequency signals outside the solid.

5. *Electron gas effect.* A semiclassical model of elastic metallic conductor imagines a lattice of metal ions in a free-electron gas. At low accelerations the electrons and ions move together thus producing zero net electric field outside the solid. However, at high frequencies, or high accelerations, the electrons and ions can have different motions and produce a net electric field outside the body as illustrated in Figure 9-13 (Moon, 1971). This effect has been observed for torsional waves propagating in a rod (Kennedy and Curtis, 1967). The conductor does not have to be ferromagnetic to exhibit this effect.

6. *Triboluminescence.* Intense dynamic plastic strains in solids have been observed to produce measurable radiation outside the solid under impact-type loads.

Of course, similar phenomena occur in nonconductors such as piezoelectric and ferroelectric solids, but we do not deal with structures of these materials in this book.

Ferromagnetic structures

A magnetized solid acts as a magnetic dipole which when vibrated produces time-varying magnetic fields outside the solid. Far from the solid, the dynamic signals behave as electromagnetic waves. However, near the vibrating solid the magnetic fields behave as a quasistatic dipole field (or the sum of dipole fields). The magnetic dipole field is derived from a scalar potential ϕ, (Eq. 3-3.15)

$$\mathbf{B} = -\mu_0 \nabla \phi \tag{9-6.1}$$

The magnetic potential for a point dipole of strength $\mathbf{m}$ is

$$\phi = \frac{\mathbf{m} \cdot \mathbf{r}}{4\pi r^3} \tag{9-6.2}$$

or

$$\mathbf{B} = \frac{\mu_0}{4\pi}\left(\frac{-\mathbf{m}}{r^3} + \frac{3(\mathbf{m} \cdot \mathbf{r})\mathbf{r}}{r^5}\right) \tag{9-6.3}$$

where $\mathbf{r}$ is the position vector from the dipole to the field point. When the structure vibrates, both $\mathbf{m}$ and $\mathbf{r}$ will vary with time. If $\mathbf{r}_0$ is the initial position and $\mathbf{u}(t)$ the displacement of dipole then

$$\mathbf{r}(t) = \mathbf{r}_0 + \mathbf{u}(t) \tag{9-6.4}$$

The magnetization can change magnitude as well as direction. If permanently magnetized, $\mathbf{m}$ can change by the rotation vector of the dipole $\boldsymbol{\Omega}(t)$, that is,

$$\mathbf{m}(t) = \mathbf{m}_0 + \boldsymbol{\Omega} \times \mathbf{m}_0 \tag{9-6.5}$$

The motions $\mathbf{u}(t)$ and $\boldsymbol{\Omega}(t)$ will generate a dynamic field $\mathbf{B}_1$, linear in $\mathbf{u}$ and $\boldsymbol{\Omega}$, and a nonlinear contribution $\mathbf{B}_2(t)$,

$$\mathbf{B} = \mathbf{B}_0 + \mathbf{B}_1(t) + \mathbf{B}_2(t) \tag{9-6.6}$$

where

$$\mathbf{B}_1(t) = \frac{\mu_0}{4\pi}\left(\frac{\mathbf{m}_0 \times \boldsymbol{\Omega}(t)}{r_0^3} + \frac{3\mathbf{r}_0(\mathbf{r}_0 \cdot \boldsymbol{\Omega} \times \mathbf{m}_0)}{r_0^5} + \frac{3}{r_0^5}[\mathbf{u}(\mathbf{m}_0 \cdot \mathbf{r}_0) + \mathbf{r}_0(\mathbf{m}_0 \cdot \mathbf{u}) + \mathbf{m}_0(\mathbf{r}_0 \cdot \mathbf{u})] + \frac{15}{r_0^7}\mathbf{r}_0(\mathbf{m}_0 \cdot \mathbf{r}_0)\mathbf{r}_0 \cdot \mathbf{u}\right) \tag{9-6.7}$$

The first two terms on the right-hand side of Eq. (9-6.7) depend on $1/r_0^3$, while the next two terms depend on $1/r_0^4$. The dynamic field $\mathbf{B}_1(t)$ is not only linear in $\boldsymbol{\Omega}$ and $\mathbf{u}$, but is also linear in the initial magnetization $\mathbf{m}_0$.

Many thin structures such as plates and shells have sharp edges which act as concentrators of magnetization. Thus for a plate with an edge concentration of magnetization $\mathbf{m}_0$ per unit length along the x axis, integration on the field contribution of the dipoles along the edge, considering the first term in $\mathbf{B}_1(t)$, gives

$$\int_{-\infty}^{\infty} \frac{\mathbf{m}_0 \times \boldsymbol{\Omega}\, dx}{(x^2 + y^2 + z^2)^{3/2}} = \frac{2\mathbf{m}_0 \times \boldsymbol{\Omega}}{\rho_0^2} \tag{9-6.8}$$

where $\rho_0 = (y^2 + z^2)^{1/2}$. [Note $\mathbf{m}_0$ and $\boldsymbol{\Omega}(t)$ are assumed to be constant along the edge.) We note further that for a plate or beam the rotation vector $\boldsymbol{\Omega}$ is related to the lateral displacement of the plate $\mathbf{u}$. Thus if a plate lies in the x-y plane,

$$\mathbf{u} = (0, 0, w(x, y, t)) \tag{9-6.9}$$

and

$$\boldsymbol{\Omega} = -\mathbf{n} \times \boldsymbol{\nabla} w$$

where $\mathbf{n}$ is normal to the plate and $\boldsymbol{\nabla}$ is the divergence operator in the plane of the plate.

Multiturn Coil Method

The time-varying magnetic field can be detected using a passive multiturn coil whose voltage output can be displaced graphically on an oscilloscope. If λ is the electromagnetic wavelength, then the radiation terms contribute at distances $r_0 \gg \lambda$; whereas for $r_0 \ll \lambda$, the quasistatic field dominates. Note that for $f = 10^6$ Hz, $\lambda = 300$ m. This justifies the neglect of radiation terms for fields close to the structure. An illustration of the use of this technique is shown in Figures 9-14–9-16.

A sensing coil was wound from 0.05-mm (0.002-in.)-dia copper wire onto a 2.1-cm × 9.0-cm plastic coil form with 1000 turns. In the experiments, free vibrations of the flexible steel structures were generated by deflecting the beam or plate and suddenly releasing the structure. The test beam had a first mode natural frequency of 140 Hz.

Another specimen used was a solid steel rod of circular cross section [6.35-mm (0.25-in.) diameter] bent into a circular ring 23.5 cm (9.25 in.) in diameter. One quadrant of the ring was clamped which produced a fundamental frequency of vibration of 88 Hz.

The small cantilevered beam-plate was first vibrated in a steady uniform magnetic field by placing the plate between the poles of a 10-cm-diam electromagnet. The magnetic-flux lines were normal to the face of the beam. The induced voltage in the 1000-turn search coil as a function of applied magnetic field is shown in Figure 9-14. As one might expect, the output voltage is linear in the range from 100 to 400 G. The asymptotic voltage as $B \rightarrow 0$ is due to the residual magnetization in the steel plate itself. In this test,

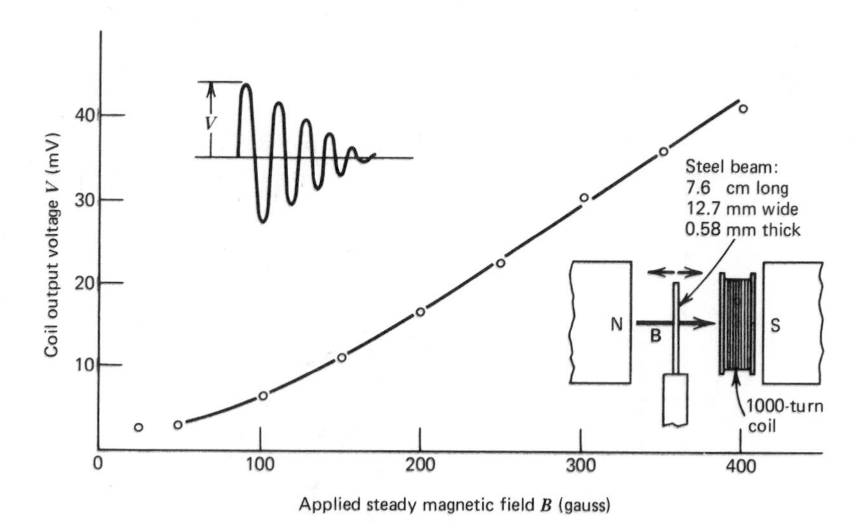

Figure 9-14 Multiturn coil vibration detection measurements—vitrating ferromagnetic beam in a steady magnetic field.

the search coil face was parallel to, and 1.3 cm from, the beam-plate. The initial deflection of the beam was 2.3 mm (0.9 in.).

The next series of experiments were conducted with the cantilevered beam out of the electromagnet in the earth's magnetic field (0.5 G). Three different residual magnetization values of the beam tip were used by magnetizing the plate: 15, 10, and 6 G at the tip of the beam. The search coil output voltage as a function of coil-beam distance is shown in Figure 9-15. The initial deflection was 6.34 mm (0.25 in.). Measurable voltages (0.1 mV) were observed at a distance of 10 cm. The coil face was again parallel to the beam face and perpendicular to the direction of vibration. No signals were obtained when the coil was rotated 90° or when its face was parallel to the direction of vibration.

The variation of output voltage as a function of distance is shown plotted on log-log scale in Figure 9-15. The behavior is of the form

$$V = \frac{K}{x^n}$$

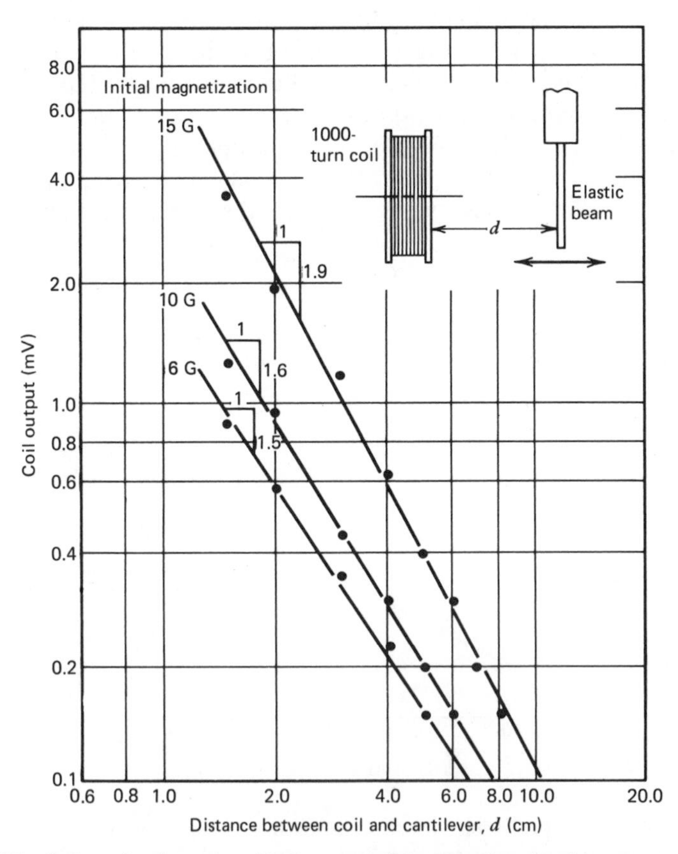

Figure 9-15 Induced voltage in a 1000-turn coil from a vibrating ferromagnetic beam in the earth's magnetic field.

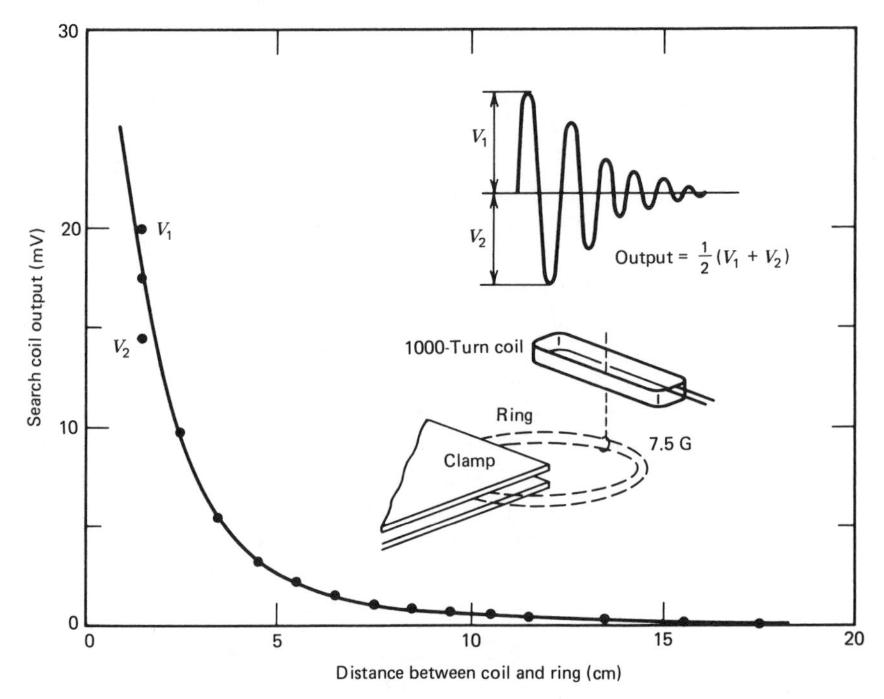

Figure 9-16 Induced voltage in a 1000-turn coil from a vibratory circular ferromagnetic ring in the earth's magnetic field.

where $1.5 < n < 1.9$. This is a smaller decrease than the simple theory would predict.

Another experiment was conducted with a solid steel rod bent into a circular ring. The free edges of the ring were clamped such that the ring could vibrate over $\frac{3}{4}$ of its circumference. This test was conducted to eliminate the effect of the sharp edges which were present in the cantilevered plate. The ring had a residual magnetization on the outside surface of 7.5 G. The ring was vibrated normal to the plane of the ring. The search coil face was placed parallel to the plane of the ring directly over the circumference of the ring. The initial deflection of the ring was 8.6 mm. The output voltage as a function of coil-ring distance is shown in Figure 9-16. The variation of voltage with distance appears to be similar to that predicted by simple theory, that is,

$$V = \frac{K}{x^3}$$

SQUID Magnetometer

The ability to measure magnetic fields due to vibrating structures may be enhanced by using more sensitive magnetic field transducers. One method might be to use a megaturn coil at cryogenic temperatures to lower the resistance. However, the most sensitive method is the "Superconducting Quantum Interference Device" or SQUID. These devices are based on the Josephson effect and are capable of measuring magnetic field changes as low as 10^{-10} T. Their drawback, however, is that at this extremely low sensitivity other types of magnetic noise external to the vibration structure will be detected. Nonetheless, it seems worthwhile to explore this technique for the detection of vibrations.

The SQUID technique for measuring small magnetic fluxes has been used to detect vibrations as low as 2×10^{-10} m. Normally, a single loop SQUID system is sensitive to external magnetic fields, such as automobiles or elevators moving in the earth's field. To avoid having to work in a shielded environment, second- and third-order magnetometers have been used. A second-order magnetometer consists of three colinear coils with equal axial separation as shown in Figure 9-17. The middle coil has twice the number of turns as the two outer coils. The coils are connected in series to the Josephson junction, such that the junction output is proportional to

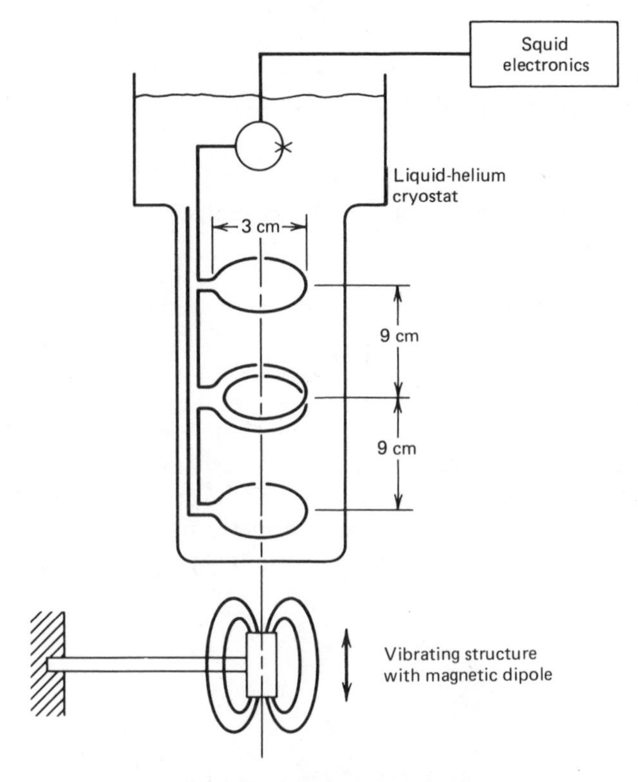

Figure 9-17 Sketch of SQUID vibration detection apparatus. (Adopted from Rutten et al., 1982, reprinted with permission. Butterworth Scientific Ltd., copyright 1982.)

$$\Phi = \Phi_1 - 2\Phi_2 + \Phi_3$$

where Φ_i is the magnetic flux thru each coil. If the magnetic field has a strong gradient in the axial direction, the flux Φ is sensitive only to the second derivative of B_z,

$$\Phi \sim A\frac{\partial^2 B_z}{\partial z^2}$$

In this way flux changes due to long-range sources such as elevators or autos are canceled out. Only changes due to the motion of magnetic dipoles near the magnetometer affect the output. This technique has been used to measure vibration amplitudes as small as 2×10^{-10} m (see Rutten et al., 1982). Applications such as measurement of eye motions and vibrations of the middle ear have employed SQUID gradient magnetometers. When the vibrating part is not magnetic, small rare-earth magnets can be attached to the vibrating object (Rutten et al., 1982).

One of the major disadvantages of this technique, however, is the relatively high cost of the electronic and cryogenic equipment.

9.8 SUPERCONDUCTING EXPERIMENTS AT CRYOGENIC TEMPERATURES

Numerous magnetic devices are designed for cryogenic environments. A few normal conducting magnets have been designed to operate at low temperatures in order to decrease the resistivity and lower the I^2R or Joule-heating losses. However, most cryogenic magnets are superconducting and to achieve high current densities they are usually operated at liquid-helium temperatures (e.g., 4.2°K at 1 atm). Specialized references on cryogenic engineering, such as Barron (1966), or a text on superconducting devices, such as Brechna (1973), should be consulted by the novitiate about to enter the realm of cryogenic experiments. This brief review may serve as a guide to the major topics needed to understand and design magnetomechanical experiments on superconducting materials or magnets.

These topics include:

1. Cryogenic fluids.
2. Dewars and transfer lines.
3. Heat transfer at low temperatures.
4. Safety issues.

Cryogenic Fluids

The most widely used cryogenic fluids for low-temperature experiments are liquid nitrogen (78°K at 1 atm) and liquid helium (4.2°K at 1 atm). A list of properties of these and other cryogenic fluids is given in Table 9-1. It should be noted that the temperature of a cryogenic fluid depends on the pressure. (Most experiments are conducted near 1 atm of pressure.) Liquid nitrogen is usually used as an intermediate thermal barrier between 4.2°K and room temperature (300°K) for superconductivity experiments. It is relatively inexpensive and readily available. Liquid neon (27°K), for example, is not readily available and has a temperature above all known superconductors, but has been proposed for cryogenic, low-resistance normal conducting magnets. Liquid hydrogen (20°K) has safety problems since it can react explosively with oxygen. Its atmospheric-pressure temperature is only marginally below the highest known superconducting temperature and is not used for superconducting experiments.

Table 9-1 Properties of Cryogenic Liquids

Liquid	Boiling Point at 1 atm (°K)	Density (g/cm^3)	Specific Heat (J/cm^3 · °K)	Thermal Conductivity (mW/cm · °K)
Helium 3	3.2	0.059	0.27	0.17
Helium 4	4.2	0.125	0.57	0.27
Hydrogen	20.4	0.071	0.69	1.18
Neon	27.1	1.205	2.22	1.30
Nitrogen	77.3	0.811	1.65	1.39

Source: Barron (1966).

Liquid helium is the most widely used cryogenic fluid environment for superconducting experiments, but is considerably more expensive than liquid nitrogen. For small-scale cryogenic environments, for example, less than 100-ℓ volume, an open loop refrigeration system is used and the gaseous helium which results from heat transfer into the cryostat is simply vented into the atmosphere. However, for larger systems, a closed loop system is employed to recover and recool the gaseous helium.

Helium Cryostats

Small-scale experiments on superconducting materials are usually performed in a pool of liquid helium. The vessel for holding the helium is called a Dewar. The standard Dewars are cylindrical in shape, with diameters ranging from around 10 cm to 1 m. The 4.2°K liquid-helium pool is shielded from the 300°K environment by both vacuum and liquid-nitrogen cylindrical annuli as shown in Figure 9-18.

The small but inevitable heat flow into the helium pool results in a boil-off of liquid into gas. Both open and closed helium supply and recovery systems are in use. In the open system the helium gas is vented to the atmosphere,

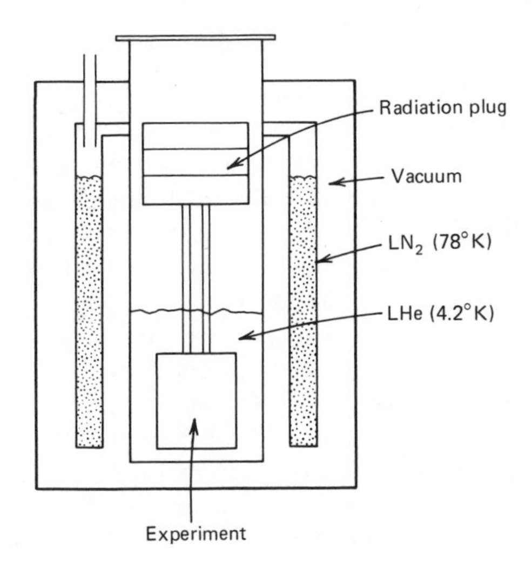

Figure 9-18 Helium cryostat for small-scale experiments on superconducting materials and structures.

whereas in the closed operation the helium gas is sent to either a gas storage tank or to a refrigeration unit which cools the gas down to the liquid state. Helium recovery and refrigeration units are relatively expensive to operate for small experiments (less than 50-ℓ Dewars). For large-scale experiments, the cost of liquid helium makes recovery economical.

To fill the Dewar with liquid helium, the inner vessel is precooled to 78°K with liquid nitrogen (LN_2). After the LN_2 is removed, liquid helium is transferred from a storage Dewar to the test Dewar via a vacuum jacketed transfer line as shown in Figure 9-19.

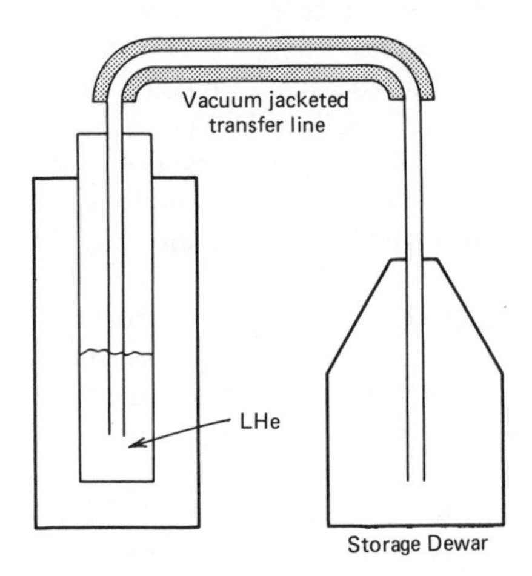

Figure 9-19 Sketch of helium transfer apparatus.

Heat Leaks

Of course, any connection between the experimental apparatus in the Dewar and the 300°K environment is a source of heat leak and a subsequent loss of liquid helium, such as current leads to a superconducting magnet or test material. One technique to minimize the heat leak through the leads is to force the evaporating helium gas to flow around the leads. Vapor-cooled current leads are standard off-the-shelf items for small-scale experiments.

Another heat loss problem is the test model support. To avoid the use of high thermal conducting materials such as steel or aluminum, one technique is to use glass-fiber epoxy structural supports between the 300°K environment and the test apparatus. For example, a glass-fiber-wound epoxy cylindrical tube provides a fairly stiff support platform for test models, especially those which will experience high magnetic forces or torques.

In addition to conduction heat leaks, one must seek to minimize radiation losses. Often a radiation shield is constructed of cylindrical styrofoam plugs separated by high reflecting foil.

9.9 EFFECTS OF MAGNETIC FIELDS ON HUMANS AND EXPOSURE LIMITS

For those who must work in magnetic field environments, there is the question of how safe is exposure to magnetic fields for humans. There has been a long history of industrial exposure to low magnetic fields (less than 0.02 T) and low frequencies (50–60 Hz) with no observed effect on health. However, there are few systematic studies under controlled conditions of the effects of electromagnetic fields on humans. Sheppard and Eisenbud (1977) and Battocletti (1976) offer two monographs which summarize biological effects of electric and magnetic fields. The first text discusses primarily biological effects, while the second surveys the extent of electric and magnetic field pollution in the modern environment.

There are three variables that must be considered when discussing limits of field exposure—magnitude, field gradient, and time rate of change. High dc magnetic fields have been observed to affect the chemical reaction rates of polymeric and biologically important molecules. DNA molecules have been observed to suffer a slight orientation in extremely high fields (>10 T) (see Sheppard and Eisenbud, 1977).

Alternating-current magnetic fields greater than 0.1 T with a frequency range of 10–100 Hz have been observed to evoke a visual response in the retina "magnet phosphene," but are not known to be harmful. Some of the effects of inhomogeneous fields include effects on tissue growth and white blood cell formation.

On the other hand, therapeutic effects of magnetic fields have been reported from the time of Mesmer in 1774. Researchers in eastern Europe claimed in 1967 to have reduced blood cholesterol, white cell count, and blood pressure by placing various parts of the body in the field of a solenoid at frequencies of 40–4000 Hz at field levels of around 0.01 T. In France, at the University of Bordeaux, experiments have been carried out on the use of electromagnetic fields to cure certain cancers in animals (see Battocletti, 1976).

In spite of some efforts to study the effects of magnetic fields on humans, the extent of risk to humans working in high-field environments has not been completely explored. Still, various exposure standards have been promulgated to eliminate what risks may be involved based on present industrial and laboratory experience. Three similar standards are compared in Table 9-2—two from U.S. laboratories and one from the Soviet Union. Note that only the third standard proscribes limits on field gradients. These standards are presumably for dc or low-frequency fields.

Table 9-2 Maximum Exposure Levels in Magnetic Fields

		LBL-DOE (United States)[a] (TESLA)	SLAC (United States)[b]	Soviet Union
Whole body or head	8 h	0.01	0.02 (8 h–1 h)	0.03 (0.2 T/m) (8 h–minutes)
	1 h	0.1		
	minutes	0.5	0.2	
Extremities (hands, etc.)	8 h	0.1	0.2 (8 h–1 h)	0.07 (0.2 T/m) (8 h–minutes)
	1 h	1		
	minutes	2	2	

[a]Lawrence Berkeley Laboratory recommendation to U.S. Department of Energy in 1979.
[b]Stanford Linear Accelerator Center.
[c]See Battocletti (1976).

REFERENCES

Abramova, K. B., Valitskii, V. P., Vandakurov, Yu. V., Zlatin, N. A., and Peregund, B. P. (1966). "Magnetohydrodynamic Instabilities during Electrical Explosion," *Sov. Phys.—Dokl.* **11**, No. 4, 301–304.

Aharoni, A. (1963). "Complete Eigenvalue Spectrum for the Nucleation in a Ferromagnetic Prolate Spheroid, " *Phys. Rev.* **131**, No. 4, 1478–1482.

Alblas, J. B. (1979). *General Theory of Electro- and Magneto-Elasticity*, Springer-Verlag, New York.

Alers, G. A., and Fleury, P. A. (1963). "Modification of the Velocity of Sound in Metals by Magnetic Fields," *Phys. Rev.* **129**, No. 6, 2425.

Al-Hassani, S. T. S (1974). "Plastic Buckling of Thin Walled Tubes Subject to Magnetomotive Force," *J. Mech. Eng. Sci.* **16**.

Al-Hassani, S. T. S., Duncan, J. L., and Johnson, W. (1974). "On the Parameters of the Magnetic Forming Process," *J. Mech. Eng. Sci.* **16**, No. 1, 1–9.

Al-Hassani, S. T. S., and Johnson, W. (1970), "The Magnetomotive Loading of Cantilevers, Beams and Frames," *Int. J. Mech. Sci.* **12**, 711–722.

Ambartsumyan, S. A. (1982). "Magnetoelasticity of Thin Plates and Shells", Review Article, *Appl. Mech. Rev.* (January).

Ambartsumyan, S. A., Bagdasaryan, G. E., and Belubekyan, M. V. (1977). *Magnetoelasticity of Thin Shells and Plates* (in Russian), Figmatgiz, Moscow.

Arp, V. (1977). "Stresses in Superconducting Solenoids," *J. App. Phys.*, **48**, No. 5, 2026–2036.

Baiko, A. V., Voevodskii, K. E., Kochetkov, V. M. (1980). "Vertical Unstable Stability of Electrodynamic Suspension of High-Speed Ground Transport," *Cryogenics* **20**, 271–276.

Baines, K., Duncan, J. L., and Johnson, W. (1965–1966). "Electromagnetic Metal Forming," *Proc. Inst. Mech. Engrs.* **180**, Paper 37, Part 31, 348–362.

Barron, R. (1966). *Cryogenics Systems*, McGraw-Hill, New York.

Battocletti, J. H. (1976). *Electromagnetics, Man and the Environment*, Westview Press, Boulder, Colo.

Bean, C. P. (1972). "Magnetization of Hard Superconductors," *Phys. Rev. Lett.* **8**, No. 6, 250–253.

Berkovsky, B., and Rosensweig, R. E. (1979). "Magneto Fluid Mechanics: A Report on an International Advanced Course and Workshop," *J. Fluid Mech.* **87**, Part 3, 521–531.

Bialek, J. (1976). *Analysis of Laminated Beams for Use in Coil Design*, Princeton Plasma Physics Department Report No. MATT-1212, Princeton, New Jersey, February 1976; Proceedings of the Sixth Symposium on Engineering Problems of Fusion Research, November 1975.

Binns, K. J., and Lawrenson, P. J. (1963). *Analysis and Computation of Electric and Magnetic Field Problems*, Macmillan, New York.

Birdsall, D. H., Ford, F. C., Furth, H. D. and Riley, R. E. (1961). "Magnetic Forming," *Am. Mach.* **105** (6), 117–121.

Bitter, F. (1959). *Magnets: The Education of a Physicist*, Doubleday, New York.

Bobrov, E. S., Marston, P. G., Kuznetsov, E. N. (1982). "Theoretical and Engineering Aspects of Momentless Structures and Coil End Turns in Superconducting MHD Magnets," *Adv. Cryogenic Eng.* **27**, 47–55.

Bobrov, E. S., and Williams, J. E. C. (1980). "Stresses in Superconducting Solenoids." In *Mechanics of Superconducting Structures*, F. C. Moon, Ed., ASME, New York, pp. 13–14.

Boley, B. A., and Tolins, I. S. (1962). "Transient Coupled Thermoelastic Boundary Value Problems in the Half Space," *J. Appl. Mech.* **29**, No. 4, 637–646.

Bonanos, P. (1981). *Bow Shaped Toroidal Field Coils*, Princeton Plasma Physics Laboratory Report No. PPPL-1790, Princeton, New Jersey.

Boyle, J. C., Mosher, E. J., and Greyerbiehl, J. M. (1964). *Experimental Determination of Eddy-Current Torques on Bodies Rotating Relative to a Uniform Magnetic Field*, NASA Report No. NASA TN D-5086, Goddard Space Flight Center, Md.

Bozorth, R. M. (1951). *Ferromagnetism*, Van Nostrand, New York.

Brechna, H. (1973). *Superconducting Magnet Systems*, Springer-Verlag, New York.

Brillouin, L. (1953). *Wave Propagation in Periodic Structures*, Dover, New York.

Brown, W. F. (1962). *Magnetostatic Principles in Ferromagnetism*, North-Holland, Amsterdam.

Brown, W. F. (1966). *Magnetoelastic Interactions*, Tracts in Natural Philosophy, Springer, Berlin, No. 9.

Burgess, T. J., Cnare, E. C., Oberkampf, W. L., Beard, S. G., and Cowan, M. (1982). "The Electromagnetic Gun and Tubular Projectiles," *IEEE Trans. Mag.* **MAG-18**, No. 1, 46–59.

Carpenter, C. J. (1977). "Comparison of Alternate Formulation of 3-Dimensional Magnetic Field and Eddy Current Problems at Power Frequencies," *Proc. IEEE* **124**, No. 11, 1026–1034.

Case, J. (1938). *The Strength of Materials*, Arnold, London.

Chadwick, P. (1956). "Elastic Wave Propagation in a Magnetic Field," Proceedings of the International Congress of Applied Mechanics, Brussels, Belgium, pp. 143–153.

Chandrasekhar, S. (1956). "On Force-Free Magnetic Fields," *Proc. Natl. Acad. Sci.* **42**, No. 1, 1–9.

Chandrasekhar, S. (1960). "The Virial Theorem in Hydrodynamics," *J. Math. Anal. Appl.* **1**, 240–252.

Chandrasekhar, S., and Fermi, E. (1953). "Problems of Gravitational Stability in the Presence of a Magnetic Field," *Astrophys. J.* **118**, 116–141.

Chari, M. V. K., and Silvester, P. P., Eds. (1980). *Finite Elements in Electrical and Magnetic Field Problems*, Wiley, New York.

Chattopadhyay, S. (1979). "Magnetoelastic Instability of Structures Carrying Electric Current," *Int. J. Solids Struct.* **15**, 467–477.

Chattopadhyay, S., and Moon, F. C. (1975). "Magnetoelastic Buckling and Vibration of a Rod Carrying Electric Current," *J. Appl. Mech.* **42**, 809–914.

Chen, P. J., and McCarthy, M. F. (1975). " The Behaviour of Plane Shock Waves in Deformable Magnetic Materials," *Acta Mech.* **23**, 91–102.

Chian, C. T., and Moon, F. C. (1981). "Magnetically Induced Cylindrical Stress Waves in a Thermoelastic Conductor," *Int. J. Solids Struct.* **17**, No. 11, 1021–1035.

Chilton, F., Hibbs, B., Kolm, H., and O'Neill, G. K. (1977). "Electromagnetic Mass-Driver," Space-Based Manufacturing from Non-Terrestrial Materials, *Progress in Astronautics and Aeronautics* **57**, AIAA, New York.

Christensen, R. M. (1979) *Mechanics of Composite Materials*, Wiley, New York.

Chu, D., and Moon, F. C. (1983). "Dynamic Instabilities in Magnetically Levitated Models," *J. Appl. Phys.* **54**, No. 3, 1619–1625.

Cockroft, J. D. (1928). "The Design of Coils for the Production of Strong Magnetic Fields," *Phil. Trans. Roy. Soc. London Ser.* **A 227**, 317.

Cornish, D. (1981). Personal Communication.

Cowen, C., Friedlander, F. J., and Joluria, R. (1976a). "Single Wire Model of High Gradient Magnetic Separation Processes I," *IEEE Trans. Mag.* **MAG-12**, No. 5, 466–470.

Cowen, C., Friedlander, F. J., and Joluria, R. (1976b). "single Wire Model of High Gradient Magnetic Separation Processes II," *IEEE Trans. Mag.* **MAG-12**, No. 6, 898–900.

Crandall, S. H., Karnopp, D. C., Kurtz, E. F. Jr., and Pridmore-Brown, D. C. (1968). *Dynamics of Mechanical and Electromechanical Systems*, McGraw-Hill, New York.

Cullity, B. D. (1972). *Introduction to Magnetic Materials*, Addison-Wesley, Reading, Mass.

Dalder, E. N. C. (1981). *Welding of Austenitic Stainless Steels for Liquid Helium Services*, Proceedings of the Conference on MHD Magnets—MIT, Cambridge, Mass. Lawrence Livermore Laboratory Preprint No. 85818, Livermore, Calif.

Dalrymple, J. M., Peach, M. O., Viegelahn, G. L. (1974). "Magnetoelastic Buckling of Thin Magnetically Soft Plates in the Cylindrical Mode," *J. Appl. Mech.* **41**, Series E, No. 1, 145–150.

Daniels, J. M. (1953). "High Power Solenoids: Stresses and Stability," *Brit. J. Appl. Phys.* **4**, No. 2, 50–54.

Davis, L. C., and Wilkie, D. F. (1971). "Analysis of Motion of Magnetic Levitation Systems: Implications," *J. Appl. Phys.* **42**, No. 12, 4779–4793.

Deadrich, F. J., Hawke, R. S., and Scudder, J. D. (1982). "MAGRAC—A Railgun Simulation Program," *IEEE Trans. Mag.* **MAG-18**, No. 1, 94–104.

DeBlois, R. W. (1967). "Ferromagnetic Properties of Single-Crystal Nickel Platelets and Submicron Whiskers," *J. Appl. Phys.* **38**, No. 3, 1291.

Deis, D. W., Cornish, D. N., Rosdahl, A. R., and Hirzel, D. G. (1977). "Mechanical Strain of Large, Multifilament Nb_3Sn Conductors for Fusion Magnets," Proceedings of the Sixth International Conference on Magnet Technology.

Deis, D. W., and McNab, I. R. (1982). "A Laboratory Demonstration Electromagnetic Launcher," *IEEE Trans. Mag.* **MAG-18**, No. 1, 16–22.

Dolbin, N. I. (1962). "Propagation of Elastic Waves in a Current-Carrying Rod," *PMTF*, No. 2 (in Russian).

Dolbin, N. I., and Morozov, A. I. (1966). "Elastic Bending Vibrations of a Rod Carrying Electric Current," *Zh. Prik. Mekh. Tek. Fiz.* No. 3, 97–103.

Dresner, L. (1979). *Eddy Current Heating of Irregularly Shaped Plates by Slow Ramped Fields*, Oak Ridge National Laboratory Report No. ORNL/TM–6968, Oak Ridge, Tenn.

Dunkin, J. W., and Eringen, A. C. (1963). "Propagation of Waves in an Electromagnetic Elastic Solid," *Int. J. Eng. Sci.* **1**, 461–495.

Dunsheath, P. (1962). *A History* of *Electrical Power Engineering*, The MIT Press, Cambridge, Mass.

Dwight, H. B. (1945). *Electrical Coils and Conductors*; *Their Electrical Characteristics and Theory*, McGraw-Hill, New York.

Earnshaw, S. (1842). "On the Nature of the Molecular Forces Which Regulate the Constitution of the Luminiferous Ether," *Trans. Cambridge Philos. Soc.* **7**, 97–114.

Ekin, J. W., and Clark, A. F. (1976). *Effect of Strain on the Critical Current of* Nb_3Sn and Nb-Ti *Multifilamentary Composite Wire*, Proceedings of the Magnetism and Magnetic Materials Conference, Pittsburg, Penn.; American Institute of Physics Publications No. 34.

Ekin, J. W., Fickett, J. R., and Clark, A. F. (1977). *Adv. Cyrogenic Eng.* **22**, 449.

Eringen, A. C. (1963). "On the Foundations of Electroelastodynamics," *Int. J. Eng. Sci.* **1**, 127–153.

Eringen, A. C. (1967). *Mechanics of Continua*, Wiley, New York.

Eringen, A. C. (1971). "Micromagnetism and Superconductivity," *J. Math. Phys.* **12**, No. 7, 1353–1358.

Eyssa, Y. M. (1980). "Design Configuration for Large Superconductive Energy Magnets," In *Mechanics of Superconducting Structures*, AMD-Vol. 41. F. C. Moon, Ed., ASME, New York.

Feynman, R. P., Leighton, R. B., and Sands, M. (1964). *The Feynman Lectures on Physics II*, Addison-Wesley, Reading, Mass.

File, J., Mills, R. G., and Sheffield, G. V. (1971). "Large Superconducting Magnet Designs for Fusion Reactors," Fourth Symposium on Engineering Problems of Fusion Research, Naval Research Laboratory Washington, D.C.

Fisher, E. S., Kim, S. H., and Linz, R. J. (1975). "Effect of Cyclic Strain on Electrical Resistivity of Cu at 4.2°K," Proceedings of the International Conference on Cryogenic Materials, Kingston, Ontario, Canada.

Flükiger, R. (1980). "A15 Superconducting Wires Produced by Discontinuous Fibre Methods," *IEEE Trans. Mag.* **MAG-16**, No. 5, 1236–1241.

Forrestal, M. J. and Overmier, D. K. (1974). "An Experiment on an Impulse Loaded Elastic Ring," *AIAA J.* **12**, No. 5, 722–724.

Frazier, R. H., Gilison, P. J., Jr., and Oberbeck, G. A. (1974). *Magnetic and Electric Suspensions*, MIT Press, Cambridge, Mass.

Frei, E. H., Shtrikman, S., and Treves, D. (1957). *Phy. Rev.* **106**, 446.

Freynik, H. S. Jr., Roach, D. R., Deis, D. W., and Hirzel, D. G. (1978). "Evaluation of Metal-Foil Strain Gages in Magnetic Fields," *Adv. Cryogenic Eng.* **24**, 473–479.

Furth, H. P. (1961). "Pulsed Magnets." In *High Magnetic Fields*, H. Kolm et al., Eds., MIT Press, Cambridge, Mass.

Furth, H. P., Levine, M. A., and Waniek, R. W. (1956). "Production and Use of High Transient Magnetic Fields, I," *Rev. Sci. Instr.* **27**, 195–203.

Furth, H. P., Levine, M. A., and Waniek, R. W. (1957). "Production and Use of High Transient Magnetic Fields II," *Rev. Sci. Instr.* **28**, 949–958.

Galkin, A. A., and Koroliuk, A. P. (1958). "Dispersion of Sound in Metals in a Magnetic Field," *J. Exptl. Theor. Phys.* (USSR) **34**, 1025–1026.

Garden, P. O. (1968). "Mechanical Stresses in Bounded Plane Helical Solenoids with Arbitrary External Field," *J. Sci. Instr.*, Ser. 2, I, 437.

Geary, P. J. (1964). *Magnetic and Electric Suspensions*, British Scientific Instrument Research Association. Report No. R314.

Georgievsky, A. V., Ziser, V. E., and Litvinenko, Yu. A. (1974). *Employment of Partly 'Force-Free' Toroidal Magnet Systems for Producing Stellarator-Type Traps*, Proceedings of the Fifth Symposium on Engineering Problems of Fusion Research, IEEE Publications No. 73CHO843-3-NPS, 511–514.

Gersdorf, R., Muller, F. A., and Roeland, L. W. (1965). "Design of High Field Magnet Coils for Long Pulses," *Rev. Sci. Instr.* **36**, No. 8, 1100–1109.

Goldstein, H. (1950). *Classical Mechanics*, Addison Wesley, Reading, Mass.

Gralnick, S. L., and Tenney, F. H. (1976). *Analytic Solutions for Constant Tension Coil Shapes*, Plasma Physics Laboratory Report No. MATT-1197 Princeton, New Jersey.

Gray, W. H., and Ballou, J. K. (1977). *Electromechanical Stress Analysis of Transversely Isotropic Solenoids*, Oak Ridge National Laboratory Report No. ORNL/TM-5528, Oak Ridge, Tenn.

Gray, W. H., and Sun, C. T. (1976). *Theoretical and Experimental Determination of Mechanical Properties of Superconducting Composite Wire*, Oak Ridge National Laboratory Report No. ORNL/TM-5331.

Grot, R. A., and Eringen, A. C. (1966). "Relativistic Continuum Mechanics, Part II—Electromagnetic Interaction with Matter," *Int. J. Eng. Sci.* **4**, 639–670.

Grover, F. W. (1946). *Inductance Calculations*, Dover, New York.

Gunn, J., and Billinghurst, E. (1957). "Magnetic Fields Affect Strain Gages," *Control Eng.*, August 1957.

Hague, B. (1929). *The Principles of Electromagnetism Applied to Electrical Machines* (originally Oxford University Press, 1929), Dover, New York.

Hashmi, S. J., and Al-Hassani, S. T. S. (1975). "Large Deflection Response of Square Frames to Distributed Impulse Loads," *Int. J. Mech. Sci.* **17**, 513–523.

Hawke, R. S., and Scudder, J. K. (1979). *Magnetic Propulsion Railguns: Their Design and Capabilities*, Second International Conference on Megagauss Magnetic Field Generation and Related Topics, Washington, D. C; Lawrence Livermore Laboratory Report No. UCRL-82677, Livermore, Calif.

Heim, J. R. (1974). Fermi National Laboratory Report No. TM-334-B.

Hein, R. A. (1974). "Superconductivity: Large-Scale Applications," *Science* **185**, No. 4147, 211–222.

Herlach, F., and McBroom, R. (1973). "Megagauss Fields in Single Turn Coils," *J. Phys. E.* **6**, 652.

Hillier, M. J., and Lal, G. K. (1968). "The Electrodynamics of Electromagnetic Forming," *Int. J. Mech. Sci.* **10**, 491–500.

Hoard, R. W. (1980). *The Effects of Strain on the Superconductivity Properties of Niobium-Tin Conductors*, Ph.D. dissertation, University of Washington; Lawrence Livermore Laboratory Report No. UCRL-53069 Livermore, Calif.

Holm. R. (1967). *Electric Contacts, Theory and Application*, 4th ed., Springer-Verlag, New York.

Horikawa, H., Dowell, E. H., and Moon, F. C. (1978). "Active Feedback Control of a Beam Subjected to a Nonconservative Force," *Int. J. Solids Struct.* **14**, 821–829.

Horvath, J. (1980). "Mechanical Behavior of the Mirror Fusion Test Facility Superconducting Magnet Coils," In *Mechanics of Superconducting Structures*, AMD-Vol. 41, F. C. Moon Ed., ASME, New York.

Huebener, R. P. (1979). *Magnetic Flux Structures in Superconductors*, Springer-Verlag, Berlin.

Hutter, K. (1973). "Electrodynamics of Deformable Continua," *Ph.D. thesis*, Cornell University, Ithaca, New York.

Hutter, K., and Pao, Y-H. (1974). "A Dynamical Theory for Magnetizable Elastic Solids with Thermal and Electrical Conduction," *J. Elas.* **4**, 89–114.

Hutter, K., and van der Ven, A. A. F. (1978). *Field Matter Interactions in Thermoelastic Solids*, Lecture Notes in Physics No. 88, Springer-Verlag, Berlin.

Iwamoto, M., Yamada, T., and Ohno, E. (1974). "Magnetic Damping Force in Electrodynamically Suspended Trains." 1974 *Digest of International Conference of the IEEE*, Toronto, Canada, Section 19, p. 14.

Jackson, J. D. (1962). *Classical Electrodynamics*, Wiley, New York.

Jahn, R. G. (1968). *Physics of Electric Propulsion*, McGraw-Hill, New York.

Jeans, J. (1925). *The Mathematical Theory of Electricity and Magnetism*, 5th ed., Cambridge University Press, Cambridge, Great Britain.

Johnson, N. E., Gray, W. H., and Weed, R. A. (1976). "Stress Analysis of Non-Homogeneous Transversely Isotropic Superconducting Solenoids," *Proceedings of the Sixth Symposium on Engineering Problems of Fusion Research*, San Diego, Calif., pp. 243–247.

Kaliski, S. (1962). "Magnetoelastic Vibrations of Perfectly Conducting Plates and Bars Assuming the Principle of Plane Sections," *Proc. Vibr. Probl.* **3**, No. 4.

Kaliski, S. (1969). "Quasi-Static Approximation to the Equation of Elastic Vibrations in a Ferromagnetic Plate under the Action of a Transverse Magnetic Field," *Bull. Acad. Pol. Sci.* **XVII**, No. 9, 411–418.

Kaliski, S., and Nowacki, W. (1965). "Combined Elastic and Electromagnetic Waves Produced by Thermal Shock in the Case of a Medium of Finite Electric Conductivity," *Bull. Acad. Sci. Ser. Sci. Tech.* **X**, No. 4, 159–168.

Kaliski, S., and Petykiewicz, J. (1959/1960). "Dynamical Equations of Motion and Solving Functions for Elastic and Inelastic Anisotropic Bodies in the Magnetic Field," *Proc. Vibr. Probl.* **1**, No. 2, 17–35.

Kalmbach, C., Dowell, E. H. and Moon F. C. (1974). "The Application of Feedback Control to the Suppression of a Dynamic Instability of an Elastic Body," *Int. J. Solids Struct.* **10**, 361–381.

Kapitza, P. L. (1927). "Further Developments of the Method of Obtaining Strong Magnetic Fields," *Proc. Roy. Soc. London Ser. A* **115**, 658–670.

Kaplan, B. Z. (1974). "A New Analysis of Tuned Circuit Levitators," *Int. J. Non-Lin. Mech.* **9**, 75–87.

Kaufman, A. (1963). "Investigation of Strain Gages for Use at Cryogenic Temperatures," *Exptl. Mech.* **3**, No. 8, 177.

Kennedy, J. D., and Curtis, C. W. (1967). "Transient Electron-Inertia Field Produced by a Strain Pulse," *J. Acoust. Soc. Am.* **41**, No. 2, 328–335.

Kittel, C. (1968). *Introduction to Solid State Physics*, 3rd ed., Wiley, New York.

Knopoff, L. (1955). "The Interaction between Elastic Wave Motions and a Magnetic Field in Electrical Conductors," *J. Geophys. Res.* **60**, 441–456.

Koch, C. C., and Easton, D. S. (1977). "A Review of Mechanical Behavior and Stress Effects in Hard Superconductors," *Cryogenics*, 391–413.

Köhler, A. (1980). "Eddy Currents; Electromagnetic Measurements of Power Density," Proceedings of the IEEE Power Engineering Society, Paper No. F80 153–7.

Kokavec, J., and Cesnak, L. (1977). "Mechanical Stresses in Cylindrical Superconducting coils," *J. Phys. D*, **10**.

Kolm, H. H., Oberteuffer, J. A., and Kelland, D. R. (1975). "High Gradient Magnetic Separation," *Sci. Am.* **233**, No. 5, 46.

Kostoff, R. N., Peaslee, A. T. Jr., and Ribe, F. L. (1982). "Possible Application of Electromagnetic Guns to Impact Fusion," *IEEE Trans. Mag.* **MAG-18**, No. 1, 194–196.

Kramer, E. J. (1975). "Scaling Laws for Flux Pinning in Hard Superconductors," *J. Appl. Phys.* **44**, No. 3, 1360–1370.

Kuznetsov, A. A. (1960). "Mechanical Stresses Produced by the Radial Electromagnetic Force in Multilayer Coil Wound with Wire of Rectangular Cross Section Carrying a Uniform Current," *Tech. Phys. USSR* **5**, 555–561 (translated from *Zh. Tekh. Fiz.* **30**, No. 5, 592–597).

Kuznetsov, A. A. (1961a). "Mechanical Stresses Produced by the Radial Force in Turns in the Central Region of a Long, Single-Layer, Plane-Helix Coil," *Tech. Phys.* (*USSR*) **6**, No. 8, 687–690 (translated from *Zh. Tekh. Fiz.* **31**, No. 8, 944–947).

Kuznetsov, A. A. (1961b). "Force-Free Coils for Magnetic Fields of Infinite Length," *Tech. Phys.* (*USSR*) **6**, No. 6, 472–475.

Laithwaite, E. R. (1977). *Transport Without Wheels*, Elek Science, London.

Lammeraner, J., and Stafl, M. (1966). *Eddy Currents*, CRC Press Illiffe Books Ltd., London.

Landau, E., and Lifshitz, M. (1960). *Electrodynamics of Continuous Media*, Pergamon Press, New York.

Lawson, W. F., Simons, W. H., and Treat, R. P. (1977). "The Dynamics of a Particle Attracted by a Magnetized Wire," *J. Appl. Phys.* **48**, No. 8, 3213–3224.

Lekhnitskii, S. G. (1963). *Theory of Elasticity of an Anisotropic Elastic Body*, Holden-Day, San Francisco.

Leontovich, M. A., and Shafronov, V. D. (1961). "The Stability of a Flexible Conductor in a Magnetic Field." In *Plasma Physics and the Problem of Controlled Thermonuclear Reactions I*, Pergamon Press, New York.

Levy, R. H. (1962). "Author's reply to Willinski's Comment on Radiation Shielding of Space Vehicles by Means of Superconducting Coils," *Am. Rocket Soc. J.* **32**, 787.

Levy, S., and Truel, R. (1953). "Ultrasonic Attenuation in Magnetic Single Crystals," *Rev. Mod. Phys.* **25**, No. 1, 140–145.

Librescu, L. (1977). "Recent Contributions Concerning the Flutter Problem of Elastic Thin Bodies in an Electrically Conducting Gas Flow, a Magnetic Field Being Present," *Solid Mech. Arch.* **2**, No. 1, 1–108.

Lippman, H. J., and Schreiner, H. (1964). "Zur Physik der Metallumforming Mit Hohen Magnetfeldimpulsen," *Zeits. Metalkd.* **55**, 737–740.

Liu, Y. A. (1979). *Industrial Applications of Magnetic Separation*, Proceedings of the International Conference on Magnetic Separation, New Hampshire: IEEE Publication No. 78CH1447–2MAG.

Lontai, L. M., and Marston, P. G. (1965). "A 100 Kilogauss Quasi-Continuous Cryogenic Solenoid–Part 1," In *Proceedings of the International Symposium on Magnet Technology*, H. Brechna and H. S. Gordon, Eds., Stanford University, Stanford, Calif. pp. 723–732.

Love, A. E. H. (1922). *A Treatise on the Mathematical Theory of Elasticity*, 4th Ed. Dover, New York.

Marshall, R. A. (1982). *Railguns*, Proceedings of the Ninth National Congress of Applied Mechanics, Cornell University, Ithaca, New York; ASME, New York.

Matteucci, C. (1850). *Ann. Chim.* **XXVIII**, 493–499.

Maugin, G. A., and Goudjo, C. (1982). "The Equations of Soft Ferromagnetic Elastic Plates," *Int. J. Solids Struct.* **18**, 889–912.

Maxwell, J. C. (1869–1972). "On Reciprocal Figures, Frames, and Diagrams of Forces," *Trans. Roy. Soc. Edinburgh* **XXVI**, 1–40.

Maxwell, J. C. (1874). "Van der Waals on the Continuity of the Gaseous and Liquid States," *Nature*, 477–480.

Maxwell, J. C. (1891). *A Treatise On Electricity and Magnetism*, Vols. I and II, Dover, New York, (reprint).

McCaig, M. (1977). *Permanent Magnets in Theory and Practice*, Pentech Press, London.

McCarthy, M. F. (1966a). "The Growth of Magnetoelastic Waves in a Cauchy Elastic Material of Finite Electrical Conductivity," *Arch. Rat. Mech. Anal.* **23**, No. 3, 191–217.

McCarthy, M. F. (1966b). "The Propagation and Growth of Plane Acceleration Waves in a Perfectly Electrically Conducting Elastic Material in a Magnetic Field," *Int. J. Eng. Sci.* **4**, 361–381.

McLachlan, N. W., (1947). *Theory and Application of Mathieu Functions*, Oxford University Press, London.

Meisenholder, S. G., and Wang, T. C. (1972). *Dynamic Analysis of an Electromagnetic Suspension System for a Suspended Vehicle System*. U. S. Federal Rail Administration Report No. FRA–RT–73–1.

Melcher, J. R. (1981). *Continuum Electromechanics*, MIT Press, Cambridge, Mass.

Mellville, D., and Mattocks, P. G. (1972). "Stress Calculations for High Magnetic Field Coils," *J. Phys. D* **5**, No. 10, 1745–1759.

Middleton, A. J., and Trowbridge, C. W. (1967). "Mechanical Stresses in Large High Field Magnet Coils," Proceedings of the Second International Conference on Magnet Technology, Oxford, England, pp. 140–149.

Miya, K. (1977). "Dynamic Analysis of Electro-Magneto-Mechanical Field by Finite Element Method," *J. Fac. Eng. Univ. Tokyo (B)*, **XXXIV**, No. 1, 137–157.

Miya, K., Hara, K., and Someya, K. (1978). "Experimental and Theoretical Study on Magnetoelastic Buckling of a Cantilever," *J. Appl. Mech.* **45**, 355–360.

Miya, K., Hara, K., and Tabata, Y. (1980). "Finite Element Analysis of Dynamic Behavior of a Cylinder Due to Electromagnetic Forces," *Nuc. Eng. Design* **59**, No. 2, 401–410.

Miya, K., and Uesaka, M. (1982). "An Application of Finite Element Method to Magnetomechanics of Superconducting Magnets for Magnetic Fusion Reactors," *Nuc. Eng. Design* **72**, 275–296.

Montgomery, B. (1980). *Solenoid Magnet Design*, Wiley, New York.

Moon, F. C. (1967). "Magnetoelastic Stability and Vibration," Ph. D. thesis, Cornell University, Ithaca, New York.

Moon, F. C. (1970a). "The Mechanics of Ferroelastic Plates in a Uniform Magnetic Field," *J. Appl. Mech.* **27**, No. 1, 153.

Moon, F. C. (1970b). "A Theory for High Current Elastic Conductors," In *Recent Advances In Engineering Science, Vol.* 5, Gordon and Breach, New York pp. 87–103.

Moon, F. C. (1971). "The Generation of Electromagnetic Radiation by Elastic Waves," *Dev. Mech.* **6**, 623–631; Proceedings of the Twelfth Midwestern Mechanics Conference.

Moon, F. C. (1974). "Laboratory Studies of Magnetic Levitation in the Thin Track Limit," *IEEE Trans. Mag.* **MAG–10**, 439–442.

Moon, F. C. (1976). *Elastic Stability and Vibrations of Superconducting Magnets for Magnetic Fusion Reactors*, ERDA Report No. C00–2780-2, Cornell University, Ithaca, New York.

Moon, F. C. (1977). "Vibration Problems in Magnetic Levitation and Propulsion," Chapter 6, In *Advances in Transport Without Wheels*, E. Laithwaite, Ed., Paul Elec. Press, London.

Moon, F. C. (1978). "Problems in Magneto-Solid Mechanics," Chapter V. In *Mechanics Today* 4, S. Nemat Nasser, Ed., Pergamon Press, New York.

Moon, F. C. (1979a). "Buckling of a Superconducting Ring in a Toroidal Magnetic Field," *J. Appl. Mech.* **46**, No. 1, 151–155.

Moon, F. C. (1979b). "Experiments on Magnetoelastic Buckling in a Superconducting Torus," *J. Appl. Mech.* **46**, No. 1, 145–150.

Moon, F. C. (1980a). "Magnetoelastic Instabilities in Superconducting Structures and Earnshaw's Theorem." In *Mechanics of Superconducting Structures*, ASME Applied Mechanics Monograph, Vol. 41, 77–90.

Moon, F. C. (1980b). "Elastic Vibrations and Waves in a Superconducting Torus", *IEEE Trans. Mag.* **MAG–16**, No. 5, 1242–1244.

Moon, F. C. (1982). "The Virial Theorem and Scaling Laws for Superconducting Magnet Systems," *J. Appl. Phys.* **53**, No. 12, 9112–9121.

Moon, F. C., and Chattopadhyay, S. (1974). "Magnetically Induced Stress Waves in a Conducting Solid-Theory and Experiment," *J. Appl. Mech.* **41**, No. 3, 641–646.

Moon, F. C., and Hara, K. (1982) "Buckling Induced Stresses in 'Martensitic' Stainless Steels for Magnetic Fusion Reactors," *Nuc. Eng. Design* **71**, 27–31.

Moon, F. C., and Kim, B. S. (1979). "Magnetoelastic Buckling of a Superconducting Energy Storage Shell," Proceedings of the Third Engineering Mechanics Conference ASCE, Austin, Tex. pp. 69–72.

Moon, F. C., and Pao, Y-H. (1968). "Magnetoelastic Buckling of a Thin Plate," *J. Appl. Mech.* **35**, No. 1, 53–58.

Moon, F. C., and Pao, Y-H. (1969). "Vibration and Dynamic Instability of a Beam-Plate in a Transverse Magnetic Field," *J. Appl. Mech.* **36**, No. 1, 1–9.

Moon, F. C., and Swanson, C. (1976). "Vibration and Stability of a Set of Superconducting Toroidal Magnets," *J. Appl. Phys.* **47**, No. 3, 914–919.

Moon, F. C., and Swanson, C. (1977). "Experiments on Buckling and Vibration of Superconducting Coils, " *J. Appl. Mech.* **44**, No. 4, 701–713.

Moon, F. C., Swanson, C., and King, S. (1977). "An Eight Coil Superconducting Torus for Elastic Buckling Studies," *Cryogenics* **17**, 341–344.

Morjaria, M., Mukherjee, S., and Moon, F. C. (1981). "Eddy Currents Around Cracks in Thin Plates due to a Current Filament," *Elec. Mach. Electromech.* **2**, 57–71.

Moses, R. N., Jr. (1976). "Configuration Design of Superconductive Energy Storage Magnets," *Adv. Cryogenic Eng.* **21**, 140–148.

Moses, R. W., Jr., and Young, W. C. (1975). "Analytic Expressions for Magnetic Forces on Sectored Toroidal Coils," Proceedings of the Sixth Symposium on Engineering Problems of Fusion Research.

Moullin, E. B. (1955). *The Principles of Electromagnetism*, 3rd ed., Oxford University Press, London.

Mozniker, R. A. (1959). "Effect of a Constant Magnetic Field on the Free Oscillations of Mechanical Systems," *Izv. Acad. Nauk Ukr. SSSR*, 847–852 (*in Russian*).

Nasar, S. A., and Boldea, I. (1976). *Linear Motion Electric Machines*, Wiley, New York.

Novitsky, V. G., and Shakhtarin, V. N. (1972). "Electrodynamic Strains and Mechanical Stresses in Superconducting Magnetic Systems," *Dokl. Aka. Nauk*

SSSR, **17**, No. 4, 50–55. (Translated from Russian: Oak Ridge National Laboratory, Report No. ORNL-tr-2902, Oak Ridge, Tenn.)

O'Neill, G. K. (1977). *The High Frontier*, Morrow, New York.

O'Neill, G. K., and Kolm, H. (1978). "Mass Driver for Lunar Transport and as a Reaction Engine," *J. Astro. Sci.* **XXV**, No. 4.

Panovko, Y. C., and Gubanova, I. I. (1965). *Stability and Oscillations of Elastic Systems* (English Translation Consultants Bureau, New York.)

Pao, Y-H. (1978). "Electromagnetic Forces in Deformable Continua," Review Article *Mech. Today*, **4**, 209–305 (S. Nemat Nasser, Ed., Pergamon Press, Oxford).

Pao, Y-H., and Hutter, K. (1973) "Magnetoelastic Waves in Soft Ferromagnetic Solids," In *Recent Advances in Engineering Science, Vol.* 6, Gordon and Breach, New York.

Pao, Y-H., and Yeh, C-S. (1973). "A Linear Theory for Soft Ferromagnetic Elastic Solids," *Int. J. Eng. Sci.* **11**, No. 4, 415.

Paria, G. (1967). "Magneto-elasticity and Magneto-thermoelasticity," In *Advances in Applied Mechanics, Vol.* 10, Academic Press, New York, 73–112.

Parker, E. N. (1958). "Reaction of Laboratory Magnetic Fields against Their Current Coils," *Phys. Rev.* **109**, 1440.

Penfield, P., and Haus, H. A. (1967). *Electrodynamics of Moving Media*, Research Monograph No. 40, MIT Press, Cambridge, Mass.

Pih, H., and Gray, W. H. (1976). *Photoelastic and Analytical Investigation of Stresses in Toroidal Magnetic Field Coils*, Proceedings of the Sixth, Symposium on Engineering Problems of Fusion Research; IEEE Publication No. 75CH1097-5-NPS.

Pipkin, A. C., and Rivlin, R. S. (1960). "Electrical Conduction in Deformed Isotropic Materials," *J. Math. Phys.* **1**, No. 2, 127–130.

Pipkin, A. C., and Rivlin, R. S. (1962). "Non-Rectilinear Current Flow in a Straight Conductor," *J. Math. Phys.* **3**, *No.* 2, 368–371.

Popelar, C. H. (1972). "Postbuckling Analysis of a Magnetoelastic Beam," *J. Appl. Mech.* **39**, 207–211.

Popelar, C. H., and Bast, C. O. (1972). "An Experimental Study of the Magnetoelastic Postbuckling Behavior of a Beam," *Exptl. Mech.* **12**, No. 12, 537–542.

Poston and Stewart (1978). *Catastrophe Theory and Its Applications*, Pitman, London.

Powell, J. D., and Batteh, J. H. (1982). "Arc Dynamics in the Rail Gun," *IEEE Tans. Mag.* **MAG-18**, No. 1, 7–10.

Rashleigh, S. C., and Marshall, R. A. (1978). "Electromagnetic Acceleration of Microparticles to High Velocities," *J. Appl. Phys.* **49**, No. 4, 2540–2542.

Rayleigh, S. C. (1894). *Theory of Sound*, Dover, New York.

Reed, R. P., Mikesell, R. P., and Clark, A. F. (1977). *Adv. Cryogenic Eng.* **22**, 463.

Reitz, J. R. (1970). "Forces on Moving Magnets due to Eddy Currents," *J. Appl. Phys.* **41**, No. 5, 2067–2071.

Reitz, J. R., and Davis, L. C. (1972). "Force on a Rectangular Coil Moving above a Conducting Slab," *J. Appl. Phys.* **43**, No. 4, 1547–1553.

Reitz, J. R., and Milford, F. J. (1960). *Foundations of Electromagnetic Theory*, Addison-Wesley, Reading, Mass.

Rhodes, R. G., and Mulhall, R. G. (1981). *Magnetic Levitation for Rail Transport*, Clarendon Press, Oxford, London.

Robey, D. H. (1953). "Magnetic Dispersion of Sound in Electrically Conducting Plates," *J. Acoust. Soc. Am.* **25**, No. 4, 603–609.

Rosensweig, R. E. (1966). "Buoyancy and Stable Levitation of a Magnetic Body Immersed in a Magnetizable Fluid," *Nature* **210**, 613–614.

Roters, H. C. (1941). *Electromagnetic Devices*, Wiley, New York.

Rutten, W. L. C., Peters, M. J., Brenkman, C. J., Mol, H., Grote, J. J., and van der Marel, L. C. (1982). "The Use of a SQUID Magnetometer for Middle Ear Research," *Cryogenics*, **22** 457–460.

Scanlan, R. M. (1979), Superconducting Materials, Lawrence Livermore Laboratory, Preprint UCRL-83492.

Scott, W. T. (1959). "Who Was Earnshaw?," *Am. J. Phys.* **27**, 418.

Sheppard, A. R., and Eisenbud, M. (1977). *Biological Effects of Electric and Magnetic Fields of Extremely Low Frequency*, New York University Press, New York.

Silvester, P., Wong, S. K., and Burke, P. E. (1971). "Modal Theory of Skin Effect in Single and Multiple Turn Coils," Proceedings of the IEEE Summer Power Meeting, Paper No. 71 TP 523-PNR, Portland, Ore.

Slemon, G. R. (1970). "Scale Factors of Physical Modelling of Magnetic Devices," *Elec. Mach. Electromech.* **1**, 1–9.

Slemon, G. R., and Straughen, A. (1980). *Electric Machines*, Addison-Wesley, Reading, Mass.

Smythe, N. R. (1968). *Static and Dynamic Electricity*, 3rd ed., McGraw-Hill, New York.

Snell, R. F., MacKallor, D. C., and Guernsey, R. (1973). "An Electromagnetic Plane Stress Wave Generator," *Exptl. Mech.*, 472–479.

Snow, W. R., Dunbar, S., Kubby, J., and O'Neill, G. K. (1982). "Mass Driver Two: A Status Report," *IEEE Trans. Mag.* **MAG-18**, No. 1, 127–134.

Snowden, A. C. (1961). "Studies of Electromagnetic Forces Occurring at Electrical Contacts," *Proc. AIEE Appl. Ind.* **24**.

Sommerfeld, A. (1952). *Electrodynamics*, Academic Press, New York.

Southwell, R. V. (1941). *An Introduction to the Theory of Elasticity*, Oxford University Press (reprinted edition, Dover, New York, 1969).

Steele, M. C., and Vural, B. (1969). *Wave Interaction in Solid State Plasmas*, McGraw-Hill, New York.

Stein, P. K. (1962). "Advanced Strain Gage Technique," Stein Engineering Services Incorporated Phoenix, Ariz. pp. 117–120.

Stekly, Z. J. J. (1963). "Feasibility of Large Superconducting Coils," AVCO–Everett Research, Vol. 160.

Stevenson, R., and Atherton, D. L. (1974). "Damage by Diamagnetic Forces in Large Superconducting Coils," Proceedings of the IEEE Intermag Meeting, Toronto, Canada.

Stoll, R. L. (1974). *The Analysis of Eddy Currents*, Clarendon Press, Oxford, London.

Stratton, J. A. (1941). *Electromagnetic Theory*, McGraw-Hill, New York.

Sun, C. T., and Gray, W. H. (1975). *Theoretical and Experimental Determination of Mechanical Properties of Superconducting Coil Composites*. Proceedings of the Sixth Symposium on Engineering Problems of Fusion Research, pp. 261–265 (also published by IEEE in 1976).

Sviatoslavsky, I. N., and Young, W. (1980). "Structural Design Features for Commercial Fusion Power Reactor Magnet System," *Nuc. Eng. Design* **58**, 207–218.

Swanson, C., and Moon, F. C. (1977). "Buckling and Vibrations in Five Coil Superconducting Partial Torus," *J. Appl. Phys.* 3110–3115.

Takaki, H., and Tsuji, T. (1958). "A Note on the Magneto-Resistance Effect of Strain Gauge Wire," *J. Phys. Soc. Jpn.* **13**, 1406.

Telinde, J. C. (1970). "Strain Gages in Cryogenic Environment," *Exptl. Mech.* **10** (9), 394-400.

Tenney, F. H. (1969). *On the Stability of Rigid Current Loops in an Axisymmetric Field*, Plasma Physics Laboratory Report No. MATT-693, Princeton, New Jersey.

Thome, R. J., and Tarrh, J. M. (1982). *MHD and Fusion Magnets; Field and Force Design Concepts*, Wiley, New York.

Thompson, J. M. T. (1982). *Instabilities and Catastrophies in Science and Engineering*, Wiley, New York.

Thompson, J. M. T., and Hunt, G. W. (1973). *A General Theory of Elastic Stability*, Wiley, London.

Thompson, W. B. (1962). *An Introduction to Plasma Physics*, Pergamon Press, New York, 106–108.

Thornton, R. D. (1973). "Design Principles for Magnetic Levitation," *Proc. IEEE* **61**, No. 5, 586–598.

Tiersten, H. F. (1964). "Coupled Magnetomechanical Equations for Magnetically Saturated Insulators," *J. Math. Phys.* **5**, No. 9, 1–21.

Tiersten, H. F. (1967). "An Extension of the London Equations of Superconductivity," *Physica* **37**, 504–538.

Tiersten, H. F. (1969). *Linear Piezoelectric Plate Vibrations*, Plenum Press, New York.

Tinkham, M. (1975). *Introduction to Superconductivity*, McGraw-Hill, New York.

Todhunter, I., and Pearson, K. (1886/1893). *A History of the Theory of Elasticity*, Vols. I, II, Dover, New York, 1960 (reprint).

Toupin, R. A. (1963). "A Dynamical Theory of Elastic Dielectrics," *Int. J. Eng. Sci.* **1**, 101–126.

Truesdell, C., and Toupin, R. (1960). *The Classical Field Theories*, Handbuch der Phy. Bd III/1, Springer, Berlin.

Tsai, S. W. and Hahn, M. T. (1980), *Introduction to Composite Materials* Technomic Publishing, Westport, Conn.

Underhill, C. R. (1924). *Magnets*, McGraw-Hill, New York.

Wakefield, K. (1964). *Design of Force-Free Toroidal Magnets*, Plasma Physics Laboratory Report No. MATT 208, Princeton University, Princeton, New Jersey.

Wallerstein, D. V., and Peach, M. O. (1972). "Magnetoelastic Buckling of Beams and Thin Plates of Magnetically Soft Material," *J. Appl. Mech.* **39**, No. 2, 451–455.

Walling, H. C., and Forrestal, M. J. (1973). "Elastic-Plastic Expansion of 6061-T6 Aluminum Rings," *AIAA J.* **11**, No. 8, 1196–1197.

Walstrom, P. L. (1975). "The Effect of High Magnetic Fields on Metal Foil Strain Gauges at 4.2°K," *Cryogenics*, 270–272.

Welch, D. O. (1980). "Alteration of the Superconducting Properties of A15 Compounds and Elementary Composite Superconductors by Nonhydrostatic Elastic Strain." In *Advances in Cryogenic Engineering Materials Vol.* 26, A. F. Clark and R. P. Reed, Eds., Plenum, New York.

Wells, D. R., and Mills, R. G. (1962). "Force Reduced Toroidal System." In *High Magnetic Fields*, H. Kohm, B. Lax, F. Bitter, and R. Mills Eds., MIT Press, Cambridge, Mass. and Wiley, New York, Chapter 5, pp. 44–47.

Werthiem, G. (1848). *Ann. Chim.* **XXVIII**, 302–327. [See also Todhunter, I. and Pearson, K. (1960). *A History of the Theory of Elasticity; and Strength of Materials*, Art 811-818, Dover, New York.]

Weston, W. F. (1975). "Low-Temperature Elastic Constants of a Superconducting Coil Composite," *J. Appl. Phys.* **46**, No. 10, 4458–4465.

Wiedemann, G. (1860). [See Todhunter, I. and Pearson, K. (1960). *A History of the Theory of Elasticity and Strength of Materials*, Art 206-715, Dover, New York.]

Wilkie, D. F. (1972). *Trans. Res.* **6**, 343.

Williams (1931). *Magnetic Phenomena*, Chapter II, McGraw-Hill, New York.

Williams, J. E. C. (1970). *Superconductivity and Its Application*, Pion Ltd., London.

Winterberg, F. (1966). "Magnetic Acceleration of a Superconducting Solenoid to Hypervelocities," *Plasma Phy.* **8**, 541–552.

Woodson, H. H., and Melcher, J. R. (1968). *Electromechanical Dynamics*, Parts I, II, and III, Wiley, New York.

Yagawa, G., and Horie, T. (1982). "Cracked Beam Under Influence of Dynamic Electromagnetic Force," *Nuc. Eng. Design* **69**, 49–55.

Yamamura, S. (1979). *Theory of Linear Induction Motors*, 2nd Ed., Wiley, New York.

Yeh, C-S. (1971). "Linear Theory of Magnetoelasticity for Soft Ferromagnetic Materials and Magnetoelastic Buckling," Ph. D. dissertation, Cornell University, Ithaca, New York.

Yuan, K. (1972) "Magneto-Thermo-Elastic Stresses in an Infinitely Long Cylindrical Conductor Carrying a Uniformly Distributed Axial Current," *Appl. Sci. Res.* **26**, 307–314.

Yuan, K. Y. (1981). "Finite Element Analysis of Magnetoelastic Plate Problems," Ph. D. thesis, Cornell University, Ithaca, New York.

Yuan, K. Y., Abel, J., and Moon, F. C. (1981). "Eddy Current Calculations in Thin Conducting Plates Using a Finite Element Stream Function Code," *IEEE Trans. Mag.* **MAG-18**, No. 2, 447–449.

Yuan, K. Y., Moon, F. C., and Abel, J. (1984). "Elastic Conducting Structures in Pulsed Magnetic Fields." In *Numerical Methods in Coupled Systems*, Ed. R. N. Lewis, J. Wiley and Sons, N.Y.

Ziegler, H. (1959). "On the Concept of Elastic Stability," *Advances in Applied Mechanics Vol.* IV, Academic Press, New York, pp. 351–403.

Ziman, J. M. (1964). *Principles of the Theory of Solids*, Cambridge University Press, London.

APPENDIX A

List of Common Symbols

A	area
$\mathbf{A}$	magnetic vector potential
B	magnetic flux density
$\mathbf{B}$	magnetic flux density vector
C	electric capacitance
c_{ijkl}	elastic constants
$\mathbf{C}$	couple vector
$\mathbf{c}$	body couple density
D	flexural stiffness
$\mathbf{D}$	electric displacement vector
$\mathbf{E}$	electric field intensity vector
e	electron charge
$\mathbf{e}_x, \mathbf{e}_y, \mathbf{e}_z$	orthogonal unit vectors
F	force
$\mathbf{F}$	force vector
$\mathbf{f}$	body force density
G	bending moments
H	twisting moment in a rod
$\mathbf{H}$	magnetic field intensity vector
I	electric current
$\mathcal{I}$	moment of area
$\mathbf{J}$	current-density vector
$\mathbf{K}$	surface current-density vector
k	wave number
L	inductance
$\mathcal{L}$	Lagrangian
$\mathbf{M}$	magnetization density vector
m	mass or mass per unit length
$\mathbf{m}$	magnetic dipole vector, magnetic moment of circuit
N	number of turns, shear force
$\mathbf{n}$	unit normal vector
$\mathbf{P}$	electric polarization density vector
P_m	magnetic pressure
Q	electric charge
q	electric charge density
$\mathbf{q}$	heat flux vector
R	electrical resistance, radius
R_H	Hall constant
$\mathcal{R}_m$	magnetic Reynold's number
$\mathcal{R}$	magnetic reluctance
$\mathbf{r}$	position vector
T	tension force
$\mathbf{T}$	stress tensor
$\mathcal{T}$	kinetic energy
t	time

t_{ij} stress tensor components

U stored energy, displacement

$\mathbf{u}$ displacement vector

V electric voltage

$\mathcal{V}$ elastic energy

$\mathbf{v}$ velocity vector

$\mathcal{W}$ magnetic energy

w transverse plate displacement

Y Young's modulus of elasticity

α linear coefficient of thermal expansion

Δ thickness

δ skin depth

ε_{ij} strain tensor components

ε_0 permittivity of vacuum

θ temperature, angle

κ thermal conductivity

λ Lamé's elastic constant

μ Lamé's elastic constant

μ_0 magnetic permeability of vacuum

μ_r relative magnetic permeability

ν Poisson's ratio

ρ mass density, electrical resistivity

σ electrical conductivity

Φ magnetic flux

ϕ electric or magnetic field potential

χ magnetic susceptibility

ψ eddy-current stream function

ω frequency

$\mathbf{\Omega}$ rotation vector

$\boldsymbol{\tau}$ stress vector

APPENDIX B

Tables of Magnetic Forces and Properties of Selected Electromagnetic Materials

Table B-1 Physical Constants

Permeability of vacuum	$\mu_0 = 4\pi \times 10^{-7}$ H/m
Permittivity of vacuum	$\varepsilon_0 = 8.854 \times 10^{-12}$ F/m
Speed of light in vacuum	$C = 1/(\mu_0 \varepsilon_0)^{1/2} \cong 3 \times 10^8$ m/sec
Mass of electron	0.91×10^{-30} kg
Charge of electron	$e = 1.6 \times 10^{-19}$ C
Electron volt	$\text{eV} = 1.6 \times 10^{-19}$ J
Planck's constant	$h = 6.626 \times 10^{-34}$ J sec
One flux quantum (fluxoid)	$\Phi_0 = h/2e = 2.07 \times 10^{-15}$ Wb

Table B-2 Magnetic Forces and Couples

F, r, F, Q_1, Q_2

$$\mathbf{F} = \frac{Q_1 Q_2}{4\pi\varepsilon_0} \frac{\mathbf{r}}{r^3} \quad \text{(N)}$$

$$\frac{1}{4\pi\varepsilon_0} = 8.99 \times 10^9 \frac{N \cdot m^2}{C^2}$$

m_2, r, m_1

$$\mathbf{F}_2 = \mathbf{m}_2 \cdot \nabla \mathbf{B}_1, \qquad \mathbf{B}_1 = \frac{\mu_0}{4\pi}\left(\frac{3(\mathbf{m}_1 \cdot \mathbf{r})\mathbf{r}}{r^5} - \frac{\mathbf{m}_1}{r^3}\right)$$

$$\mathbf{F}_2 = \frac{\mu_0}{4\pi} \frac{3}{r^5}\left((\mathbf{m}_1 \cdot \mathbf{m}_2)\mathbf{r} + (\mathbf{m}_1 \cdot \mathbf{r})\mathbf{m}_2 + (\mathbf{m}_2 \cdot \mathbf{r})\mathbf{m}_1 - \frac{5(\mathbf{m}_1 \cdot \mathbf{r})(\mathbf{m}_2 \cdot \mathbf{r})\mathbf{r}}{r^2}\right)$$

$$\mathbf{C}_2 = \mathbf{m}_2 \times \mathbf{B}_1$$

$$\mathbf{C}_2 = \frac{\mu_0}{4\pi}\left(\frac{3(\mathbf{m}_1 \cdot \mathbf{r})\mathbf{m}_2 \times \mathbf{r}}{r^5} - \frac{\mathbf{m}_2 \times \mathbf{m}_1}{r^3}\right)$$

$$\mathbf{F} = -\frac{\mu_0}{2\pi}\frac{I_1 I_2}{r^2}\mathbf{r} \quad \text{(N/m)}$$

$$F = \frac{\mu_0}{\pi}\frac{I_1 I_2}{a}\left[\tan^{-1}\left(\frac{a}{d}\right) - \frac{d}{2a}\ln\left(1 + \frac{a^2}{d^2}\right)\right] \quad \text{(N/m)}$$

$$F = \frac{\mu_0 I_1 I_2}{2\pi d_1 d_2}\int_s^{s+d_2} \ln\frac{d_1 + x}{x}\,dx \quad \text{(N/m)}$$

$$F = \frac{\mu_r - 1}{\mu_r + 1}\frac{\mu_0 I^2}{4\pi h} \quad \text{(N/m)}$$

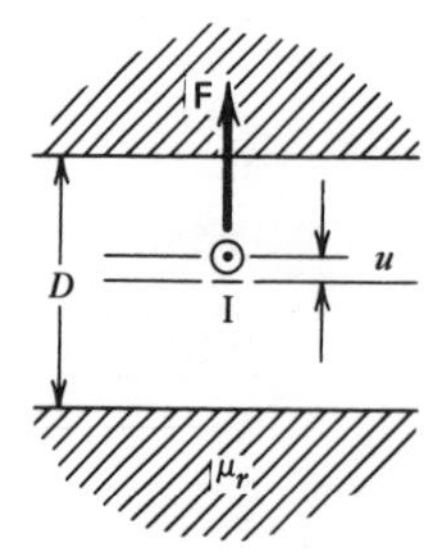

$$F = \frac{\mu_0 I^2}{4\pi}\frac{\pi}{D}\tan\frac{\pi u}{D} \quad \text{(N/m)}$$

$$(\mu_r \to \infty)$$

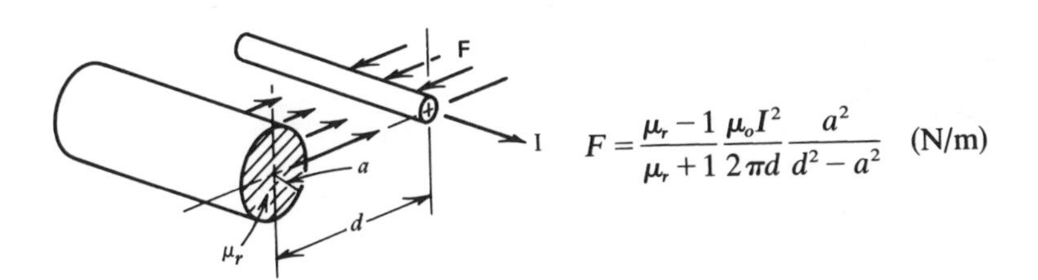

$$F = \frac{\mu_r - 1}{\mu_r + 1} \frac{\mu_o I^2}{2\pi d} \frac{a^2}{d^2 - a^2} \quad \text{(N/m)}$$

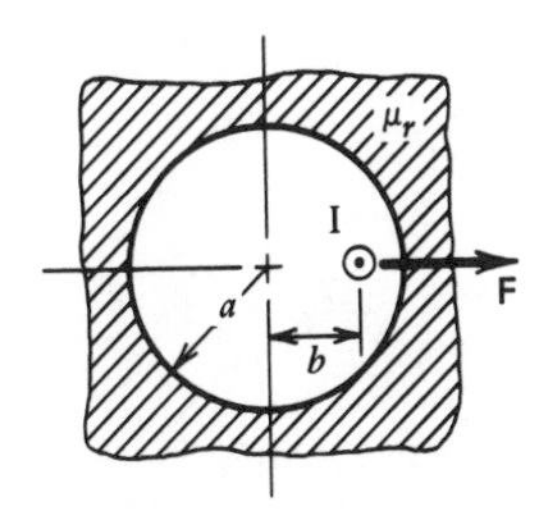

$$F = \frac{\mu_r - 1}{\mu_r + 1} \frac{\mu_o I^2}{2\pi} \frac{b}{a^2 - b^2} \quad \text{(N/m)}$$

$$F_x = \frac{\mu_0 I^2}{2\pi r}\left(1 + \frac{r^2}{r^2 + 4h^2}\right)$$

$$F_y = \frac{\mu_0 I^2}{4\pi h}\left(1 - \frac{4h^2}{r^2 + 4h^2}\right)$$

$$F_L = \frac{\mu_0 I^2}{4\pi h} \frac{v^2}{v^2 + w^2}$$

$$F_D = \frac{w}{v} F_L \quad \text{(N/m)}$$

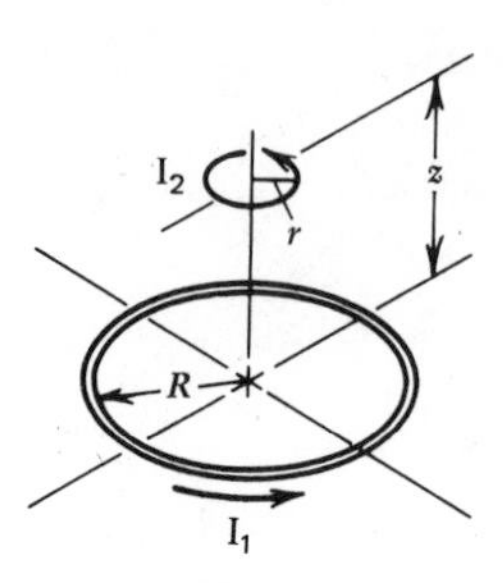

$$F_z = -\frac{3}{2}\mu_0 I_1 I_2 \frac{\pi R^2 r^2 z}{(R^2+z^2)^{5/2}} \quad \text{(N)}$$

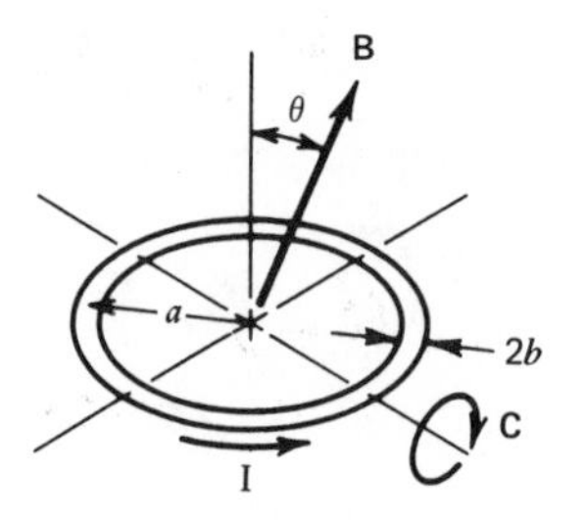

$$C = \tfrac{1}{4}\pi B I (4a^2 - b^2)\sin\theta \quad (\text{N}\cdot\text{m})$$

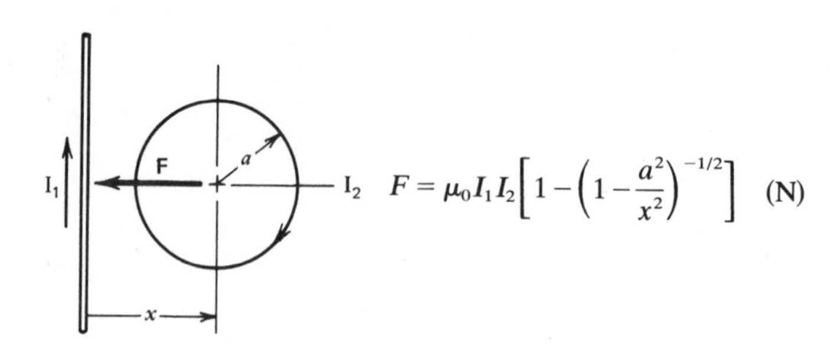

$$F = \mu_0 I_1 I_2 \left[1 - \left(1 - \frac{a^2}{x^2}\right)^{-1/2}\right] \quad \text{(N)}$$

I

2b

R

T_θ

$$T_\theta = \frac{\mu_0 I^2}{4\pi}\left[\ln\frac{8R}{b} - \frac{3}{4}\right]$$

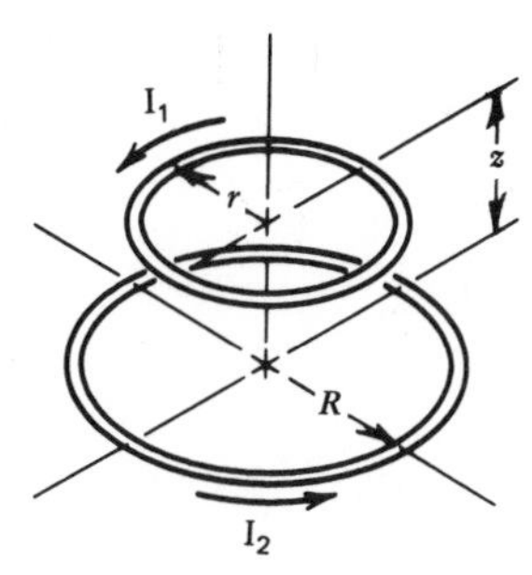

$$F = \frac{\mu_0 I_1 I_2 z}{[(r+R)^2 + z^2]^{1/2}}\left(-K(k) + \frac{r^2 + R^2 + z^2}{(r-R)^2 + z^2}E(k)\right)$$

$$k = 4rR[(r+R)^2 + z^2]^{-1}$$

K, E = complete elliptic integrals of first and second kind

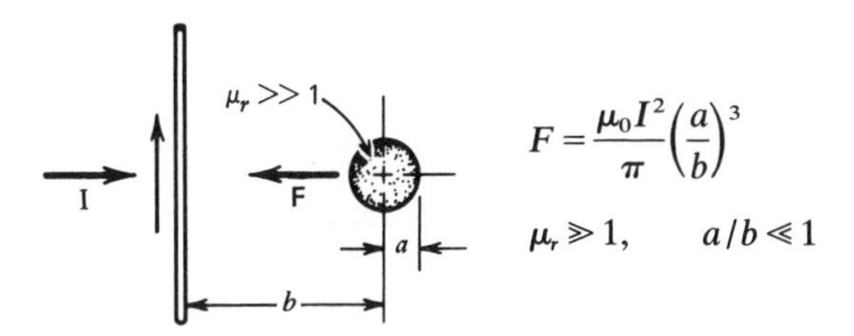

$$F = \frac{\mu_0 I^2}{\pi}\left(\frac{a}{b}\right)^3$$

$$\mu_r \gg 1, \qquad a/b \ll 1$$

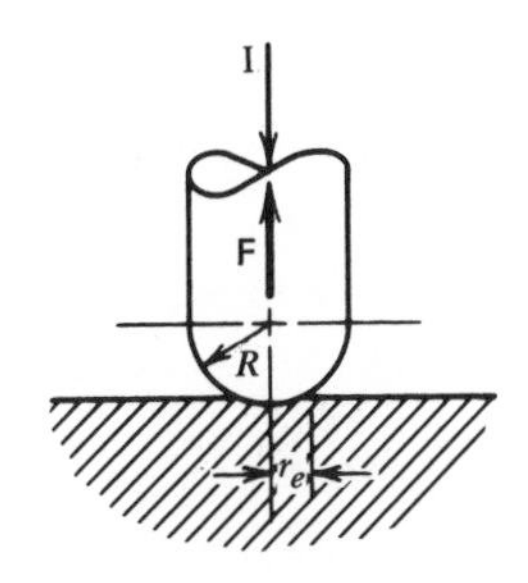

$$F = \frac{\mu_0 I^2}{4\pi}\left[\ln\frac{R}{r_c} + \frac{1}{4}\right]$$

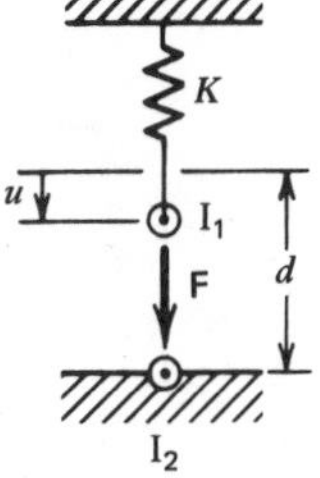

$$F = \frac{\mu_0 I_1 I_2}{\pi d[1 + (1 - 4\beta)^{1/2}]}$$

$$u = \tfrac{1}{2}d[1 - (1 - 4\beta)^{1/2}]$$

$$\beta = \frac{\mu_0 I_1 I_2}{2\pi K d^2}$$

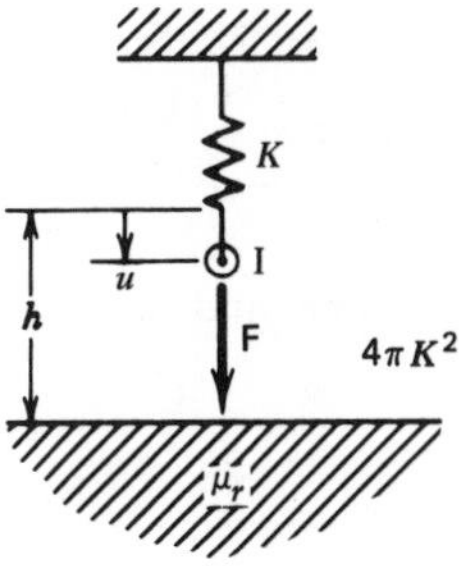

$$F = \frac{\mu_r - 1}{\mu_r + 1} \frac{\mu_0 I^2}{\pi h} \frac{1}{[1 + (1 - 4\beta)^{1/2}]}$$

$$u = \tfrac{1}{2} h [1 - (1 - 4\beta)^{1/2}]$$

$$\beta = \frac{\mu_r - 1}{\mu_r + 1} \frac{\mu_0 I^2}{4\pi K h^2}$$

Table B-3 Electromagnetic Properties of Ferromagnetic Materials

Material	Curie Temperature (°C)	μ_r Initial	μ_r Maximum	Electricity Resistivity ($10^{-8}\ \Omega \cdot m$)	Density (kg/m^3)	Saturation Field (tesla)
Fe (purified)	770	10^4	2×10^5	10	7800	2.15
Co	1131					
Ni	358			7.8		
Grain-oriented Fe-Si (3% Si)	740	7500	55,000	47	7670	2.0
78 Permalloy (78.5% Ni)	600	8000	10^5	16	8600	1.08
Supermalloy (5% Mo, 79% Ni)	400	10^5	10^6	60	8770	0.79
45 Permalloy (45% Ni)	400	2500	25,000	45	8170	1.6
Fe Commercial (0.2% impurities)	770	250	9,000	10	7800	2.16

Source: *American Institute of Physics Handbook*, McGraw-Hill, New York. Bozorth (1951).

Table B-4 Nominal Properties of Selected Superconducting Materials[a]

Material	T_c (°K)	B_{c1} (tesla)	B_{c2} (tesla)	J_c at 4 T (A/cm^2)
Ti	0.4	0.002	—	
Al	1.2	0.01	—	
Sn	3.7	0.03	—	
Nb	9.2	0.13	0.25	
Nb-Ti	9.5–10.5	0.04–0.06	9.0–12	$\sim 0.8\text{–}2.5 \times 10^5$
Nb-25%Zr	10.8	0.04	7	$\sim 0.2\text{–}0.4 \times 10^5$
V_3Ga	14.5		21	$\sim 10^5$
V_3Si	16.9		14	2×10^5
Nb_3Sn	18.1	~0.02	20	$0.7\text{–}3 \times 10^5$
Nb_3Al	18.8	~0.02	31	
Nb_3Ge	22.3			

Source: Brechna (1973) and Hein (1974).
[a] All values at 4.2°K (liquid helium) unless noted.

Table B-5 Elastic and Electrical Properties of Selected Metals[a]

	Density (g/cm^3)	Young's modulus (10^{10} N/m^2)	Poisson's ratio	Dilatational speed (km/sec)	Shear speed (km/sec)	Electrical resistivity (10^{-8} $\Omega \cdot m$)	Hall constant 10^{-10} m^3/C	$\mu_0 \sigma$ (mks)
Aluminum	2.7	6.8–7.1	0.355	6.42	3.04	2.83	−0.30	44.4
Brass (70% Cu, 30% Zn)	8.5–8.7	10.4	0.374	4.70	2.11			
Copper	8.3–8.93	12.1–12.8	0.37	5.01	2.27	1.69	−0.55	74.4
Lead	11.4	1.5–1.7	0.43	1.96	0.690	19.8	0.09	6.35
Stainless steel	7.91	19.6	0.30	5.79	3.10	10–20		6.3–12.6
Tin	7.3	5.5	0.34	3.32	1.67	11.5	−0.04	10.9
Titanium	4.50	36.2	0.35	5.41	2.64	43.1		2.92
Zinc	7.04–7.18	10.5	0.25	4.21	2.44	5.75	0.33	21.9

Source: *American Institute of Physics Handbook*, 2nd ed., McGraw-Hill, New York, 1951.
[a] All data at room temperature.

Table B-6 Ratio of Magnetic to Thermal Skin Depths

Material	δ_m / δ_{th} at 0°C
Cu	10
Al	15
Brass	38
Fe (at $\mu_r = 1$)	55
BeCu	55
Stainless steel	350

Table B-7 Properties of Selected Materials at 4.2°K

Material	Young's modulus 10^9 N/m^2	Yield Stress 10^6 N/m^2	Electrical Resistivity ($\Omega \cdot$ m)
Austenitic stainless steels	200	400–1200	5×10^{-7}
Fe- and Ni-base superalloys	200	750–1400	10×10^{-7}
Ferritic steels	200	1400–2200	3×10^{-5}
Low expansion Ni alloys	140	850–1300	5×10^{-7}
Copper alloys	140	60–1000	1×10^{-11} 6×10^{-8}
Aluminum alloys	70	65–570	5.6×10^{-11} 2×10^{-8}
Glass-reinforced composites	70[a]	20[a]	Very high

Source: C. J. Long, *Structural Materials for Large Superconducting Magnets for Tokamaks*, Oak Ridge National Laboratory Report No. ORNL/TM-5632, Oak Ridge, Tennessee, December 1976.

[a]Longitudinal direction.

Author Index

Page numbers in *italics* refer to references

Subject Index